'A well-written and superbly illustrated book on the ever-fascinating theme of the descent of man . . . this extraordinarily detailed, serious and engaging book'   George Walden, *Daily Mail*

'Amazing and brilliant . . . [Dawkins is] the most temperate and invigorating of persuaders, one of the most cultured and humane . . . a work of immense erudition, engaging geniality and originality of conception and composition'   James Grieve, *Canberra Times*

'To read *The Ancestor's Tale* is to be amazed at the multiplicity and ingenuity that results . . . Dawkins's capacity for clear explanation, assisted by excellent and often beautiful illustrations, is formidable'
Quentin de la Bédoyère, *Catholic Herald*

'I believe I speak for most of us when I say that there's not a huge demand to know more about protosomes, sauropsids and sea squirts. But when you realize that these are things we simply had to be in order to become human, they are rather more alluring, and in Dawkins' clear, measured prose they become reliably and unexpectedly absorbing'   Bill Bryson, *Daily Express*

D0259843

'Beautifully written . . . Dawkins's account cites a stunning array of biologists past and present. No other book gives such an impression of sheer intellectual vitality and pluralism among the past century's evolutionary scientists. Virtually every page exemplifies a memorable insight into the strangeness and prodigality of nature, its culs-de-sac and its extraordinary leaps' John Cornwell, *Sunday Times*

'As a contribution to the history of ideas this book is well worthy of Britain's top public intellectual. The arguments are as sharply honed as we have come to expect from Dawkins'

Matt Ridley, *Guardian*

'One of the richest accounts of evolution ever written . . . the tales of the pilgrims dart around with a delightful unpredictability, propelled like a firecracker by Dawkins's wonderful way with words. He is so good at explaining scientific issues that readers will learn painlessly about matters well outside the author's field of evolutionary biology from maths to cosmology . . . we have no right to expect [another] *magnum opus* on the scale of *The Ancestor's Tale*'

Clive Cookson, *Financial Times*

'Dawkins is one of the outstanding scientific inventors of our time – an inventor of concepts, that is, rather than experiments or machines. Over the past 30 years he has established himself as a great explainer too, combining serious research with zestful popular writing . . . In 500 pages crammed with lucid prose and lovely pictures, it sketches the story of life on Earth since its origin four billion years ago. And it does it with punch . . . A book which tries, with much brilliance and some success, to treat our vaunted human-ity as no more than a tiny episode in a vast drama, equivalent to a couple of seconds of madness at the end of a very long day'

Jonathan Ree, *Evening Standard*

'The result is not just a wealth of ideas about how living things evolved, but a strong sense of the urgency and absorption with which science is done' Marek Kohn, *Independent*

'*The Ancestor's Tale* is a pilgrimage. Dawkins's subject here is the history of life . . . you never lose sight of the fact that it is our family tree we are discussing . . . No other book I have read has given me such a dizzyingly immediate sense of the vastness and strangeness of the changes brought about by evolution over the eons, or how intimately all life is bound together – far more intimately than we could have conceived a few years ago . . . the meticulous clarity of Dawkins's prose and the absolute absence of fuzziness offer assurance that, given time and thought, the knottiest passages will yield'

Robert Hanks, *Daily Telegraph*

'This pilgrimage, this "backwards history", is a wonderfully informed enquiry into how we got to be here, as well as a reminder that, in spite of our arrogant belief in "man as evolution's last word", evolution does not end with us . . . *The Ancestor's Tale* is just as satisfying: as intellectually stimulating as its predecessors, but also more generous, more warm-blooded (as it were). As always with Dawkins, the writing is beautiful: economical, vivid and, often, both elegant and witty'

John Burnside, *Scotsman*

'A new chronicle of life, wonderfully illustrated, from this great evolutionist'

Economist

'Should be given to all intelligent young persons starting out on their exploration of the world. It will excite their curiosity and awe and prove to them that the world is inexhaustible in its fascination'

Anthony Daniels, *Sunday Telegraph*

'This is epic stuff indeed and Mr Dawkins carries it off with triumphant skill, never sacrificing the complexity of his argument to the voracious god of dumbing down'

Dan Colwell, *Wall Street Journal*

'*The Ancestor's Tale* achieves the almost impossible: it makes biology (not biochemistry, brain science, or bird-watching, but biology as a whole) interesting again'

Steve Jones, *Lancet*

'Its central philosophy is well stated on the last page. Pilgrimage implies reverence, and such reverence should go to "the sublime grandeur of the real world"'

Crispin Tickell, *Literary Review*

'*The Ancestor's Tale* is an audacious book, a monumental work that takes us back in time to the origins of life on this planet'

Dick Ahlstrom, *Irish Times*

Professor Richard Dawkins is a world-renowned evolutionary biologist and author. He is a Fellow of the Royal Society and holds the Charles Simonyi Chair of Public Understanding of Science at Oxford University. His first book, *The Selfish Gene* (1976), was an instant international bestseller, and has become an established classic work of modern evolutionary biology. *The Blind Watchmaker* (1986), too, has become world-famous. His other works for the general public have each been highly successful.

By Richard Dawkins

The Selfish Gene
The Extended Phenotype
The Blind Watchmaker
River Out of Eden
Climbing Mount Improbable
Unweaving the Rainbow
A Devil's Chaplain
The Ancestor's Tale
The God Delusion

THE
ANCESTOR'S TALE

*A Pilgrimage to
the Dawn of Life*

Richard Dawkins

with additional research by Yan Wong

PHOENIX

A PHOENIX PAPERBACK

First published in Great Britain in 2004
by Weidenfeld & Nicolson
This paperback edition published in 2005
by Phoenix,
an imprint of Orion Books Ltd,
Orion House, 5 Upper St Martin's Lane,
London WC2H 9EA

An Hachette Livre UK company

10 9

A CIP catalogue record for this book
is available from the British Library.

ISBN 978-0-7538-1996-8

Typeset at The Spartan Press Ltd,
Lymington, Hants

Printed and bound in Great Britain by
Clays Ltd, St Ives plc

The Orion Publishing Group's policy is to use papers that
are natural, renewable and recyclable products and
made from wood grown in sustainable forests. The logging
and manufacturing processes are expected to conform to
the environmental regulations of the country of origin.

www.orionbooks.co.uk

John Maynard Smith
(1920–2004)

He saw a draft and graciously accepted the dedication,
which now, sadly, must become

In Memoriam

'Never mind the lectures or the "workshops"; be blowed to the
motor coach excursions to local beauty spots; forget your fancy
visual aids and radio microphones; the only thing that really
matters at a conference is that John Maynard Smith must be in
residence and there must be a spacious, convivial bar. If he can't
manage the dates you have in mind, you must just reschedule the
conference . . . He will charm and amuse the young research
workers, listen to their stories, inspire them, rekindle enthusiasms
that might be flagging, and send them back to their laboratories or
their muddy fields, enlivened and invigorated, eager to try out the
new ideas he has generously shared with them.'

It isn't only conferences that will never be the same again.

ACKNOWLEDGMENTS

I was persuaded to write this book by Anthony Cheetham, founder of Orion Books. The fact that he had moved on before the book was published reflects my unconscionable delay in finishing it. Michael Dover tolerated that delay with humour and fortitude, and always encouraged me by his swift and intelligent understanding of what I was trying to do. The best of his many good decisions was to engage Latha Menon as a freelance editor. As with *A Devil's Chaplain*, Latha's support has been beyond all estimation. Her grasp of the big picture simultaneously with the details, her encyclopaedic knowledge, her love of science and her selfless devotion to promoting it have benefited me, and this book, in more ways than I can count. Others at the publishers helped greatly, but Jennie Condell and the designer, Ken Wilson, went beyond the call of duty.

My research assistant Yan Wong has been intimately involved at every stage of the planning, researching and writing of the book. His resourcefulness and detailed familiarity with modern biology have been matched only by his green fingers with computers. If, here, I have gratefully assumed the role of apprentice, it could be said that he was my apprentice before I was his, for I was his tutor at New College. He then did his doctorate under the supervision of Alan Grafen, once my own graduate student, so I suppose Yan could be called my grandstudent as well as my student. Apprentice or master, Yan's contribution has been so great that, for certain tales, I have insisted on adding his name as joint author. When Yan left to cycle across Patagonia, the book in its final stages benefited greatly from Sam Turvey's extraordinary knowledge of zoology and his conscientious care in deploying it.

Advice and help of various kinds were willingly given by Michael Yudkin, Mark Griffith, Steve Simpson, Angela Douglas, George McGavin, Jack Pettigrew, George Barlow, Colin Blakemore, John Mollon, Henry Bennet-Clark, Robin Elisabeth Cornwell, Lindell Bromham, Mark Sutton, Bethia Thomas, Eliza Howlett, Tom Kemp, Malgosia Nowak-Kemp, Richard Fortey, Derek Siveter, Alex Freeman,

Nicky Warren, A. V. Grimstone, Alan Cooper, and especially Christine DeBlase-Ballstadt. Others are acknowledged in the Notes at the end.

I am deeply grateful to Mark Ridley and Peter Holland, who were engaged by the publishers as critical readers and gave me exactly the right kind of advice. The routine authorial claim of responsibility for the remaining shortcomings is more than usually necessary in my case.

As always, I gratefully acknowledge the imaginative generosity of Charles Simonyi. And my wife, Lalla Ward, has once again been my help and strength.

RICHARD DAWKINS

CONTENTS

THE
ANCESTOR'S TALE

THE CONCEIT OF HINDSIGHT

History doesn't repeat itself, but it rhymes.
MARK TWAIN

History repeats itself; that's one of the things that's wrong with history.
CLARENCE DARROW

History has been described as one damn thing after another. The remark can be seen as a warning against a pair of temptations but, duly warned, I shall cautiously flirt with both. First, the historian is tempted to scour the past for patterns that repeat themselves; or at least, following Mark Twain, to seek reason and rhyme for everything. This appetite for pattern affronts those who insist that, as Mark Twain will also be found to have said, 'History is usually a random, messy affair', going nowhere and following no rules. The second connected temptation is the vanity of the present: of seeing the past as aimed at our own time, as though the characters in history's play had nothing better to do with their lives than foreshadow us.

Under names that need not trouble us, these are live issues in human history and they arise with greater force, and no greater agreement, on the longer timescale of evolution. Evolutionary history can be represented as one damn species after another. But many biologists will join me in finding this an impoverished view. Look at evolution that way and you miss most of what matters. Evolution rhymes, patterns recur. And this doesn't just happen to be so. It is so for well-understood reasons: Darwinian reasons mostly, for biology, unlike human history or even physics, already has its grand unifying theory, accepted by all informed practitioners, though in varying versions and interpretations. In writing evolutionary history I do not shrink from seeking patterns and principles, but I try to be careful about it.

What of the second temptation, the conceit of hindsight, the idea that the past works to deliver our particular present? The late Stephen Jay Gould rightly pointed out that a dominant icon of evolution in popular mythology, a caricature almost as ubiquitous as lemmings jumping over cliffs (and that myth is false too), is a shambling file of simian ancestors, rising progressively in the wake of the erect, striding, majestic figure of *Homo sapiens sapiens:* man as evolution's last word (and in this context it always is man rather than woman); man as what the whole enterprise is pointing towards; man as a magnet, drawing evolution from the past towards his eminence.

There is a physicist's version which is less obviously vainglorious and which I should mention in passing. This is the 'anthropic' notion that the very laws of physics themselves, or the fundamental constants of the universe, are a carefully tuned put-up job, calculated to bring humanity eventually into existence. It is not necessarily founded on vanity. It doesn't have to mean that the universe was deliberately made in order that we should exist. It need mean only that we are here, and we could not be in a universe that lacked the capability of producing us. As physicists have pointed out, it is no accident that we see stars in our sky, for stars are a necessary part of any universe capable of generating us. Again, this does not imply that stars exist in order to make us. It is just that without stars there would be no atoms heavier than lithium in the periodic table, and a chemistry of only three elements is too impoverished to support life. Seeing is the kind of activity that can go on only in the kind of universe where what you see is stars.

But there is a little more that needs to be said. Granted the trivial fact that our presence requires physical laws and constants capable of producing us, the existence of such potent ground rules may still seem tantalisingly improbable. Depending upon their assumptions, physicists may reckon that the set of possible universes vastly outnumbers that subset whose laws and constants allowed physics to mature, via stars into chemistry and via planets into biology. To some, this means that the laws and constants must have been deliberately premeditated from the start (although it baffles me why anybody regards this as an explanation for anything, given that the problem so swiftly regresses to the larger one of explaining the existence of the equally fine-tuned and improbable Premeditator).

Other physicists are less confident that the laws and constants were free to vary in the first place. When I was little it was not obvious to me why five times eight had to give the same result as eight times five. I accepted it as one of those facts that grownups assert. Only later did I understand, perhaps through visualising rectangles, why such pairs of multiplications are not free to vary independently of one another. We understand that the circumference and the diameter of a circle are not independent, otherwise we might feel tempted to postulate a plethora of possible universes, each with a different value of π. Perhaps, argue some physicists such as the Nobel Prize-winning theorist Steven Weinberg, the fundamental constants of the universe, which at present we treat as independent of one another, will in some Grand Unified fullness of time be understood to have fewer degrees of freedom than we now imagine. Maybe there is only one way for a universe to be. That would undermine the appearance of anthropic coincidence.

Other physicists, including Sir Martin Rees, the present Astronomer Royal, accept that there is a real coincidence in need of explanation, and explain it by postulating many actual universes existing in parallel, mutually incommunicado, each with its own set of laws and constants.* Obviously we, who find ourselves reflecting upon such things, must be in one of those universes, however rare, whose laws and constants are capable of evolving us.

The theoretical physicist Lee Smolin added an ingenious Darwinian spin which reduces the apparent statistical improbability of our existence. In Smolin's model, universes give birth to daughter universes, which vary in their laws and constants. Daughter universes are born in black holes produced by a parent universe, and they inherit its laws and constants but with some possibility of small random change – 'mutation'. Those daughter universes that have what it takes to reproduce (last long enough to make black holes, for instance) are, of course, the universes that pass on their laws and constants to their daughters. Stars are precursors to black holes which, in the Smolin model, are the birth events. So universes that have what it takes to make stars are favoured in this cosmic Darwinism. The properties of a universe that furnish this gift to the future are the self-same properties that incidentally lead to the manufacture of large atoms, including vital carbon atoms. Not only do we live in a universe that is capable of producing life. Successive generations of universes progressively evolve to become increasingly the sort of universe that, as a by-product, is capable of producing life.

---

* This 'many universes' idea is not to be confused (though it often is) with Hugh Everett's 'many worlds' interpretation of quantum theory, brilliantly advocated by David Deutsch in *The Fabric of Reality*. The resemblance between the two theories is superficial and meaningless. Both theories could be true, or neither, or one, or the other. They were proposed to answer completely different problems. In the Everett theory, the different universes don't differ in their fundamental constants. But it is the entire point of the theory we are here considering that the different universes have different fundamental constants.

The logic of the Smolin theory is bound to appeal to a Darwinian, indeed to anyone of imagination, but as for the physics I am not qualified to judge. I cannot find a physicist to condemn the theory as definitely wrong – the most negative thing they will say is that it is superfluous. Some, as we saw, dream of a final theory in whose light the alleged fine-tuning of the universe will turn out to be a delusion anyway. Nothing we know rules out Smolin's theory, and he claims for it the merit – which scientists rate more highly than many laymen appreciate – of testability. His book is *The Life of the Cosmos* and I recommend it.

But that was a digression about the physicist's version of the conceit of hindsight. The biologist's version is easier to dismiss since Darwin, though harder before him, and it is our concern here. Biological evolution has no privileged line of descent and no designated end. Evolution has reached many millions of interim ends (the number of surviving species at the time of observation), and there is no reason other than vanity – human vanity as it happens, since we are doing the talking – to designate any one as more privileged or climactic than any other.

This doesn't mean, as I shall continue to argue, that there is a total dearth of reasons or rhymes in evolutionary history. I believe there are recurring patterns. I also believe, though this is more controversial today than it once was, that there are senses in which evolution may be said to be directional, progressive and even predictable. But progress is emphatically not the same thing as progress towards humanity, and we must live with a weak and unflattering sense of the predictable. The historian must beware of stringing together a narrative that seems, even to the smallest degree, to be homing in on a human climax.

A book in my possession (in the main a good book, so I shall not name and shame it) provides an example. It is comparing *Homo habilis* (a human species, probably ancestral to us) with its predecessors the australopithecines.* What the book says is that *Homo habilis* was 'considerably more evolved than the Australopithecines'.

---

* The laws of zoological nomenclature follow strict precedence, and I fear there is no hope of changing the name *Australopithecus* to something less confusing to the contemporary majority who lack a classical education. It has nothing to do with Australia. No member of the genus has ever been found outside Africa. *Australo* simply means southern. Australia is the great southern continent, the Aurora australis is the southern equivalent of the Aurora borealis (boreal means northern), and *Australopithecus* was first found in *south* Africa, in the person of the Taung child.

More evolved? What can this mean but that evolution is moving in some pre-specified direction? The book leaves us in no doubt of what the presumed direction is. 'The first signs of a chin are apparent.' 'First' encourages us to expect second and third signs, towards a 'complete' human chin. 'The teeth start to resemble ours . . .' As if those teeth were the way they were, not because it suited the habiline diet but because they were embarking upon the road towards becoming our teeth. The passage ends with a telltale remark about a later species of extinct human, *Homo erectus:*

> Although their faces are still different from ours, they have a much more human look in their eyes. They are like sculptures in the making, 'unfinished' works.

In the making? Unfinished? Only with the unwisdom of hindsight. In excuse of that book it is probably true that, were we to meet a *Homo erectus* face to face, it might well look to our eyes like an unfinished sculpture in the making. But that is only because we are looking with human hindsight. A living creature is always in the business of surviving in its own environment. It is never unfinished – or, in another sense, it is always unfinished. So, presumably, are we.

The conceit of hindsight tempts us at other stages in our history. From our human point of view, the emergence of our remote fish ancestors from water to land was a momentous step, an evolutionary rite of passage. It was undertaken in the Devonian Period by lobe-finned fish a bit like modern lungfish. We look at fossils of the period with a pardonable yearning to gaze upon our forebears, and are seduced by a knowledge of what came later: drawn into seeing these Devonian fish as 'half way' towards becoming land animals; everything about them earnestly transitional, bound into an epic quest to invade the land and initiate the next big phase of evolution. That is not the way it was at the time. Those Devonian fish had a living to earn. They were not on a mission to evolve, not on a quest towards the distant future. An otherwise excellent book about vertebrate evolution contains the following sentence about fish which

> ventured out of the water on to the land at the end of the Devonian Period and jumped the gap, so to speak, from one vertebrate class to another to become the first amphibians . . .

The 'gap' comes from hindsight. There was nothing resembling a gap at the time, and the 'classes' that we now recognise were no more separate, in those days, than two species. As we shall see again, jumping gaps is not what evolution does.

It makes no more sense (and no less) to aim our historical narrative towards *Homo sapiens* than towards any other modern species – *Octopus vulgaris,* say, or *Panthera leo* or *Sequoia sempervirens.* A historically minded swift, understandably proud of flight as self-evidently the premier accomplishment of life, will regard swift-kind – those spectacular flying machines with their swept-back wings, who stay aloft for a year at a time and even copulate in free flight – as the acme of evolutionary progress. To build on a fancy of Steven Pinker, if elephants could write history they might portray tapirs, elephant shrews, elephant seals and proboscis monkeys as tentative beginners along the main trunk road of evolution, taking the first fumbling steps but each – for some reason – never quite making it: so near yet so far. Elephant astronomers might wonder whether, on some other world, there exist alien life forms that have crossed the nasal rubicon and taken the final leap to full proboscitude.

We are not swifts nor elephants, we are people. As we wander in imagination through some long-dead epoch, it is humanly natural to reserve a special warmth and curiosity for whichever otherwise ordinary species in that ancient landscape is our ancestor (it is an intriguingly unfamiliar thought that there is always one such species). It is hard to deny our human temptation to see this one species as 'on the main line' of evolution, the others as supporting cast, walk-on parts, sidelined cameos. Without succumbing to that error, there is one way to indulge a legitimate human-centrism while respecting historical propriety. That way is to do our history backwards, and it is the way of this book.

Backward chronology in search of ancestors really can sensibly aim towards a single distant target. The distant target is the grand ancestor of all life, and we can't help converging upon it no matter where we start – elephant or eagle, swift or salmonella, wellingtonia or woman. Backward chronology and forward chronology are each good for different purposes. Go backwards and, no matter where you start, you end up celebrating the unity of life. Go forwards and you extol diversity. It works on small timescales as well as large. The forward chronology of the mammals, within their large but still limited timescale, is a story of branching diversification, uncovering the richness of that group of hairy warmbloods. Backward chronology, taking any modern mammal as our starting point, will always converge upon the same unique ur-mammal: shadowy, insectivorous, nocturnal contemporary of the dinosaurs. This is a local convergence. A yet more local one converges on the most recent ancestor of all rodents, who lived somewhere around the time the dinosaurs went extinct. More local still is the backward convergence of all apes (including humans) on their shared ancestor, who lived about 18 million years ago. On a larger scale, there is a comparable convergence to be found if we work backwards from any vertebrate, an even larger convergence working backwards from any animal to the ancestor of all animals. The largest convergence of all takes us from any modern creature – animal, plant, fungus or bacterium – back to the universal progenitor of all surviving organisms, probably resembling some kind of bacterium.

I used 'convergence' in the last paragraph, but I really want to reserve that word for a completely different meaning in forward chronology. So for the present purpose I shall substitute 'confluence' or, for reasons that will make sense in a moment, 'rendezvous'. I could have used 'coalescence', except that, as we shall see, geneticists have already adopted it in a more precise sense, similar to my 'confluence' but concentrating on genes rather than species. In a backward chronology, the ancestors of any set of species must eventually meet at a particular geological moment. Their point of rendezvous is the last common ancestor that they all share, what I shall call their 'Concestor':* the focal rodent or the focal mammal or the focal vertebrate, say. The oldest concestor is the grand ancestor of all surviving life.

We can be very sure there really is a single concestor of all surviving life forms on this planet. The evidence is that all that have ever been examined share (exactly in most cases, almost exactly in the rest) the same genetic code; and the genetic code is too detailed, in arbitrary aspects of its complexity, to have been invented twice. Although not every species has been examined, we already have enough coverage to be pretty certain that no surprises – alas – await us. If we now were to discover a life form sufficiently alien to have a completely different genetic code, it would be the most exciting biological discovery in my adult lifetime, whether it lives on this planet or another. As things stand, it appears that all known life

* I am grateful to Nicky Warren for suggesting this word.

forms can be traced to a single ancestor which lived more than 3 billion years ago. If there were other, independent origins of life, they have left no descendants that we have discovered. And if new ones arose now they would swiftly be eaten, probably by bacteria.

The grand confluence of all surviving life is not the same thing as the origin of life itself. This is because all surviving species presumably share a concestor who lived after the origin of life: anything else would be an unlikely coincidence, for it would suggest that the original life form *immediately* branched and more than one of its branches survive to this day. Current textbook orthodoxy dates the oldest bacterial fossils at about 3.5 billion years ago, so the origin of life must at least be earlier than that. If we accept a recent disputation* of these apparently ancient fossils, our dating of the origin of life might be a bit more recent. The grand confluence – the last common ancestor of all surviving creatures – could pre-date the oldest fossils (it didn't fossilise) or it could have lived a billion years later (all but one of the other lineages went extinct).

Given that all backward chronologies, no matter where they start, culminate in the one grand confluence, we can legitimately indulge our human preoccupation and concentrate upon the single line of our own ancestors. Instead of treating evolution as aimed towards us, we *choose* modern *Homo sapiens* as our arbitrary, but forgivably preferred, starting point for a reverse chronology. We choose this route, out of all possible routes to the past, because we are curious about our own great grancestors. At the same time, although we need not follow them in detail, we shall not forget that there are other historians, animals and plants belonging to other species, who are independently walking backwards from their separate starting points, on separate pilgrimages to visit their own ancestors, including eventually the ones they share with us. If we retrace our own ancestral steps, we shall inevitably meet these other pilgrims and join forces with them in a definite order, the order in which their lineages rendezvous with ours, the order of ever more inclusive cousinship.

---

* J. W. Schopf's much-cited evidence for 3.5 billion-years-old bacteria has been sharply criticised by my Oxford colleague Martin Brasier. Brasier may be right about Schopf's evidence, but new evidence, published when this book was in proof, may reinstate 3.5 billion years as the date of the oldest fossils. The Norwegian scientist Harald Furnes and his coworkers found tiny holes in volcanic glass of that age in South Africa, which they believe were etched by micro-organisms. These 'burrows' contain carbon, which the discoverers claim is of biological origin. No trace of the micro-organisms themselves remains.

Pilgrimages? Join forces with pilgrims? Yes, why not? Pilgrimage is an apt way to think about our journey to the past. This book will be cast in the form of an epic pilgrimage from the present to the past. All roads lead to the origin of life. But because we are human, the path we shall follow will be that of our own ancestors. It will be a human pilgrimage to discover human ancestors. As we go, we shall greet other pilgrims who will join us in strict order, as we reach the common ancestors we share with them.

The first fellow pilgrims we shall greet, some 5 million years ago, deep in Africa where Stanley memorably shook hands with Livingstone, are the chimpanzees. The chimpanzee and bonobo pilgrims will already have joined forces with each other 'before' we greet them. And here we have a little linguistic trickiness which I must face at the outset, before it dogs us any further. I placed 'before' in inverted commas because it could confuse. I used it to mean before in the backwards sense – 'before, in the course of the pilgrimage to the past'. But that of course means *after* in the chronological sense, the exact opposite meaning! My guess is that no reader was confused in this particular case, but there will be other instances where the reader's patience may be tested. While writing this book I tried the experiment of coining a new preposition, tailored to the peculiar needs of a backward historian. But it didn't fly. Instead, I shall adopt the convention of 'before' in inverted commas. When you see 'before', remember that it really means after! When you see before, it really means before. And the same for 'after' and after, *mutatis mutandis.*

The next pilgrims with whom we shall rendezvous as we push back along our journey are gorillas, then orang utans (quite a lot deeper into the past, and probably no longer in Africa). Next we shall greet gibbons, then Old World monkeys, then New World monkeys, then various other groups of mammals . . . and so on until eventually all the pilgrims of life are marching together in one single backward quest for the origin of life itself. As we push on back, there will come a time when it is no longer meaningful to name the continent in which a rendezvous takes place: the map of the world was so different, because of the remarkable phenomenon of plate tectonics. And further back still, all rendezvous take place in the sea.

It is a rather surprising fact that we human pilgrims pass only about 40 rendezvous points in all, before we hit the origin of life itself. At each of the 40 steps we shall find one particular shared ancestor, the Concestor, which will bear the same labelling number as the Rendezvous. For example, Concestor 2, whom we meet at Rendezvous 2, is the most recent common ancestor of gorillas on the one hand and {humans + {chimpanzees + bonobos}} on the other. Concestor 3 is the most recent common ancestor of orang utans and {{humans + {chimpanzees + bonobos}} + gorillas}. Concestor 39 is the grand ancestor of all surviving life forms. Concestor 0 is a special case, the most recent ancestor of all surviving humans.

We shall be pilgrims, then, sharing fellowship ever more inclusively with other pilgrim bands, which also have been swelling on their own way to their rendezvous with us. After each meeting, we continue together on the high road back to our shared Archaean goal, our 'Canterbury'. There are other literary allusions, of course, and I almost made Bunyan my model and *Pilgrim's Regress* my title. But it was to Chaucer's *Canterbury Tales* that I and my research assistant Yan Wong kept returning in our discussions, and it seemed increasingly natural to think of Chaucer throughout this book.

Unlike (most of) Chaucer's pilgrims, mine do not all set out together, although they do set off at the same time, the present. These other pilgrims aim towards their ancient Canterbury from different starting points, joining our human pilgrimage at various rendezvous along the road. In this respect, my pilgrims are unlike those who gathered in London's Tabard Inn. Mine are more like the sinister canon and his understandably disloyal yeoman, who joined Chaucer's pilgrims at Boughton-under-Blee, five miles short of Canterbury. Following Chaucer's lead, my pilgrims, which are all the different species of living creature, will have the opportunity to tell tales along the way to their Canterbury which is the origin of life. It is these tales that form the main substance of this book.

Dead men tell no tales, and extinct creatures such as trilobites are deemed not to be pilgrims capable of telling them, but I shall make exceptions of two special classes. Animals such as the dodo, which survived into historical times and whose DNA is still available to us, are treated as honorary members of the modern fauna setting off on pilgrimage at the same time as us, and joining us at some particular rendezvous. Since we are responsible for their so recent extinction, it seems the least we can do. The other honorary pilgrims, exceptions to the rule that dead men tell no tales, really are men (or women). Since we human pilgrims are directly seeking our own ancestors, fossils that might plausibly be considered candidates for *being* our ancestors are deemed members of our human pilgrimage and we shall hear tales from some of these 'shadow pilgrims', for example the Handyman, *Homo habilis*.

I decided it would be twee to let my animal and plant tale-tellers speak in the first person singular, and I shall not do so. Save for occasional asides and prefatory remarks, Chaucer's pilgrims don't either. Many of Chaucer's Tales have their own Prologue, and some have an Epilogue too, all written in Chaucer's own voice as narrator of the pilgrimage. I shall occasionally follow his example. As with Chaucer, an epilogue may serve as a bridge from one tale to the next.

Before his Tales begin, Chaucer has a long General Prologue in which he sets out his cast list: the professions and in some cases the names of the pilgrims who are about to set off from the tavern. Instead, I shall introduce new pilgrims as they join us. Chaucer's jovial host offers to guide the pilgrims, and encourages them to tell their tales to while away the journey. In my role as host I shall use the General Prologue for some preparatory remarks about methods and problems of reconstructing evolutionary history, which must be faced and solved whether we do our history backwards or forwards.

Then we shall embark on our backwards history itself. Although we shall concentrate on our own ancestors, noting other creatures usually only when they join us, we shall from time to time look up from our road and remind ourselves that there are other pilgrims on their own more or less independent routes to our ultimate destination. The numbered rendezvous milestones, plus a few intermediate markers necessary to consolidate the chronology, will provide the scaffolding for our narrative. Each will mark a new chapter, where we halt to take stock of our pilgrimage, and maybe listen to a tale or two. On rare occasions, something important happens in the world around us, and then our pilgrims may pause briefly to reflect on it. But, for the most part, we shall mark our progress to the dawn of life by the measure of those 40 natural milestones, the trysts that enrich our pilgrimage.

THE GENERAL PROLOGUE

How shall we know the past, and how date it? What aids to our vision will help us peer into theatres of ancient life and reconstruct the scenes and the players, their exits and their entrances, of long ago? Conventional human history has three main methods, and we shall find their counterparts on the larger timescale of evolution. First there is archaeology, the study of bones, arrowheads, fragments of pots, oystershell middens, figurines and other relics that survive as hard evidence from the past. In evolutionary history, the most obvious hard relics are bones and teeth, and the *fossils* that they eventually become. Second, there are *renewed relics,* records that are not themselves old but which contain or embody a copy or representation of what is old. In human history these are written or spoken accounts, handed down, repeated, reprinted or otherwise duplicated from the past to the present. In evolution, I shall propose DNA as the main renewed relic, equivalent to a written and recopied record. Third, there is *triangulation.* This name comes from a method of judging distances by measuring angles. Take a bearing on a target. Now walk a measured distance sideways and take another. From the intercept of the two angles, calculate the distance of the target. Some camera rangefinders use the principle, and map surveyors traditionally relied upon it. Evolutionists can be said to 'triangulate' an ancestor by comparing two (or more) of its surviving descendants. I shall take the three kinds of evidence in order, beginning with hard relics and, in particular, fossils.

Fossils

Bodies or bones may survive for our attention, having somehow escaped that of hyenas, burying beetles and bacteria. The 'Ice Man' of the Italian Tyrol was preserved in his glacier for 5,000 years. Insects have become embalmed in amber (petrified gum from trees) for 100 million years. Without benefit of ice or amber, hard parts like teeth, bones and shells stand the best chance of being preserved. Teeth last longest of all because, to do their job in life, they had to be harder than anything their owner was likely to eat. Bones and shells need to be hard for different reasons, and they too can last a long time. Such hard parts and, under exceptionally lucky circumstances, soft parts too, occasionally become petrified as stone fossils that last for hundreds of millions of years.

In spite of the fascination of fossils, it is surprising how much we would still know about our evolutionary past without them. If every fossil were magicked away, the comparative study of modern organisms, of how their patterns of resemblances, especially of their genetic sequences, are distributed among species, and of how species are distributed among continents and islands, would still demonstrate, beyond all sane doubt, that our history is evolutionary, and that all living creatures are cousins. Fossils are a bonus. A welcome bonus, to be sure, but not an essential one. It is worth remembering this when creationists go on (as they tediously do) about 'gaps' in the fossil record. The fossil record could be one big gap, and the evidence for evolution would still be overwhelmingly strong. At the same time, if we had *only* fossils and no other evidence, the fact of evolution would again be overwhelmingly supported. As things stand, we are blessed with both.

The word fossil is conventionally used to mean any relic dating back more than 10,000 years: not a helpful convention, for there is nothing special about a round number like 10,000. If we had fewer or more than ten fingers, we'd recognise a different set of numbers as round.* When we speak of a fossil, we normally mean that the original material has been substituted or infiltrated by a mineral of a different chemical composition and therefore given, as one might say, a new lease of death. An imprint of the original form may be preserved in stone for a very long time indeed, perhaps mixed with some of the original material. There are various ways in which this can happen. I leave the details – what is technically called taphonomy – for the Ergast's Tale.

When fossils were first discovered and mapped, their ages were unknown. The most we could hope for was a rank ordering of oldness. Age ranking depends upon the assumption known as the Law of Superposition. For obvious reasons, younger strata lie atop older ones, unless the circumstances are exceptional. Such exceptions, though they sometimes cause temporary puzzlement, are

---

* If we had eight (or sixteen) fingers, we'd think naturally in octal (or hexadecimal) arithmetic, binary logic would be easier to understand, and computers might have been invented much earlier.

usually pretty obvious. A lump of old rock, complete with fossils, may be thrown on top of a younger stratum, say by a glacier. Or a series of strata may be turned over wholesale, and its vertical ordering exactly reversed. These anomalies can be taken care of by comparing equivalent rocks in other parts of the world. Once this is done, the palaeontologist can piece together the true sequence of the whole fossil record, in a jigsaw of overlapping sequences from different parts of the world. The logic is complicated in practice, though not in principle, by the fact (see the Elephant Bird's Tale) that the map of the world itself changes as the ages go by.

Why is the jigsaw necessary? Why can't we just dig down as far as we like, and treat this as equivalent to digging steadily backwards through time? Well, time itself may flow smoothly, but this doesn't mean that anywhere in the world there is a single sequence of sediment deposited smoothly and continuously from start to finish through geological time. Fossil beds are laid down in fits and starts, when the conditions are right.

In any one location, at any one time, it is rather likely that no sedimentary rocks, and no fossils, are being laid down. But it is quite likely that, in *some* part of the world, fossils are being deposited at any given time. By hopping around the world, from site to site where different strata happen to be accessibly near the surface, the palaeontologist can aspire to piece together something approaching a continuous record. Of course individual palaeontologists don't hop from site to site. They hop from museum to museum looking at specimens in drawers, or from journal to journal in university libraries looking at written descriptions of fossils whose site of discovery has been carefully labelled, and they use these descriptions to piece together the fragments of the puzzle from different parts of the world.

The task is eased by the fact that particular strata, with recognisably characteristic rock properties, and consistently housing the same kinds of fossils, keep turning up in different regions. Devonian rock, so-called because it was first recognised as the 'Old Red Sandstone' of the beautiful county of Devon, crops up in various other parts of the British Isles, in Germany, Greenland, North America and elsewhere. Devonian rocks are recognisable as Devonian wherever they may be found, partly because of the quality of the rock but also because of the internal evidence of the fossils that they contain. This sounds like a circular argument but it really isn't: no more so than when a scholar recognises a Dead Sea Scroll, from internal evidence, as a fragment of the First Book of Samuel. Devonian rocks are reliably labelled by the presence of certain characteristic fossils.

The same goes for rocks from other geological periods, right back to the time of the earliest hard-bodied fossils. From the ancient Cambrian through to the present Holocene, the geological periods listed in the chart on plate 1 were mostly separated on the basis of changes in the fossil record. And as a result, the end of one period and the start of another is often delimited by extinctions that conspicuously interrupt the continuity of the fossils. As Stephen Jay Gould has put it, no palaeontologist has any trouble identifying whether a lump of rock lies before or after the great end-Permian mass extinction. There is almost no overlap in animal types. Indeed, fossils (especially microfossils) are so useful in labelling and dating rocks that the oil and mining industries are among their principal users.

Such 'relative dating', then, has long been possible by vertical piecing together of the jigsaw of rocks. The geological periods were named for purposes of relative dating, before absolute dating became possible. And they are still useful. But relative dating is more difficult for rocks with scarce fossils – and that includes all rocks older than the Cambrian: the first eight-ninths of Earth's history (see plate 1).

Absolute dating had to wait for recent developments in physics, especially the physics of radioactivity. This needs some explaining, and the details must wait for the Redwood's Tale. For now, it is enough to know that we have a range of reliable methods for putting an absolute age on fossils, or the rocks that contain or surround them. Moreover, different methods in this range provide sensitivity across the whole spectrum of ages from hundreds of years (tree rings), through thousands of years (carbon 14), millions, hundreds of millions (uranium-thorium-lead) to billions of years (potassium-argon).

Renewed Relics

Fossils, like archaeological specimens, are more-or-less direct relics of the past. We turn now to our second category of historical evidence, *renewed* relics, copied successively down the generations. For historians of human affairs this might mean eyewitness accounts, handed down by oral tradition or in written documents. We cannot ask any

living witnesses what it was like to live in fourteenth-century England, but we know about it thanks to written documents, including
Chaucer's. They contain information that has been copied, printed,
stored in libraries, reprinted and distributed for us to read today.
Once a story gets into print or, nowadays, a computer medium of
some kind, copies of it have a fair chance of being perpetuated into
the distant future.

Written records are more reliable than oral tradition, by a disconcerting margin. You might think that each generation of children,
knowing their parents as well as most children do, would listen to
their detailed reminiscences and relay them to the next generation.
Five generations on, a voluminous oral tradition should, one might
think, have survived. I remember my four grandparents clearly, but
of my eight great-grandparents I know a handful of fragmentary
anecdotes. One great-grandfather habitually sang a certain nonsense
rhyme (which I can sing), but only while lacing his boots. Another
was greedy for cream, and would knock the chess board over when
losing. A third was a country doctor. That is about my limit. How
have eight entire lives been so reduced? How, when the chain of
informants connecting us back to the eyewitness seems so short, and
human conversation so rich, could all those thousands of personal
details that made up the lifetimes of eight human individuals be so
fast forgotten?

Frustratingly, oral tradition peters out almost immediately, unless
hallowed in bardic recitations like those that were eventually written
down by Homer, and even then the history is far from accurate. It
decays into nonsense and falsehood after amazingly few generations.
Historical facts about real heroes, villains, animals and volcanoes
rapidly degenerate (or blossom, depending upon your taste) into
myths about demigods, devils, centaurs and fire-breathing dragons.*
But oral traditions and their imperfections needn't detain us because,
in any case, they have no equivalent in evolutionary history.

Writing is a huge improvement. Paper, papyrus and even stone
tablets may wear out or decay, but written records have the potential

---

* John Reader, in his *Man on Earth,* notes that the Incas, who had no written language
(unless, as has been recently suggested, their knotted strings were used for language as
well as for counting), made a perhaps compensatory effort to improve the accuracy of
their oral tradition. Official historians were 'obliged to memorise vast amounts of
information and repeat it for the benefit of administrators as required. Not surprisingly,
the role of historian passed from father to son.'

to be copied accurately for an indefinite number of generations, although in practice the accuracy is not total. I should explain the special sense in which I mean accuracy and, indeed, the special sense in which I mean generations. If you handwrite me a message and I copy it and pass it on to a third person (the next copying 'generation'), it will not be an exact replica, for my handwriting is different from yours. But if you write with care, and if I painstakingly match each of your squiggles with exactly one from our shared alphabet, your message has a good chance of being copied by me with total accuracy. In theory this accuracy could be preserved through an indefinite number of 'generations' of scribes. Given that there is a discrete alphabet agreed by writer and reader, copying lets a message survive the destruction of the original. This property of writing can be called 'self-normalising'. It works because letters of a true alphabet are discontinuous. The point, reminiscent of the distinction between analogue and digital codes, needs a little more explanation.

There exists a consonant sound which is intermediate between the English hard c and g (it is the French hard c in *comme*). But nobody would think of trying to represent this sound by writing a character which looked intermediate between c and g. We all understand that a written character in English must be one, and only one, member of our 26-letter alphabet. We understand that French uses the same 26 letters for sounds that are not exactly the same as ours and which may be intermediate between ours. Each language, indeed each local accent or dialect, separately uses the alphabet for self-normalising on different sounds.

Self-normalisation fights against the 'Chinese Whispers'* degrading of messages over generations. The same protection is not available to a drawing, copied and recopied along a line of imitative artists, unless the drawing style incorporates ritual conventions as its own version of 'self-normalisation'. An eyewitness record of some event, which is written down, as opposed to drawn as a picture, has a good chance of still being accurately reproduced in history books centuries later. We have what is probably an accurate account of the destruction of Pompeii in 79 AD because a witness, Pliny the

---

* In the game of Chinese Whispers (American children call it 'Telephone'), a number of children stand in a line. A story is whispered to the first child, who whispers it to the second, and so on until the last child, whose finally revealed version of the story turns out to be an amusingly garbled and degraded version of the original.

Younger, wrote down what he saw, in two epistles to the historian
Tacitus, and some of Tacitus's writings survived, by successive copying and eventually printing, for us to read them today. Even in pre-
Gutenberg days when documents were duplicated by scribes, writing
represented a great advance in accuracy compared with memory and
oral tradition.

It is only a theoretical ideal that repetitive copying retains perfect
accuracy. In practice scribes are fallible, and not above massaging
their copy to make it say things that they think (no doubt sincerely)
the original document ought to have said. The most famous example
of this, painstakingly documented by nineteenth-century German
theologians, is the doctoring of New Testament history to make it
conform to Old Testament prophecies. The scribes concerned were
probably not wilfully mendacious. Like the gospel-makers, who
themselves lived long after Jesus's death, they genuinely believed he
had been the incarnation of Old Testament messianic prophecies. He
'must', therefore, have been born in Bethlehem, and descended from
David. If the documents unaccountably failed to say so, it was the
scribe's conscientious duty to rectify the deficiency. A sufficiently
devout scribe would, I suppose, no more have regarded this as
falsification than we do when we automatically correct a spelling
mistake or a grammatical infelicity.

Quite apart from positive massaging, all repeated copying is subject to straightforward errors like skipping a line, or a word in a list.
But in any case writing cannot take us back beyond its invention,
which was only about 5,000 years ago. Identification symbols, count-
ing-marks and pictures go back a bit further, perhaps some tens of
thousands of years, but all such periods are chickenfeed compared
with evolutionary time.

Fortunately, when we turn to evolution there is another kind of
duplicated information which goes back an almost unimaginably
large number of copying generations and which, with a little poetic
licence, we can regard as the equivalent of a written text: a historical
record that renews itself with astounding accuracy for hundreds of
millions of generations precisely because, like our writing system, it
has a self-normalising alphabet. The DNA information in all living
creatures has been handed down from remote ancestors with
prodigious fidelity. The individual atoms in DNA are turning over
continually, but the information that they encode in the pattern of
their arrangement is copied for millions, sometimes hundreds of

millions, of years. We can read this record directly, using the arts of modern molecular biology to spell out the actual DNA letter sequences or, slightly more indirectly, the amino acid sequences of protein into which they are translated. Or, much more indirectly as through a glass darkly, we can read it by studying the embryological products of the DNA: the shapes of bodies and their organs and chemistries. We don't need fossils to peer back into history. Because DNA changes very slowly through the generations, history is woven into the fabric of modern animals and plants, and inscribed in its coded characters.

DNA messages are written in a true alphabet. Like the Roman, Greek and Cyrillic writing systems, the DNA alphabet is a strictly limited repertoire of symbols with no self-evident meaning. Arbitrary symbols are chosen and combined to make meaningful messages of unlimited complexity and size. Where the English alphabet has 26 letters and the Greek one 24, the DNA alphabet is a four-letter alphabet. Most useful DNA spells out three-letter words from a dictionary limited to 64 words, each word called a 'codon'. Some of the codons in the dictionary are synonymous with others, which is to say that the genetic code is technically 'degenerate'.*

The dictionary maps 64 code words onto 21 meanings – the 20 biological amino acids, plus one all-purpose punctuation mark. Human languages are numerous and changing, and their dictionaries contain tens of thousands of distinct words, but the 64-word DNA dictionary is universal and unchanging (with very minor variations in a few rare cases). The 20 amino acids are strung into sequences of typically a few hundred, each sequence a particular protein molecule. Whereas the number of letters is limited to four and the number of codons to 64, there is no theoretical limit to the number of proteins that can be spelled out by different sequences of codons. It is beyond all counting. A 'sentence' of codons specifying one protein molecule is an identifiable unit often called a gene. The genes are not separated from their neighbours (whether other genes or repetitive nonsense)

---

* 'Redundant' is sometimes mistakenly used instead of degenerate, but it means something different. The genetic code is, as it happens, redundant too, in that either strand of the double helix could be decoded to yield the same information. Only one of them is actually decoded, but the other is used for correcting errors. Engineers, too, use redundancy – repetitiousness – to correct errors. The degeneracy of the genetic code is something different, and it is what we are talking about here. A degenerate code contains synonyms and could therefore accommodate a larger range of meanings than it actually does.

by any delimiters apart from what can be read from their sequence. In this respect they resemble TELEGRAMS THAT LACK PUNCTU-ATION MARKS COMMA AND HAVE TO SPELL THEM OUT AS WORDS COMMA ALTHOUGH EVEN TELEGRAMS HAVE THE AD-VANTAGE OF SPACES BETWEEN WORDS COMMA WHICH DNA LACKS STOP

DNA differs from written language in that islands of sense are separated by a sea of nonsense, never transcribed. 'Whole' genes are assembled, during transcription, from meaningful 'exons' separated by meaningless 'introns' whose texts are simply skipped by the reading apparatus. And even meaningful stretches of DNA are in many cases never read – presumably they are superseded copies of once useful genes that hang around like early drafts of a chapter on a cluttered hard disk. Indeed, the image of the genome as an old hard disk, badly in need of a spring clean, is one that will serve us from time to time during the book.

It bears repeating that the DNA molecules of long-dead animals are not themselves preserved. The *information* in DNA can be preserved for ever, but only by dint of frequent re-copying. The plot of *Jurassic Park,* though not silly, falls foul of practical facts. Conceivably, for a short while after becoming embalmed in amber, a bloodsucking insect could have contained the instructions needed to reconstruct a dinosaur. But unfortunately, after an organism is dead, the DNA in its body, and in blood that it has sucked, doesn't survive intact longer than a few years – only days in the case of some soft tissues. Fossilisation doesn't preserve DNA either.

Even deep freezing doesn't preserve it for very long. As I write this, scientists are excavating a frozen mammoth from the Siberian permafrost in the hope of extracting enough DNA to grow a new mammoth, cloned in the womb of a modern elephant. I fear this is a vain hope, though the mammoth is only a few thousand years dead. Among the oldest corpses from which readable DNA has been extracted is a Neanderthal man. Imagine the kerfuffle if somebody managed to clone him. But alas, only disjointed fragments of his 30,000-year-old DNA can be recovered. For plants in permafrost, the record is about 400,000 years.

The important point about DNA is that, as long as the chain of reproducing life is not broken, its coded *information* is copied to a new molecule before the old molecule is destroyed. In this form, DNA information far outlives its molecules. It is renewable – copied

– and since the copies are literally perfect for most of its letters on any one occasion, it can potentially last an indefinitely long time. Large quantities of our ancestors' DNA information survives completely unchanged, some even from hundreds of millions of years ago, preserved in successive generations of living bodies.

Understood in this way, the DNA record is an almost unbelievably rich gift to the historian. What historian could have dared hope for a world in which every single individual of every species carries, within its body, a long and detailed text: a written document handed down through time? Moreover, it has minor random changes, which occur seldom enough not to mess up the record yet often enough to furnish distinct labels. It is even better than that. The text is not just arbitrary. In *Unweaving the Rainbow*, I made a Darwinian case for regarding an animal's DNA as a 'Genetic Book of the Dead': a descriptive record of ancestral worlds. It follows from the fact of Darwinian evolution that everything about an animal or plant, including its bodily form, its inherited behaviour and the chemistry of its cells, is a coded message about the worlds in which its ancestors survived: the food they sought; the predators they escaped; the climates they endured; the mates they beguiled. The message is ultimately scripted in the DNA that fell through the succession of sieves that is natural selection. When we learn to read it properly, the DNA of a dolphin may one day confirm what we already know from the telltale giveaways in its anatomy and physiology: that its ancestors once lived on dry land. Three hundred million years earlier, the ancestors of all land-dwelling vertebrates, including the land-dwelling ancestors of dolphins, came out of the sea where they had lived since the origin of life. Doubtless our DNA records this fact if we could read it. Everything about a modern animal, especially its DNA, but its limbs and its heart, its brain and its breeding cycle too, can be regarded as an archive, a chronicle of its past, even if that chronicle is a palimpsest, many times overwritten.

The DNA chronicle may be a gift to the historian, but it is a hard one to read, demanding deeply informed interpretation. It is made more powerful if combined with our third method of historical reconstruction, triangulation. It is to this that we now turn, and again we start with the analogous case of human history, specifically the history of languages.

Triangulation

Linguists often wish to trace languages back through history. Where written records survive it is rather easy. The historical linguist can use the second of our two methods of reconstruction, tracing back renewed relics, in this case words. Modern English goes back via Middle English to Anglo-Saxon using the continuous literary tradition, through Shakespeare, Chaucer and *Beowulf*. But speech obviously goes back long before the invention of writing, and many languages have no written form anyway. For the earlier history of dead languages, linguists resort to a version of what I am calling triangulation. They compare modern languages and group them hierarchically into families within families. Romance, Germanic, Slavic, Celtic and other European language families are in turn grouped with some Indian language families into Indo-European. Linguists believe that 'Proto-Indo-European' was an actual language, spoken by a particular tribe around 6,000 years ago. They even aspire to reconstruct many of its details by extrapolating back from the shared features of its descendants. Other language families in other parts of the world, of equivalent rank to Indo-European, have been traced back in the same way, for instance Altaic, Dravidian and Uralic-Yukaghir. Some optimistic (and controversial) linguists believe they can go back even further, uniting such major families in an even more all-embracing family of families. In this way they have persuaded themselves that they can reconstruct elements of a hypothetical ur-language which they call Nostratic, and which they believe was spoken between 12,000 and 15,000 years ago.

Many linguists, while happy about Proto-Indo-European and other ancestral languages of equivalent rank, doubt the possibility of reconstructing a language as ancient as Nostratic. Their professional scepticism reinforces my own amateur incredulity. But there is no doubt at all that equivalent triangulation methods – various techniques for comparing modern organisms – work for evolutionary history, and can be used for penetrating back hundreds of millions of years. Even if we had no fossils, a sophisticated comparison of modern animals would permit a fair and plausible reconstruction of their ancestors. Just as a linguist penetrates the past to Proto-Indo-European, triangulating from modern languages and from already reconstructed dead languages, we can do the same with modern organisms, comparing either their external characteristics or their protein or DNA sequences. As the libraries of the world accumulate long and exact DNA listings from more and more modern species, the reliability of our triangulations will increase, particularly because DNA texts have such a large range of overlaps.

Let me explain what I mean by 'range of overlaps'. Even when taken from extremely distant relations, for example humans and bacteria, large sections of DNA still unequivocally resemble each other. And very close relations, such as humans and chimpanzees, have much more DNA in common. If you choose your molecules judiciously, there is a complete spectrum of steadily increasing proportions of shared DNA, all the way in between. Molecules can be chosen which, between them, span the gamut of comparison, from remote cousins like humans and bacteria, to close cousins like two species of frogs. Resemblances between languages are harder to discern, all except close pairs of languages like German and Dutch. The chain of reasoning that leads some hopeful linguists to Nostratic is tenuous enough to make the links the subject of scepticism on the part of other linguists. Would the DNA equivalent of triangulating to Nostratic be triangulation between, say, humans and bacteria? But humans and bacteria have some genes that have hardly changed at all since the common ancestor, their equivalent of Nostratic. And the genetic code itself is virtually identical in all species and must have been the same in the shared ancestors. One could say that the resemblance between German and Dutch is comparable to that between any pair of mammals. Human and chimpanzee DNA are so similar, they are like English spoken in two slightly different accents. The resemblance between English and Japanese, or between Spanish and Basque, is so slight that no pair of living organisms can be chosen for analogy, not even humans and bacteria. Humans and bacteria have DNA sequences which are so similar that whole paragraphs are word-for-word identical.

I have been talking about using DNA sequences for triangulation. In principle it works for gross morphological characters as well but, in the absence of molecular information, distant ancestors are about as elusive as Nostratic. With morphological characters, as with DNA, we assume that features shared by many descendants of an ancestor are likely (or at least slightly more likely than not) to have been inherited from that ancestor. All vertebrates have a backbone and we assume that they inherited it (strictly inherited the genes for growing it) from a remote ancestor which lived, the fossils suggest, more than half a billion years ago and also had a backbone. It is this sort of morphological triangulation that has been used to help imagine the bodily forms of concestors in this book. I would have preferred to rely more heavily upon triangulation using DNA directly, but our ability to predict how a change in a gene will change the morphology of an organism is inadequate to the task.

Triangulation is even more effective if we include many species. But for this we need sophisticated methods which rely on having an accurately constructed family tree. These methods will be explained in the Gibbon's Tale. Triangulation also lends itself to a technique for calculating the date of any evolutionary branch point you like. This is the 'Molecular Clock'. Briefly, the method is to count discrepancies in molecular sequences between surviving species. Close cousins with recent common ancestors have fewer discrepancies than distant cousins, the age of the common ancestor being – or so it is hoped – proportional to the number of molecular discrepancies between their two descendants. Then we calibrate the arbitrary timescale of the molecular clock, translating it into real years, by using fossils of known date for a few key branch-points where fossils happen to be available. In practice it isn't as simple as that, and the complications, difficulties and associated controversies will occupy the Epilogue to the Velvet Worm's Tale.

Chaucer's General Prologue introduced the complete cast of his pilgrimage, one by one. My cast list is much too large for that. In any case, the narrative itself is a long sequence of introductions – at the 40 rendezvous points. But one preliminary introduction is necessary, in a way that it wasn't for Chaucer. His cast list was a set of individuals. Mine is a set of groupings. The way we group animals and plants needs introducing. At Rendezvous 10, our pilgrimage is joined by some 2,000 species of rodents, plus 87 species of rabbits, hares and pikas, collectively called Glires. Species are grouped in hierarchically inclusive ways, and each grouping has a name of its own (the family of mouse-like rodents is called Muridae, and of squirrel-like rodents Sciuridae). And each category of grouping has a name. Muridae is a family, so is Sciuridae. Rodentia is the name of the order to which both belong. Glires is the superorder that unites rodents with rabbits and their kind. There is a hierarchy of such category names, family and order being somewhere in the middle of the hierarchy. Species lies near the bottom of the hierarchy. We work up through genus (plural genera), family, order, class, and phylum (plural phyla), with prefixes like sub- and super- offering scope for interpolation.

Species has a particular status, as we shall learn in the course of various tales. Every species has a unique scientific binomial, consisting of its genus name with an initial capital letter, followed by its species name with no initial capital, both printed in italics. The leopard ('panther'), lion and tiger are all members of the genus *Panthera*: respectively *Panthera pardus, Panthera leo* and *Panthera tigris,* within the cat family, Felidae, which in turn is a member of the order Carnivora, the class Mammalia, the subphylum Vertebrata and the phylum Chordata. I shan't expatiate on the principles of taxonomy any further here, but will mention them, as necessary, during the book.

THE PILGRIMAGE BEGINS

It is time to set off on our pilgrimage to the past, which we can think of as a journey in a time machine in quest of our ancestors. Or more accurately, for reasons to be explained in the Neanderthal's Tale, in quest of our ancestral genes. For the first few tens of thousands of years of our backwards quest, our ancestral genes reside in individuals who look the same as us. Well, that is obviously not literally true, because we don't look exactly the same as each other. Let me rephrase it. For the first tens of thousands of years of our pilgrimage, the people we meet as we step outside our time machine will be no more different from us than we today are different from each other. Bear in mind that 'we today' includes Germans and Zulus, Pygmies and Chinese, Berbers and Melanesians. Our genetic ancestors of 50,000 years ago would have fallen within the same envelope of variability as we see around the world today.

If not biological evolution, then, what changes shall we see, as we go back through tens of millennia, as opposed to hundreds or thousands of millennia? There is an evolution-like process, orders of magnitude faster than biological evolution, which, in the early stages of our time machine's journey, will dominate the view from the porthole. This is variously called cultural evolution, exo-somatic evolution or technological evolution. We notice it in the 'evolution' of the motor car, or of the necktie or of the English language. We mustn't over-estimate its resemblance to biological evolution, and it will in any case not detain us long. We have a 4-billion-year road to run, and we shall soon have to set the time machine into a gear too high to allow us more than a fleeting glimpse of events on the scale of human history.

But first, while our time machine is still in bottom gear, travelling on the timescale of human rather than evolutionary history, a pair of tales about two major cultural advances. The Farmer's Tale is the story of the Agricultural Revolution, arguably the human innovation that has had the greatest repercussions for the rest of the world's organisms. And the Cro-Magnon's Tale is about the 'Great Leap Forward', that flowering of the human mind which, in a special sense, provided a new medium for the evolutionary process itself.

THE FARMER'S TALE

The agricultural revolution began at the wane of the last Ice Age, about 10,000 years ago, in the so-called Fertile Crescent between the Tigris and the Euphrates. This is the cradle of human civilisation whose irreplaceable relics in the Baghdad Museum were vandalised in 2003, during the chaos that attended the American invasion of Iraq. Agriculture also arose, probably independently, in China and along the banks of the Nile, and completely independently in the New World. An interesting case can be made for yet another independent cradle of agricultural civilisation in the astonishingly isolated highland interior of New Guinea. The Agricultural Revolution dates the start of the new stone age, the Neolithic.

The transition from wandering hunter-gatherers to a settled agricultural lifestyle may represent the first time people had a concept of a home. Contemporaries of the first farmers, in other parts of the world, were unreconstructed hunter-gatherers who wandered more-or-less continuously. Indeed, the hunter-gatherer lifestyle ('hunter' can include fisher) has not died out. It is still practised in pockets around the world: by Australian Aborigines, by San and related tribes in Southern Africa (called 'bushmen'), by various Native American tribes (called 'Indians' after a navigational error), and by the Inuit of the Arctic (who prefer not to be called Eskimos). Hunter-gatherers typically do not cultivate plants and do not keep livestock. In practice all intermediates between pure hunter-gatherers and pure agriculturalists or pastoralists are found. But, earlier than about 10,000 years ago, all human populations were hunter-gatherers. Soon, probably none will be. Those not extinct will be 'civilised' – or corrupted, depending on your point of view.

Colin Tudge, in his little book *Neanderthals, Bandits and Farmers: How Agriculture Really Began,* agrees with Jared Diamond (*The Third Chimpanzee*) that the switch to agriculture from hunting and gathering was by no means the improvement we, in our complacent hindsight, might think. The Agricultural Revolution did not, in their view, increase human happiness. Agriculture supported larger populations than the hunter-gatherer lifestyle that it superseded, but not in obviously improved health or happiness. In fact, larger populations generally harbour more vicious diseases, for sound evolutionary reasons (a parasite is less concerned to prolong the life of its present host if it can easily find new victims to infect).

Nevertheless, our situation as hunter-gatherers cannot have been a Utopia either. It has lately become fashionable to regard hunter-gatherers and primitive* agricultural societies as more 'in balance' with nature than us. This is probably a mistake. They may well have had greater knowledge of the wild, simply because they lived and survived in it. But, like us, they seem to have used their knowledge to exploit (and often overexploit) the environment to the best of their abilities at the time. Jared Diamond emphasises overexploitation by early agriculturalists leading to ecological collapse, and the demise of their society. Far from being in balance with nature, pre-agricultural hunter-gatherers were probably responsible for widespread extinctions of many large animals around the globe. Just prior to the Agricultural Revolution, the colonisation of remote areas by hunter-gatherer peoples is suspiciously often followed in the archaeological record by the wiping out of many large (and presumably palatable) birds and mammals.

We tend to regard 'urban' as the antithesis of 'agricultural' but, in the longer perspective that this book must adopt, city dwellers should be lumped in with farmers as opposed to hunter-gatherers. Almost all the food of a town comes from owned and cultivated land – in ancient times from fields round about the town, in modern times from anywhere in the world, transported and sold on through middlemen before being consumed. The Agricultural Revolution soon led to specialisation. Potters, weavers and smiths traded their skills for food which others grew. Before the Agricultural Revolution, food was not cultivated on owned land but captured or gathered on unowned commons. Pastoralism, the herding of animals on common land, may have been an intermediate stage.

Whether it was a change for better or worse, the Agricultural Revolution was presumably not a sudden event. Husbandry was not the overnight brainwave of some genius, the neolithic equivalent of Turnip Townshend. To begin with, hunters of wild animals in open and unowned country might have guarded hunting territories against rival hunters, or guarded the herds themselves while following them about. From there it was a natural progression to herding

---

* Throughout this book I use 'primitive' in the technical sense, to mean 'more like the ancestral state'. No implication of inferiority is intended.

them; then feeding them, and finally corralling and housing them. I dare say none of these changes would have seemed revolutionary when they happened.

Meanwhile the animals themselves were evolving – becoming 'domesticated' by rudimentary forms of artificial selection. The Darwinian consequences on the animals would have been gradual. Without any deliberate intention to breed 'for' domestic tractability, our ancestors inadvertently changed the selection pressures on the animals. Within the gene pools of the herds, there would no longer be a premium on fleetness or other survival skills of the wild. Successive generations of domestic animals became tamer, less able to fend for themselves, more apt to flourish and grow fat under feather-bedded domestic conditions. There are alluring parallels in the domestication, by social ants and termites, of aphid 'cattle' and fungus 'crops'. We shall hear about these in the Leaf Cutter's Tale, when the ant pilgrims join us at Rendezvous 26.

Unlike modern plant and animal breeders, our forebears of the Agricultural Revolution would not knowingly have practised artificial selection for desirable characteristics. I doubt if they realised that, in order to increase milk yield, you have to mate high-yielding cows with bulls born to other high-yielding cows, and discard the calves of low-yielders. Some idea of the accidental genetic consequences of domestication is given by some interesting Russian work on silver foxes.

D. K. Belyaev and his colleagues took captive silver foxes, *Vulpes vulpes*, and set out systematically to breed for tameness. They succeeded, dramatically. By mating together the tamest individuals of each generation, Belyaev had, within 20 years, produced foxes that behaved like Border collies, actively seeking human company and wagging their tails when approached. That is not very surprising, although the speed with which it happened may be. Less expected were the by-products of selection for tameness. These genetically tamed foxes not only behaved like collies, they looked like collies. They grew black-and-white coats, with white face patches and muzzles. Instead of the characteristic pricked ears of a wild fox, they developed 'lovable' floppy ears. Their reproductive hormone balance changed, and they assumed the habit of breeding all the year round instead of in a breeding season. Probably associated with their lowered aggression, they were found to contain higher levels of the neurally active chemical serotonin. It took only 20 years to turn foxes into 'dogs' by artificial selection.*

I put 'dogs' in inverted commas, because our domestic dogs are not descended from foxes, they are descended from wolves. Incidentally, Konrad Lorenz's well-known speculation that only some breeds of dog (his favourites such as chow chows) are derived from wolves, the rest from jackals is now known to be wrong. He supported his theory with insightful anecdotes on temperament and behaviour. But molecular taxonomy trumps human insight, and molecular evidence clearly shows that all modern breeds of dog are descended from the grey wolf, *Canis lupus*. The next closest relatives to dogs (and wolves) are coyotes, and Simien 'jackals' (which it now seems should be called Simien wolves). True jackals (golden, side-striped and black-backed jackals) are more distantly related, although they are still placed in the genus *Canis*.

No doubt the original story of the evolution of dogs from wolves was similar to the new one simulated by Belyaev with foxes, with the difference that Belyaev was breeding for tameness deliberately. Our ancestors did it inadvertently, and it probably happened several times, independently in different parts of the world. Perhaps initially, wolves took to scavenging around human encampments. Humans may have found such scavengers a convenient means of refuse disposal, and they may also have valued them as watchdogs, and even as warm sleep comforters. If this amicable scenario sounds surprising, reflect that the medieval legend of wolves as mythic symbols of terror coming out of the forest was born of ignorance. Our wild ancestors, living in more open country, would have known better. Indeed, they evidently did know better, because they ended up domesticating the wolf, thereby making the loyal, trusted dog.

From the wolf's point of view human camps provided rich pickings for a scavenger, and the individuals most likely to benefit were those whose serotonin levels and other brain characteristics ('propensity to tameness') happened to make them feel at home with humans. Several writers have speculated, plausibly enough, about orphaned cubs being adopted as pets by children. Experiments have shown that domestic dogs are better than wolves at 'reading' the expressions on human faces. This is presumably an inadvertent

---

* The Canadian archaeologist Susan Crockford has attributed such changes to changing levels of two thyroid hormones.

consequence of our mutualistic evolution over many generations. At the same time we read their faces, and dog facial expressions have become more human-like than those of wolves, because of inadvertent selection by humans. This is presumably why we think wolves look sinister while dogs look loving, guilty, soppy and so on.

A distant parallel is the case of the Japanese 'samurai crabs'. These wild crabs have a pattern on their back which resembles the face of a Samurai warrior. The Darwinian theory to account for this is that superstitious fishermen tossed back into the sea individual crabs that slightly resembled a Samurai warrior. Over the generations, as genes for resembling a human face were more likely to survive in the bodies of 'their' crabs, the frequency of such genes increased in the population until today it is the norm. Whether that story of wild crabs is true or not, something like it surely went on in the evolution of truly domesticated animals.

Back to the Russian fox experiment, which demonstrates the speed with which domestication can happen, and the likelihood that a train of incidental effects would follow in the wake of selection for tameness. It is entirely probable that cattle, pigs, horses, sheep, goats, chickens, geese, ducks and camels followed a course which was just as fast, and just as rich in unexpected side-effects. It also seems plausible that we ourselves evolved down a parallel road of domestication after the Agricultural Revolution, towards our own version of tameness and associated by-product traits.

In some cases, the story of our own domestication is clearly written in our genes. The classic example, meticulously documented by William Durham in his book *Coevolution,* is lactose tolerance. Milk is baby food, not 'intended' for adults and, originally, not good for them. Lactose, the sugar in milk, requires a particular enzyme, lactase, to digest it. (This terminological convention is worth remembering, by the way. An enzyme's name will often be constructed by adding '-ase' to the first part of the name of the substance on which it works.) Young mammals switch off the gene that produces lactase after they pass the age of normal weaning. It isn't that they lack the gene, of course. Genes needed only in childhood are not removed from the genome, not even in butterflies, which must carry large numbers of genes needed only for making caterpillars. But lactase production is switched off in human infants at the age of about four, under the influence of other, controlling genes. Fresh milk makes adults feel ill, with symptoms ranging from flatulence and intestinal cramps to diarrhoea and vomiting.

All adults? No, of course not. There are exceptions. I am one of them, and there is a good chance that you are too. My generalisation concerned the human species as a whole and, by implication, the wild *Homo sapiens* from which we are all descended. It is as if I had said 'Wolves are big, fierce carnivores that hunt in packs and bay at the moon', knowing full well that Pekineses and Yorkshire terriers belie it. The difference is that we have a separate word, dog, for domestic wolf, but not for domestic human. The genes of domestic animals have changed as a result of generations of contact with humanity, inadvertently following the same sort of course as the genes of the silver fox. The genes of (some) humans have changed as a result of generations of contact with domestic animals. Lactose tolerance seems to have evolved in a minority of tribes including the Tutsi of Rwanda (and, to a lesser extent, their traditional enemies the Hutu), the pastoral Fulani of West Africa (though, interestingly, not the sedentary branch of the Fulani), the Sindhi of North India, the Tuareg of West Africa, the Beja of Eastern North Africa, and some European tribes from which I, and possibly you, are descended. Significantly, what these tribes have in common is a history of pastoralism.

At the other end of the spectrum, peoples who have retained the normal human intolerance of lactose as adults include Chinese, Japanese, Inuit, most Native Americans, Javanese, Fijians, Australian Aborigines, Iranians, Lebanese, Turks, Tamils, Singhalese, Tunisians, and many African tribes including the San, and the Tswanas, Zulus, Xhosas and Swazis of southern Africa, the Dinkas and Nuers of North Africa, and the Yorubas and Igbos of West Africa. In general, these lactose-intolerant peoples do not have a history of pastoralism. There are instructive exceptions. The traditional diet of the Masai of East Africa consists of little else besides milk and blood, and you might think they'd be particularly tolerant of lactose. This is not the case, however, probably because they curdle their milk before consuming it. As with cheese, the lactose is largely removed by bacteria. That's one way of getting rid of its bad effects – get rid of the stuff itself. The other way is to change your genes. This happened in the other pastoral tribes listed above.

Of course nobody deliberately changes their genes. Science is only now beginning to work out how to do that. As usual, the job was done for us by natural selection, and it happened millennia ago. I don't know exactly by what route natural selection produced adult lactose tolerance. Perhaps adults resorted to baby food in times of desperation, and the individuals that were most tolerant of it survived better. Perhaps some cultures postponed weaning, and selection for survival of children under these conditions spilled over gradually into adult tolerance. Whatever the details, the change, though genetic, was culture-driven. The evolution of tameness and increasing milk yields in cattle, sheep and goats paralleled that of lactose tolerance in the tribes that herded them. Both were true evolutionary trends in that they were changes in gene frequencies in populations. But both were driven by non-genetic cultural changes.

Is lactose tolerance just the tip of the iceberg? Are our genomes riddled with evidences of domestication, affecting not just our biochemistry but our minds? Like Belyaev's domesticated foxes, and like the domesticated wolves that we call dogs, have we become tamer, more lovable, with the human equivalents of floppy ears, soppy faces and wagging tails? I leave you with the thought, and move hastily on.

While hunting was sliding into herding, gathering presumably followed a similar slide into cultivation of plants. Again, it was probably mostly inadvertent. No doubt there were moments of creative discovery, as when people first noticed that if you put seeds in the ground they make plants like those from which they came. Or when somebody first observed that it helps to water them, weed them and manure them. It was probably more difficult to work out that it might be a good idea to keep back the best seed for planting, rather than follow the obvious course of eating the best and planting the dross (my father, as a young man fresh out of college, taught agriculture to peasant farmers in central Africa in the 1940s, and he tells me that this was one of the hardest lessons to get across). But mostly the transition from gatherer to cultivator passed unnoticed by those concerned, like the transition from hunter to herder.

Many of our staple food crops, including wheat, oats, barley, rye and maize, are members of the grass family which have become greatly modified since the dawn of agriculture by inadvertent and later deliberate human selection. It is possible that we too have become genetically modified over the millennia to increase our tolerance of cereals, in a way parallel to our evolution of tolerance to milk. Starchy cereals such as wheat and oats cannot have featured prominently in our diets before the Agricultural Revolution. Unlike oranges and strawberries, cereal seeds do not 'want' to be eaten. Passing through an animal's digestive tract is no part of their dispersal strategy, as it is of plum and tomato seeds. On our side of the relationship, the human digestive tract is not able, unaided, to absorb much nutriment from seeds of the grass family, with their meagre starch reserves and hard, unsympathetic husks. Some aid comes from milling and cooking, but it also seems conceivable that, in parallel with the evolution of tolerance to milk, we might have evolved an increased physiological tolerance to wheat, compared to our wild ancestors. Wheat intolerance is a known problem for a substantial number of unfortunate individuals who discover, by painful experience, that they are happier if they avoid it. A comparison of the incidence of wheat intolerance in hunter-gatherers such as the San, and other peoples whose agricultural ancestors have long eaten wheat, might be revealing. If there has been a large comparative study of wheat tolerance, like the one that has been made of lactose tolerance across different tribes, I am unaware of it. A systematic comparative study of alcohol intolerance, too, would be interesting. It is known that certain genetic alleles make our livers less capable of breaking down alcohol than we might wish.

In any case, co-evolution between animals and their food plants was nothing new. Grazing animals had been exerting a kind of benevolent Darwinian selection on grasses, guiding their evolution towards mutualistic co-operation, for millions of years before we started domesticating wheat, barley, oats, rye and maize. Grasses flourish in the presence of grazers, and they probably have been doing so for most of the 20 million years since their pollens first announce them in the fossil record. It is not, of course, that individual plants actually benefit by being eaten, but that grasses can withstand being cropped better than rival plants can. My enemy's enemy is my friend, and grasses, even when grazed, thrive when herbivores eat (along with the grasses themselves) other plants that would compete for soil, sun and water. Grasses became ever more able to thrive in the presence of wild cattle, antelopes, horses and other grazers (and eventually lawnmowers), as the millions of years went by. And the herbivores became better equipped, for example with specialised teeth, and complicated digestive tracts including fermentation vats with cultures of micro-organisms, to flourish on a diet of grass.

This isn't what we ordinarily mean by domestication, but in effect it is not far from it. When, starting about 10,000 years ago, wild grasses of the genus *Triticum* were domesticated by our ancestors into what we now call wheat, it was, in a way, a continuation of what herbivores of many kinds had been doing to the ancestors of *Triticum* for 20 million years. Our ancestors accelerated the process, especially when we later switched from inadvertent, accidental domestication to deliberate, planned selective breeding (and very recently scientific hybridisation and genetically engineered mutations).

That is all I want to say about the origins of agriculture. Now, as our time machine leaves the 10,000-year mark and heads for Rendezvous 0, we briefly pause, one more time, around 40,000 years ago. Here human society, entirely consisting of hunter-gatherers, underwent what may have been an even larger revolution than the agricultural one, the 'cultural Great Leap Forward'. The tale of the Great Leap Forward will be told by Cro-Magnon Man, named after the cave in the Dordogne where fossils of this race of *Homo sapiens* were first discovered.

THE CRO-MAGNON'S TALE

Archaeology suggests that something very special began to happen to our species around 40,000 years ago. Anatomically, our ancestors who lived before this watershed date were the same as those who came later. Humans sampled earlier than the watershed would be no more different from us than they were from their own contemporaries in other parts of the world, or indeed than we are from our contemporaries. That's if you look at their anatomy. If you look at their culture, there is a huge difference. Of course there are also huge differences between the cultures of different peoples across the world today, and probably then too. But this wasn't true if we go back much more than 40,000 years. Something happened then – many archaeologists regard it as sudden enough to be called an 'event'. I like Jared Diamond's name for it, the Great Leap Forward.

Earlier than the Great Leap Forward, man-made artefacts had hardly changed for a million years. The ones that survive for us are almost entirely stone tools and weapons, quite crudely shaped. Doubtless wood (or, in Asia, bamboo) was a more frequently worked material, but wooden relics don't easily survive. As far as we can tell, there were no paintings, no carvings, no figurines, no grave goods, no ornamentation. After the Leap, all these things suddenly appear in the archaeological record, together with musical instruments such as bone flutes, and it wasn't long before stunning creations like the Lascaux Cave murals were created by Cro-Magnon people (see plate 2). A disinterested observer taking the long view from another planet might see our modern culture, with its computers, supersonic planes and space exploration, as an afterthought to the Great Leap Forward. On the very long geological timescale, all our modern achievements, from the Sistine Chapel to Special Relativity, from the *Goldberg Variations* to the Goldbach Conjecture, could be seen as almost contemporaneous with the Venus of Willendorf and the Lascaux Caves, all part of the same cultural revolution, all part of the blooming cultural upsurge that succeeded the long Lower Palaeolithic stagnation. Actually I'm not sure that our extra-planetary observer's uniformitarian view would stand up to much searching analysis, but it could be at least briefly defended.

David Lewis-Williams's *The Mind in the Cave* considers the whole question of Upper Palaeolithic cave art, and what it can tell us about the flowering of consciousness in *Homo sapiens*.

Some authorities are so impressed by the Great Leap Forward that they think it coincided with the origin of language. What else, they ask, could account for such a sudden change? It is not as silly as it sounds to suggest that language arose suddenly. Nobody thinks writing goes back more than a few thousand years, and everyone agrees that brain anatomy didn't change to coincide with anything so recent as the invention of writing. In theory, speech could be another example of the same thing. Nevertheless, my hunch, supported by the authority of linguists such as Steven Pinker, is that language is older than the Leap. We'll come back to the point a million years further into the past, when our pilgrimage reaches *Homo ergaster* (*erectus*).

If not language itself, perhaps the Great Leap Forward coincided with the sudden discovery of what we might call a new software technique: maybe a new trick of grammar, such as the conditional clause, which, at a stroke, would have enabled 'what if' imagination to flower. Or maybe early language, before the leap, could be used to talk only about things that were there, on the scene. Perhaps some forgotten genius realised the possibility of using words referentially as tokens of things that were not immediately present. It is the difference between 'That waterhole which we can both see' and 'Suppose there was a waterhole the other side of the hill'. Or perhaps representational art, which is all but unknown in the archaeological record before the Leap, was the bridge to referential language. Perhaps people learned to draw bison, before they learned to talk about bison that were not immediately visible.

Much as I would like to linger around the heady time of the Great Leap Forward, we have a long pilgrimage to accomplish and we must press on backwards. We are approaching the point where we can start looking for Concestor 0, the most recent ancestor of all surviving humans.

ALL HUMANKIND

The human genome project has reached completion, hailed by a justly proud humanity. We might pardonably wonder *whose* genome has been sequenced. Has an illustrious dignitary been singled out for the honour, or is it a random nobody pulled off the street, or even an anonymous clone of cells from a tissue culture lab? It makes a difference because we vary. I have brown eyes while you, perhaps, have blue. I can't curl my tongue into a tube, whereas it's 50/50 that you can. Which version of the tongue-curling gene makes it into the published human genome? What is the canonical eye colour?

I raise the question only to draw a parallel. This book traces 'our' ancestors back through time, but *whose* ancestors are we talking about: yours or mine, a Bambuti Pygmy's or a Torres Strait Islander's? I shall come to the question presently. But first, having raised the analogous question about the Human Genome Project, I can't just leave it dangling. Whose genome is chosen for analysis? In the case of the 'official' Human Genome Project the answer is that, for the low percentage of DNA letters that vary, the canonical genome is the majority 'vote' among a couple of hundred people chosen to give a good spread of racial diversity. In the case of the rival project initiated by Dr Craig Venter, the genome analysed was mostly that of . . . Dr Craig Venter. This was announced by the man himself,* to the mild consternation of the ethics committee which had recommended, for all sorts of warm and worthy reasons, that the donors should be anonymous and drawn from a spread of different races. There are other projects for the study of human genetic diversity itself, which, bizarrely, come under recurrent political attack as though it were somehow improper to admit that humans vary. Thank goodness we do, if not very much.

But now, to our backwards pilgrimage. Whose ancestors are we going to trace? If we go sufficiently far back, everybody's ancestors are shared. All your ancestors are mine, whoever you are, and all

---

* When his team went on to decipher the dog genome, it was no surprise to discover that the individual honoured was Dr Venter's own poodle, Shadow.

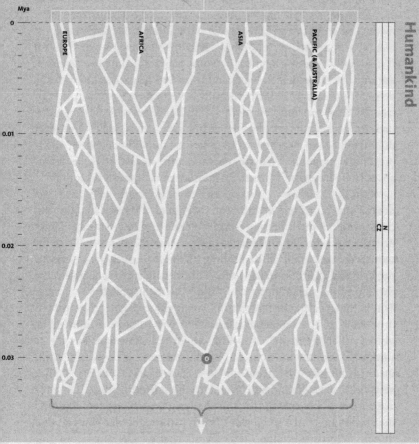

Humankind. A stylised impression of the human family tree. It is not intended as an accurate depiction – the real tree would be unmanageably dense. Moving down the page means going back in time, with the geological timescale (see plate 1) given by the bar on the right. White lines illustrate patterns of interbreeding, with lots of it within continents and occasional migration between them. The numbered circle marks Concestor 0, the most recent common ancestor of all living humans. Verify this by following routes upwards from Concestor 0: you can reach any of the modern-day-human end points.

mine are yours. Not just approximately but literally. This is one of those truths that turns out, on reflection, to need no new evidence. We prove it by pure reason, using the mathematician's trick of *reductio ad absurdum*. Take our imaginary time machine absurdly far back, say 100 million years, to an age when our ancestors resembled shrews or opossums. Somewhere in the world at that ancient date, at least one of my personal ancestors must have been living, or I wouldn't be here. Let us call this particular little mammal Henry (it happens to be a family name). We seek to prove that if Henry is my ancestor he must be yours too. Imagine, for a moment, the contrary: I am descended from Henry and you are not. For this to be so, your lineage and mine would have to have marched, side by side yet never touching, through 100 million years of evolution to the present, never interbreeding yet ending up at the same evolutionary destination – so alike that your relatives are still capable of interbreeding with mine. This *reductio* is clearly absurd. If Henry is my ancestor he has to be yours too. If not mine, he cannot be yours.

Without specifying how ancient is 'sufficiently', we have just proved that a sufficiently ancient individual with any human descendants at all must be an ancestor of the entire human race. Long-distance ancestry, of a particular group of descendants such as the human species, is an all-or-nothing affair. Moreover, it is perfectly possible that Henry is my ancestor (and necessarily yours, given that you are human enough to be reading this book) while his brother Eric is the ancestor of, say, all the surviving aardvarks. Not only is it possible. It is a remarkable fact that there *must* be a moment in history when there were two animals in the same species, one of whom became the ancestor of all humans and no aardvarks, while the other became the ancestor of all aardvarks and no humans. They may well have met, and may even have been brothers. You can cross out aardvark and substitute any other modern species you like, and the statement must still be true. Think it through, and you will find that it follows from the fact that all species are cousins of one another. Bear in mind when you do so that the 'ancestor of all aardvarks' will also be the ancestor of lots of very different things besides aardvarks (in this case, the entire major group called Afrotheria which we shall meet at Rendezvous 13, and which includes elephants and dugongs, hyraxes and Madagascan tenrecs).

My reasoning was constructed as a *reductio ad absurdum*. It assumed that 'Henry' lived long enough ago for it to be obvious that he begat either all living humans, or none. How long is long enough? That's a harder question. A hundred million years is more than enough to assure the conclusion we seek. If we go back only a hundred years, no individual can claim the entire human race as direct descendants. Between the obvious cases of 100 years and 100 million, what can we say about unobvious intermediates such as 10,000, 100,000 or 1 million years? The precise calculations were beyond me when I explained this *reductio* in *River Out of Eden* but, happily, a Yale University statistician called Joseph T. Chang has now made a start on them. His conclusions and their implications form the Tasmanian's Tale, a tale of particular relevance to this rendezvous because Concestor 0 is the most recent common ancestor of all living humans. It is more elaborate versions of calculations like Chang's that we need to do in order to date Rendezvous 0.

Rendezvous 0 is the time when, on our backwards pilgrimage, we *first* meet a common human ancestor. But according to our *reductio* there is a point further in the past when *every* individual that we encounter with our time machine is either a common ancestor or no ancestor at all. And although no one ancestor can be singled out for attention at this more distant milestone, it is worth a nod as we go by, because it marks the point where we can stop worrying about whether it is your ancestors we trace or mine: from that milestone on, all my readers march, shoulder to shoulder, in a phalanx of pilgrims towards the past.

THE TASMANIAN'S TALE
Written with Yan Wong

Tracing ancestors is a beguiling pastime. As with history itself, there are two methods. You can go backwards, listing your two parents, four grandparents, eight great-grandparents, and so on. Or you can pick a distant ancestor and go forwards, listing his children, grandchildren, great-grandchildren, until you end up with yourself. Amateur genealogists do both, going back and forth between generations until they have filled in the tree as far as parish registers and family Bibles allow. This tale, like the book as a whole, uses the backwards method.

Pick any two people and go backwards and, sooner or later, we hit a most recent common ancestor – MRCA. You and me, the plumber and the queen, any set of us must converge on a single concestor (or couple). But unless we pick close relatives, finding the concestor requires a vast family tree, and most of it will be unknown. This applies *a fortiori* to the concestor of all humans alive today. Dating Concestor 0, the most recent common ancestor of all living humans, is not a task that can be undertaken by a practising genealogist. It is a task in estimation: a task for a mathematician.

An applied mathematician tries to understand the real world by setting up a simplified version of it – a 'model'. The model eases thought, while not losing all power to illuminate reality. Sometimes a model gives us a baseline, departures from which elucidate the real world.

In framing a mathematical model to date the common ancestors of all surviving humans, a good simplifying assumption – a sort of toy world – is a breeding population of fixed and constant size, living on an island with no immigration or emigration. Let it be an idealised population of Tasmanian aboriginals, in happier times before they were exterminated as agricultural vermin by nineteenth-century settlers. The last pure-bred Tasmanian, Truganinni, died in 1876, soon after her friend 'King Billy' whose scrotum was made into a tobacco pouch (shades of Nazi lamps). The Tasmanian aboriginals were isolated some 13,000 years ago when land bridges to Australia were flooded by rising sea levels, and they then saw no outsiders until they saw them with a vengeance in their nineteenth-century holocaust. For our modelling purposes, we consider Tasmania to be perfectly isolated from the rest of the world for 13,000 years until 1800. Our notional 'present', for modelling purposes, will be defined as 1800 AD.

The next step is to model the mating pattern. In the real world people fall in love, or into arranged marriages, but here we are modellers, ruthlessly replacing human detail by tractable mathematics. There's more than one mating model we could imagine. The random diffusion model has men and women behaving as particles diffusing outwards from their birthplace, more likely to bump into near than distant neighbours. An even simpler and less realistic model is the random mating model. Here, we forget about distance altogether and simply assume that, strictly within the island, mating between any male and any female is equally likely.

Of course neither model is remotely plausible. Random diffusion assumes that people walk in any direction from their starting point. In reality there are paths or roads which guide their feet: narrow gene conduits through the island's forests and grasslands. The random mating model is even more unrealistic. Never mind. We set up models to see what happens under ideally simplified conditions. It can be surprising. Then we have to consider whether the real world is more surprising or less, and in which directions.

Joseph Chang, following a long tradition of mathematical geneticists, opted for random mating. His model ignored population size by assuming it constant. He did not deal with Tasmania in particular but we shall assume, again as a calculated oversimplification, that our toy population remained constant at 5,000, which is one estimate for Tasmania's aboriginal population in 1800 before the massacres began. I must repeat that such simplifications are of the essence in mathematical modelling: not a weakness of the method but, for certain purposes, a strength. Chang of course doesn't believe people mate at random, any more than Euclid believed lines have no breadth. We follow abstract assumptions to see where they lead, and then decide whether the detailed differences from the real world matter.

So, how many generations would you have to go back, in order to be reasonably sure of finding an individual who was ancestor to everybody alive in the present? The calculated answer from the abstract model is the logarithm (base 2) of the population size. The base 2 logarithm of a number is the number of times you have to multiply 2 by itself to get that number. To get 5,000, you need to multiply 2 by itself about 12.3 times so, for our Tasmanian example, theory tells us to go back 12.3 generations to find the concestor. Assuming four generations per century, this is less than four centuries. It's even less if people reproduce younger than 25.

I give the name 'Chang One' to the date of the most recent common ancestor of some specified population. Continuing backwards from Chang One, it doesn't take long before we hit the point – I shall call it 'Chang Two' – at which *everybody* is either a common ancestor or has no surviving descendants. Only during the brief interregnum between Chang One and Chang Two does there exist an intermediate category of people who have some surviving descendants but are not common ancestors of everybody. A surprising deduction, whose rationale I won't spell out, is that at Chang Two a large number of people are universal ancestors: about 80 per cent of individuals in any generation will in theory be ancestors of everybody alive in the distant future.

As for the timing, well, the mathematics yield the result that Chang Two is approximately 1.77 times older than Chang One. 1.77 times 12.3 gives just under 22 generations, between five and six centuries. As we ride our time machine backwards in Tasmania, therefore, around the time of Geoffrey Chaucer in England we enter 'all or nothing' territory. From there on backwards, to the time when Tasmania was joined to Australia and all bets are off, everyone our time machine encounters will have either the entire population as descendants or no descendants at all.

I don't know about you, but I find these calculated dates astonishingly recent. What's more, the conclusions don't change much if you assume a larger population. Taking a model population the size of Britain's today, 60 million, we still need to go back only 23 generations to reach Chang One and our youngest universal ancestor. If the model applied to Britain, Chang Two, when everybody is either the ancestor of all modern British people or of none, is only about 40 generations ago, or about 1000 AD. If the assumptions of the model are true (of course they aren't) King Alfred the Great is the ancestor of either all today's British or none.[*]

I must repeat the cautions with which I began. There are all sorts of differences between 'model' and 'real' populations, in Britain or Tasmania or anywhere else. Britain's population has climbed steeply in historical time to reach its present size, and that completely changes the calculations. In any real population, people don't mate at random. They favour their own tribe, language group or local area, and of course they all have individual preferences. Britain's history adds the complication that, although a geographical island, its population is far from isolated. Waves of external immigrants have swept in from Europe over the centuries: Romans, Saxons, Danes, and Normans among them.

If Tasmania and Britain are islands, the world is a larger 'island' since it has no immigration or emigration (give or take alien abductions in flying saucers). But it is imperfectly subdivided into continents and smaller islands, with not just seas but mountain ranges,

---

[*] With characteristic prescience, the great statistician and evolutionary geneticist Sir Ronald Fisher (1890–1962) wrote the following, in a letter dated 15 January 1929 to Major Leonard Darwin (1850–1943, Charles's second youngest son): 'King Solomon lived 100 generations ago, and his line may be extinct; if not, I wager he is in the ancestry of all of us, and in nearly equal proportions, however unequally his wisdom may be distributed.' In J. H. Bennett, (Ed. 1983) *Natural Selection, Heredity and Eugenics.* Oxford: Clarendon Press, p. 95.

rivers and deserts impeding the movement of people to varying degrees. Complicated departures from random mating confound our calculations, not just slightly but grossly. The present population of the world is 6 billion, but it would be absurd to look up the logarithm of 6 billion and swallow the resulting medieval date for Rendezvous o! The real date is older, if only because pockets of humanity have been separated far longer than the orders of magnitude we are now calculating. If an island has been isolated for 13,000 years, as Tasmania was, it is impossible for the human race as a whole to have a universal ancestor younger than 13,000 years. Even partial isolation of sub-populations plays havoc with our all-too-tidy calculations, as does any kind of non-random mating.

The date when the most isolated island population in the world became isolated sets a lower bound on the date of Rendezvous o. But to take this lower bound seriously, isolation must be absolute. This follows from the calculated figure of 80 per cent that we met earlier. A single migrant to Tasmania, once he has been sufficiently accepted into society to reproduce normally, has an 80 per cent chance of eventually becoming a common ancestor to all Tasmanians. So even tiny amounts of migration are enough to graft the family tree of an otherwise isolated population to that of the mainland. The timing of Rendezvous o is likely to depend on the date at which the most isolated pocket of humanity became completely isolated from its neighbour, plus the date at which its neighbour then became completely isolated from *its* neighbour, and so on. A few island hops may be needed before we can join all the family trees together, but it is then an insignificant number of centuries back until we tumble upon Concestor o. That would put Rendezvous o some few tens of thousands of years ago, conceivably somewhere in the high tens of thousands, no more.

As to where Rendezvous o took place, this is almost as surprising. You might be inclined to think of Africa, as was my initial reaction. Africa houses the deepest genetic divides within humankind, so it seems a logical place to look for a common ancestor of all living humans. It has been well said that if you wiped out sub-Saharan Africa you would lose the great majority of human genetic diversity, whereas you could wipe out everywhere except Africa and nothing much would change. Nevertheless Concestor o may well have lived outside Africa. Concestor o is the most recent common ancestor that unites the most geographically isolated population – Tasmania for the sake of argument – with the rest of the world. If we assume that populations throughout the rest of the world, including Africa, indulged in at least some interbreeding during a long period when Tasmania was totally isolated, the logic of Chang's calculations could lead us to suspect that Concestor 0 lived outside Africa, near the take-off point for the migrants whose offspring became Tasmanian immigrants. Yet African groups still retain most of humanity's genetic diversity. This seeming paradox is resolved in the next tale, when we explore family trees of genes rather than of people.

Our surprising conclusion is that Concestor 0 probably lived tens of thousands of years ago, and very possibly not even in Africa.[*] Other species too may generally have quite recent common ancestors. But this is not the only part of the Tasmanian's Tale that forces us to examine biological ideas in a new light. To professional Darwinian specialists, it seems a paradox that 80 per cent of a population will become universal ancestors. Let me explain. We are used to thinking of individual organisms as striving to maximise a quantity called 'fitness'. Exactly what fitness means is disputed. One favoured approximation is 'total number of children'. Another is 'total number of grandchildren', but there is no obvious reason to stop at grandchildren, and many authorities prefer to say something like 'total number of descendants alive at some distant date in the future'. But we seem to have a problem if, in our theoretically idealised population in the absence of natural selection, 80 per cent of the population can expect to have the maximum possible 'fitness': that is, they can expect to claim the entire population as their descendants! This matters for Darwinians because they widely presume that 'fitness' is what all animals constantly struggle to maximise.

I have long argued that the only reason an organism behaves as a quasi-purposeful entity at all – an entity capable of maximising anything – is that it is built by genes that have survived through past generations. There is a temptation to personify and impute intention: to turn 'gene survival in the past' into something like 'intention to reproduce in the future'. Or 'individual intention to have lots of descendants in the future'. Such personification can also apply to genes: we are tempted to see genes as influencing individual

[*] Just after the first print run of this book, a paper by Rohde, Olsen and Chang was published in *Nature* magazine (Vol. 431, p. 562) suggesting that Rendezvous 0 occurred a mere 3,500 years ago, even more recently than I dared hope. They too come to the conclusion that Concestor 0 was probably Asian.

bodies to behave in such a way as to increase the number of future copies of those same genes.

Scientists who use such language, whether at the level of the individual or the gene, know very well that it is only a figure of speech. Genes are just DNA molecules. You'd have to be barking mad to think that 'selfish' genes *really* have deliberate intentions to survive! We can always translate back into respectable language: the world becomes full of those genes that have survived in the past. Because the world has a certain stability and doesn't change capriciously, the genes that have survived in the past tend to be the ones that are going to be good at surviving in the future. That means good at programming bodies to survive and make children, grandchildren and long-distance descendants. So, we have arrived back at our individual-based definition of fitness looking into the future. But we now recognise that individuals matter only as vehicles of gene survival. Individuals having grandchildren and distant descendants is only a means to the end of gene survival. And this brings us again to our paradox. 80 per cent of reproducing individuals seem to be crammed up against the ceiling – saturated out at maximum fitness!

To resolve the paradox, we return to the theoretical bedrock: the genes. We neutralise one paradox by erecting another, almost as if two wrongs could make a right. Think on this: an individual organism can be a universal ancestor of the entire population at some distant time in the future, and yet not a single one of his genes survives into that future! How can this be?

Every time an individual has a child, *exactly* half his genes go into that child. Every time he has a grandchild, a quarter of his genes *on average* go into that child. Unlike the first generation offspring where the percentage contribution is exact, the figure for each grandchild is statistical. It could be more than a quarter, it could be less. Half your genes come from your father, half from your mother. When you make a child, you put half of your genes into her. But which half of your genes do you give to the child? On average they will be drawn equally from the ones you originally got from the child's grandfather and the ones you originally got from the child's grandmother. But, by chance, you could *happen* to give all your mother's genes to your child, and none of your father's. In this case, your father would have given no genes to his grandchild. Of course such a scenario is highly unlikely, but as we go down to more distant descendants, total non-contribution of genes becomes more possible. On average you can expect one-eighth of your genes to end up in each great-grandchild, one-sixteenth in each great-great-grandchild, but it could be more or it could be less. And so on until the likelihood of a literally zero contribution to a given descendant becomes significant.

In our hypothetical Tasmanian population, the Chang Two date is 22 generations back. So when we say that 80 per cent of the population can expect to be ancestors of all surviving individuals, we are talking about their 22-greats-grandchildren. The fraction of an ancestor's genome which, on average, we can expect to find in a particular one of his 22-greats-grandchildren is one four-millionth part. Since the human genome has only tens of thousands of genes, it would appear that one four-millionth part is going to be fairly thinly spread! It won't be quite like that, of course, because the population of our hypothetical Tasmania is only 5,000. Any individual may be descended from a particular ancestor through many different routes. But still, it could easily happen by chance that some universal ancestors happen to end up contributing none of their genes to distant posterity.

Perhaps I am biased, but I see this as yet another reason to return to the gene as the focus of natural selection: to think backwards about the genes that have survived up to the present, rather than forwards about individuals, or indeed genes, trying to survive into the future. The 'forward intentional' style of thought can be helpful if used carefully and not misunderstood, but it is not really necessary. 'Backwards gene' language is just as vivid when you get used to it, is closer to the truth, and is less likely to yield the wrong answer.

In the Tasmanian's Tale we have talked about genealogical ancestors: historical individuals who are ancestors of modern ones in the conventional genealogist's sense: 'people ancestors'. But what you can do for people you can do for genes. Genes too have parent genes, grandparent genes, grandchild genes. Genes too have pedigrees, family trees, 'Most Recent Common Ancestors' (MRCAs). Genes too have their own Rendezvous 0 and here we really can say that, for the majority of genes, their own Rendezvous 0 was in Africa. This apparent contradiction will take some explaining, and this is the purpose of Eve's Tale.

Before proceeding, I must clear up a possible confusion over the meaning of the word gene. It can mean lots of things to different people, but the particular confusion that threatens here is the following. Some biologists, especially molecular geneticists, strictly reserve the word gene for a location on a chromosome ('locus'), and they use the word 'allele' for each of the alternative versions of the gene that might sit at that locus. To take an oversimplified example, the gene for eye colour comes in different versions or alleles, including a blue allele and a brown allele. Other biologists, especially the kind to which I belong, who are sometimes called sociobiologists, behavioural ecologists or ethologists, tend to use the word gene to mean the same as allele. When we want a word for the slot in the chromosome which could be filled by any of a set of alleles, we tend to say 'locus'. People like me are apt to say 'Imagine a gene for blue eyes, and a rival gene for brown eyes'. Not all molecular geneticists like that, but it is a well-established habit with my kind of biologist and I shall occasionally follow them.

EVE'S TALE
Written with Yan Wong

There's a telling difference between 'gene trees' and 'people trees'. Unlike a person who is descended from two parents, a gene has one parent only. Each one of your genes must have come from either your mother or your father, from one and only one of your four grandparents, from one and only one of your eight great-grandparents, and so on. But when whole people trace their ancestors in the conventional way, they descend equally from two parents, four grandparents, eight great-grandparents and so on. This means that a 'people genealogy' is much more mixed up than a 'gene genealogy'. In a sense, a gene takes a single path chosen from the maze of crisscrossing routes mapped by the (people) family tree. Surnames behave like genes, not like people. Your surname picks out a thin line through your full family tree. It highlights your male to male to male ancestry. DNA, with two notable exceptions which I shall come to later, is not so sexist as a surname: genes trace their ancestry through males and females with equal likelihood.

Some of the best-recorded human pedigrees are of European royal families. In the family tree of the house of Saxe-Coburg (overleaf), look at the princes Alexis, Waldemar, Heinrich, and Rupert. The 'gene tree' of one of their genes is easy to trace because, unfortunately for them but fortunately for us, the gene concerned was defective. It gave the four princes, and many others of their ill-favoured family, the easily recognised blood disease haemophilia: their blood

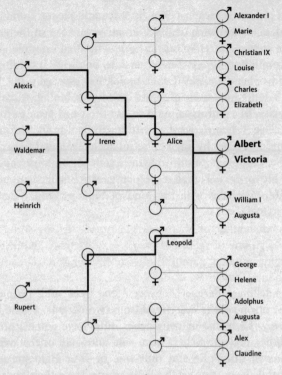

Bloodlines in the ill-fated House of Saxe-Coburg.

wouldn't clot properly. Haemophilia is inherited in a special manner: it is carried on the X chromosome. Males have only one X chromosome which they inherit from their mother. Females have two X chromosomes, one inherited from each parent. They suffer from the disease only if they have inherited the defective version of the gene from both their mother and their father (i.e. haemophilia is 'recessive'). Males suffer from the disease if their single 'unguarded' X chromosome bears the defective gene. Extremely few females suffer from haemophilia, therefore, but lots of females are 'carriers'. They have one copy of the faulty gene, and a 50 per cent chance of passing it on to each child. Carrier females who are pregnant always hope for a daughter, but they still have a substantial risk of haemophiliac grandsons. If a haemophiliac male lives long enough to have children, he cannot pass the gene on to a son (males never receive their X chromosome from their father), but he must pass it on to a daughter (females always receive their father's only X chromosome). Knowing these rules, and knowing which royal males had haemophilia, we can trace the faulty gene. Here is the backwards family tree, with the path the haemophilia gene must have taken in bold.

It seems that Queen Victoria herself was the mutant. It wasn't Albert, because his son, Prince Leopold, was haemophiliac, and sons don't get their X chromosome from their father. None of Victoria's collateral relatives suffered from haemophilia. She was the first royal individual to carry the gene. The miscopying must have occurred either in an egg of her mother, Victoria of Saxe-Coburg, or, which is more likely for reasons explained by my colleague Steve Jones in *The Language of the Genes,* 'in the august testicles of her father, Edward Duke of Kent'.

Although neither of Victoria's parents carried or suffered from haemophilia, one of them did have a gene (strictly an allele) which was the pre-mutated 'parent' of the royal haemophilia gene. We can think about (though we cannot detect) the ancestry of Victoria's haemophilia gene, back before it mutated to become a haemophilia gene. For our purposes it is irrelevant, except as a matter of diagnostic convenience, that Victoria's copy of the gene was diseased while its predecessors were not. As we trace back the family tree of the gene we ignore its effects, except insofar as they render it visible. The gene's lineage must go back before Victoria, but the visible trail goes cold when it wasn't a haemophilia gene. The lesson is that every gene has one parent gene even if, through mutation, it is not identical to that parent gene. Similarly it has only one grandparent gene, only one great-grandparent gene, and so on. This may seem an odd way to think, but remember that we are on an ancestor-hunting pilgrimage. The present exercise is to see what an ancestor-hunting pilgrimage would look like from a gene's point of view, instead of an individual's.

In the Tasmanian's Tale we encountered the acronym MRCA (Most Recent Common Ancestor) as an alternative to 'concestor'. I want to reserve 'concestor' for the most recent common ancestor in an entire (people or organism) genealogy. So when talking about genes I shall use 'MRCA'. Two or more alleles in different individuals (or even, as we shall see, in the same individual) certainly do have an MRCA. It is the ancestral gene of which they are each a (possibly mutated) copy. The MRCA of the haemophilia genes of Princes Waldemar and Heinrich of Prussia sat on one of the two X chromosomes of their mother, Irene von Hesse und bei Rhein. When she was still a foetus, two copies of the one haemophilia gene she carried were peeled off and passed successively into two of her egg cells, the progenitors of her luckless sons. These genes in turn share an MRCA with the haemophilia gene of Tsarevitch Alexis of Russia (1904–1918), in the form of a gene carried by their grandmother, Princess Alice of Hesse. Finally, the MRCA of the haemophilia gene in all four of our chosen princes is the very one that flagged itself up for attention in the first place, the mutant gene of Victoria herself.

Geneticists have a word for this sort of backwards tracing of a gene: it is called the coalescent. Looking backwards in time, two gene lineages can be said to coalesce into one at the point where, looking forwards again, a parent runs off two copies of the gene for two successive children. The point of coalescence is the MRCA. Any gene tree has many coalescence points. The haemophilia genes of Waldemar and Heinrich coalesce into the MRCA gene carried by their mother, Irene. That then coalesces with the lineage heading backwards from Tsarevitch Alexis. And, as we've seen, the grand coalescence of all the royal haemophilia genes occurs in Queen Victoria. Her genome holds the MRCA haemophilia gene for the whole dynasty.

In my example, the coalescence of the haemophilia genes of all four princes occurs in the very individual (Victoria) who happens also to be their most recent common *genealogical* ('people') ancestor, their concestor. But that is just coincidence. If we were to choose another gene (say for eye colour), then the path it took through the family tree would be quite different, and the genes would coalesce in a more distant ancestor than Victoria. If we picked a gene for brown eyes in Prince Rupert and one for blue eyes in Prince Heinrich, then the coalescence must be at least as far off as the separation of an ancestral eye-colour gene into two forms, brown and blue, an event buried in prehistory. Each piece of DNA has a genealogy which may be traced in a way that is separate but parallel to the sort of genealogy where we follow surnames through records of Births, Marriages and Deaths.

We can even do this for two identical genes in the same person. Prince Charles has blue eyes, which means, since blue is recessive, that he has two blue-eyed alleles. Those two alleles must coalesce somewhere in the past, but we can't tell when or where. It could be centuries or millennia ago, but in the special case of Prince Charles it is possible that the two blue-eyed alleles coalesce in as recent an individual as Queen Victoria. This is because, as it happens, Prince Charles is descended from Victoria twice: once via King Edward VII and once via Princess Alice of Hesse. On this hypothesis, a single blue-eyed gene of Victoria made two copies of itself at different times. These two copies of the same gene came down to the present Queen (Edward VII's great-granddaughter) and to her husband, Prince Philip (Princess Alice's great-grandson) respectively. Two copies of one Victorian gene could therefore have met again, on two different chromosomes, in Prince Charles. In fact, that almost certainly has happened for some of his genes, whether for blue eyes or not. And regardless of whether his two blue-eyed genes coalesce in Queen Victoria or in somebody farther back, those two genes must have had an MRCA at some specific point in the past. It doesn't matter whether we are talking about two genes in one person (Charles) or in two people (Rupert and Heinrich): the logic is the same. Any two alleles, in different people or in the same person, are fair game for the question: When, and in whom, do these genes coalesce as we look back? And, by extension, we can ask the same question of any three genes, or any number of genes in the population, at the same genetic location ('locus').

Looking much further back still, we can ask the same question for pairs of genes at different loci, because genes give rise to genes at different loci by the process of 'gene duplication'. We shall meet this phenomenon again in the Howler Monkey's Tale, and in the Lamprey's Tale.

Individual people who are closely related share a large number of *gene trees*. We share the majority of our gene trees with our close kin. But some gene trees deliver a 'minority vote', placing us closer to our otherwise more distant relatives. We can think of closeness of kinship among *people* as a kind of majority vote among genes. Some of your genes vote for, say, the Queen, as a close cousin. Others argue that you are closer to seemingly much more distant individuals (as we shall see, even members of other species). When quizzed, each piece of DNA has a different view of what history is all about, because each has blazed a different path through the generations. We can hope to gain a comprehensive view only by questioning a large number of genes. But at this point we must be suspicious of genes situated close to each other on a chromosome. To see why this is, we need to know something about the phenomenon of recombination, which happens every time a sperm or an egg is made.

In recombination, randomly chosen sections of matching DNA are swapped between chromosomes. On average, only one or two swaps are seen per human chromosome (fewer when making sperm, more when making eggs: it is not known why). But over numerous generations, many different parts of the chromosome will eventually be swapped around. So, generally speaking, the nearer two pieces of DNA are on a chromosome, the lower is the chance of a swap occurring between them, and the more likely they are to be inherited together.

When taking 'votes' from genes, therefore, we have to remember that the nearer a pair of genes are to each other on a chromosome, the more likely they are to experience the same history. And this motivates genes which are close colleagues to back up each other's vote. At the extreme are sections of DNA so tightly bound together that the entire chunk has travelled through history as a single unit. Such fellow-travelling chunks are known as 'haplotypes', a word that we shall meet again. Among such caucuses within the genetic parliament, two stand out, not because their view of history is more valid, but because they have been extensively used to settle biological debates. Both hold sexist views, because one has come down entirely through female bodies, and the other has never been outside a male body. These are the two major exceptions to unbiased gene inheritance that I previously mentioned.

Like a surname, the (non-recombining portion of the) Y chromosome always passes through the male line only. Together with a few other genes, the Y chromosome contains the genetic material that actually switches an embryo into the male pattern of development rather than the female one. Mitochondrial DNA, on the other hand, passes exclusively down the female line (although in this case it is not responsible for making the embryo develop as a female: males have mitochondria, it is just that they don't pass them on). As we shall see in the Great Historic Rendezvous, mitochondria are tiny bodies inside cells, relics of once-free bacteria who, probably about 2 billion years ago, took up exclusive residence inside cells where they have been reproducing, nonsexually by simple division, ever since. They have lost many of their bacterial qualities and most of their DNA, but they retain enough to be useful to geneticists. Mitochondria constitute an independent line of genetic reproduction inside our bodies, unconnected with the main nuclear line which we think of as our 'own' genes.

Because of their mutation rate, Y chromosomes are most useful for studies of recent populations. One neat study took samples of Y-chromosomal DNA in a straight line across modern Britain. The results showed that Anglo-Saxon Y chromosomes moved west across England from Europe, stopping rather abruptly at the Welsh border. It is not hard to imagine reasons why this male-carried DNA is unrepresentative of the rest of the genome. To take a more obvious example, Viking ships carried cargoes of Y chromosomes (and other genes) and spread them among widely scattered populations. The distribution of Viking Y-chromosome genes today presumably shows them to be slightly more 'travelled' than other Viking genes, which were statistically more likely to favour home-acre over Widow-maker:

> *What is a woman that you forsake her,*
> *And the hearth-fire and the home-acre,*
> *To go with the old grey Widow-maker?*
>
> RUDYARD KIPLING
> "Harp Song of the Dane Women"

Mitochondrial DNA too can be revealing, particularly for very ancient patterns. If we compare your mitochondrial DNA with mine, we can tell how long ago they shared an ancestral mitochondrion. And, since we all get our mitochondria from our mothers, and hence maternal grandmothers, maternal great-grandmothers, etc., mitochondrial comparison can tell us when our most recent female-line ancestor lived. The same can be done for Y chromosomes, to tell us when our most recent male-line ancestor lived but, for technical reasons, it is not so easy. The beauty of Y-chromosomal and mitochondrial DNA is that neither of them is contaminated by sexual mixing. This makes tracing these particular classes of ancestor easy.

The mitochondrial MRCA of all humanity, which pinpoints the 'people' common ancestor in the all-female line, is sometimes called Mitochondrial Eve – she whose tale this is. And of course the equivalent in the all-male line might as well be called Y-chromosome Adam. All human males have Adam's Y chromosome (creationists please refrain from deliberate misquotation). If surnames had always been strictly inherited by modern Western rules we'd all have Adam's surname too, which would rather lose the point of having a surname.

Eve is a great temptress to error and it is good to be forearmed.

The errors are quite instructive. First, it is important to understand that Eve and Adam are only two out of a multitude of MRCAs that we could reach if we traced our way back through different lines. They are the special-case common ancestors that we reach if we travel up the family tree from mother to mother to mother, or father to father to father respectively. But there are many, many other ways of going up the family tree: mother to father to father to mother, mother to mother to father to father, and so forth. Each of these possible pathways will have a different MRCA.

Second, Eve and Adam were not a couple. It would be a major coincidence if they ever met, and they could well have been separated by tens of thousands of years. As a subsidiary point, there are independent reasons to believe that Eve preceded Adam. Males are more variable in reproductive success than females: where some females have five times as many children as other females, the most successful males could have hundreds of times as many children as unsuccessful males. A male with a large harem finds it easy to become a universal ancestor. A female, since she is less likely to have a large family, needs a larger number of generations to achieve the same feat. And indeed, today's best 'molecular clock' estimates for their respective dates are about 140,000 years ago for Eve and only about 60,000 for Adam.

Third, Adam and Eve are shifting honorific titles, not names of particular individuals. If, tomorrow, the last member of some outlying tribe were to die, the baton of Adam, or of Eve, could abruptly be thrown forward several thousand years. The same is true of all the other MRCAs defined by different gene trees. To see why this is so, suppose Eve had two daughters, one of whom eventually gave rise to the Tasmanian aborigines and the other of whom spawned the rest of humanity. And suppose, entirely plausibly, that the female-line MRCA uniting 'the rest of humanity' lived 10,000 years later, all other collateral lines descending from Eve having gone extinct apart from the Tasmanians. When Truganinni, the last Tasmanian, died, the title of Eve would instantly have jumped forward 10,000 years.

Fourth, there was nothing to single out either Adam or Eve for particular notice in their own times. Despite their legendary namesakes, Mitochondrial Eve and Y-chromosome Adam were not particularly lonely. Both would have had plenty of companions, and each may well have had many sexual partners, with whom they may also have surviving descendants. The only thing that singles them out is that Adam eventually turned out to be hugely endowed with descendants down the male line, and Eve with descendants down the female line. Others among their contemporaries may have left as many descendants all told.

While I was writing this, somebody sent me a videotape of a BBC television documentary called *Motherland,* hyped as 'an incredibly poignant film', and as 'truly beautiful, a really memorable piece'. The heroes of the film were three 'black'* people whose families had immigrated to Britain from Jamaica. Their DNA was matched up against worldwide databases, in an attempt to trace the part of Africa from which their ancestors were taken as slaves. The production company then staged lachrymose 'reunions' between our heroes and their long-lost African families. They used Y-chromosomal and mitochondrial DNA because, for the reasons we have seen, they are more traceable than genes in general. But unfortunately, the producers never really came clean about the limitations this imposed. In particular, no doubt for sound televisual reasons, they came close to actively deceiving these individuals, and also their long-lost African 'relatives', into becoming far more emotional about the reunions than they had any right to be.

Let me explain. When Mark, later given the tribal name Kaigama, visited the Kanuri tribe in Niger, he believed he was 'returning' to the land of 'his people'. Beaula was welcomed as a long-lost daughter by eight women of the Bubi tribe on an island off the coast of Guinea, whose mitochondria matched hers. Beaula said,

> It was like blood touching blood . . . It was like family . . . I was just crying, my eyes were just filled with tears, my heart was pounding. All I just kept thinking was: 'I'm going to my motherland.'

Sentimental rubbish, and she should never have been deceived into thinking this. All that she, or Mark, were really visiting – at least as far as there was any evidence to suppose – were individuals who shared their mitochondria. As a matter of fact, Mark had already been told that his Y chromosome came from Europe (which upset him and he was later palpably relieved to discover respectable African roots for his mitochondria!). Beaula, of course, has no Y chromosome, and apparently they didn't bother to look at her father's although that would have been interesting, for she was quite light-skinned. But it

---

* For explanation of the inverted commas around 'black', see the Grasshopper's Tale.

was explained to neither Beaula nor Mark, nor the television audience, that genes outside their mitochondria almost certainly came from a huge variety of 'homelands', nowhere near those identified for purposes of the documentary. If their other genes had been traced, they could have had equally emotional 'reunions' in hundreds of different sites, all over Africa, Europe and very probably Asia too. That would have spoiled the dramatic impact, of course.

As I have been continually reiterating, reliance on a single gene can be misleading. But the combined evidence from many genes gives us a powerful tool for reaching back into history. The gene trees of a population, and the coalescence points which define them, reflect the events of the past. Not only can we identify these coalescence points, we can also guess at their dates because of the molecular clock. And herein lies the key, because the pattern of branchings through time tells a story. Random mating, the assumption made in the Tasmanian's Tale, generates a very different pattern of coalescence from various kinds of non-random mating – each of which, in turn, imprints its own shape on the coalescence tree. Fluctuations in population size, too, leave their own characteristic signature. So we can work backwards from today's patterns of gene distribution and make inferences about population sizes, and about the timings of migrations. For example, when a population is small, coalescence events will occur more frequently. An expanding population is signified by trees with long end branches, so coalescence points will be concentrated near the base of the tree, back when the population was small. With the aid of the molecular clock, this effect can be used to work out when the population expanded, and when it contracted in 'bottlenecks'. (Although unfortunately, by wiping out genetic lineages, severe bottlenecks tend to erase the traces of what happened before them.)

Coalescent gene trees have helped resolve a long-standing debate over human origins. The 'Out of Africa' theory holds that all surviving peoples outside Africa are descended from a single exodus around a hundred thousand years ago, more or less. At the other extreme are the 'Separate Origins' theorists or 'Multiregionalists', who believe that the races still living in, say, Asia, Australia and Europe are anciently divided, separately descended from regional populations of the earlier species, *Homo erectus*. Both names are misleading. 'Out of Africa' is unfortunate because everybody agrees that our ancestors are from Africa if you go back far enough.

'Separate Origins' is also not an ideal name because, again if you go back far enough, the separation must disappear on any theory. The disagreement concerns the date when we came out of Africa. It might be better to call the two theories 'Young Out of Africa' (YOOA) and 'Old Out of Africa' (OOOA). This has the added advantage of emphasising the continuum between them.

If today's non-Africans all stem from a single recent emigration from that continent, we would expect modern gene distributions to demonstrate a recent, Africa-centred, small-population 'bottleneck'. Coalescence points would be concentrated around the time of the exodus. If we are separately descended from regional *H. erectus*, however, then genes should instead show evidence of anciently separated genetic lineages in each region. At the time when YOOA supporters claim an exodus, we would instead see a dearth of coalescence points. Which is it?

By expecting a single answer to this question we have fallen into the same trap as the *Motherland* television documentary. Different genes tell different stories. It is perfectly possible for some of our genes to have recently come out of Africa, while others have been passed to us from separate *H. erectus* populations. Or to put it another way, we can be both descendants of a recent African exodus, and simultaneously descendants of regional *H. erectus*, because at any given time in the past we have a huge number of genealogical ancestors. Some could have recently left Africa. Some could have been resident in, say, Java for thousands of years. And we could have inherited African genes from some and Javan genes from others. A single chunk of DNA, such as from a mitochondrion or Y chromosome, gives as impoverished a view of the past as a single sentence from a history book. Yet the YOOA position is often supported on the basis of the placement of Mitochondrial Eve. What happens if we quiz the other members of the parliament of genes?

This is, in effect, what the evolutionary biologist Alan Templeton did, and he came up with his engagingly titled theory 'Out of Africa Again and Again'. Templeton used a type of coalescence theory, similar to that in our haemophilia discussion, but he did it for lots of separate genes instead of just one. This enabled him to reconstruct the history and geography of genes over the whole world and over hundreds of thousands of years. At the moment, I favour Temple-ton's 'Out of Africa Again and Again' theory, because he seems to me to use all the available information in a way that maximises its power to generate inferences; and because he bent over backwards, at every step of his work, to guard against overreaching the evidence.

Here is what Templeton did. He looked through the genetic literature, using strict criteria to skim the cream: he wanted only large studies of human genetics, where samples had been taken from different parts of the world, including Europe, Asia and Africa. The genes examined belonged to long-lived 'haplotypes'. A haplotype, as we have seen, is a chunk of genome which is either impervious to being broken up by sexual recombination (as with Y-chromosomal and mitochondrial DNA), or (as with certain smaller parts of the genome) can be recognised intact through enough generations to cover the timescale of interest. A haplotype is a long-lived, recognisable chunk of genome. You don't go too far wrong if you think of it as a large 'gene'.

Templeton zeroed in on 13 haplotypes. For each of them, he calculated their 'gene tree', and dated the various coalescence points using the molecular clock which is ultimately calibrated with fossils. From these dates, and from the geographical distribution of the samples, he was able to pull out inferences about the genetic history of our species over the past couple of million years. He summarised his conclusions in a helpful diagram, reproduced on page 61.

Templeton's main conclusion is that there were not two major migrations out of Africa but three. In addition to the OOOA (*Homo erectus*) exodus around 1.7 million years ago (which everyone accepts and for which the evidence is mostly from fossils) and the recent migration as promoted by the YOOA theory, there was another Great Trek from Africa to Asia between 840,000 and 420,000 years ago. This middle emigration – shall we call it MOOA? – is supported by extant 'signals' from three of the 13 haplotypes. The YOOA emigration is supported by mitochondrial and Y-chromosomal evidence. Other genetic 'signals' betray a major back-migration from Asia to Africa about 50,000 years ago. A little later, mitochondrial DNA and various smaller genes disclose other migrations: from southern to northern Europe, from southern Asia to northern Asia, across the Pacific and to Australia. Finally, as shown by mitochondrial DNA and archaeological evidence, North America was colonised across what was then the Bering land bridge from north-east Asia, around 14,000 years ago. Colonisation of South America through the Isthmus of Panama rapidly followed. The suggestion, by the way, that

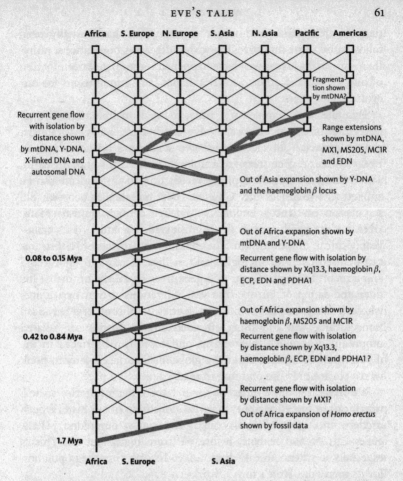

Out of Africa again and again. Templeton's summary of major human migrations, based on the study of 13 haplotypes. Vertical lines represent genetic descent; diagonal lines represent gene flow. The major human migrations indicated by genetic data are shown by the thick arrows. Adapted from Templeton [284] (square brackets refer to sources in the Bibliography).

either Christopher Columbus or Leif Ericsson 'discovered' America is nothing short of racist. Equally distasteful, in my view, is relativist 'respect' for Native American oral histories which ignorantly deny that their ancestors ever lived outside America.

Between Templeton's three major migrations out of Africa, other genetic signals reveal continual eddies of gene flow back and forth between Africa, southern Europe and southern Asia. His evidence suggests that major and minor immigrations have usually been followed by some interbreeding with indigenous populations, rather than – as might just as well have happened – complete extermination of one side or the other. Clearly this has large implications for our evolutionary ancestry.

This tale, and Templeton's study, focused on humans and their genes. But of course all species have family trees. All species inherit genetic material. All species with two sexes have an Adam and an Eve. Genes and gene trees are a ubiquitous feature of life on Earth. The techniques that we apply to recent human history can also be applied to the rest of life. Cheetah DNA reveals a 12,000-year-old population bottleneck important to feline conservationists. Maize DNA has stamped upon it the unmistakable signature of its 9,000-year Mexican domestication. The coalescence patterns of HIV strains can be used by epidemiologists and medical doctors to understand and contain the virus. Genes and gene trees reveal the history of the flora and fauna of Europe: the vast migrations driven by ice ages whose waxing pushed temperate species into southern-European refuges, and whose waning stranded Arctic species on isolated mountain ranges. All these events and more can be traced in the distribution of DNA around the globe, a historical reference book which we are only just learning to read.

We have seen how different genes have different stories to tell, which can be pieced together to reveal something of our history, both modern and ancient. How ancient? Amazingly, our oldest MRCA genes can even date back before we were human at all. This is especially so when natural selection favours variety in the population for its own sake. Here's how it works.

Suppose there are two blood types called A and B, which confer immunity to different diseases. Each blood type is susceptible to the disease against which the other type has immunity. Diseases flourish when the blood type that they can attack is abundant, because an epidemic can get going. So if B people, say, happen to be common in the population, the disease that hurts them will enjoy an epidemic. Consequently, B people will die until they cease to be common, and the A people increase – and vice versa. Whenever we have two types, the rarer of which is favoured because it is rare, it is a recipe for *polymorphism*: the positive maintenance of variety for variety's sake. The ABO blood group system is a famous polymorphism which has probably been maintained for this kind of reason.

Some polymorphisms can be quite stable – so stable that they span the change from an ancestral to a descendant species. Astonishingly, our ABO polymorphism is present in chimpanzees. It could be that we and chimps have independently 'invented' the polymorphism, and for the same reason. But it is more plausible that we have both inherited it from our shared ancestor, and independently kept it going during our six million years of separate descent, because the relevant diseases have been continuously at large throughout that time. This is called trans-specific polymorphism, and it may apply to far more distant cousins than chimpanzees are to us.

A stunning conclusion is that, for particular genes, you are more closely related to some chimpanzees than to some humans. And I am closer to some chimpanzees than to you (or to 'your' chimpanzees). Humans as a species, as well as humans as individuals, are temporary vessels containing a mix of genes from different sources. Individuals are temporary meeting points on the crisscrossing routes that genes take through history. This is a tree-based way to express the central message of *The Selfish Gene,* my first book. As I put it there, 'When we have served our purpose we are cast aside. But genes are denizens of geological time: genes are forever.' At the concluding banquet to a conference in America, I recited the same message in verse:

> *An itinerant selfish gene*
> *Said 'Bodies a-plenty I've seen.*
> *You think you're so clever*
> *But I'll live for ever.*
> *You're just a survival machine.'*

And, as the body's immediate reply to the gene, I parodied the very same *Harp Song of the Dane Women* quoted previously:

> *What is a body that first you take her,*
> *Grow her up, and then forsake her,*
> *To go with the old blind watchmaker?*

We estimated the date of Rendezvous 0 as probably tens of thousands of years ago, and at most hundreds of thousands. We have not travelled far on our backward pilgrimage. The next rendezvous, our meeting with the chimpanzee pilgrims at Rendezvous 1, is millions of years away, and most of our rendezvous are hundreds of millions beyond that. To stand a chance of completing our pilgrimage, we shall need to speed up, and begin the move into 'deep time'. We must accelerate past the rest of the 30 or so ice ages that punctuate the last three million years, past such drastic events as the drying and refilling of the Mediterranean that occurred between 4.5 and 6 million years ago. To ease this initial acceleration, I shall take the otherwise unusual liberty of stopping at a few intermediate milestones *en route,* and allowing dead fossils to tell tales. The fossilised 'shadow' pilgrims we shall meet, and the tales they tell, will help satisfy our natural preoccupation with our direct ancestors.

ARCHAIC *HOMO SAPIENS*

Our first milestone on the way back to Rendezvous 1 is in the depths of the ice age before last, about 160,000 years ago. I have chosen this way station to look at fossil finds from Herto in the Afar depression of Ethiopia.* The Herto humans are intriguing because, in the words of their discoverers, Tim White and his colleagues, they are from a 'population that is on the verge of anatomical modernity but not yet fully modern'. The distinguished palaeoanthropologist Christopher Stringer regards 'the Herto material as the oldest definite record of what we currently think of as modern *H. sapiens*', a record previously held by younger Middle Eastern fossils dating from about 100,000 years ago. Regardless of hair-splitting distinctions between 'modern' and 'nearly modern', it is clear that the Herto people are on the cusp between modern humans and those predecessors that we know by the catch-all name of 'Archaic *Homo sapiens*'. Certain authorities use this name back to about 900,000 years ago where it grades into an earlier species, *Homo erectus*. As we shall see, others prefer to give various Latin names to the bridging archaic forms. I shall sidestep the disputes by using anglicisms in the style of my colleague Jonathan Kingdon: 'Moderns', 'Archaics', 'Erects', and others that I'll mention as we come to them. We should not expect to draw a neat line between early Archaics and the Erects from whom they evolved, or between Archaics and the earliest Moderns who evolved from them. Don't be confused, incidentally, by the fact that the Erects were even more archaic (with a small a) than the Archaics (with a large A), and that all three types were erect with a small e!

Archaic forms persisted alongside Modern forms until at least 100,000 years ago (longer still if we include the Neanderthals, of whom more in a moment). Archaic fossils are found all around the world, dating from various times during the last few hundred thousand years: examples are the German 'Heidelberg man', 'Rhodesian man' from Zambia (which used to be called Northern Rhodesia), and

---

* The same 'Afar' after which the much older *Australopithecus afarensis*, or Lucy, is named.

the Chinese 'Dali man'. Archaics had big brains like us, averaging 1,200 to 1,300 cubic centimetres. This is a little smaller than our average of 1,400 cubic centimetres but the range comfortably overlaps with ours. Their bodies were more robust than ours, their skulls were thicker, and they had more pronounced brow ridges and less pronounced chins. They looked more like Erects than we do, and hindsight justly sees them as intermediate. Some taxonomists recognise them as a subspecies of *Homo sapiens* called *Homo sapiens heidelbergensis* (where we would be *Homo sapiens sapiens*). Others do not recognise the Archaics as *Homo sapiens* at all, but call them *Homo heidelbergensis.* Yet others divide the Archaics into more than one species, for instance *Homo heidelbergensis, Homo rhodesiensis,* and *Homo antecessor.* If you think about it, we should be worried if there was *not* disagreement over the divisions. On the evolutionary view of life, a continuous range of intermediates is to be expected.

Modern *Homo sapiens sapiens* are not the only offshoot of the Archaics. Another species of advanced humans, the so-called Neanderthals, were our contemporaries for much of our prehistory. They resembled the Archaics more than we do in some respects, and they seem to have emerged from an Archaic root between about one and two hundred thousand years ago – in this case not in Africa but in Europe and the Middle East. Fossils from these regions show a gradual transition from Archaics to Neanderthals with the first unequivocal Neanderthal fossils found just before the beginning of the last Ice Age, about 130,000 years ago. They then persisted in Europe for most of this cold period, vanishing about 28,000 years ago. In other words, for their entire existence Neanderthals were contemporaries of European Modern émigrés from Africa. Some people believe that Moderns were responsible for their extinction, either by killing them directly or by competing with them.

Neanderthal* anatomy was sufficiently different from ours that some people prefer to give them a separate species name, *Homo neanderthalensis.* They retained some features of Archaics such as large brow ridges which Moderns did not (which is why some

---

* Pedants' Corner: *Thal* or, in modern German, *Tal,* means valley. The Neander Valley is where the first fossil of this type was discovered. When German spelling was reformed at the end of the nineteenth century, the valley changed from a *Thal* to a *Tal,* but the Latin name, *Homo neanderthalensis,* was left high and dry, trapped by the laws of zoological terminology. To conform with tradition and with the Latin, I prefer to leave the English spelling in its original form and stick with the h.

authorities classify them as just another type of Archaic). Adaptations to their cold environment include stockiness, short limbs and enormous noses, and they surely must have been warmly clothed, presumably in animal furs. Their brains were as big as ours or even bigger. Much is made of slight indications that they ceremonially buried their dead. Nobody knows whether they could speak, and opinions differ on this important question. Archaeology hints that technological ideas may have passed both ways between Neanderthals and Moderns, but this could have been by imitation rather than by language.

The rules for the pilgrimage stated that only modern animals setting off from the present were entitled to tell tales. We are making an exception for the dodo and the elephant bird, because they lived in recent historical times. And the fossils *Homo erectus* and *Homo habilis* qualify as 'shadow pilgrims' because a plausible case could be made that they are our direct ancestors. Do the Neanderthals, too, qualify under this rubric? Are we descended from them? Well, as it happens, that very question is the topic of the tale that the Neanderthals want to tell. Think of the Neanderthal's Tale as a plea to be allowed to tell it.

THE NEANDERTHAL'S TALE
Written with Yan Wong

Are we descended from Neanderthals? If so, they would have to have interbred with *Homo sapiens sapiens*. But did they? They overlapped for a long time in Europe, and there was surely contact between them. But did it go beyond contact? Do modern Europeans inherit any Neanderthal genes? This is a hotly debated issue, recently reignited by a remarkable extraction of DNA from late Neanderthal bones. So far, we have extracted only the maternally inherited mitochondrial DNA, but this is enough for a tentative verdict. Neanderthal mitochondria are quite distinct from those of all surviving humans, suggesting that Neanderthals are no closer to Europeans than to any other modern peoples. In other words, the female-line common ancestor of Neanderthals and all surviving humans long pre-dates Mitochondrial Eve: about 500,000 years as opposed to 140,000. This genetic evidence suggests that successful interbreeding between Neanderthals and Moderns was rare. And so it is often said that they died out without leaving any descendants.

But don't let's forget that '80 per cent' argument which so surprised us in the Tasmanian's Tale. A single immigrant who managed to break into the Tasmanian breeding population had an 80 per cent chance of joining the set of universal ancestors: the set of individuals who could call themselves ancestors of all surviving Tasmanians in the distant future. By the same token, if only one Neanderthal male, say, bred into a *sapiens* population, that gave him a reasonable chance of being a common ancestor to all Europeans alive today. This can be true even if Europeans contain no Neanderthal genes at all. A striking thought.

So although few, if any, of our genes come from Neanderthals, it is possible that some people have many Neanderthal ancestors. This was the distinction we met in Eve's Tale between gene trees and people trees. Evolution is governed by the flow of genes, and the moral of the Neanderthal's Tale, if we allow him to tell it, is that we cannot, should not, look at evolution in terms of pedigrees of individuals. Of course individuals are important in all sorts of other ways, but if we are talking pedigrees it is gene trees that count. The words 'evolutionary descent' refer to gene ancestors, not genealogical ancestors.

Fossil changes too are a reflection of gene pedigrees, not (or only incidentally) genealogical pedigrees. Fossils indicate that Modern anatomy passed to the rest of the world via young out-of-Africa migrations. But Alan Templeton's work (described in Eve's Tale) suggests that we are also partly 'descended from' non-African Archaics, possibly even non-African *Homo erectus.* The description is both simpler and more powerful if we switch from people talk to gene talk. The genes that determine our Modern anatomy were carried out of Africa by the YOOA migrants, leaving fossils in their wake. At the same time, Templeton's evidence suggests that other genes we now possess were flowing around the world by different routes, but left little anatomical evidence to show for it. Most of our genes probably took the young out-of-Africa route, while just a few came to us through other routes. What could be a more powerful way to express it?

So, have the Neanderthals established their right to tell a tale? Maybe a tale of genealogy if not a tale of genes.

ERGASTS

Moving deeper in time, we touch down again at one million years in search of ancestors. The only likely candidates of this age are of the type usually called *Homo erectus,* although some would call the African ones *Homo ergaster* and I shall follow them. In seeking an anglicised form for these creatures, I shall call them Ergasts rather than Erects, partly because I believe the majority of our genes trace back to the African form, and partly because, as I've already remarked, they were no more erect than their predecessors (*Homo habilis*) or their successors (us). Whatever name we prefer, the Ergast type persisted from about 1.8 million until about a quarter of a million years ago. They are widely accepted as the immediate predecessors, and partial contemporaries, of the Archaics who are in turn the predecessors of us Moderns.

The Ergasts were noticeably different from modern *Homo sapiens,* and, unlike the Archaic *sapiens* people, they differed from us in some respects that show no overlap. Fossil finds show they lived in the Middle East and Far East including Java, and represent an ancient migration out of Africa. You may have heard them referred to by their old names of Java Man and Peking Man. In Latin, before they were admitted into the *Homo* fold, they had the generic names *Pithecanthropus* and *Sinanthropus.* They walked on two legs like us, but had smaller brains (900 cc in early specimens to 1,100 cc in late ones), housed in lower, less domed, more 'swept-back' skulls than ours, and they had receding chins. Their jutting brow ridges made a pronounced horizontal ledge above the eyes, set in wide faces, with a pinching in of the skull behind the eyes.

Hair doesn't fossilise, so there is no natural place in our history to discuss the obvious fact that at some point in our evolution we lost most of our body hair, with the luxuriant exception of the tops of our heads. Very likely the Ergasts were hairier than us, but we can't rule out the possibility that Ergasts had already lost their body hair by a million years ago. They could have been as hairless as we are. Equally, nobody should complain of an imaginative reconstruction as hairy as a chimpanzee, or any intermediate level of shagginess.

Modern people, males at least, remain quite variable in how hairy they are. Hairiness is one of those characteristics that can increase or decrease in evolution again and again. Vestigial hairs, with their associated cellular support structures, lurk in even the barest-seeming skin, ready to evolve into a full coat of thick hair at short notice (or shrink again) should natural selection at any time call them out of retirement. Look at the woolly mammoths and woolly rhinoceroses that rapidly evolved in response to the recent ice ages in Eurasia. We shall return to the evolutionary loss of human hair in – strangely enough – the Peacock's Tale.

Subtle evidence of repeatedly used hearths suggests that at least some groups of Ergasts discovered the use of fire – with hindsight a momentous event in our history. The evidence is less conclusive than we might hope. Blackening from soot and charcoal does not survive immense timespans, but fires leave other traces that last longer. Modern experimenters have systematically constructed fires of various kinds and then examined them afterwards for their trace effects. It emerges that deliberately built campfires magnetise the soil in a way that distinguishes them from bushfires and from burnt-out tree stumps – I don't know why. But such signs provide evidence that Ergasts, both in Africa and Asia, had campfires nearly one and a half million years ago. This doesn't have to mean that they knew how to light fires. They could have begun by capturing and tending naturally occurring fire, feeding it and keeping it alive as one might look after a Tamagochi pet. Maybe, before they began to cook food, they used fires to scare away dangerous animals and provide light, heat, and a social focus.

The Ergasts also shaped and used stone tools, and presumably wooden and bone ones too. Nobody knows whether they could speak, and evidence is hard to come by. You might think that 'hard to come by' is an understatement, but we have now reached a point in our backward journey when fossil evidence starts to tell. Just as campfires leave traces in the soil, so the needs of speech call forth tiny changes in the skeleton: nothing so dramatic as the hollow bony box in the throat with which the howler monkeys of the South American forests amplify their stentorian voices, but still telltale signs such as one might hope to detect in a few fossils. Unfortunately, the signs that have been unearthed are not telltale enough to settle the matter, and it remains controversial.

There are two parts of the modern human brain which seem to go with speech. When in our history did these parts – Broca's Area and Wernicke's Area – enlarge? The nearest approach we have to fossil brains is endocasts, to be described in the Ergast's Tale. Unfortunately the lines dividing different regions of the brain do not fossilise very clearly, but some experts think they can say that the speech areas of the brain were already enlarged before two million years ago. Those who want to believe that Ergasts possessed the power of speech are encouraged by this evidence.

They are discouraged, however, when they move down the skeleton. The most complete *Homo ergaster* we know is the Turkana Boy, who died near Lake Turkana, in Kenya, about 1.5 million years ago. His ribs, and the small size of the portholes in the vertebrae through which the nerves pass, suggest that he lacked the fine control over breathing that seems to be associated with speech. Other scientists, studying the base of the skull, have concluded that even Neanderthals, as recently as 60,000 years ago, were speechless. The evidence is that their throat shape would not have allowed the full range of vowels that we deploy. On the other hand, as the linguist and evolutionary psychologist Steven Pinker has remarked, 'e lengeege weth e smell nember ef vewels cen remeen quete expresseve'. If written Hebrew can be intelligible without vowels, I don't see why spoken Neander or even Ergaster couldn't too. The veteran South African anthropologist Phillip Tobias suspects that language may pre-date even *Homo ergaster*, and he may just possibly be right. As we have seen, there are a few who go to the opposite extreme and date the origin of language to the Great Leap Forward, just a few tens of thousands of years ago.

This may be one of those disagreements that can never be resolved. All considerations of the origin of language begin by citing the Linguistic Society of Paris which, in 1866, banned discussion of the question because it was deemed unanswerable and futile. It may be difficult to answer, but it is not in principle unanswerable like some philosophical questions. Where scientific ingenuity is concerned, I am an optimist. Just as continental drift is now sewn up beyond all doubt, with multiple threads of convincing evidence, and just as DNA fingerprinting can establish the exact source of a bloodstain with a confidence that forensic experts could once only dream of, I guardedly expect that scientists will one day discover some ingenious new method of establishing when our ancestors started to speak.

Even I, however, have no hope that we shall ever know what they said to each other, or the language in which they said it. Did it begin with pure words and no grammar: the equivalent of an infant babbling nounspeak? Or did grammar come early and – which is not impossible and not even silly – suddenly? Perhaps the capacity for grammar was already deep in the brain, being used for something else like mental planning. Is it even possible that grammar, as applied to communication at least, was the sudden invention of a genius? I doubt it, but in this field I wouldn't rule anything out with confidence.

As a small step towards finding out the date at which language arose, some promising genetic evidence has appeared. A family code-named KE suffers from a strange hereditary defect. Out of approximately 30 family members spread over three generations, about half are normal, but fifteen show a curious linguistic disorder, which seems to affect both speech and understanding. It has been called verbal dyspraxia, and it first shows itself as an inability to articulate clearly in childhood. Other authorities think the trouble stems from 'feature blindness', meaning an inability to grasp certain grammatical features such as gender, tense and number. What is clear is that the abnormality is genetic. Individuals either have it or they don't, and it is associated with a mutation of an important gene called FOXP2, which the rest of us have in unmutated form. Like most of our genes, a version of FOXP2 is present in mice and other species, and it probably does various things in the brain and elsewhere.* The evidence of the KE family suggests that in humans FOXP2 is important for the development of some part of the brain that is involved in language.

So, we naturally want to compare human FOXP2 with the same gene in animals that lack language. You can compare genes either by looking at the DNA sequences themselves, or by looking at the amino acid sequences in the proteins that they encode. There are times when it makes a difference, and this is one of them. FOXP2 codes for a protein chain 715 amino acids long. The mouse and chimpanzee versions of the gene differ in only one amino acid. The human version differs from both these animals in an additional two amino acids. You see what this might mean? Although humans and chimpanzees share the great majority of their evolution and their genes, the FOXP2 gene is one place where humans seem to have evolved

* Many genes have more than one effect: a phenomenon known as pleiotropism.

rapidly in the short time since we split from them. And one of the most important respects in which we differ from chimpanzees is that we have language and they don't. A gene that changed somewhere along the line towards us, but after the separation from chimpanzees, is exactly the sort of gene we should be looking for if we are trying to understand the evolution of language. And it is the very same gene that has mutated in the unfortunate KE family (and also, in a different way, in a wholly unrelated individual with the same kind of language defect). Perhaps it was changes in *FOXP2* that made humans, as opposed to chimpanzees, capable of language. Did the Ergasts have the mutated *FOXP2* gene?

Wouldn't it be wonderful if we could use this genetic hypothesis to date the origin of language in our ancestors? While we can't do it with certainty, we can do something quite suggestive, along these lines. The obvious approach would be to triangulate backwards from variants among modern humans, and try to calculate the antiquity of the *FOXP2* gene. But with the exception of rare unfortunates like the members of the KE family, there is no variation among humans in any of the *FOXP2* amino acids. So there isn't enough variation there to triangulate from. Luckily, however, there are other parts of the gene which are never translated into protein and which are therefore free to mutate without natural selection 'noticing': they are 'silent' code letters, in those parts of the gene that are never transcribed and are called introns (as opposed to 'exons' which are 'expressed' and therefore 'seen' by natural selection). The silent letters, unlike the expressed ones, are quite variable among individual humans, and between humans and chimpanzees. We can get some understanding of the evolution of the gene if we look at the patterns of variation in the silent areas. Even though the silent letters are not subject to natural selection themselves, they can be swept along by selection of neighbouring exons. Even better, the mathematically analysed pattern of variation in the silent introns gives a good indication of *when* the sweeps of natural selection occurred. And the answer for *FOXP2* is less than 200,000 years ago. A naturally selected change to the human version of *FOXP2* seems roughly to coincide with the change from archaic *Homo sapiens* to anatomically modern *Homo sapiens*. Could this be when language was born? The margin of error in this sort of calculation is wide, but this ingenious genetic evidence counts as a vote against the theory that *Homo ergaster* could talk. More importantly for me, the unexpected new method boosts my optimism that one day science will find a way to confound the pessimists of the Linguistic Society of Paris.

Homo ergaster is the first fossil ancestor we have met on our pilgrimage who is unequivocally of a different species from ourselves. We are about to embark on a portion of the pilgrimage in which fossils provide the most important evidence, and they will continue to bulk large – though they will never overwhelm molecular evidence – until we reach extremely ancient times and relevant fossils start to peter out. It is a good moment to look in more detail at fossils, and how they are formed. The Ergast will tell the tale.

THE ERGAST'S TALE

Richard Leakey movingly describes the discovery, by his colleague Kimoya Kimeu on 22 August 1984, of the Turkana Boy (*Homo ergaster*), at 1.5 million years the oldest near-complete hominid skeleton ever found. Equally moving is Donald Johanson's description of the older, and unsurprisingly less complete, australopithecine familiarly known as Lucy. The discovery of 'Little Foot', yet to be fully described, is just as remarkable (see page 92). Whatever freak conditions blessed Lucy, 'Little Foot', and the Turkana Boy with their version of immortality, would we not wish it for ourselves when our time comes? What hurdles must we cross to achieve this ambition? How does any fossil come to be formed? This is the subject of the Ergast's Tale. To begin, we need a small digression into geology.

Rocks are built of crystals, though these are often too small for the unaided eye to see. A crystal is a single giant molecule, its atoms arranged in an orderly lattice with a regular spacing pattern repeated billions of times until, eventually, the edge of the crystal is reached. Crystals grow when atoms come out of the liquid state and build up on the expanding edge of an existing crystal. The liquid is usually water. On other occasions, it is not a solvent at all but the molten mineral itself. The shape of the crystal, and the angles at which its plane facets meet, is a direct rendition, in the large, of the atomic lattice. The lattice shape is sometimes projected very large indeed, as in a diamond or amethyst whose facets betray to the naked eye the three-dimensional geometry of the self-assembled atomic arrays. Usually, however, the crystalline units of which rocks are made are too small for the eye to detect them, which is one reason why most rocks are not transparent. Among important and common rock crystals are quartz (silicon dioxide), feldspars (mostly silicon dioxide again, but some of the silicon atoms are replaced by aluminium atoms), and calcite (calcium carbonate). Granite is a densely packed mixture of quartz, feldspar and mica, crystallised out of molten magma. Limestone is mostly calcite, sandstone mostly quartz, in both cases ground small and then compacted from sediments of sand or mud.

Igneous rocks begin as cooled lava (which in turn is molten rock). Often, as with granite, they are crystalline. Sometimes their shape may be visibly that of a glass-like solidified liquid and, with great good fortune, molten lava may sometimes be cast in a natural mould, such as a dinosaur's footprint or an empty skull. But the main usefulness of igneous rock to historians of life is in dating. As we shall see in the Redwood's Tale, the best dating methods are available for igneous rocks alone. Fossils usually cannot be precisely dated themselves, but we can look for igneous rocks in the vicinity. We then either assume that the fossil is contemporaneous, or we seek two datable igneous samples that sandwich our fossil and fix upper and lower bounds to its date. This sandwich dating is open to the slight risk that a corpse has been carried by floodwater, or by hyenas or their dinosaur equivalents, to an anachronistic site. With luck this will usually be obvious; otherwise we have to fall back on consistency with a general statistical pattern.

Sedimentary rocks such as sandstone and limestone are formed from tiny fragments, ground by wind or water from earlier rocks or other hard materials such as shells. They are carried in suspension, as sand, silt or dust, and deposited somewhere else, where they settle and compact themselves over time into new layers of rock. Most fossils lie in sedimentary beds.

It is in the nature of sedimentary rock that its materials are continually being recycled. Old mountains such as the Scottish Highlands have been slowly ground down by wind and water, yielding materials which later settle into sediments and may ultimately push up again somewhere else as new mountains like the Alps, and the cycle resumes. In a world of such recycling, we have to curb our importunate demands for a continuous fossil record to bridge every gap in evolution. It isn't just bad luck that fossils are often missing, but an inherent consequence of the way sedimentary rocks are made. It would be positively worrying if there were no gaps in the fossil record. Old rocks, with their fossils, are actively being destroyed by the very process that goes to make new ones.

Often fossils are formed when mineral-charged water penetrates the fabric of a buried creature. In life, bone is porous and spongy, for good engineering and economic reasons. When water seeps through the interstices of a dead bone, minerals are slowly deposited as the ages pass. I say slowly almost as a ritual, but it isn't always slow. Think how fast a kettle furs up. On an Australian beach I once found a bottletop embedded in stone. But the process usually is slow. Whatever the speed, the stone of a fossil eventually takes on the shape of the original bone, and that shape is revealed to us millions of years later, even if – which doesn't always happen – every atom of the original bone has disappeared. The petrified forest in the Painted Desert of Arizona consists of trees whose tissues were slowly replaced by silica and other minerals leached out of ground water. Two hundred million years dead, the trees are now stone through and through, but many of their microscopic cellular details can still be clearly seen in petrified form.

I've already mentioned that sometimes the original organism, or a part of it, forms a natural mould or imprint from which it is subsequently removed, or dissolved. I fondly recall two happy days in Texas in 1987 spent wading through the Paluxy River examining, and even putting my feet in, the dinosaur footprints preserved in its smooth limestone bed. A bizarre local legend grew up that some of these are giant manprints contemporary with undoubted dinosaur prints, and in consequence the nearby town of Glen Rose became home to a thriving cottage industry, artlessly faking giant manprints in blocks of cement (for sale to gullible creationists who know, all too well, that 'There were giants in the earth in those days': Genesis 6:4). The story of the real footprints has been carefully worked out, and is fascinating. The obviously dinosaurian ones are three-toed. The ones that look faintly like a human foot have no toes, and were made by dinosaurs walking on the back of the foot rather than running on their toes. Also, the viscous mud would have tended to ooze back in at the sides of the footprint, obscuring the side toes of the dinosaurs.

More poignant for us, at Laetoli in Tanzania are the companionable footprints of three real hominids, probably *Australopithecus afarensis*, walking together 3.6 million years ago in what was then fresh volcanic ash (see plate 3). Who does not wonder what these individuals were to each other, whether they held hands or even talked, and what forgotten errand they shared in a Pliocene dawn?

Sometimes, as I mentioned when discussing lava, the mould may become filled with a different material, which subsequently hardens to form a cast of the original animal or organ. I am writing this on a table in the garden whose top is a six-inch thick, seven-foot square slab of Purbeck sedimentary limestone, of Jurassic age, perhaps 150 million years old.* Along with lots of fossil mollusc shells, there is an alleged (by the distinguished and eccentric sculptor who procured it for me) dinosaur footprint on the underside of the table, but it is a footprint in relief, standing out from the surface. The original footprint (if indeed it is genuine, for it looks pretty nondescript to me) must have served as a mould, into which the sediment later settled. The mould then disappeared. Much of what we know about ancient brains comes to us in the form of such casts: 'endocasts' of the insides of skulls, often imprinted with surprisingly full details of the brain surface itself.

Less frequently than shells, bones or teeth, soft parts of animals sometimes fossilise. The most famous sites are the Burgess Shale of the Canadian Rockies, and the slightly older Chengjiang in South China which we shall meet again in the Velvet Worm's Tale. At both these sites, fossils of worms and other soft, boneless and toothless creatures (as well as the usual hard ones) wonderfully record the Cambrian Period, more than half a billion years ago. We are outstandingly lucky to have Chengjiang (see plate 4) and the Burgess Shale. Indeed, as I have already remarked, we are pretty lucky to have fossils at all, anywhere. It has been estimated that 90 per cent of all species will never be known to us as fossils. If that is the figure for whole species, just think how few individuals can ever hope to achieve the ambition with which the tale began, and end up as fossils. One estimate puts the odds at one in a million among vertebrates. That sounds high to me, and the true figure must be far less among animals with no hard parts.

---

* A journalist interviewed me at this two-tonne megalith for over an hour and then described it in his newspaper as a 'white wrought-iron table': my favourite example of the fallibility of eyewitness evidence.

HABILINES

Back another million years from *Homo ergaster,* 2 million years ago there is no longer any doubt in which continent our genetic roots lie. Everyone agrees, 'multiregionalists' included, that Africa is the place. The most compelling fossil bones at this age are normally classified as *Homo habilis.* Some authorities recognise a second, very similar contemporary type, which they call *Homo rudolfensis.* Others equate it with *Kenyapithecus,* described by the Leakey team in 2001. Yet others cautiously refrain from giving these fossils a species name at all, and just call them all 'Early *Homo*'. As usual I shan't take a stand on names. What matters is the real flesh and bone creatures themselves, and I shall use 'Habilines' as an anglicism for all of them. Habiline fossils, being older, are understandably less plentiful than Ergasts. The best-preserved skull bears the reference number KNM-ER 1470 and is widely known as Fourteen Seventy. It lived about 1.9 million years ago.

The Habilines were about as different from Ergasts as Ergasts from us, and, as we should expect, there were intermediates which are hard to classify. Habiline skulls are less robust than Ergast skulls, and lack the pronounced brow ridges. In this respect, Habilines were more like us. This should cause no surprise. Robustness and brow ridges are peculiarities that, possibly like hair, hominids seem able to acquire and lose again at the drop of an evolutionary hat.

Habilines mark the place in our history where the brain, that most dramatic of human peculiarities, starts to expand. Or more accurately, starts to expand beyond the normal size of the already large brains of other apes. This distinction, indeed, is the rationale for placing the Habilines in the genus *Homo* at all. For many palaeontologists, the large brain is the distinguishing feature of our genus. Habilines, with their brains pushing the 750 cc barrier, have crossed the rubicon and are human.

As readers may soon become tired of hearing, I am not a lover of rubicons, barriers and gaps. In particular, there is no reason to expect an early Habiline to be separated from its predecessor by a bigger gap than from its successor. It might seem tempting because the predecessor has a different generic name (*Australopithecus*) whereas the successor *(Homo ergaster)* is 'merely' another *Homo.* It is true that when we look at living species, we expect members of different genera to be less alike than members of different species within the same genus. But it can't work like that for fossils, if we have a continuous historical lineage in evolution. At the borderline between any fossil species and its immediate predecessor, there must be some individuals about whom it is absurd to argue, since the *reductio* of such an argument must be that parents of one species gave birth to a child of the other. It is even more absurd to suggest that a baby of the genus *Homo* was born to parents of a completely different genus, *Australopithecus.* These are evolutionary regions into which our zoological naming conventions were never designed to go.*

Setting names to one side frees us for a more constructive discussion about *why* the brain suddenly started to enlarge. How would we measure the enlargement of the hominid brain and plot a graph of average brain size against geological time? There is no problem about the units in which we measure time: millions of years. Brain size is harder. Fossil skulls and endocasts allow us to estimate brain size in cubic centimetres, and it is easy enough to convert this to grams. But absolute brain size is not necessarily the measure you want. An elephant has a bigger brain than a person, and it isn't just vanity that makes us think we are brainier than elephants. *Tyrannosaurus*'s brain was not much smaller than ours, but all dinosaurs are regarded as small-brained, slow-witted creatures. What makes us cleverer is that we have bigger brains *for our size* than dinosaurs. But what, more precisely, does 'for our size' mean?

There are mathematical methods of correcting for absolute size, and expressing an animal's brain size as a function of how big it 'ought to be' given its body size. This is a topic worthy of a tale in its own right, and *Homo habilis,* handyman, from his uneasy vantage point straddling the brain-size 'rubicon', will tell it.

---

* The 750 cc rubicon for the definition of *Homo* was originally chosen by Sir Arthur Keith. As Richard Leakey tells us in *The Origin of Humankind,* when Louis Leakey first described *Homo habilis* his specimen had a brain capacity of 650 cc, and Leakey actually moved the rubicon to accommodate it. Later specimens of *Homo habilis* retrospectively vindicated him by turning in figures closer to 800 cc. All grist to my anti-rubicon mill.

THE HANDYMAN'S TALE

We want to know whether the brain of a particular creature such as *Homo habilis* is larger or smaller than it 'ought' to be, given that animal's body size. We accept (slightly unwillingly in my case but I'll let it pass) that large animals just have to have large brains and small animals small brains. Making allowance for this, we still want to know whether some species are 'brainier' than others. So, how do we make allowance for body size? We need a reasonable basis for calculating the expected brain size of an animal from its body size, so that we can decide whether the actual brain of a particular animal is larger or smaller than expected.

In our pilgrimage to the past, we happen to have met the problem in connection with brains, but similar questions can arise with respect to any part of the body. Do some animals have larger (or smaller) hearts, or kidneys, or shoulder-blades than they 'ought' to have for their size? If so, this might suggest that their way of life makes special demands on the heart (kidney or shoulder-blade). How do we know what size any bit of an animal 'ought' to be, given that we know its total body size? Note that 'ought to be' doesn't mean 'needs to have for functional reasons'. It means 'would be expected to have, knowing what comparable animals have'. Since this is the Handyman's Tale, and since the Handyman's most surprising feature is his brain, we'll go on using brains for the sake of discussion. The lessons we learn will be more general.

We begin by making a scatter plot of brain mass against body mass for a large number of species. Each symbol in the graph on the opposite page (from my colleague the distinguished anthropologist Robert Martin) represents one species of living mammal – 309 of them, ranging from the smallest to the largest. In case you are interested, *Homo sapiens* is the point with the arrow, and the one immediately next to us is a dolphin. The heavy black line drawn through the middle of the points is the straight line that, according to statistical calculation, gives the best fit to all the points.*

A slight complication, which will make sense in a moment, is that things work better if we make the scales of both axes logarithmic, and that is how this graph was made. We plot the logarithm of

---

* It is the line that minimises the sum of the squares of the distances of the points from it.

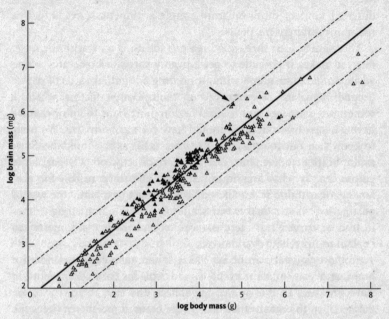

Log-log plot of brain mass against body mass for different species of placental mammal, with primates as filled triangles. Adapted from Martin [185].

an animal's brain mass against the logarithm of its body mass. Logarithmic means that equal steps along the bottom of the graph (or equal steps up the side) represent *multiplications* by some fixed number, say ten, rather than additions of a number, as in an ordinary graph. The reason ten is convenient is that we can then think of a logarithm as a count of the number of noughts. If you have to multiply a mouse's mass by a million to get an elephant's, this means you have to add six noughts to the mouse's mass: you have to add six to the logarithm of the one, to get the logarithm of the other. Half way between them on the logarithmic scale – three noughts – lies an animal that weighs a thousand times as much as a mouse, or a thousandth of an elephant: a person, perhaps. Using round numbers like a thousand and a million is just to make the explanation easy. 'Three and a half noughts' means somewhere between a thousand and ten thousand. Note that 'half way between' when we are counting noughts is a very different matter from half way between when we are counting grams. This is all taken care of automatically by looking up the logarithms of the numbers. Logarithmic scales call on a different kind of intuition from simple arithmetic scales, which is useful for different purposes.

There are at least three good reasons for using a logarithmic scale. First, it makes it possible to get a pygmy shrew, a horse and a blue whale on the same graph without needing a hundred yards of paper. Second, it makes it easy to read off multiplicative factors, which is sometimes what we want to do. We don't just want to know that we have a bigger brain than we should have for our body size. We want to know that our brain is, say, six times as big as it 'should' be. Such multiplicative judgements can be read directly off a logarithmic graph: that is what logarithmic means. The third reason for preferring logarithmic scales takes a little longer to explain. One way of putting it is that it makes our scatterpoints fall along straight lines instead of curves, but there is more to it than that. Let me try to explain to my fellow dysnumerics.

Suppose you take an object like a sphere or a cube, or indeed a brain, and you inflate it evenly so it is still the same shape but ten times the size. In the case of the sphere, this means ten times the diameter. In the case of the cube, or the brain, it means ten times the width (and height and depth). In all these cases of proportionate scaling up, what will happen to the volume? It will not be ten times as great – it will be a thousand times as great! You can prove it for cubes if you imagine stacking sugar lumps. The same applies to uniformly inflating any shape you like. Multiply length by ten and, provided the shape doesn't change, you automatically multiply volume by a thousand. In the special case of a tenfold inflation, this is equivalent to adding three noughts. More generally, volume is proportional to the third power of length, and the logarithm is multiplied by three.

We can do the same sort of calculation for area. But area increases in proportion to the second power of length rather than the third power. Not for nothing is raising to the second power called squaring while raising to the third power is called cubing. The volume of a sugar lump determines how much sugar there is, and what it costs. But how fast it dissolves will be determined by its surface area (not a simple calculation because, as it dissolves, the remaining surface area will shrink more slowly than the volume of sugar remaining). When you uniformly inflate an object by doubling its length (width, etc.), you multiply the surface area by $2 \times 2 = 4$. Multiply its length by ten, and you multiply the surface area by $10 \times 10 = 100$ or add two noughts to the number. The logarithm of area increases as double the logarithm of length, while the logarithm of volume increases as treble the logarithm of length. A two-centimetre sugar lump will contain eight times as much sugar as a one-centimetre lump, but it will release that sugar into the tea only four times as fast (at least initially), because it is the surface of the lump that is exposed to the tea.

Now imagine that we make a scatter plot of sugar lumps of a wide range of sizes, with mass of lump (proportional to volume) along the bottom axis, and (initial) rate of dissolving up the side of the graph (assumed proportional to area). In a non-logarithmic graph, the points will fall along a *curved* line, which will be quite hard to interpret and not very helpful. But if we plot the logarithm of mass against the logarithm of initial dissolving rate, we shall see something much more informative. For every threefold increment of log mass, we shall see a doubling of log surface. On the log-log scale, the points will not fall along a curve, they will fall along a straight line. What is more, the *slope* of the straight line will mean something very precise. It will be a slope of two-thirds: for every two steps along the area axis, the line takes three steps along the volume axis. For every doubling of the logarithm of area, the logarithm of volume is tripled. Two-thirds is not the only informative slope of line we might see in a log-log plot. Plots of this kind are informative because the slope of the line gives us an intuitive feel for what is going on vis-à-vis such things as volumes and areas. And volumes and areas and the complicated relationships between them are extremely important in understanding living bodies and their parts.

I am not particularly mathematical – that's putting it mildly – but even I can see the fascination of this. And it gets better, because the same principle works for all shapes, not just tidy ones like cubes and spheres, but complicated shapes like animals and bits of animals such as kidneys and brains. All that is required is that size change should come about by simple inflation or deflation without a change of shape. This gives us a sort of null-expectation, against which to compare real measurements. If one species of animal is 10 times the length of another, its mass will be 1,000 times as great, *but only if the shapes are the same.* In fact, shape is very likely to have evolved to be systematically different as you go from small animals to large, and we can now see why.

Big animals need to be a different shape from small animals, if only because of the area/volume scaling rules we have just seen. If you turned a shrew into an elephant just by inflating it, retaining the same shape, it wouldn't survive. Because it is now about a million times heavier, a whole lot of new problems arise. Some of the problems an animal faces depend upon volume (mass). Others depend on area. Still others depend on some complicated function of the two, or on some different consideration altogether. Like a sugar lump's rate of dissolving, an animal's rate of losing heat, or of losing water through the skin, will be proportional to the area that it presents to the outside world. But its rate of generating heat is probably more related to the number of cells in the body, which is a function of volume.

A shrew scaled up to elephant size would have spindly legs that would break under the strain, and its slender muscles would be too weak to work. The strength of a muscle is proportional not to its volume but to its cross-sectional area. This is because muscular movement is the summed movement of millions of molecular fibres, sliding past each other in parallel. The number of fibres you can pack into a muscle depends upon the area of its cross-section (second power of linear size). But the task that the muscle has to perform – supporting an elephant, say – is proportional to the mass of the elephant (third power of linear size). So, the elephant *needs* proportionately more muscle fibres than a shrew, in order to support its mass. Therefore the cross-sectional area of elephant muscles needs to be larger than you'd expect from simple scaling up, and the volume of muscle in an elephant must be more than you'd expect from simple scaling up. For different particular reasons, the conclusion is similar for bones. This is why large animals like elephants have massive tree-trunk shaped legs.

Suppose an elephant-sized animal is 100 times as long as a shrew-sized animal. With no change of shape, the area of its outer skin would be 10,000 times as great as the shrew's and its volume and mass a million times as great. If touch-sensitive cells are equally spaced through the skin, the elephant will need 10,000 times as many of them, and the part of the brain that services them will perhaps need to be scaled in proportion. The total number of cells in the elephant's body will be a million times as great as in the shrew, and they'll all have to be serviced by capillary blood vessels. What does this do to the number of miles of blood vessel that we expect in a large animal, as distinct from a small one? That's a complicated calculation, and one that we'll return to in a later tale. For the moment, it is enough for us to understand that when we calculate it we cannot ignore these scaling rules for volumes and areas. And the logarithmic plot is a good method for getting intuitive clues to such things. The main conclusion is that, as animals get larger or smaller in evolution, we positively expect their shape to change in predictable directions.

We got into this through thinking about brain size. We can't just compare our brains with those of *Homo habilis, Australopithecus* or any other species without making allowance for body size. We need some index of brain size which makes allowance for body size. We can't divide brain size by body size, though that would be better than just comparing absolute brain sizes. A better way is to make use of the logarithmic plots we have just been discussing. Plot the logarithm of brain mass against the logarithm of body mass for lots of species of different sizes. The points will probably fall around a straight line, as indeed they do in the graph on page 81. If the slope of the line is $\frac{1}{1}$ (brain size exactly proportional to body size) it will suggest that each brain cell is capable of servicing some fixed number of body cells. A slope of $\frac{2}{3}$ would suggest that brains are like bones and muscles: a given volume of body (or number of body cells) demands a certain surface area of brain. Some other slope would need yet a different interpretation. So, what is the actual slope of the line?

It is neither $\frac{1}{1}$ nor $\frac{2}{3}$ but something in between. To be exact, it is a remarkably good fit to $\frac{3}{4}$. Why $\frac{3}{4}$? Well, that is a tale in itself, which will be told, as you will no doubt have guessed, by the cauliflower (well, a brain does *look* a bit like a cauliflower). Without pre-empting the Cauliflower's Tale, I will just say that the $\frac{3}{4}$ slope is not special to brains, but crops up all over the place in all sorts of living creatures, including plants like cauliflowers. Applied to brain size, and with the intuitive rationale that must wait for the Cauliflower's Tale, this observed line, with its $\frac{3}{4}$ slope, is the meaning we are going to attach to the word 'expect' as it was used in the opening paragraphs of this tale.

Although the points cluster about the 'expected' straight line of slope $\frac{3}{4}$, not all the points fall exactly on the line. A 'brainy' species is one whose point on the graph falls above the line. Its brain is larger than 'expected' for its body size. A species whose brain is smaller than 'expected' falls below the line. The distance above, or below, the line, is our measure of how *much* bigger than 'expected', or smaller, it is. A point that falls exactly on the line represents a species whose brain is exactly the size expected for its body size.

Expected on what assumption? On the assumption that it is typical of the set of species whose data contributed to calculating the line. So, if the line was calculated from a representative range of land vertebrates, from geckos to elephants, the fact that all mammals fall above the line (and all reptiles below) means that mammals have bigger brains than you would 'expect' of a typical vertebrate. If we calculate a separate line from a representative range of mammals, it will be parallel to the vertebrate line, still with a slope of $\frac{3}{4}$, but its absolute height will be higher. A separate line calculated from a representative range of primates (monkeys and apes) will be higher again, but still parallel with a slope of $\frac{3}{4}$. And *Homo sapiens* is higher than any of them.

The human brain is 'too' big, even by the standards of primates, and the average primate brain is too big by the standards of mammals generally. For that matter, the average mammal brain is too big by the standards of vertebrates. Another way to say all this is that the scatter of points in the vertebrate graph is wider than the scatter of points on the mammal graph, which is in turn wider than the primate scatter which it includes. The xenarthran scatter of points on the graph (xenarthrans are an order of South American mammals, including sloths, anteaters and armadillos) sits below the average of mammals, of which the xenarthran scatter forms a part.

Harry Jerison, the father of fossil brain size studies, proposed an index, the Encephalisation Quotient or EQ, as a measure of how much bigger, or smaller, the brain of a particular species is than it 'should' be for its size, given that it is a member of some larger grouping, such as the vertebrates or the mammals. Notice that the EQ requires us to specify the larger group which is being used as the baseline for comparison. The EQ of a species is its distance above, or below, the average line for the specified larger grouping. Jerison thought the slope of the line was $\frac{2}{3}$, whereas modern studies agree that it is $\frac{3}{4}$, so Jerison's own estimates of EQ have to be amended accordingly, as was pointed out by Robert Martin. When this is done, it turns out that the modern human brain is about six times as big as it should be, for a mammal of equivalent size (the EQ would be larger, if calculated against the standard of the vertebrates as a whole, rather than the mammals as a whole. And it would be smaller if calculated against the standard of primates as a whole).* A modern chimpanzee's brain is about twice the size it should be for a typical mammal, and so are the brains of australopithecines. *Homo habilis* and *Homo erectus,* the species that are probably intermediate in evolution between *Australopithecus* and ourselves, are also intermediate in brain size. Both have an EQ of about 4, meaning that their brains were about four times as big as they should have been for a mammal of equivalent size.

The graph on the following page shows an estimate of EQ, the 'braininess index', for various fossil primates and ape-men, as a function of the time at which they lived. With considerable pinches of salt you could read it as a rough graph of decreasing braininess as we go backwards in evolutionary time. At the top of the graph is modern *Homo sapiens* with an EQ of 6, meaning our brain is six times as heavy as it 'should' be for a typical mammal of our size. At the bottom of the graph are fossils who might possibly represent something like Concestor 5, our common ancestor with the Old World monkeys. Their estimated EQ was about 1, meaning they had a brain which would be 'about right' for a typical mammal of their size today. Intermediate on the graph are various species of *Australopithecus* and *Homo* who might be close to our ancestral line at the time they lived. The drawn line is, once again, the straight line which best fits the points on the graph.

I advised pinches of salt, and let me raise that to ladles of salt. The EQ 'braininess index' is calculated from two measured quantities, the brain mass and the body mass. In the case of fossils, both these quantities have to be estimated from the fragments that have come down to us, and there is a huge margin of error, especially in the estimation of body mass. The point on the graph for *Homo habilis* shows it as 'brainier' than *Homo erectus.* I don't believe this. The absolute brain size of *H. erectus* is undeniably larger. The inflation of the *H. habilis* EQ comes from the much lower estimated body mass. But to get an idea of the margin of error, think of the enormous range of body mass in modern humans. EQ as a measure is extremely

* Much the same applies to IQ. It is *not* an absolute measure of intelligence. Rather, your IQ reflects how much more (or less) intelligent you are than the average for a particular population, that average being standardised at 100. My IQ if standardised against the background population of Oxford University would be lower than if standardised against the background population of England. Hence the joke about the politician lamenting the fact that half the population has an IQ less than 100.

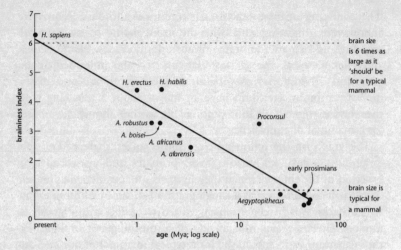

Plot of EQ or 'braininess index' for various fossil species, against time in millions of years on a log scale. The results have been corrected for a slope of $\frac{3}{4}$ for the reference baseline (see text).

sensitive to error in measuring body mass, which is raised, remember, to a power in the EQ formula. So, the scatter of points about the line largely reflects erratic estimation of body mass. On the other hand, the trend over time, as represented by the line, is probably real. The methods explained in this tale, in particular the estimates of EQ in the graph at the end, bear out our subjective impression that one of the most important things that has happened during the last 3 million years of our evolution was the ballooning of our already large primate brain. The next obvious question is why. What Darwinian selection pressure drove the enlargement of the brain during the past three million years?

Because it happened after we rose up on our hind legs, some people have suggested that brain inflation was driven by the freeing of the hands and the opportunity this offered for precision-controlled manual dexterity. In a general way I find this a plausible idea, though no more than several others that have been offered. But the enlargement of the human brain looks, as evolutionary trends go, explosive. I think inflationary evolution demands a special kind of inflationary explanation. In *Unweaving the Rainbow*, in the chapter called 'The Balloon of the Mind', I developed this inflationary theme in a general theory of what I called 'software-hardware co-evolution'.

The computer analogy is with software innovations and hardware innovations triggering each other in an escalating spiral. Software innovations demand an escalation in hardware, which in turn provokes an escalation in software, and so the inflation gathers pace. In the brain, my candidates for the *kind* of thing I meant by a software innovation were language, spoor-tracking, throwing, and memes. One theory of brain inflation that I didn't do justice to in my earlier book was sexual selection, and it is for this reason alone that I shall give it special prominence later in this book.

Could the enlarged human brain, or rather its products such as body painting, epic poetry and ritual dances, have evolved as a kind of mental peacock's tail? I have long had a soft spot for the idea, but nobody developed it into a proper theory until Geoffrey Miller, a young American evolutionary psychologist working in England, wrote his book, *The Mating Mind*. We shall hear this idea in the Peacock's Tale, after the bird pilgrims join us at Rendezvous 16.

APE-MEN

The popular literature on human fossils is hyped up with alleged ambition to discover the 'earliest' human ancestor. This is silly. You can ask a specific question like 'Which was the earliest human ancestor to walk habitually on two legs?' Or 'Which was the first creature to be our ancestor and not the ancestor of a chimpanzee?' Or 'Which was the earliest human ancestor to have a brain volume larger than 600 cc?' Those questions at least mean something in principle, although they are hard to answer in practice and some of them suffer from the vice of erecting artificial gaps in a seamless continuum. But 'Who was the earliest human ancestor?' means nothing at all.

More insidiously, the competition to find human ancestors means that new fossil discoveries are touted as on the 'main' human line whenever remotely possible. But as the ground yields up more and more fossils it becomes increasingly clear that, during most of hominid history, Africa housed several species of hominid simultaneously. This has to mean that many fossil species now thought of as ancestral will turn out to be our cousins.

At various times since *Homo* first appeared in Africa, it shared the continent with more robust hominids, perhaps several different species of them. As usual their affinities, and the exact number of species, are hotly disputed. Names that have been attached to various of these creatures (we met them in the graph at the end of the Handyman's Tale) are *Australopithecus* (or *Paranthropus*) *robustus*, *Australopithecus* (or *Paranthropus* or *Zinjanthropus*) *boisei*, and *Australopithecus* (or *Paranthropus*) *aethiopicus*. They seem to have evolved from more 'gracile' apes (gracile being the opposite of robust). The gracile apes are also placed in the genus *Australopithecus*, and we too almost certainly emerged from among gracile australopithecine ranks. Indeed, it is often difficult to distinguish early *Homo* from gracile australopithecines – which prompted my diatribe on the naming conventions that place them in separate genera.

The immediate ancestors of *Homo* would be classified as some kind of gracile australopithecine. Let's look at some of the gracile fossils. Mrs Ples is one for whom I have had special affection ever since the Transvaal Museum in Pretoria presented me with a beautiful cast of her skull, on the fiftieth anniversary of her discovery at Sterkfontein nearby, when I gave the Robert Broom Memorial Lecture in honour of her discoverer. She lived about 2.5 million years ago. Her nickname comes from the genus *Plesianthropus*, to which she was originally assigned before people decided to incorporate her into *Australopithecus;* and from the fact that she was thought (perhaps erroneously as is now suspected) to be female. Individual fossil hominids often pick up pet names like this. 'Mr Ples', naturally, is a more recently discovered fossil from Sterkfontein who is in the same species as Mrs Ples, *Australopithecus africanus*. Other fossils with nicknames include 'Dear Boy', a robust australopithecine also known as 'Zinj' because he was originally named *Zinjanthropus boisei*, 'Little Foot' (see below) and the famous Lucy, to whom we now turn.

We meet Lucy as our time machine's odometer touches 3.2 million years. Another gracile australopithecine, she is often mentioned because her species, *Australopithecus afarensis*, is a hot contender for a human ancestor. Her discoverers, Donald Johanson and his colleagues, also found fossils of 13 similar individuals in the same area, known as the 'First Family'. Other 'Lucys' have since been found between about 3 and 4 million years ago in other parts of East Africa. The 3.6-million-year-old footprints discovered by Mary Leakey at Laetoli (page 76) are attributed to *A. afarensis*. Whatever the Latin name, evidently somebody was walking bipedally at that time. Lucy is not greatly different from Mrs Ples, and some people think of Lucys as an earlier version of Mrs Ples. They are anyway more like each other than either is like the robust australopithecines. Early East African Lucys are said to have a slightly smaller brain than later South African Mrs Pleses, but there isn't much in it. Their brains were no more different from each other than some modern human brains are from other modern human brains.

As we have come to expect, the more recent *afarensis* individuals such as Lucy are slightly different from the earliest 3.9-million-year-old *afarensis* forms. Differences collect over time and, as we emerge from our time machine 4 million years ago, we find more creatures who might well be ancestral to Lucy and her kin, but who are sufficiently different, in the direction of being more chimpanzee-like, to merit a different species name. Discovered by Meave Leakey and her team, these *Australopithecus anamensis* consist of more than 80 fossils from two different sites near Lake Turkana. No intact skull has been found, but there is a splendid lower jaw which plausibly could belong to an ancestor of ours.

But the most exciting discovery from this time period, and a good reason for calling a temporary halt here, is a fossil yet to be fully described in print. Affectionately known as Little Foot, this skeleton from the Sterkfontein caves of South Africa was originally dated to about three million years ago, but has recently been redated to just over four million. Its discovery is a piece of detective work worthy of a Conan Doyle story. Bits of Little Foot's left foot were dug up from Sterkfontein in 1978, but the bones were stored away, unremarked and unlabelled, until 1994 when the palaeontologist Ronald Clarke, working under the direction of Phillip Tobias, accidentally rediscovered them in a box in the shed used by workers at the Sterkfontein cave. Three years later, Clarke chanced upon another box of bones from Sterkfontein, in a store room at Witwatersrand University. This box was labelled 'Cercopithecoids'. Clarke had an interest in this kind of monkey, so he looked in the box and was delighted to notice a hominid foot bone in amongst the monkey bones. Several foot and leg bones in the box seemed to match the bones previously found in the Sterkfontein shed. One was half a right shinbone, broken across. Clarke gave a cast of the shinbone to two African assistants, Nkwane Molefe and Stephen Motsumi, and asked them to return to Sterkfontein and look for the other half.

> The task I had set them was like looking for a needle in a haystack as the grotto is an enormous, deep, dark cavern with breccia exposed on the walls, floor and ceiling. After two days of searching with the aid of hand-held lamps, they found it on 3 July 1997.

Molefe and Motsumi's jigsaw feat was the more astonishing because the bone that fitted their cast was

> at the opposite end to where we had previously excavated. The fit was perfect, despite the bone having been blasted apart by lime workers 65 or more years previously. To the left of the exposed end of the right tibia could be seen the section of the broken-off shaft of the left tibia, to which the lower end of the left tibia with foot bones could be joined. To the left of that could be seen the broken-off shaft of the left fibula.

From their positions with the lower limbs in correct anatomical relationship, it seemed that the whole skeleton had to be there, lying face downwards.

Actually, it wasn't quite there but, after pondering the geological collapses in the area, Clarke deduced where it must be and, sure enough, Motsumi's chisel found it there. Clarke and his team were indeed lucky, but here we have a first-class example of that maxim of scientists since Louis Pasteur: 'Fortune favours the prepared mind.'

Little Foot is still to be fully excavated, described and formally named, but preliminary reports suggest a spectacular find, rivalling Lucy in completeness but older. Although more human-like than chimpanzee-like, the big toe is more divergent than our toes. This might suggest that Little Foot grasped tree boughs with its feet in a way that we cannot. Although it almost certainly walked bipedally, it probably climbed too and walked with a different gait from us. Like other australopithecines, it may have spent time in trees, perhaps bivouacking in them at night like modern chimpanzees.

Having paused at the 4-million-year milestone, let's take a quick peek at the journey yet to unfold. There are some fragmentary remains of a possibly bipedal *Australopithecus*-like creature even further back in time, about 4.4 million years ago. Tim White and his colleagues discovered it in Ethiopia, quite close to Lucy's last resting place. They named it *Ardipithecus ramidus,** although some prefer to keep it in the genus *Australopithecus*. No skull of *Ardipithecus* has so far been found, but its teeth suggest that it was more chimpanzee-like than any later humans. Its tooth enamel was thicker than that of chimpanzees, but not as thick as ours. A few isolated cranial bones have been found, and these indicate that the skull rested on top of the vertebral column, as in us, rather than in front of it, as in chimpanzees. This suggests a vertical stance, and such foot bones as have been found support the idea that *Ardipithecus* was bipedal.

Bipedality separates humans from the rest of the mammals so dramatically that I feel it deserves a tale to itself. And who better fitted to tell it than Little Foot?

---

* Some people distinguish a second species, *Ardipithecus kadabba*.

LITTLE FOOT'S TALE

It isn't particularly helpful to dream up reasons why walking on two legs might be generally a good thing. If it were, the chimps would do it too, to say nothing of other mammals. There is no obvious reason for saying that either bipedal or quadrupedal running is faster or more efficient than the other. Galloping mammals can be astonishingly fleet, using the up-and-down flexibility of the backbone to achieve – among other benefits – a lengthened effective stride. But ostriches show that a man-like bipedal gait can be a match for a quadrupedal horse. Indeed a top human sprinter, though noticeably slower than a horse or dog (or ostrich or kangaroo, for that matter), is not disgracefully slow. Quadrupedal monkeys and apes are generally undistinguished runners, perhaps because their bodily designs have to compromise with the needs of a climber. Even baboons, which normally forage and run on the ground, resort to the trees to sleep and as a defence against predators, but baboons can run fast when they need to.

So, when we ask why our ancestors rose up on their hind legs, and when we imagine the quadrupedal alternative that we forsook, it is unfair to 'think cheetah', or anything like it. When our ancestors first stood up, there was no overwhelmingly strong advantage in efficiency or speed. We should look elsewhere for the natural selection pressure which drove us to this revolutionary change in gait.

Like some other quadrupeds, chimpanzees can be trained to walk bipedally, and they often do it anyway over short distances. So it probably wouldn't be insuperably difficult for them to make the switch if there were strong benefits to doing so. Orang utans are even better at it. Wild gibbons, whose fastest method of locomotion is brachiation – swinging under the boughs by their arms – also run across clearings on their hind legs. Some monkeys rise upright, to peer over long grass or to wade through water. A lemur, Verreaux's sifaka, although it lives mainly in trees where it is a spectacular acrobat, 'dances' across the ground between trees on its hind legs, the arms held up with balletic grace.

Doctors sometimes ask us to run on the spot in a mask, so they can measure our oxygen consumption and other metabolic indices when we are exerting ourselves. In 1973 two American biologists, C. R. Taylor and V. J. Rowntree, did this with trained chimpanzees and capuchin monkeys, running on a treadmill. By making the animals run the treadmill either on four legs or on two (they were given something to hold on to), the researchers could compare the oxygen consumption and efficiency of the two gaits. They expected that quadrupedal running would be more efficient. This, after all, is what both species naturally do, and it is what their anatomy fits them for. Maybe bipedalism was helped by the fact that they had something to hold on to. In any case, the result was otherwise. There was no significant difference between the oxygen consumption of the two gaits. Taylor and Rowntree concluded that:

> The relative energy cost of bipedal versus quadrupedal running should not be used in arguments about the evolution of bipedal locomotion in man.

Even if this is an exaggeration, it should at least encourage us to look elsewhere for possible benefits of our unusual gait. It arouses the suspicion that, whatever non-locomotor benefits of bipedality we might propose as drivers of its evolution, they probably did not have to fight against strong locomotor costs.

What might a non-locomotor benefit look like? A stimulating suggestion is the sexual selection theory of Maxine Sheets-Johnstone, of the University of Oregon. She thinks we rose on our hind legs as a means of showing off our penises. Those of us that have penises, that is. Females, in her view, were doing it for the opposite reason: concealing their genitals which, in primates, are more prominently displayed on all fours. This is an appealing idea but I don't carry a torch for it. I mention it only as an example of the *kind* of thing I mean by a non-locomotor theory. As with so many of these theories, we are left wondering why it would apply to our lineage and not to other apes or monkeys.

A different set of theories stresses the freeing of the hands as the really important advantage of bipedality. Perhaps we rose on our hind legs, not because that is a good way of getting about, but because of what we were then able to do with our hands – carry food, for instance. Many apes and monkeys feed on plant matter that is widely available but not particularly rich or concentrated, so you must eat as you go, more or less continuously like a cow. Other kinds of food such as meat or large underground tubers are harder to acquire but, when you do find them, they are valuable – worth carrying home in greater quantity than you can eat. When a leopard makes a kill, the first thing it normally does is drag it up a tree and hang it over a branch, where it will be relatively safe from marauding scavengers and can be revisited for meals. The leopard uses its powerful jaws to hold the carcass, needing all four legs to climb the tree. Having much smaller and weaker jaws than a leopard, did our ancestors benefit from the skill of walking on two legs because it freed their hands for carrying food – perhaps back to a mate or children, or to trade favours with other companions, or to keep in a larder for future needs?

Incidentally the latter two possibilities may be closer to each other than they appear. The idea (I attribute this inspired way of expressing it to Steven Pinker) is that before the invention of the freezer the best larder for meat was a companion's belly. How so? The meat itself is no longer available, of course, but the goodwill it buys is safe in long-term storage in a companion's brain. Your companion will remember the favour and repay it when fortunes are reversed.[*] Chimpanzees are known to share meat for favours. In historic times, this kind of I.O.U. became tokenised as money.

A particular version of the 'carrying food home' theory is that of the American anthropologist Owen Lovejoy. He suggests that females would often have been hampered in their foraging by nursing infants, therefore unable to travel far and wide looking for food. The consequent poor nutrition and poor milk production would have delayed weaning. Suckling females are infertile. Any male who feeds a nursing female accelerates the weaning of her current child and brings her into receptiveness earlier. When this happens, she might make her receptiveness especially available to the male whose provisioning accelerated it. So, a male who can bring lots of food home might gain a direct reproductive advantage over a rival male who just eats where he finds. Hence the evolution of bipedalism to free the hands for carrying.

Other hypotheses of bipedal evolution invoke the benefits of height, perhaps standing upright to look over the long grass; or to keep the head above water while wading. This last is the imaginative 'aquatic ape' theory of Alister Hardy, ably championed by Elaine Morgan. Another theory, favoured by John Reader in his fascinating biography of Africa, suggests that upright posture minimises exposure to the

---

[*] There is a well-developed theory of reciprocal altruism in Darwinism, beginning with the pioneering work of Robert Trivers and continuing with the modelling of Robert Axelrod and others. Trading favours, with delayed repayment, really works. My own exposition of it is in *The Selfish Gene*, especially the second edition.

sun, limiting it to the top of the head which is consequently furnished with protective hair. Moreover, when the body is not hunched close to the ground, it can lose heat more rapidly.

My colleague the distinguished artist and zoologist Jonathan Kingdon has centred a whole book, *Lowly Origin,* around the question of the evolution of human bipedality. After a lively review of 13 more-or-less distinct hypotheses, including the ones I have mentioned, Kingdon advances his own sophisticated and multifaceted theory. Rather than seek an immediate benefit of walking upright, Kingdon expounds a complex of quantitative anatomical shifts which arose for some other reason, but which then made it easier to become bipedal (the technical term for this kind of thing is pre-adaptation). The pre-adaptation that Kingdon proposes is what he calls squat feeding. Squat feeding is familiar from baboons in open country, and Kingdon visualises something similar in our ape ancestors in the forest, turning over stones or leaf litter for insects, worms, snails and other nutritious morsels. To do this effectively they would have had to undo some of their adaptations to living up trees. Their feet, previously hand-like for gripping branches, would have become flatter, forming a stable platform for squatting on the haunches. You will already be getting a glimmering of where the argument is going. Flatter, less hand-like feet for squatting are later going to serve as pre-adaptations for upright walking. And you will, as usual, understand that this apparently purposeful way of writing – they had to 'undo' their tree-swinging adaptations, etc. – is a shorthand which is easily translated into Darwinian terms. Those individuals whose genes happened to make their feet more suitable for squat feeding survived to pass on those genes because squat feeding was efficient and aided their survival. I shall continue to employ the shorthand because it chimes with the way humans naturally think.

A tree-swinging, 'brachiating' ape could fancifully be said to walk upside down under the branches – run and leap in the case of an athletic gibbon – using the arms as its 'legs' and the shoulder girdle as its 'pelvis'. Our ancestors probably passed through a brachiating phase, and the true pelvis consequently became rather inflexibly bound to the trunk by long blades of bone, which form a substantial part of a rigid trunk that can be swung as a single unit. Much of this, according to Kingdon, would have needed to change, to make an efficient squat feeder out of an ancestral brachiator. Not all, however. The arms could have remained long. Indeed, long brachiating arms would have been a positively beneficial 'pre-adaptation', increasing the reach of the squat feeder and decreasing the frequency with which it had to shuffle to a new squatting position. But the massive, inflexible, top-heavy ape trunk would have been a disadvantage in a squat feeder. The pelvis would have needed to free itself and become less rigidly tied to the trunk, and its blades would have shrunk – to more human proportions. This, to anticipate the later stages of the argument again (you might say that anticipation is what a pre-adaptation argument is all about) just happens to make a better pelvis for bipedal walking. The waist became more flexible, and the spine was held more vertically, to allow the squat-feeding animal to search all around with its arms, turning on the platform of the flat feet and the squatting haunches. The shoulders became lighter and the body less top-heavy. And the point is that these subtle quantitative changes, and the balancing and compensating shifts that went with them, incidentally had the effect of 'preparing' the body for bipedal walking.

Not for a moment is Kingdon proposing any kind of anticipation of the future. It is just that an ape whose ancestors were tree-swingers, but which has switched to squat feeding on the forest floor, now has a body which feels relatively *comfortable* walking on its hind feet. And it would have begun to do this while squat feeding, shuffling to a new squatting position as the old one became depleted. Without realising what was happening, squat feeders were, over the generations, preparing their bodies to feel more comfortable when upright and on two legs; to feel more awkward on four. I use the word comfortable deliberately. It is not a trivial consideration. We are capable of walking on all fours like a typical mammal, but it is uncomfortable: hard work, because of our altered body proportions. Those proportional changes which now make us feel comfortable on two legs originally came about, Kingdon suggests, in the service of a minor shift in food habits – to squat feeding.

There is much more in Jonathan Kingdon's subtle and complex theory, but I will now recommend his book, *Lowly Origin,* and move on. My own slightly wayout theory of bipedality is very different but not incompatible with his. Indeed, most of the theories of human bipedality are mutually compatible, with the potential to assist rather than oppose one another. As in the case of the enlargement of the human brain, my tentative suggestion is that bipedality may have evolved through sexual selection, so again I postpone the matter to the Peacock's Tale.

Whatever theory we believe about the evolutionary origins of human bipedality, it subsequently turned out to be an extremely important event. In former times it was possible to believe, as respected anthropologists did up to the 1960s, that the decisive evolutionary event that first separated us from the other apes was the enlargement of the brain. Rising up on the hind legs was secondary, driven by the benefits of freeing the hands to do the kind of skilled work which the enlarged brain was now capable of controlling and exploiting. Recent fossil finds point decisively towards the reverse sequence. Bipedality came first. Lucy, who lived long after Rendezvous 1, was bipedal, nearly or completely as bipedal as we are, yet her brain was approximately the same size as a chimpanzee's. The enlargement of the brain could still have been associated with the freeing of the hands, but the sequence of events was reversed. If anything it would be the freeing of the hands by bipedal walking that drove the enlargement of the brain. The manual hardware came first, then the controlling brainware evolved to take advantage of it, rather than the other way around.

EPILOGUE TO LITTLE FOOT'S TALE

Whatever the reason for the evolution of bipedality, recent fossil discoveries seem to indicate that hominids were already bipedal at a date which is pushing disconcertingly close to Rendezvous 1, the fork between ourselves and chimpanzees (disconcerting because it seems to leave little time for bipedality to evolve). In the year 2000, a French team led by Brigitte Senut and Martin Pickford announced a new fossil from the Tugen Hills, east of Lake Victoria in Kenya. Dubbed 'Millennium Man', dated at 6 million years and given yet another new generic name, *Orrorin tugenensis* was also, according to its discoverers, bipedal. Indeed, they claim that the top of its femur, near the hip joint, was more human-like than that of *Australopithecus*. This evidence, supplemented by fragments of skull bones, suggested to Senut and Pickford that orrorins are ancestral to later hominids and that Lucys are not. These French workers go further and suggest that *Ardipithecus* might be ancestral to modern chimpanzees rather than to us. Clearly we need more fossils to settle these arguments. Other scientists are sceptical of these French claims, and some doubt that there is enough evidence to show whether *Orrorin* was or was not bipedal. If it was, since 6 million years is

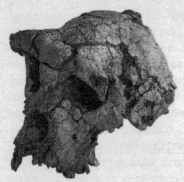

Hope of Life. Skull of *Sahelanthropus tchadensis*, or 'Toumai', discovered in the Sahel region of Chad by Michel Brunet and colleagues in 2001.

approximately the time of the split from chimpanzees according to molecular evidence, this raises difficult questions about the speed with which bipedality must have arisen.

If a bipedal *Orrorin* pushes back alarmingly close to Rendezvous 1, a newly discovered skull from Chad in southern Sahara, found by another French team led by Michel Brunet, is even more disturbing to accepted ideas. This is partly because it is so old, and partly because the site is far to the west of the Rift Valley (as we shall see, many authorities had thought early hominid evolution confined to the east of the Rift). Nicknamed Toumai (Hope of Life in the local Goran language) its official name is *Sahelanthropus tchadensis*, after the Sahel region of the Sahara in Chad where it was found. It is an intriguing skull, looking rather human from in front (lacking the protruding face of a chimpanzee or gorilla) but chimpanzee-like from behind, with a chimpanzee-sized braincase. It has an extremely well-developed brow-ridge, even thicker than a gorilla's, which is the main reason for thinking Toumai was male. The teeth are rather human-like, especially the thickness of the enamel which is intermediate between a chimpanzee's and our own. The foramen magnum (the big hole through which the spinal cord passes) is placed further forward than in a chimpanzee or gorilla, suggesting to Brunet himself, though not to some others, that Toumai was bipedal. Ideally, this should be confirmed by pelvis and leg bones but, unfortunately, nothing but a skull has so far been found.

There are no volcanic remains in the area to provide radiometric dates, and Brunet's team had to use other fossils in the area as an indirect clock. These are compared with already known faunas from other parts of Africa which can be dated absolutely. The comparison yields a date for Toumai of between 6 and 7 million years. Brunet and his colleagues claim it as older than *Orrorin*, which has predictably elicited indignant ripostes from *Orrorin*'s discoverers. One of them, Brigitte Senut, of the Natural History Museum in Paris, has said that Toumai is 'a female gorilla', while her colleague Martin Pickford described Toumai's canine teeth as typical 'of a large female monkey'. These were the two, remember, who (perhaps rightly) wrote off the human credentials of *Ardipithecus*, another threat to the priority of their own baby, *Orrorin*. Other authorities have hailed Toumai more generously: 'Astonishing.' 'Amazing.' 'This will have the impact of a small nuclear bomb.'

If their discoverers are right that *Orrorin* and Toumai were bipedal, this poses problems to any tidy view of human origins. The naive expectation is that evolutionary change spreads itself uniformly to fill the time available for it. If 6 million years elapsed between Rendezvous 1 and modern *Homo sapiens*, the quantity of change ought to be spun out, *pro rata* one might naively think, through the 6 million years. But *Orrorin* and Toumai both lived very close to the date identified from molecular evidence as that of Concestor 1, the split between our line and that of chimpanzees. These fossils even pre-date Concestor 1 according to some datings.

Assuming that the molecular and fossil dates are correct, there seem to be four ways (or some combination from among the four) in which we might respond to *Orrorin* and Toumai.

1. *Orrorin* and/or Toumai walked on all fours. This is not unlikely, but the remaining three possibilities assume, for the sake of argument, that it is wrong. If we accept option 1, the problem just goes away.
2. An extremely rapid burst of evolution occurred immediately after Concestor 1, which itself walked on all fours like a chimpanzee. The more humanoid Toumai and *Orrorin* evolved their bipedality so swiftly after Concestor 1 that the separation in dates cannot easily be resolved.
3. Humanoid features such as bipedality have evolved more than once, maybe many times. *Orrorin* and Toumai could represent earlier occasions when African apes experimented with bipedality, and perhaps other human features too. On this hypothesis, they could indeed pre-date Concestor 1 while being bipedal, and our own lineage would constitute a later foray into bipedality.

4. Chimpanzees and gorillas descend from more human-like, even bipedal ancestors, and have reverted to all fours more recently. On this hypothesis, Toumai, say, could actually be Concestor 1.

The last three hypotheses all have difficulties, and many authorities are driven to doubt either the dating, or the supposed bipedality, of Toumai and *Orrorin*. But if we accept these for the moment and look at the three hypotheses that assume ancient bipedality, there is no strong theoretical reason to favour or disfavour any particular one of them. We shall learn from the Galapagos Finch's Tale and the Lungfish's Tale that evolution can be extremely rapid or can be extremely slow. So Theory 2 is not implausible. The Marsupial Mole's Tale will teach us that evolution can follow the same path, or strikingly parallel paths, on more than one occasion. There's nothing particularly implausible, then, about Theory 3. Theory 4, at first sight, seems the most surprising. We are so used to the idea that we have risen 'up' from the apes that Theory 4 seems to put the cart before the horse, and may even insult human dignity into the bargain (often good for a laugh in my experience). Also there is a so-called law, Dollo's Law, which states that evolution never reverses itself, and it might seem that Theory 4 violates it.

The Blind Cave Fish's Tale, which is about Dollo's Law, will reassure us that this last is not the case. There is nothing in principle wrong with Theory 4. Chimpanzees really could have passed through a more humanoid, bipedal stage before reverting to quadrupedal apehood. As it happens, this very suggestion has been revived by John Gribbin and Jeremy Cherfas, in their two books, *The Monkey Puzzle* and *The First Chimpanzee*. They go so far as to suggest that chimpanzees are descended from gracile australopithecines (like Lucy), and gorillas from robust australopithecines (like 'Dear Boy'). For such an in-your-face radical suggestion, they make a surprisingly good case. It centres on an interpretation of human evolution which has long been widely accepted, although not without controversy: people are juvenile apes who have become sexually mature. Or, putting it another way, we are like chimpanzees who have never grown up.

The Axolotl's Tale explains the theory, which is known as neoteny. To summarise, the axolotl is an overgrown larva, a tadpole with sex organs. In a classic experiment by Vilém Laufberger in Germany, hormone injections persuaded an axolotl to grow into a fully adult salamander of a species that nobody had ever seen. More famously in the English-speaking world, Julian Huxley later repeated the experiment, not knowing it had already been done. In the evolution of the axolotl, the adult stage had been chopped off the end of the life cycle. Under the influence of experimentally injected hormone, the axolotl finally grew up, and an adult salamander was recreated, presumably never before seen. The missing last stage of the life cycle was restored.

The lesson was not lost on Julian's younger brother, the novelist Aldous Huxley. His *After Many a Summer** was one of my favourite novels when I was a teenager. It is about a rich man, Jo Stoyte, who resembles William Randolph Hearst and collects *objets d'art* with the same voracious indifference. His strict religious upbringing has left him with a terror of death, and he employs and equips a brilliant but cynical biologist, Dr Sigismund Obispo, to research how to prolong life in general and Jo Stoyte's life in particular. Jeremy Pordage, a (very) British scholar, has been hired to catalogue some eighteenth-century manuscripts recently acquired as a job lot for Mr Stoyte's library. In an old diary kept by the Fifth Earl of Gonister, Jeremy makes a sensational discovery which he imparts to Dr Obispo. The old Earl was hot on the trail of everlasting life (you have to eat raw fish guts), and there is no evidence that he ever died. Obispo takes the increasingly fretful Stoyte to England in quest of the Fifth Earl's remains . . . and finds him still alive at 200. The catch is that he has finally matured from the juvenile ape which all the rest of us are into a fully adult ape: quadrupedal, hairy, repellent, urinating on the floor while humming a grotesquely distorted vestige of a Mozart aria. The diabolical Dr Obispo, beside himself with gleeful laughter and evidently acquainted with Julian Huxley's work, tells Stoyte he can start on the fish guts tomorrow.

Gribbin and Cherfas are in effect suggesting that modern chimpanzees and gorillas are like the Earl of Gonister. They are humans (or australopithecines, orrorins or sahelanthropes) who have grown up and become quadrupedal apes again, like their, and our, more distant ancestors. I never thought the Gribbin/Cherfas theory was obviously silly. The new findings of very ancient hominids like *Orrorin* and Toumai, whose dates push up against our split with chimpanzees, could almost justify them in a *sotto voce* 'We told you so'.

---

* The American edition rounds off the Tennyson quotation: 'Dies the Swan.'

Even if we accept *Orrorin* and Toumai as bipedal, I would not choose with confidence between Theories 2, 3 and 4. And we mustn't forget Theory 1, that they walked on all fours and the problem goes away, which many people think is the most plausible. But of course these different theories make predictions about Concestor 1, our next stopping point. Theories 1, 2, and 3 agree in assuming a chimpanzee-like Concestor 1, walking on all fours, but occasionally rising on the hind legs. Theory 4 by contrast differs in assuming a more humanoid Concestor 1. In narrating Rendezvous 1, I have been forced to make a decision between the theories. Somewhat reluctantly, I'll go with the majority, and assume a chimpanzee-like concestor. On to meet it.

CHIMPANZEES

Between 5 and 7 million years ago, somewhere in Africa, we human pilgrims enjoy a momentous encounter. It is Rendezvous 1, our first meeting with pilgrims from another species. Two other species to be precise, for the common chimpanzee pilgrims and the pygmy chimpanzee or bonobo pilgrims have already joined forces with each other some 4 million years 'before' their rendezvous with us. The common ancestor we share with them, Concestor 1, is our 250,000-greats-grandparent – an approximate guess this, of course, like the comparable estimates that I shall be making for other concestors.

As we approach Rendezvous 1, then, the chimpanzee pilgrims are approaching the same point from another direction. Unfortunately we don't know anything about that other direction. Although Africa has yielded up some thousands of hominid fossils or fragments of fossils, not a single fossil has ever been found which can definitely be regarded as along the chimpanzee line of descent from Concestor 1. This may be because they are forest animals, and the leaf litter of forest floors is not friendly to fossils. Whatever the reason, it means that the chimpanzee pilgrims are searching blind. Their equivalent contemporaries of the Turkana Boy, of 1470, of Mrs Ples, Lucy, Little Foot, Dear Boy, and the rest of 'our' fossils – have never been found.

Nevertheless, in our fantasy the chimpanzee pilgrims meet us in some Pliocene forest clearing, and their dark brown eyes, like our less predictable ones, are fixed upon Concestor 1: their ancestor as well as ours. In trying to imagine the shared ancestor, an obvious question to ask is, is it more like modern chimpanzees or modern humans, is it intermediate, or completely different from either?

Notwithstanding the pleasing speculation that ended the previous section – which I would by no means rule out – the prudent answer is that Concestor 1 was more like a chimpanzee, if only because chimpanzees are more like the rest of the apes than humans are. Humans are the odd ones out among apes, both living and fossil. Which is only to say that more evolutionary change has occurred along the human line of descent from the common ancestor, than

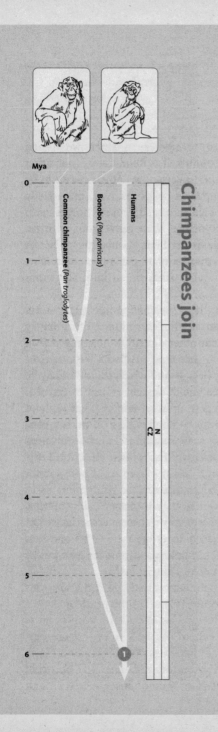

Chimpanzees join

Mya

0

1

2

3

4

5

6

①

Common chimpanzee (*Pan troglodytes*)

Bonobo (*Pan paniscus*)

Humans

CZ

N

Chimpanzees join. White lines depict the evolutionary tree (or 'phylogeny') of chimps and humans, branching apart at Concestor 1 (marked by a numbered circle). The vertical right branch represents the current set of pilgrims: in this case, only humans. The left branch shows chimps splitting into two species about 2 million years ago.

If we were to zoom in on any of the lines, we would find them not solid, but crisscrossing networks of interbreeding, as depicted in the humankind diagram at Rendezvous 0. From now on we'll continue to use this solid line representation.

Images, left to right: common chimpanzee (*Pan troglodytes*); bonobo (*Pan paniscus*).

along the lines leading to the chimpanzees. We must not assume, as many laymen do, that our ancestors *were* chimpanzees. Indeed, the very phrase 'missing link' is suggestive of this misunderstanding. You still hear people saying things like, 'Well, if we are descended from chimpanzees, why are there still chimpanzees around?'

So, when we and the chimpanzee/bonobo pilgrims meet at the rendezvous point, the likelihood is that the shared ancestor that we greet in that Pliocene clearing was hairy like a chimpanzee, and had a chimpanzee-sized brain. Reluctantly to set aside the speculations of the previous chapter, it probably walked on its hands (knuckles) like a chimp, as well as its feet. It probably spent some time up trees, but also lots of time on the ground, maybe squat feeding as Jonathan Kingdon would say. All available evidence suggests that it lived in Africa, and only in Africa. It probably used and made tools, following local traditions as modern chimpanzees still do. It probably was omnivorous, sometimes hunting, but with a preference for fruit.

Bonobos have been seen to kill duikers, but hunting is more frequently documented for common chimpanzees, including highly co-ordinated group pursuits of colobus monkeys. But meat is only a supplement to fruit, which is the main diet of both species. Jane Goodall, who first discovered hunting and intergroup warfare in chimpanzees, was also the first to report their now famous habit of termite fishing, using tools of their own construction. Bonobos have not been seen to do this, but that may be because they have been studied less. Captive bonobos readily use tools. Common chimpanzees in different parts of Africa develop local traditions of tool use. Where Jane Goodall's animals on the east side of the range fish for termites, other groups to the west have developed local traditions of cracking nuts using stone or wood hammers and anvils. Some skill is required. You have to hit hard enough to break the kernel but not so hard as to pulp the nut itself.

Although often spoken of as a new and exciting discovery, by the way, nut cracking was mentioned by Darwin in Chapter 3 of *The Descent of Man* (1871):

> It has often been said that no animal uses any tool; but the chimpanzee in a state of nature cracks a native fruit, somewhat like a walnut, with a stone.

The evidence cited by Darwin (a report by a missionary in Liberia in the 1843 issue of the *Boston Journal of Natural History*) is brief and non-specific. It simply states that 'the *Troglodytes niger,* or Black Orang of Africa' is fond of a species of unidentified nut, which 'they crack with stones precisely in the manner of human beings'.

The especially interesting thing about nut cracking, termite fishing and other such chimpanzee habits is that local groups have local customs, handed down locally. This is true culture. Local cultures extend to social habits and manners. For example, one local group in the Mahale Mountains in Tanzania has a particular style of social grooming known as the grooming hand clasp. The same gesture has been seen in another population in the Kibale forest in Uganda. But it has never been seen in Jane Goodall's intensively studied population at Gombe Stream. Interestingly, this gesture also spontaneously arose and spread among a captive group of chimpanzees.

If both species of modern chimpanzee used tools in the wild as we do, this would encourage us to think that Concestor 1 probably did too. I think it probably did – even though bonobos have not been seen using tools in the wild, they are adept tool-users in captivity. The fact that common chimpanzees use different tools in different areas, following local traditions, suggests to me that lack of such a tradition in a particular area should not be taken as negative evidence. After all, Jane Goodall's Gombe Stream chimpanzees haven't been seen to crack nuts. Presumably they would, if the West African nut-cracking tradition were introduced to them. I suspect that the same might be true of bonobos. Maybe they just haven't been studied enough in the wild. In any case, I think the indications are strong enough that Concestor 1 made and used tools. This idea is strengthened by the fact that tool use also occurs in wild orang utans, local populations again differing in ways that suggest local traditions.[*]

The present-day representatives of the chimpanzee lineage are both forest apes, whereas we are savannah apes, more like baboons except, of course, that baboons are not apes at all but monkeys. Bonobos today are confined to the forests south of the great curve of the River Congo and north of its tributary the Kasai. Common chimpanzees inhabit a wider belt of the continent, north of the Congo, westward to the coast, and extending as far as the Rift Valley in the east.

As we shall see in the Cichlid's Tale, current Darwinian orthodoxy

---

[*] Tool use is, in any case, widespread among mammals and birds, as Jane Goodall herself (among others) has documented.

suggests that usually, in order for an ancestral species to split into two daughter species, there is an initial, accidental geographical separation between them. Without the geographical barrier, sexual mixing of the two gene pools keeps them together. It is plausible that the great Congo river provided the barrier to gene flow which assisted the evolutionary divergence of the two chimpanzee species from each other, two or three million years ago. In the same way, it has been suggested that the Rift Valley, in the throes of its formation at the time, may have provided the barrier to gene flow which, further in the past, allowed our line to separate from that which gave rise to the chimpanzees.

This Rift Valley theory was proposed and supported by the distinguished Dutch primatologist Adriaan Kortlandt. It became better known when it was later espoused by the French palaeontologist Yves Coppens, and it is now widely called by the name Coppens gave it, East Side Story. Incidentally, I don't know what to make of the fact that, in his native France, Yves Coppens is widely cited as the discoverer of Lucy, even as the 'father' of Lucy. In the English-speaking world, this important discovery is universally attributed to Donald Johanson. East Side Story has a hard time dealing with *Sahelanthropus* ('Toumai') from Chad, thousands of miles to the west of the Rift Valley. *Australopithecus bahrelghazali,* a poorly known australopithecine also discovered in Chad, adds to the problem, although it is younger.

Whatever I say on this matter will soon be out of date when new fossils are discovered, so I'll hand over at this point to the bonobo and his tale.

THE BONOBO'S TALE

The bonobo, *Pan paniscus,* looks pretty much like a common chimpanzee, *Pan troglodytes,* and before 1929 they were not recognised as separate species. The bonobo, despite its other name of pygmy chimpanzee, which should be abandoned, is not noticeably smaller than the common chimpanzee. Its body proportions are slightly different, and so are its habits, and that is the cue for its brief tale. The primatologist Frans de Waal put it neatly: 'The chimpanzee resolves sexual issues with power; the bonobo resolves power issues with sex . . .' Bonobos use sex as a currency of social interaction, somewhat as we use money. They use copulation, or copulatory gestures, to appease, to assert dominance, to cement bonds with other troop members of any age or sex, including small infants. Paedophilia is not a hang-up with bonobos; all kinds of philia seem fine to them. De Waal describes how, in a group of captive bonobos that he watched, the males would develop erections as soon as a keeper approached at feeding time. He speculates that this is in preparation for sexually mediated food-sharing. Female bonobos pair off to practise so-called GG (genital–genital) rubbing.

> One female facing another clings with arms and legs to a partner that, standing on both hands and feet, lifts her off the ground. The two females then rub their genital swellings laterally together, emitting grins and squeals that probably reflect orgasmic experiences.

The 'Haight-Ashbury' image of free-loving bonobos has led to a piece of wishful thinking among nice people, who perhaps came of age in the 1960s – or maybe they are of the 'medieval bestiary' school of thought, in which animals exist only to point moral lessons to us. The wishful thinking is that we are more closely related to bonobos than to common chimpanzees. The Margaret Mead in us feels closer to this gentle role-model than to the patriarchal, monkey-butchering chimpanzee. Unfortunately, however, like it or not, we are exactly equally close to both species. This is simply because *P. troglodytes* and *P. paniscus* share a common ancestor which lived more recently than the ancestor they share with us. By the same token, molecular evidence suggests that chimpanzees and bonobos are more closely related to humans than they are to gorillas. From this it follows that humans are exactly as close to gorillas as chimpanzees and bonobos are. And we are exactly as close cousins of orang utans as chimpanzees, bonobos and gorillas are.

It does not follow from this that we *resemble* chimpanzees and bonobos equally. If chimpanzees have changed more than bonobos since the shared ancestor, Concestor 1, we might be more like bonobos than chimpanzees, or vice versa – and we shall probably find different things in common with both our *Pan* cousins, perhaps in roughly equal measure. They are equally closely related to us because they are linked to us via the same shared ancestor. This is the moral of the Bonobo's Tale, a simple moral and a very general one, which we shall meet again and again at other junctures of our pilgrimage.

GORILLAS

The molecular clock tells us that Rendezvous 2, where the gorillas join us, again in Africa, is only a million years further into our pilgrimage than Rendezvous 1. Seven million years ago, North and South America were not joined, the Andes had not undergone their major uplift and the Himalayas only just so. Nevertheless the continents would have looked pretty much as now and the African climate, while less seasonal and slightly wetter, would have been similar. Africa was more thoroughly forested then than now – even the Sahara would have been wooded savannah at the time.

Unfortunately there are no fossils to bridge the gap between Concestors 2 and 1, nothing to guide us in deciding whether Concestor 2, which is perhaps our 300,000-greats-grandparent, was more like a gorilla or more like a chimpanzee or, indeed, more like a human. My guess would be chimpanzee, but this is only because the huge gorilla seems more extreme, and less like the generality of apes. Don't let's exaggerate the unusualness of gorillas, however. They are not the largest apes that have ever lived. The Asian ape *Gigantopithecus*, a sort of giant orang utan, would have stood head and massive shoulders over the largest gorilla. It lived in China, and went extinct only recently, about half a million years ago, overlapping with *Homo erectus* and archaic *Homo sapiens*. This is so recent that some enterprising fantasists have gone so far as to suggest that the Yeti or Abominable Snowman of the Himalayas . . . but I digress. *Gigantopithecus* presumably walked like a gorilla, probably on the knuckles of its hands and the soles of its feet as gorillas and chimpanzees do, and as orang utans, committed as they are to life up trees, do not.

It is a reasonable guess that Concestor 2 was also a knuckle-walker but that, like chimpanzees, it spent time in trees as well, especially at night. Natural selection under a tropical sun favours dark pigmentation as protection against ultraviolet rays, so if we had to guess at Concestor 2's colour we would presumably say black or dark brown. All apes except humans are hairy, so it would be surprising if Concestors 1 and 2 were not. Since chimpanzees, bonobos and

Mya

0

Gorillas join

Gorillas (*Gorilla*)

Already joined

1

2

3

N
CZ

4

5

6 •

7 ②

Gorillas join. Phylogeny showing the gorillas diverging from the other African apes around 7 million years ago, as suggested by genetics. The right branch now represents the chimpanzees and humans (Concestor 1 is marked on the branch with a dot at 6 million years ago). The left branch represents the single genus of gorillas, now thought to comprise two species.

Image: western gorilla (*Gorilla gorilla*).

gorillas are inhabitants of deep forest, it is plausible to locate Rendezvous 2 in a forest, in Africa, but there is no strong reason to guess any particular part of Africa.

Gorillas are not just giant chimpanzees, they are different in other respects which we need to think about in trying to reconstruct Concestor 2. Gorillas are entirely vegetarian. The males have harems of females. Chimpanzees are more promiscuous, and the differences in breeding systems have interesting consequences on the size of their testes as we shall learn from the Seal's Tale. I suspect that breeding systems are evolutionarily labile, meaning easily changed. I don't see any obvious way to guess where Concestor 2 stood in this respect. Indeed, the fact that different human cultures today show a large range of breeding systems, from faithful monogamy to potentially very large harems, reinforces my reluctance to speculate about such matters for Concestor 2, and persuades me to bring my speculations as to its nature to a swift end.

Apes, perhaps especially gorillas, have long been potent generators – and victims – of human myths. The Gorilla's Tale considers our changing attitudes to our closest cousins.

THE GORILLA'S TALE

The rise of Darwinism in the nineteenth century polarised attitudes towards the apes. Opponents who might have stomached evolution itself balked with visceral horror at cousinship with what they perceived as low and revolting brutes, and desperately tried to inflate our differences from them. This was nowhere more true than with gorillas. Apes were 'animals'; we were set apart. Worse, where other animals such as cats or deer could be seen as beautiful in their own way, gorillas and other apes, precisely because of their similarity to ourselves, seemed like caricatures, distortions, grotesques.

Darwin never missed an opportunity to put the other side, sometimes in little asides such as his charming observation in *The Descent of Man* that monkeys 'smoke tobacco with pleasure'. T. H. Huxley, Darwin's formidable ally, had a robust exchange with Sir Richard Owen, the leading anatomist of the day, who claimed (wrongly as Huxley showed) that the 'hippocampus minor' was uniquely diagnostic of the human brain. Nowadays, scientists not only think we resemble apes. We include ourselves within the apes, specifically the African apes. We emphasise, by contrast, the distinctness of apes, including humans, from monkeys. To call a gorilla or a chimpanzee a monkey is a solecism.

It has not always been so. In former times, apes were frequently lumped with monkeys, and some of the early descriptions confused apes with baboons, or with Barbary macaques, which indeed are still known as Barbary apes. More surprisingly, long before people thought in terms of evolution at all, and before apes were clearly distinguished from each other or from monkeys, great apes were often confused with humans. Agreeable as it would be to approve this apparent prescience of evolution, it unfortunately may owe more to racism. Early white explorers in Africa saw chimpanzees and gorillas as close kin only to black humans, not to themselves. Interestingly, tribes in both South East Asia and Africa have traditional legends suggesting a reversal of evolution as conventionally seen: their local great apes are regarded as humans who fell from grace. Orang utan means 'man of the woods' in Malay.

A picture of an 'Ourang Outang' by the Dutch doctor Bontius in 1658 is, in T. H. Huxley's words, 'nothing but a very hairy woman of rather comely aspect and with proportions and feet wholly human'. Hairy she is except, oddly, in one of the few places where a real woman is: her pubic region is conspicuously naked. Also very human are the pictures made, a century later, by Linnaeus's pupil Hoppius (1763). One of his creatures has a tail, but is otherwise wholly human, bipedal, and carries a walking stick. Pliny the Elder says that 'the tailed species have even been known to play at draughts' (American 'checkers').

One might have thought such a mythology would have prepared our civilisation for the idea of evolution when it arrived in the nineteenth century, and might even have accelerated its discovery. Apparently not. Instead, the picture is one of confusion between apes, monkeys and humans. This makes it hard to date the scientific discovery of each species of great ape, and it is often unclear which one is being discovered. The exception is the gorilla, which became known to science the most recently.

In 1847 an American missionary, Dr Thomas Savage, saw in the house of another missionary on the Gaboon river 'a skull represented by the natives to be a monkey-like animal, remarkable for its size, ferocity and habits'. The unjust reputation for ferocity, later to be hyperbolised in the story of King Kong, comes through loud and clear in an article about the gorilla in the *Illustrated London News* published in the same year as the *Origin of Species*. This piece is replete with falsehoods of a quantity and magnitude that try even the high standards set by travellers' tales of the time:

> . . . a close inspection is almost an impossibility, especially as the moment it sees a man it attacks him. The strength of the adult male being prodigious, and the teeth heavy and powerful, it is said to watch, concealed in the thick branches of the forest trees, the approach of any of the human species, and, as they pass under the tree, let down its terrible hind feet, furnished with an enormous thumb, grasp its victim round the throat, lift him from the earth, and, finally, drop him on the ground dead. Sheer malignity prompts the animal to this course, for it does not eat the dead man's flesh, but finds a fiendish gratification in the mere act of killing.

Savage believed the skull in the missionary's possession belonged 'to a new species of Orang'. He later decided that his new species was none other than the 'Pongo' of earlier travellers' tales in Africa. In naming it formally, Savage, with his anatomist colleague Professor Wyman, avoided *Pongo* and revived *Gorilla,* the name used by an ancient Carthaginian admiral for a race of wild hairy people which he claimed to have found on an island off the African coast. Gorilla has survived as both the Latin and common name for Savage's animal, while *Pongo* is now the Latin name of the orang utan of Asia.

Judging from its location, Savage's species must have been the western gorilla, *Gorilla gorilla.* Savage and Wyman put it in the same genus as the chimpanzees, and called it *Troglodytes gorilla.* By the rules of zoological nomenclature, *Troglodytes* had to be relinquished by both chimpanzee and gorilla because it had already been used for – of all things – the tiny wren. It survived as the specific name of the common chimpanzee, *Pan troglodytes,* while the former specific name of Savage's gorilla was promoted to become its generic name, *Gorilla.* The 'mountain gorilla' was 'discovered' – he shot it! – by the German Robert von Beringe as late as 1902. As we shall see, it is now regarded as a subspecies of the eastern gorilla, and the whole eastern species now – unfairly, one might think – bears his name: *Gorilla beringei.*

Savage did not believe his gorillas really were the race of islanders reported by the Carthaginian sailor. But the 'pygmies', originally mentioned by Homer and Herodotus as a legendary race of very small humans, were later assumed by seventeenth- and eighteenth-century explorers to be none other than the chimpanzees then being discovered in Africa. Tyson (1699) shows a drawing of a 'Pygmie' which, as Huxley says, is plainly a young chimpanzee although it, too, is depicted walking upright and carrying a walking stick. Now, of course, we use the word pygmy for small humans again.

This leads us back to the racism which, until relatively late in the twentieth century, was endemic in our culture. Early explorers often assigned the native peoples of the forests a closer affinity with chimpanzees, gorillas or orangs than with the explorers themselves. In the nineteenth century, after Darwin, evolutionists often regarded African peoples as intermediate between apes and Europeans, on the upward path to white supremacy. This is not only factually wrong. It violates a fundamental principle of evolution. Two cousins are always exactly equally related to any outgroup, because they are connected to that outgroup via a shared ancestor. For the reasons given in the Bonobo's Tale, all humans are *exactly* equally close cousins to all gorillas. Racism and speciesism, and our perennial confusion over how inclusively we wish to cast our moral and ethical net, are brought into sharp and sometimes uncomfortable focus in the history of our attitudes to our fellow humans, and our attitudes to apes – our *fellow* apes.[*]

[*] The Great Ape Project, dreamed up by the distinguished moral philosopher Peter Singer, goes to the heart of the matter by proposing that great apes should be granted, as far as is practically possible, the same moral status as humans. My own contribution to the book *The Great Ape Project* is one of the essays reprinted in *A Devil's Chaplain*.

ORANG UTANS

Molecular evidence puts Rendezvous 3 – where our ancestral pilgrimage is joined by the orang utans – at 14 million years ago, right in the middle of the Miocene Epoch. Although the world was starting to enter its current cool phase, the climate was warmer and the sea levels higher than at present. Coupled with minor differences in the positions of the continents, this led to the land between Asia and Africa, as well as much of south-east Europe, being intermittently submerged by sea. This bears, as we shall see, on our calculation of where Concestor 3, perhaps our two-thirds-of-a-million-greats-grandparent, might have lived. Did it live in Africa like 1 and 2, or Asia? As the common ancestor of ourselves and an Asian ape, we should be prepared to find it in either continent, and partisans of both are not hard to find. In favour of Asia is its richness of plausible fossils from around the right time, the mid-to-late Miocene. Africa, on the other hand, seems to be where the apes originated, before the beginning of the Miocene. Africa witnessed a great flowering of ape life in the early Miocene, in the form of proconsulids (several species of the early ape genus *Proconsul*) and others such as *Afropithecus* and *Kenyapithecus*. Our closest living relatives today, and all our post-Miocene fossils, are African.

But our special relationship to chimpanzees and gorillas has been known only for a few decades. Before that, most anthropologists thought we were the sister group to all the apes, and therefore equally close to African and Asian apes. The consensus favoured Asia as the home of our late Miocene ancestors, and some authorities even picked out a particular fossil 'ancestor', *Ramapithecus*. This animal is now thought to be the same as one previously called *Sivapithecus* which therefore, by the laws of zoological nomenclature, takes precedence. *Ramapithecus* should no longer be used – a pity because the name had become familiar. Whatever one feels about *Sivapithecus/Ramapithecus* as a human ancestor, many authorities agree that it is close to the line that gave rise to the orang utan and might even be the orang utan's direct ancestor. *Gigantopithecus* could be regarded as a kind of giant, ground-dwelling version of *Sivapithecus*. Several

Mya

0 —

Orang utans (*Pongo*)

Already joined

Orang utans join

5 —

CZ

N

10 —

14 —

③

Orang utans join. The two species of Asian orang utan are generally accepted to have diverged from the rest of the great apes approximately 14 million years ago. As with all our rendezvous phylogenies, the right branch represents the species which have already joined the pilgrimage, with the positions of previous concestors marked with dots.

Image: Borneo orang utan (*Pongo pygmaeus*).

other Asian fossils occur from about the right time. *Ouranopithecus* and *Dryopithecus* seem almost to be jostling for the title of most plausible human ancestor of the Miocene. If only, it is tempting to remark, they were in the right continent. As we shall see, this 'if only' just might turn out to be true.

If only the late Miocene apes were in Africa instead of Asia, we'd have a smooth series of plausible fossils linking the modern African apes all the way back to the early Miocene and the rich proconsulid ape fauna of Africa. When molecular evidence established beyond any doubt our affinities with the African chimpanzees and gorillas, rather than with the Asian orangs, seekers of human ancestors reluctantly turned their backs on Asia. They assumed, in spite of the plausibility of the Asian apes themselves, that our ancestral line must lie in Africa right through the Miocene and concluded that, for some reason, our African ancestors had not fossilised after the early burgeoning of proconsulid apes in the early Miocene.

That's where things stood until 1998, when an ingenious piece of lateral thinking appeared in a paper called 'Primate evolution – in and out of Africa' by Caro-Beth Stewart and Todd R. Disotell. This tale, of back and forth traffic between Africa and Asia, will be told by the orang utan. Its conclusion will be that Concestor 3 probably lived in Asia after all.

But never mind, for the moment, where it lived. What did Concestor 3 look like? It is the common ancestor of the orang utans and all today's African apes, so it might resemble either or both of them (see plate 5). Which fossils might give us helpful clues? Well, looking at the family tree, the fossils known as *Lufengpithecus, Oreopithecus, Sivapithecus, Dryopithecus* and *Ouranopithecus* all lived around the right time or slightly later. Our best-guess reconstruction of Concestor 3 might combine elements of all five of these Asian fossil genera – but it would help if we could accept Asia as the location of the concestor. Let's listen to the Orang Utan's Tale and see what we think.

THE ORANG UTAN'S TALE

Perhaps we have been too ready to assume that our links with Africa go back a very long way. What if, instead, our ancestral lineage hopped sideways out of Africa around 20 million years ago, flourished in Asia until around 10 million years ago, and then hopped back to Africa?

On this view, all the surviving apes, including the ones that ended up in Africa, are descended from a lineage that migrated out of Africa into Asia. Gibbons and orang utans are descendants of these migrants who stayed in Asia. Later descendants of the migrants returned to Africa, where the earlier Miocene apes had gone extinct. Back in their old ancestral home of Africa, these migrants then gave rise to gorillas, chimpanzees and bonobos, and us.

The known facts about the drifting of the continents and the fluctuations of sea levels are compatible. There were land bridges available across Arabia at the right times. The positive evidence in favour of the theory depends upon 'parsimony': an economy of assumptions. A good theory is one that needs to postulate little, in order to explain lots. (By this criterion, as I have often remarked elsewhere, Darwin's theory of natural selection may be the best theory of all time.) Here we are talking about minimising our assumptions about migration events. The theory that our ancestors stayed in Africa all along (no migrations) seemed, on the face of it, more economical with its assumptions than the theory that our ancestors moved from Africa into Asia (a first migration) and later moved back to Africa (a second migration).

But that parsimony calculation was too narrow. It concentrated on our own lineage and neglected all the other apes, especially the many fossil species. Stewart and Disotell did a recount of the migration events, but they counted those that would be needed to explain the distribution of all the apes including fossils. In order to do this, you first have to construct a family tree on which you mark all the species about which you have sufficient information. The next step is to indicate, for each species on the family tree, whether it lived in Africa or Asia. In the diagram on the opposite page, which is taken from Stewart and Disotell's paper, Asian fossils are highlighted in black, African ones are in white. Not all the known fossils are there, but Stewart and Disotell did include all whose position on the family tree could be clearly worked out. They also drew in the Old World monkeys, who diverged from the apes around 25 million years ago (the most obvious difference between monkeys and apes, as we shall see, is that the monkeys retained their tails). Migration events are indicated by arrows.

Taking into account the fossils, the 'hop to Asia and back again' theory is now more parsimonious than the 'our ancestors were in Africa all along' theory. Leaving out the monkeys which, on both

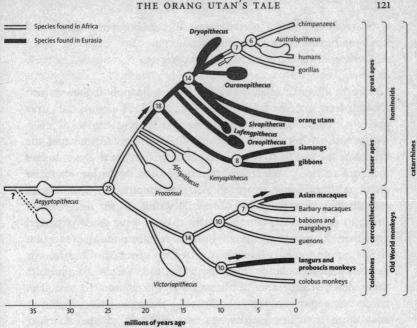

Species found in Africa

Species found in Eurasia

In and out of Africa. Stewart and Disotell's family tree of African and Asian apes. Swollen areas represent dates known from fossils, while the lines linking these to the tree are inferred from parsimony analysis. Arrows represent inferred migration events. Adapted from Stewart and Disotell [273].

theories, account for two migration events from Africa to Asia, it need postulate only two ape migrations, as follows:

1. A population of apes migrated from Africa to Asia around 20 million years ago and became all the Asian apes including the living gibbons and orang utans.
2. A population of apes migrated back from Asia to Africa and became today's African apes including us.

Conversely, the 'our ancestors were in Africa all along' theory demands six migration events to account for ape distributions, all from Africa to Asia, by ancestors of the following:

1. Gibbons, around 18 million years ago
2. *Oreopithecus*, around 16 million years ago
3. *Lufengpithecus*, around 15 million years ago

4. *Sivapithecus* and orang utans, around 14 million years ago
5. *Dryopithecus*, around 13 million years ago
6. *Ouranopithecus*, around 12 million years ago

Of course all these migration counts are valid only if Stewart and Disotell have got the family tree right, based on anatomical comparisons. They think, for example, that among the fossil apes, *Ouranopithecus* is the closest cousin to the modern African apes (its branch is the last to come off the family tree in the diagram before the African apes). The next closest cousins, according to their anatomical assessments, are all Asian (*Dryopithecus*, *Sivapithecus*, etc.). If they have got the anatomy all wrong: if, for instance, the African fossil *Kenyapithecus* is actually closest to the modern African apes, then the migration counts would have to be done all over again.

The family tree was itself constructed on grounds of parsimony. But it is a different kind of parsimony. Instead of trying to minimise the number of geographical migration events we need to postulate, we forget about geography and try to minimise the number of anatomical coincidences (convergent evolution) we need to postulate. Having got our family tree without regard to geography, we then superimpose the geographical information (the black and white coding on the diagram) to count migration events. And we conclude that it is most likely that the 'recent' African apes, that is gorillas, chimpanzees and humans, arrived from Asia.

Now here's an interesting little fact. A leading textbook of human evolution, by Richard G. Klein of Stanford University, gives a fine description of what is known of the anatomy of the main fossils. At one point Klein compares the Asian *Ouranopithecus* and the African *Kenyapithecus* and asks which most resembles our own close cousin (or ancestor) *Australopithecus*. Klein concludes that *Australopithecus* resembles *Ouranopithecus* more than it resembles *Kenyapithecus*. He goes on to say that, if only *Ouranopithecus* had lived in Africa, it might even make a plausible human ancestor. 'On combined geographic-morphologic grounds', however, *Kenyapithecus* is a better candidate. You see what is going on here? Klein is making the tacit assumption that African apes are unlikely to be descended from an Asian ancestor, even if the anatomical evidence suggests that they were. Geographical parsimony is being subconsciously allowed to pull rank over anatomical parsimony. Anatomical parsimony suggests that *Ouranopithecus* is a closer cousin to us than *Kenya-*

pithecus is. But, without being explicitly so called, geographical parsimony is assumed to trump anatomical parsimony. Stewart and Disotell argue that, when you take into account the geography of *all* the fossils, anatomical and geographical parsimony *agree* with each other. Geography turns out to agree with Klein's initial anatomical judgement that *Ouranopithecus* is closer to *Australopithecus* than *Kenyapithecus* is.

This argument may not be settled yet. It is a complicated business juggling anatomical and geographical parsimony. Stewart and Disotell's paper has unleashed a flourishing correspondence in the scientific journals, both for and against. As the available evidence stands at present, I think we should on balance prefer the 'hop to Asia and back' theory of ape evolution. Two migration events is more parsimonious than six. And there really do seem to be some telling resemblances between the late Miocene apes in Asia and our own line of African apes such as *Australopithecus* and chimpanzees. It is only a preference 'on balance', but it leads me to locate Rendezvous 3 (and Rendezvous 4) in Asia rather than Africa.

The moral of the Orang Utan's Tale is twofold. Parsimony is always in the forefront of a scientist's mind when choosing between theories, but it isn't always obvious how to judge it. And possessing a good family tree is often an essential first prerequisite to powerful further reasoning in evolutionary theory. But building a good family tree is a demanding exercise in itself. The ins and outs of it will be the concern of the gibbons, in the tale that they will tell us in melodious chorus after they join our pilgrimage at Rendezvous 4.

Rendezvous 4

GIBBONS

Rendezvous 4, where we are joined by the gibbons, occurs around 18 million years ago, probably in Asia, in the warmer and more wooded world of the early Miocene. Depending on which authority you consult, there are up to twelve modern species of gibbons. All live in South East Asia, including Indonesia and Borneo. Some authorities place them all in the genus *Hylobates*. The siamang used to be separated off, and people spoke of 'gibbons and siamangs'. With the realisation that they divide into four groups, not two, this distinction has become obsolete, and I shall call them all gibbons.*

Gibbons are small apes, and perhaps the finest arboreal acrobats that have ever lived. In the Miocene there were lots of small apes. Getting smaller and getting larger are easy changes to achieve in evolution. Just as *Gigantopithecus* and *Gorilla* got large independently of each other, plenty of apes, in the Miocene golden age of apes, got small. The pliopithecids, for instance, were small apes which flourished in Europe in the early Miocene and probably lived in a similar way to gibbons, without being ancestral to them. I suppose, for example, that they 'brachiated'.

Brachia is the Latin for 'arm'. Brachiation means using your arms rather than your legs to get about, and gibbons are spectacularly good at it. Their big grasping hands and powerful wrists are like upside-down seven-league boots, spring-loaded to slingshot the gibbon from branch to branch and from tree to tree. A gibbon's long arms, perfectly in tune with the physics of pendulums, are capable of hurling it across a sheer ten-metre gap in the canopy.

---

* Siamangs were separated off because they are larger, and they have a throat sac for amplifying their calls.

Opposite: **Gibbons join.** The 12 species of gibbon are now generally thought to fall into four groups. The order of branching between these four is controversial, as discussed in detail in the Gibbon's Tale.

Images, left to right: hoolock gibbon (*Bunopithecus hoolock*); agile gibbon (*Hylobates agilis*); siamang (*Symphalangus syndactylus*); golden-cheeked gibbon (*Nomascus gabriellae*).

Gibbons join

Mya

0

Hoolock (*Bunopithecus*)

Other gibbons (*Hylobates*)

Siamang (*Symphalangus*)

Crested gibbons (*Nomascus*)

Already joined

5

10

15

18

4

N
CZ

My imagination finds high-speed brachiation more exciting even than flying, and I like to dream of my ancestors enjoying what must surely have been one of the great experiences life could offer. Unfortunately, current thinking doubts that our ancestry ever went through a fully gibbon-like stage, but it is reasonable to conjecture that Concestor 4, approximately our 1-million-greats-grandparent, was a small tree-dwelling ape with at least some proficiency in brachiation.

Among the apes, gibbons are also second only to humans in the difficult art of walking upright. Using its hands only to steady itself, a gibbon will use bipedal walking to travel along the length of a branch, whereas it uses brachiation to travel across from branch to branch. If Concestor 4 practised the same art and passed it on to its gibbon descendants, could some vestige of the skill have persisted in the brain of its human descendants too, waiting to resurface again in Africa? That is no more than a pleasing speculation, but it is true that apes in general have a tendency to walk bipedally from time to time. We can also only speculate on whether Concestor 4 shared the vocal virtuosity of its gibbon descendants, and whether this might have presaged the unique versatility of the human voice, in speech and in music. Then again, gibbons are faithfully monogamous, unlike the great apes which are our closer relatives. Unlike, indeed, the majority of human cultures, in which custom and in several cases religion encourages (or at least allows) polygyny. We do not know whether Concestor 4 resembled its gibbon descendants, or its great ape descendants in this respect.*

Let's summarise what we can guess about Concestor 4, making the usual weak assumption that it had a good number of the features shared by all its descendants, which means all the apes including us. It was probably more dedicated to life in the trees than Concestor 3, and smaller. If, as I suspect, it hung and swung from its arms, its arms were probably not so extremely specialised for brachiation as those of modern gibbons, and not so long. It probably had a gibbon-like face, with a short snout. It didn't have a tail. Or, to be more

---

* Perhaps the good old-fashioned family values of the gibbons, and the pious hope that our evolutionary ancestors once shared them, should be drawn to the attention of the right-wing 'moral majority', whose ignorant and single-minded opposition to the teaching of evolution endangers educational standards in several backward North American States. Of course, to draw any moral would be to commit the 'naturalistic fallacy' but fallacies are what these people do best.

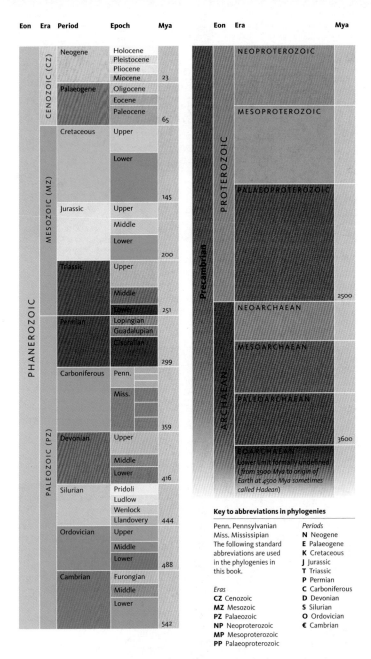

Eon	Era	Period	Epoch	Mya
PHANEROZOIC	CENOZOIC (CZ)	Neogene	Holocene Pleistocene Pliocene Miocene	23
		Palaeogene	Oligocene Eocene Paleocene	65
	MESOZOIC (MZ)	Cretaceous	Upper Lower	145
		Jurassic	Upper Middle Lower	200
		Triassic	Upper Middle Lower	251
	PALEOZOIC (PZ)	Permian	Lopingian Guadalupian Cisuralian	299
		Carboniferous	Penn. Miss.	359
		Devonian	Upper Middle Lower	416
		Silurian	Pridoli Ludlow Wenlock Llandovery	444
		Ordovician	Upper Middle Lower	488
		Cambrian	Furongian Middle Lower	542

Eon	Era	Mya
PROTEROZOIC (Precambrian)	NEOPROTEROZOIC	
	MESOPROTEROZOIC	
	PALAEOPROTEROZOIC	2500
ARCHAEAN	NEOARCHAEAN	
	MESOARCHAEAN	
	PALEOARCHAEAN	3600
	EOARCHAEAN Lower limit formally undefined (from 3900 Mya to origin of Earth at 4500 Mya sometimes called Hadean)	

Key to abbreviations in phylogenies

Penn. Pennsylvanian
Miss. Mississippian
The following standard abbreviations are used in the phylogenies in this book.

Eras
CZ Cenozoic
MZ Mesozoic
PZ Palaeozoic
NP Neoproterozoic
MP Mesoproterozoic
PP Palaeoproterozoic

Periods
N Neogene
E Palaeogene
K Cretaceous
J Jurassic
T Triassic
P Permian
C Carboniferous
D Devonian
S Silurian
O Ordovician
€ Cambrian

1. Simplified version of the timescale published by the International Commission on Stratigraphy (*www.stratigraphy.org*). The timescale is divided into eons, eras, periods, and epochs. Time is measured in 'millions of years' (Mya) (see page 15).

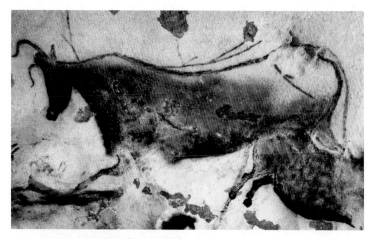

2. Something very special began to happen ...
This painting of a bull is from the Lascaux Caves in the Dordogne, France. Discovered in 1940, the paintings are over 16,000 years old. They show a deep understanding of animal forms and movement, and a fine artistic sense. The purpose of the paintings is unknown (see page 36).

3. Did they hold hands?
The 3.6-million-year-old hominid footprints at Laetoli, Tanzania, were discovered by Mary Leakey in 1978. They were fossilised in volcanic ash. The trail extends for some 70 metres and was probably made by *Australopithecus afarensis* (see page 76).

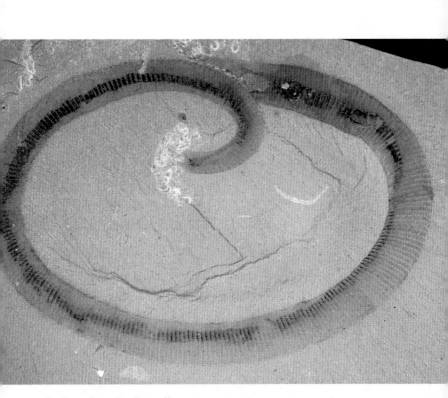

4. Lucky to have fossils at all
A fossil worm, *Palaeoscolex sinensis*, from the Chengjiang fossil beds,
showing fine details of soft body parts. The Chengjiang fossils date back to
the Lower Cambrian, about 525 million years ago (see page 77).

6. Would a Martian feel at home in Madagascar?
Avenue of baobabs, Morondava, Madagascar. This species of baobab, *Adansonia grandidieri*, is one of six unique to Madagascar (see page 173).

Opposite
5. Concestor 3
Imaginative reconstruction of Concestor 3, a large quadrupedal ape which probably spent much of its time up in the trees. Like all great apes, it would have displayed considerable intelligence. Artistic reconstruction by Malcolm Godwin (see page 119).

7. Swimming in its extended phenotype
European beaver (*Castor fiber*) (see page 198).

8. A whale of a surprise
Hippopotamus amphibius in its element. Two species of hippo survive today in Africa (the other is the pygmy hippo, *Hexaprotodon liberiensis*), but fossil remains indicate that perhaps three species of hippo lived in Madagascar right up into the Holocene (see page 203).

9. No new thing under the sun
Detail from Ernst Haeckel's evolutionary tree of mammals, published in 1866 [119], showing hippos closely related to whales (see page 209).

10. Size is at a premium
Male and female southern elephant seals (*Mirounga leonina*) (see page 214).

11. One of the small five
Cape elephant shrew (*Elephantulus edwardii*) (see page 224).

12. Leading with a spade
Artist's impression of a 'shovel tusker' or *Amebelodon* (see page 228).

precise, its tail vertebrae were, as in all the apes, joined together in a short internal tail, the coccyx (pronounced koxix).

I don't know why we apes lost our tail. It is a subject that biologists discuss surprisingly little. A recent exception is Jonathan Kingdon in *Lowly Origin*, but even he reaches no satisfactory closure. Zoologists faced with this kind of conundrum often think comparatively. Look around the mammals, note where taillessness (or a very short tail) has independently cropped up, and try to make sense of it. I don't think anyone has done this systematically, and it would be a nice thing to undertake. Apart from apes, tail loss is found in moles, hedgehogs, the tailless tenrec *Tenrec ecaudatus*, guinea pigs, hamsters, bears, bats, koalas, sloths, agoutis and several others. Perhaps most interesting for our purposes, there are tailless monkeys, or monkeys with a tail so short it might as well not be there, as in a Manx cat. Manx cats have a single gene that makes them tailless. It is lethal when homozygous (present twice) so is unlikely to spread in evolution. But it has crossed my mind to wonder whether the first apes were 'Manx monkeys'. If so, the mutation would presumably be in a Hox gene (see the Fruit Fly's Tale). My bias is against such 'hopeful monster' theories of evolution, but could this be an exception? It would be interesting to examine the skeleton of tailless mutants of normally tailed 'Manx' mammals, to see whether they 'do' taillessness in the same kind of way as apes.

The Barbary macaque *Macaca sylvanus* is a tailless monkey and, perhaps in consequence, is often miscalled the Barbary ape. The 'Celebes ape' *Macaca nigra* is another tailless monkey. Jonathan Kingdon tells me it looks and walks just like a miniature chimpanzee. Madagascar has some tailless lemurs, such as the indri, and several extinct species including 'koala lemurs' (*Megaladapis*) and 'sloth lemurs', some of which were gorilla-sized.

Any organ which is not used will, other things being equal, shrink for reasons of economy if nothing else. Tails are used for a surprisingly wide variety of purposes among mammals. Sheep keep a fat reserve in the tail. Beavers use it as a paddle. The spider monkey tail has a horny gripping pad and is used as a 'fifth limb' in the treetops of South America. The massive tail of a kangaroo is spring loaded to assist bounding. Hoofed animals use the tail as a fly whisk. Wolves and many other mammals use it for signalling, but this is likely to be secondary 'opportunism' on natural selection's part.

But here we must be especially concerned with animals who live up trees. Squirrel tails catch the air, so a 'leap' is almost like flying. Tree-dwellers often have long tails as counterweights, or as rudders for leaping. Lorises and pottos, whom we shall meet at Rendezvous 8, creep about the trees, slowly stalking their prey, and they have extremely short tails. Their relatives the bushbabies, on the other hand, are energetic leapers, and they have long feathery tails. Tree sloths are tailless, like the marsupial koalas who might be regarded as their Australian equivalents, and both move slowly in the trees like lorises.

In Borneo and Sumatra, the long-tailed macaque lives up trees, while the closely related pig-tailed macaque lives on the ground and has a short tail. Monkeys that are active in trees usually have long tails. They run along the branches on all fours, using the tail for balance. They leap from branch to branch with the body in a horizontal position and the tail held out as a balancing rudder behind. Why, then, do gibbons, who are as active in trees as any monkey, have no tail? Maybe the answer lies in the very different way in which they move. All apes, as we have seen, are occasionally bipedal, and gibbons, when not brachiating, run along branches on their hind legs, using their long arms to steady themselves. It is easy to imagine a tail being a nuisance for a bipedal walker. My colleague Desmond Morris tells me that spider monkeys sometimes walk bipedally, and the long tail is obviously a major encumbrance. And when a gibbon projects itself to a distant branch it does so from a vertically hanging position, unlike the monkey's horizontal leaping posture. Far from being a steadying rudder streaming out behind, a tail would be a positive drag for a vertical brachiator like a gibbon or, presumably, Concestor 4.

That is the best I can do. I think zoologists need to give more attention to the puzzle of why we apes lost our tail. The *a posteriori* counterfactual engenders pleasing speculations. How would the tail have sat with our habit of wearing clothes, especially trousers? It gives a different urgency to the classic tailor's question, 'Does Sir hang to the left or to the right?'

THE GIBBON'S TALE
Written with Yan Wong

Rendezvous 4 is the first time we greet a pilgrim band of more than a couple of already united species. Any more than that, and there can be problems with deducing relationships. These problems will

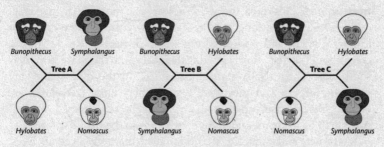

become worse as our pilgrimage advances. How to solve them is the topic of the Gibbon's Tale.*

We have seen that there are 12 species of gibbons, falling into four major groups. They are *Bunopithecus* (a group consisting of a single species, commonly known as the hoolock), *Hylobates* (six species, of which the best-known is the white-handed gibbon *Hylobates lar*), *Symphalangus* (the siamang), and *Nomascus* (four species of 'crested' gibbons). This tale explains how to build an evolutionary relationship, or phylogeny, relating the four groups.

Family trees can be 'rooted' or 'unrooted'. When we draw a rooted tree, we know where the ancestor is. Most of the tree diagrams in this book are rooted. Unrooted trees, by contrast, have no sense of direction. They are often called star diagrams, and there is no arrow of time. They don't start at one side of a page and end on the other. Above are three examples, which exhaust the possibilities for relating four entities.

At every fork in a tree, it makes no difference which is the left and which the right branch. And so far (though that will change later in the tale) no information is conveyed by the lengths of the branches. A tree diagram whose branch lengths are meaningless is known as a cladogram (an unrooted cladogram in this case). The order of branching is the only information conveyed by a cladogram: the rest is cosmetic. Try, for example, rotating either of the side forks about the horizontal line in the middle. It will make no difference to the pattern of relationships.

---

* The subject matter of this tale inevitably makes it tougher than other parts of the book. Readers should either don thinking caps for the next thirteen pages, or skip now to page 143 and return to the tale when they want their neurons exercised. Incidentally, I have often wondered what a 'thinking cap' actually is. I wish I had one. My benefactor Charles Simonyi, one of the world's greatest computer programmers, is said to wear a special 'debugging suit' which may help to account for his formidable success.

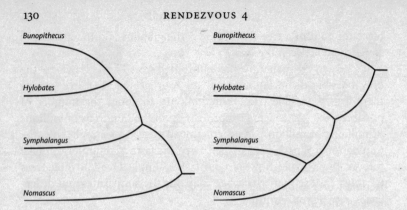

These three unrooted cladograms represent the only possible ways of connecting four species, as long as we restrict ourselves to connections via branches that only ever split in two (dichotomies). As with rooted trees, it is conventional to discount three-way splits (trichotomies) or more (polytomies) as temporary admissions of ignorance – 'unresolved'.

Any unrooted cladogram turns into a rooted one the moment we specify the oldest point (the 'root') of the tree. Certain researchers – those we have relied upon for the tree at the start of this tale – have suggested the rooted cladogram of gibbons shown above, on the left. However, other researchers have suggested the rooted cladogram on the right.

In the first tree the crested gibbons, *Nomascus,* are distant relatives of all the other gibbons. In the second, it is the hoolock gibbon, *Bunopithecus,* who holds this distinction. Despite their differences, both derive from the same unrooted tree (Tree A, on page 129). The cladograms differ only in their rooting. The first is found by dangling the root of Tree A off the branch leading to *Nomascus,* the second by placing the root on the branch leading to *Bunopithecus.*

How do we 'root' a tree? The usual method is to extend the tree to include at least one – and preferably more than one – 'outgroup': a member of a group that is universally agreed beforehand to be only distantly related to all the others. In the gibbon tree, for example, orang utans or gorillas – or indeed elephants or kangaroos – could do duty as the outgroup. However uncertain we may be about relationships *among* gibbons, we know that the common ancestor of *any* gibbon with a great ape or an elephant is older than the common ancestor of any gibbon with any other gibbon: it is uncontroversial to place the root of a tree that includes the gibbons and the great apes somewhere between the two.

It's easy to verify that the three unrooted trees I have drawn are the only possible dichotomous trees for four groups. For five groups there are 15 possible trees. But don't try to count the number of possible trees for, say, 20 groups. It is up in the hundreds of millions of millions of millions.* As the count of trees rises steeply with the number of groups to be classified, even the fastest computer can take forever. In principle, however, our task is simple. Of all possible trees we must choose that which best explains the similarities and differences between our groups.

How do we judge 'best explains'? Infinitely rich similarities and differences present themselves when we look at a set of animals. But they are harder to count than you might think. Often one 'feature' is an inextricable part of another. If you count them as separate, you've really counted the same one twice. As an extreme example suppose there are four millipede species, A, B, C, and D. A and B resemble each other in all respects except that A has red legs and B has blue legs. C and D are the same as each other and very different from A and B, except that C has red legs while D has blue legs. If we count leg colour as a single 'feature' we correctly group AB apart from CD. But if we naively count each of 100 legs as separate, their colours will give a hundredfold boost to the number of features supporting the alternative grouping of AC as against BD. Everyone would agree that we have spuriously counted the same feature 100 times. It is 'really' only one feature, because a single embryological 'decision' determined the colour of all 100 legs simultaneously.

The same goes for left-right symmetry: embryology works in such a way that, with few exceptions, each side of an animal is a mirror image of the other. No zoologist would count each mirrored feature twice in making a cladogram, but non-independence isn't always so obvious. A pigeon needs a deep breastbone to attach the flight muscles. A flightless bird like a kiwi does not. Do we count deep breastbone and flapping wings as two separate features by which pigeons differ from kiwis? Or do we count them as only a single feature, on the grounds that the state of one character determines the other, or at least reduces its freedom to vary? In the case of the

* The actual number is $(3 \times 2 - 5) \times (4 \times 2 - 5) \times (5 \times 2 - 5) \times \ldots \times (n \times 2 - 5)$ where n is the number of groups.

millipedes and the mirroring, the sensible answer is pretty obvious. In the case of the breastbones it isn't. Reasonable people can be found arguing on opposite sides.

That was all about visible resemblances and differences. But visible features evolve only if they are manifestations of DNA sequences. Nowadays we can compare DNA sequences directly. As an added benefit, being long strings, DNA texts provide a lot more items to count and compare. Problems of the wing-and-breastbone variety are likely to be drowned out in the flood of data. Even better, many DNA differences will be invisible to natural selection and so provide a 'purer' signal of ancestry. As an extreme example, some DNA codes are synonymous: they specify exactly the same amino acid. A mutation that changes a DNA word to one of its synonyms is invisible to natural selection. But to a geneticist, such a mutation is no less visible than any other. The same goes for 'pseudogenes' (usually accidental duplicates of real genes) and for many other 'junk DNA' sequences, which sit in the chromosome but are never read and never used. Freedom from natural selection leaves DNA free to mutate in ways that leave highly informative traces for taxonomists. None of this alters the fact that some mutations do have real and important effects. Even if these are only the tips of icebergs, it is those tips that are visible to natural selection and account for all the visible and familiar beauties and complexities of life.

DNA too is far from immune to the problem of multiple counting – the molecular equivalent of the millipedes' legs. Sometimes a sequence is duplicated many times throughout the genome. About half of human DNA consists of multiple copies of meaningless sequences, 'transposable elements', which may be parasites that hijack the machinery of DNA replication to spread themselves about the genome. Just one of these parasitic elements, *Alu*, is present in over a million copies in most individuals, and we shall meet it again in the Howler Monkey's Tale. Even in the case of meaningful and useful DNA, there are a few cases where genes are present in dozens of identical (or near-identical) copies. But in practice multiple counting tends not to be a problem because duplicate DNA sequences are usually easy to spot.

As a better reason for caution, extensive regions of DNA occasionally show up enigmatic resemblances between comparatively unrelated creatures. Nobody doubts that birds are more closely related to turtles, lizards, snakes and crocodiles than to mammals (see Rendezvous 16). Nevertheless, the DNA sequences of birds and mammals have resemblances greater than one might expect given their distant relationship. Both have an excess of G-C pairings in their non-coding DNA. The G-C pairing is chemically stronger than the A-T one, and it may be that warm-blooded species (birds and mammals) need more tightly bound DNA. Whatever the reason, we should beware of allowing this G-C bias to persuade us of a close relationship between all warm-blooded animals. DNA seems to promise a Utopia for biological systematists, but we must be aware of such dangers: there is a lot that we still don't understand about genomes.

So, having taken the necessary invocation of caution, how can we use the information present in DNA? Fascinatingly, literary scholars use the same techniques as evolutionary biologists in tracing the ancestries of texts. And – almost too good to be true – one of the best examples happens to be the work of the *Canterbury Tales* Project. Members of this international syndicate of literary scholars have used the tools of evolutionary biology to trace the history of 85 different manuscript versions of *The Canterbury Tales*. These ancient manuscripts, hand-copied before the advent of printing, are our best hope of reconstructing Chaucer's lost original. As with DNA, Chaucer's text has survived through repeated copyings, with accidental changes perpetuated in the copies. By meticulously scoring the accumulated differences, scholars can reconstruct the history of copying, the evolutionary tree – for it really is an evolutionary process, consisting of a gradual accumulation of errors over successive generations. So similar are the techniques and difficulties in DNA evolution and literary text evolution, that each can be used to illustrate the other.

So, let's temporarily turn from our gibbons to Chaucer, and in particular four of the 85 manuscript versions of *The Canterbury Tales:* the 'British Library', 'Christ Church', 'Egerton', and 'Hengwrt' versions.* Here are the first two lines of the General Prologue:

---

* The 'British Library' manuscript belonged to Henry Dene, Archbishop of Canterbury in 1501, and, together with the Egerton manuscript and others, is now kept at the British Library in London. The 'Christ Church' manuscript now resides close to where I am writing, in the library of Christ Church, Oxford. The earliest record of the 'Hengwrt' manuscript shows it belonging to Fulke Dutton in 1537. Damaged by rats gnawing at the sheepskin on which it is written, it is now in the National Library of Wales.

BRITISH LIBRARY: Whan that Aprylle / wyth hys showres soote
The drowhte of Marche / hath pcede to the rote
CHRIST CHURCH: Whan that Auerell w^t his shoures soote
The droght of Marche hath pced to the roote
EGERTON: Whan that Aprille with his showres soote
The drowte of marche hath pced to the roote
HENGWRT: Whan that Aueryll w^t his shoures soote
The droghte of March / hath pced to the roote

The first thing that we must do with either DNA or literary texts is to locate the similarities and differences. For this we have to 'align' them – not always an easy task, for texts can be fragmentary or jumbled and of unequal length. A computer is a great help when the going gets tough, but we don't need it to align the first two lines of Chaucer's General Prologue, which I have highlighted at the fourteen points where the scripts disagree (see opposite page).

Two places, the second and the fifth, have three variants rather than two. That makes a total of sixteen 'differences'. Having compiled a list of differences we now work out which tree best explains them. There are many ways of doing this, and all can be used for animals as well as for literary texts. The simplest is to group the texts on the basis of overall similarity. This usually relies upon some variant on the following method. First we locate the pair of texts that are the most similar. We then treat this pair as a single averaged text, and put it alongside the remaining texts while we look for the next most similar pair. And so on, forming successive, nested groups until a tree of relationships is built up. These sorts of techniques – one of the most common is known as 'neighbour-joining' – are quick to calculate, but do not incorporate the logic of the evolutionary process. They are purely measures of similarity. For this reason, the 'cladist' school of taxonomy, which is deeply evolutionary in its rationale (although not all its members realise it) prefers other methods, of which the earliest to be devised was the parsimony method.

Parsimony, as we saw in the Orang Utan's Tale, here means economy of explanation. In evolution, whether of animals or manuscripts, the most parsimonious explanation is the one that postulates the least quantity of evolutionary change. If two texts share a common feature, the parsimonious explanation is that they have jointly inherited it from a shared ancestor rather than that each evolved it independently. It is very far from an invariable rule, but it

```
Bl:  Whan that A██ryll█ | w█th h█s sho█res soote    the dro█nt█ of █arch█ / hath pced█ to the r█ote
Ch:  Whan that A█ervll█ | w██ h█s sho█res soote    the dro█nt█ of █arch█   hath pced█ to the r█ote
Eg:  Whan that A█ryll█ | with h█s sho█res soote    the dro█nt█ of █arch█   hath pced█ to the r█ote
Hg:  Whan that A█ryll█ | w██ h█s sho█res soote    the dro█nt█ of █arch█ / hath pced█ to the r█ote
```

is at least more likely to be true than the opposite. The method of parsimony – at least in principle – looks over all possible trees and chooses the one that minimises the quantity of change.

When we are choosing trees for their parsimony, certain types of difference can't help us. Differences that are unique to a single manuscript, or a single species of animal, are *uninformative*. The neighbour-joining method uses them, but the method of parsimony ignores them completely. Parsimony relies upon *informative* changes: ones that are shared by more than one manuscript. The preferred tree is the one that uses shared ancestry to explain as many informative differences as possible. In our Chaucerian lines there are five informative differences to account for. Four split the manuscripts into

{British Library plus Egerton} *versus* {Christ Church plus Hengwrt}.

These are the differences highlighted by the first, third, seventh, and eighth vertical black lines. The fifth, the virgule (diagonal stroke) highlighted by the twelfth grey line, splits the manuscripts differently, into

{British Library plus Hengwrt} *versus* {Christ Church plus Egerton}.

These splits conflict with each other. We can draw no tree in which each change happens just once. The best we can do is the following (note that it is an unrooted tree) which minimises the conflict, requiring only the virgule to appear or disappear twice.

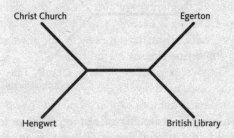

Actually, in this case I haven't much confidence in our guess. Convergences or reversions are common in texts, especially when the

meaning of the verse is not changed. A medieval scribe might have little compunction in changing a spelling, and even less in inserting or removing a punctuation mark such as a virgule. Better indicators of relationship would be changes such as the reordering of words. The genetic equivalents are 'rare genomic changes': events such as large insertions, deletions, or duplications of DNA. We can explicitly acknowledge these by giving more or less weight to different types of change. Changes known to be common or unreliable are down-weighted when counting up extra changes. Changes known to be rare, or reliable indicators of kinship, are given increased weighting. Heavy weighting to a change means we especially don't want to count it twice. The most parsimonious tree, then, is the one with the lowest overall weight.

The parsimony method is much used to find evolutionary trees. But if convergences or reversions are common – as with many DNA sequences and also in our Chaucerian texts – parsimony can be misleading. It is the notorious bugbear known as 'long branch attraction'. Here's what this means.

Cladograms, whether rooted or unrooted, convey only the order of branching. *Phylograms,* or phylogenetic trees (Greek *phylon* = race/tribe/class), are similar but also use the length of branches to convey information. Typically branch lengths represent evolutionary distance: long branches represent a lot of change, short ones little change. The first line of *The Canterbury Tales* yields the following phylogram:

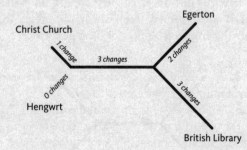

In this phylogram, the branches are not too different in length. But imagine what would happen if two of the manuscripts changed a lot, compared to the other two. The branches leading to these two would be drawn very long. And a proportion of the changes would not be unique. They would just happen to be identical to changes elsewhere on the tree, but (and now here is the point) *especially* to those on the other long branch. This is because long branches are where the most changes are anyway. With enough evolutionary changes, the ones that spuriously link the two long branches will drown out the true signal. Based upon a simple count of the number of changes, parsimony erroneously groups together the termini of especially long branches. The method of parsimony makes long branches spuriously 'attract' one another.

The problem of long branch attraction is an important headache for biological taxonomists. It rears its head whenever convergences and reversions are common, and unfortunately we cannot hope to avoid it by looking at more text. On the contrary, the more text we look at, the more erroneous similarities we find, and the stronger our conviction in the wrong answer. Such trees are said to lie in the dangerous-sounding 'Felsenstein zone', named after the distinguished American biologist Joe Felsenstein. Unfortunately, DNA data are particularly vulnerable to long branch attraction. The main reason is that there are only four letters in the DNA code. If the majority of differences are single letter changes, independent mutation to the same letter by accident is extremely likely. This sets up a minefield of long branch attraction. Clearly we need an alternative to parsimony in these cases. It comes in the form of a technique known as likelihood analysis, which is increasingly favoured in biological taxonomy.

Likelihood analysis burns even more computer power than parsimony, because now the lengths of the branches matter. So we have vastly more trees to contend with because, in addition to looking at all possible branching patterns, we must also look at all possible branch lengths – a Herculean task. This means that, despite clever shortcuts, today's computers can only cope with likelihood analysis involving small numbers of species.

'Likelihood' is not a vague term. On the contrary, it has a precise meaning. For a tree of a particular shape (remembering to include branch lengths), of all the possible evolutionary paths that could produce a phylogenetic tree of the same shape, only a tiny number would generate precisely those texts that we now see. The 'likelihood' of a given tree is the vanishingly small probability of ending up with the actual existing texts, rather than any of the other texts that could possibly have been generated by such a tree. Although the likelihood

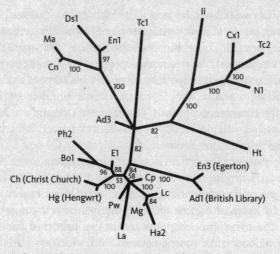

'By me was nothyng added ne mynusshyd' (Caxton's Preface). Unrooted phylogenetic tree of the first 250 lines of 24 different manuscript versions of *The Canterbury Tales*. This represents a subset of the manuscripts studied by the *Canterbury Tales* Project, whose abbreviations for the manuscripts are used here. The tree was constructed by parsimony analysis, and bootstrap values are shown on the branches. The four versions discussed are named in full.

value for a tree is tiny, we can still compare one tiny value with another as a means of judgement.

Within likelihood analysis, there are various alternative methods of obtaining the 'best' tree. The simplest is to search for the single one that has the highest likelihood: the tree which is the most likely. Not unreasonably, this goes under the name 'maximum likelihood', but just because it is the single most likely tree doesn't mean that other possible trees aren't almost as likely. More recently it has been suggested that instead of believing in a single most likely tree, we should look at all possible trees, but give proportionally more credence to the more likely ones. This approach, an alternative to maximum likelihood, is known as Bayesian phylogenetics. If many likely trees agree on a particular branch point, then we calculate that it has a high probability of being correct. Of course, just as in maximum likelihood, we can't look at all possible trees, but there are computational shortcuts and they work pretty well.

Our *confidence* in the tree we finally choose will depend on our certainty that its various branches are correct, and it is common to place measures of this beside each branch point. Probabilities are automatically calculated when using the Bayesian method, but for others such as parsimony or maximum likelihood, we need alternative measures. A commonly used one is the 'bootstrap' method, which resamples different parts of the data repeatedly to see how much difference it makes to the final tree – how robust the tree is, in other words, to error. The higher the 'bootstrap' value, the more trustworthy the branch point, but even experts struggle to interpret exactly what a particular bootstrap value tells us. Similar methods are the 'jackknife', and the 'decay index'. All are measures of how much we should believe each branch point on the tree.

Before we leave literature and return to biology, on the opposite page is a summary diagram of the evolutionary relationships between the first 250 lines of 24 Chaucer manuscripts. It is a phylogram, in which not just the branching pattern but the lengths of the lines are meaningful. You can immediately read off which manuscripts are minor variants of each other, which are aberrant outliers. It is unrooted – it doesn't commit itself as to which of the 24 manuscripts is closest to the 'original'.

It's time to return to our gibbons. Over the years, many people have tried to work out gibbon relationships. Parsimony suggested four groups of gibbons. On the next page is a rooted cladogram based on physical characteristics.

This cladogram shows convincingly that the *Hylobates* species group together, as do the *Nomascus* ones. Both groupings have reasonably high bootstrap values (the numbers on the lines). But in several places the order of branching is unresolved. Even though it looks as though *Hylobates* and *Bunopithecus* form a group, the bootstrap value, 63, is unconvincing to those trained to read such runes. Morphological features do not suffice to resolve the tree.

For this reason, Christian Roos and Thomas Geissmann of Germany turned to molecular genetics, specifically to a section of mitochondrial DNA called the 'control region'. Using DNA from six gibbons, they deciphered the sequences, lined them up letter-for-letter, and carried out neighbour-joining, parsimony, and maximum likelihood analyses on them. Maximum likelihood, which is the best of the three methods at coping with long branch attraction, gave the most convincing result. Their final verdict on the gibbons is shown on page 141, and you can see that it resolves the relationship between the four groups. The bootstrap values were enough to convince me that this was the tree to use for the phylogeny at the start of this chapter.

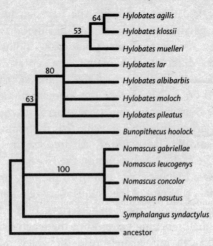

Rooted cladogram of gibbons, based on morphology. Adapted from Geissmann [100].

Gibbons 'speciated' – branched into their separate species – relatively recently. But as we look at more and more distantly related species, separated by longer and longer branches, even the sophisticated techniques of maximum likelihood and Bayesian analysis start to fail us. There can come a point where an unacceptably large proportion of similarities are coincidental. The DNA differences are then said to be saturated. No fancy techniques can recover the signal of ancestry, because any vestiges of relationship have been overwritten by the ravages of time. The problem is especially acute with neutral DNA differences. Strong natural selection keeps genes on the straight and narrow. In extreme cases, important functional genes can stay literally identical over hundreds of millions of years. But, for a pseudogene that never does anything, such lengths of time are enough to lead to hopeless saturation. In such cases, we need different data. The most promising idea is to use the rare genomic changes that I mentioned before – changes that involve DNA reorganisation rather than single letter changes. These being rare, indeed usually unique, coincidental resemblance is much less of a problem. And once found, they can reveal remarkable relationships, as we shall learn when our swelling pilgrim band is joined by the hippo, and we are bowled over by its whale of a surprising tale.

And now, an important afterthought on evolutionary trees, drawing in lessons from Eve's Tale and the Neanderthal's Tale. We might

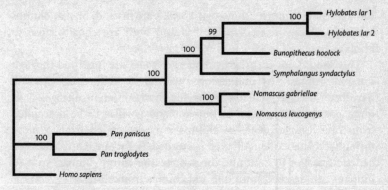

Cladogram of gibbons, based on maximum likelihood analysis of DNA. Adapted from Roos and Geissmann [246].

call it the gibbon's decline and fall of the species tree. We normally assume that we can draw a single evolutionary tree for a set of species. But Eve's Tale told us that different parts of DNA (and thus different parts of an organism) can have different trees. I think this poses an inherent problem with the very idea of species trees. Species are composites of DNA from many different sources. As we saw in Eve's Tale and reiterated in the Neanderthal's Tale, each gene, in fact each DNA letter, takes its own path through history. Each piece of DNA, and each aspect of an organism, can have a different evolutionary tree.

An example of this comes up every day, but familiarity leads us to overlook its message. A Martian taxonomist shown only the genitals of a male human, a female human, and a male gibbon would have no hesitation in classifying the two males as more closely related to each other than either is to the female. Indeed, the gene determining maleness (called *SRY*) has never been in a female body, at least since long before we and the gibbons diverged. Traditionally, morphologists plead a special case for sexual characteristics, to avoid 'nonsensical' classifications. But identical problems arise elsewhere. We saw it previously with ABO blood groups, in Eve's Tale. My B-group gene relates me more closely to a B-group chimpanzee than an A-group human. And it is not just sex genes or blood groups, but *all* genes and characteristics which are susceptible to this effect, under certain circumstances. The majority of both molecular and morphological characteristics show chimps as our closest relatives.

But a sizeable minority show that gorillas are instead, or that chimps are most closely related to gorillas and both are equally close to humans.

This should not surprise us. Different genes are inherited through different routes. The population ancestral to all three species will have been diverse – each gene having many different lineages. It is quite possible for a gene in humans and gorillas to be descended from one lineage, while in chimps it is descended from a more distantly related one. All that is needed is for anciently diverged genetic lineages to continue through to the chimp–human split so humans can descend from one and chimps from another.[*]

So we have to admit that a single tree is not the whole story. Species trees *can* be drawn, but they must be considered a simplified summary of a multitude of gene trees. I can imagine interpreting a species tree in two different ways. The first is the conventional genealogical interpretation. One species is the closest relative of another if, out of all the species considered, it shares the most recent common genealogical ancestor. The second is, I suspect, the way of the future. A species tree can be seen as depicting the relationships among a democratic majority of the genome. It represents the result of a 'majority vote' among gene trees.

The democratic idea – the genetic vote – is the one that I prefer. In this book, all relationships between species should be interpreted in this way. All the phylogenetic trees I present should be viewed in this spirit of genetic democracy, from the relationships between apes to the relationships between the animals, plants, fungi and bacteria.

[*] The longer the time between species splits (or the smaller the population size), the more ancestral lineages are lost by genetic drift. So tidy-minded taxonomists, who hope that species trees coincide with gene trees, will find it easier to deal with animals whose divergences are well spaced out in time, unlike African apes. But there are always genes, such as *SRY*, for which separate lineages are systematically maintained by natural selection over huge spans of time.

OLD WORLD MONKEYS

As we near this rendezvous and prepare to greet Concestor 5 – approximately our 1.5-million-greats-grandparent – we cross a momentous (if somewhat arbitrary) boundary. For the first time in our journey we leave one geological period, the Neogene, to enter an earlier one, the Palaeogene. The next time we do this will be to burst into the Cretaceous world of the dinosaurs. Rendezvous 5 is scheduled at about 25 million years ago, in the Palaeogene. More specifically it is in the Oligocene Epoch of that Period, the last stop on our backward journey when the climate and vegetation of the world are recognisably similar to today's. Much further back, and we shall not find any evidence of the open grasslands that so typify our own Neogene Period, or the wandering herds of grazers that accompanied their spread. Twenty-five million years ago, Africa was completely isolated from the rest of the world, separated from the nearest piece of land – Spain – by a sea as wide as that which separates it from Madagascar today. It is on that gigantic island of Africa that our pilgrimage is about to be invigorated by a new influx of spirited and resourceful recruits, the Old World monkeys – the first pilgrims to arrive bearing tails.

Today, the Old World monkeys number just under 100 species, some of which have migrated out of their mother continent into Asia (see the Orang Utan's Tale). They are divided into two main groups: on the one hand are the colobus monkeys of Africa together with the langurs and proboscis monkeys of Asia; on the other hand are the mostly Asian macaques plus the baboons and guenons, etc. of Africa.

The last common ancestor of all surviving Old World monkeys lived some 11 million years later than Concestor 5, probably around 14 million years ago. The most helpful fossil genus for illuminating the period is *Victoriapithecus,* which is now known from more than a thousand fragments, including a splendid skull, from Maboko Island in Lake Victoria. All the Old World monkey pilgrims join hands around 14 million years ago to greet their own concestor, perhaps *Victoriapithecus* itself, or something like it. They then march on backwards to join the ape pilgrims at our own Concestor 5, 25 million years ago.

And what was Concestor 5 like? Perhaps a bit like the fossil genus *Aegyptopithecus,* which actually lived about 7 million years earlier. Concestor 5 itself, according to our usual rule of thumb, is more likely than not to have had the characteristics shared by its descendants, the catarrhines, defined as consisting of the apes and the Old World monkeys. For example (it's the feature that gives the catarrhines their name) Concestor 5 probably had narrow, downwards-facing nostrils, unlike the wide, sideways-facing nostrils of the New World monkeys, the platyrrhines. The females probably showed full menstruation, as is common among apes and Old World monkeys but not New World monkeys. It probably had an ear tube formed by the tympanic bone, unlike New World monkeys whose ear lacks a bony tube.

Did it have a tail? Almost certainly yes. Given that the most obvious difference between apes and monkeys is the presence or absence of the tail, we are tempted by the *non sequitur* that the divide of 25 million years ago corresponds to the moment at which the tail was lost. In fact, Concestor 5 was presumably tailed like virtually all other mammals, and Concestor 4 was tailless like all its descendants the modern apes. But we don't know at what point along the road leading from Concestor 5 to Concestor 4 the tail was lost. Nor is there any particular reason for us suddenly to start using the word 'ape' to signify the loss of the tail. The African fossil genus *Proconsul,* for example, can be called an ape rather than a monkey, because it lies on the ape side of the fork at Rendezvous 5. But the fact that it lies on the ape side of the fork tells us nothing about whether it had a tail. As it happens, the balance of the evidence suggests that, to quote the title of an authoritative recent paper, '*Proconsul* did not have a tail.' But that in no way follows from the fact that it is on the ape side of the rendezvous divide.

Opposite: **Old World monkeys join.** This phylogeny of the 100 or so species of Old World monkey is generally accepted. The circles now visible at the tips of the branches indicate the number of known species in each group as an order of magnitude: no circle means 1–9 known species, a small width circle means 10–99, a larger circle, 100–999, etc.; each of the four groups shown contain between 10 and 99 species.

Images, left to right: mandrill (*Mandrillus sphinx*); redtail monkey (*Cercopithecus ascanius*); proboscis monkey (*Nasalis larvatus*); Angolan black-and-white colobus (*Colobus angolensis*).

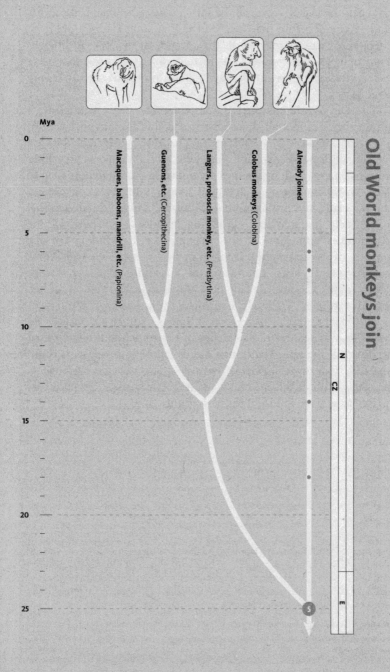

Mya

Old World monkeys join

Macaques, baboons, mandrill, etc. (Papionina)

Guenons, etc. (Cercopithecina)

Langurs, proboscis monkey, etc. (Presbytina)

Colobus monkeys (Colobina)

Already joined

0

5

10

15

20

25

N

CZ

E

5

What, then, should we call the intermediates between Concestor 5 and *Proconsul* before they lost their tail? A strict cladist would call them apes, because they lie on the ape side of the fork. A different kind of taxonomist would call them monkeys because they were tailed. Not for the first time, I say it is silly to become too worked up over names.

The Old World monkeys, Cercopithecidae, are a true clade, a group that includes all descendants of a single common ancestor. However 'monkeys' as a whole are not, because they include the New World monkeys, Platyrrhini. The Old World monkeys are closer cousins to apes, with whom they are united in the Catarrhini, than to New World monkeys. All apes and monkeys together constitute a natural clade, the Anthropoidea. 'Monkeys' constitutes an artificial (technically 'paraphyletic') grouping because it includes all the platyrrhines plus some of the catarrhines but excluding the ape portion of the catarrhines. It might be better to call the Old World monkeys tailed apes. Catarrhine, as I mentioned earlier, means 'down nose': the nostrils face downwards – in this respect we are ideal catarrhines. Voltaire's Dr Pangloss observed that 'the nose is formed for spectacles, therefore we come to wear spectacles'. He could have added that our catarrhine nostrils are beautifully directed to keep out the rain. Platyrrhine means flat or broad nose. It is not the only diagnostic difference between these two great groups of primates, but it is the one that gives them their names. Let's press on to Rendezvous 6, and meet the platyrrhines.

NEW WORLD MONKEYS

Rendezvous 6, where the New World platyrrhine 'monkeys' meet us and our approximately 3-million-greats-grandparent, Concestor 6, the first anthropoid, is some 40 million years ago. It was a time of lush tropical forests – even Antarctica was at least partly green in those days. Although all platyrrhine monkeys now live in South or Central America, the rendezvous itself almost certainly did not take place there. My guess is that Rendezvous 6 is somewhere in Africa. A group of African primates with flat noses, who have left no surviving African descendants, somehow managed, in the form of a small founding population, to get across to South America. We don't know when this happened, but it was before 25 million years ago (when the first monkey fossils appear in South America) and after 40 million years ago (Rendezvous 6). South America and Africa were closer to each other than they are now, and sea levels were low, perhaps exposing a chain of islands across the gap from West Africa, convenient for island-hopping. The monkeys probably rafted across, perhaps on fragments of mangrove swamps that could support life as floating islands for a short while. Currents were in the right direction for inadvertent rafting. Another major group of animals, the hystricognath rodents, probably arrived in South America around the same time. Again probably they came from Africa, and indeed they are named after the African porcupine, *Hystrix*. Probably the monkeys rafted across the same island chain as the rodents, using the same favourable currents, though presumably not the same rafts.

Are all the New World primates descended from a single immigrant? Or was the island-hopping corridor used* more than once by primates? What would constitute positive evidence for a double immigration? In the case of the rodents, there are still hystricognath rodents in Africa, including African porcupines, mole rats, dassie rats and cane rats. If it turned out that some of the South American rodents were close cousins of some African ones (say porcupines)

---

* 'Used' is, of course, unfortunate if it implies anything more than inadvertence. As we shall see in the Dodo's Tale, no animal ever tries to colonise brand new territory. But when it accidentally happens, the evolutionary consequences can be momentous.

while other South American rodents were closer cousins to other African ones (say mole rats) this would be good evidence that rodents more than once drifted to South America. That this is not the case is compatible with the view that rodents dispersed to South America only once, though it is not strong evidence. The South American primates, too, are all closer cousins to each other than they are to any African primate. Again this is compatible with the hypothesis of a single dispersal event, but again the evidence is not strong.

This is a good moment to repeat that the improbability of a rafting event is very far from being a reason for doubting that it happened. This sounds surprising. Usually, in everyday life, massive improbability is a good reason for thinking that something won't happen. The point about intercontinental rafting of monkeys, or rodents or anything else, is that it only had to happen once, and the time available for it to happen, in order to have momentous consequences, is way outside what we can grasp intuitively. The odds against a floating mangrove bearing a pregnant female monkey and reaching landfall in any one year may be ten thousand to one against. That sounds tantamount to impossible by the lights of human experience. But given 10 million years it becomes almost inevitable. Once it happened, the rest was easy. The lucky female gave birth to a family, which eventually became a dynasty, which eventually branched to become all the species of New World monkeys. It only had to happen once: great things then grew from small beginnings.

In any case, accidental rafting is not nearly so rare as you might think. Small animals are often seen on flotsam. And the animals aren't always small. The green iguana is typically a metre long and can be up to two metres. I quote from a note to *Nature* by Ellen J. Censky and others:

> On 4 October 1995, at least 15 individuals of the green iguana, *Iguana iguana*, appeared on the eastern beaches of Anguilla in the Caribbean.

Opposite: **New World monkeys join.** The phylogeny of the 100 or so species of New World monkeys is somewhat disputed, but here we follow the modern consensus.

Images, left to right: golden lion tamarin (*Leontopithecus rosalia*); owl monkey (*Aotus trivirgatus*); squirrel monkey (*Saimiri sciureus*); black howler monkey (*Alouatta caraya*); monk saki (*Pithecia monachus*).

New World monkeys join

Mya

0

10

20

30

40

Tamarins and marmosets (Callitrichinae)

Night or owl monkeys (Aotinae)

Capuchin and squirrel monkeys (Cebinae)

Spider, howler, woolly monkeys (Atelinae)

Sakis, uakaris, titis (Pitheciinae)

Already joined

N

CZ

E

6

This species did not previously occur on the island. They arrived on a mat of logs and uprooted trees, some of which were more than 30 feet long and had large root masses. Local fishermen say the mat was extensive and took two days to pile up on shore. They reported seeing iguanas on both the beach and on logs in the bay.

The iguanas were presumably roosting in trees on some other island, which were uprooted and sent to sea by a hurricane: either Luis, which had raged through the Eastern Caribbean on 4–5 September, or Marilyn, a fortnight later. Neither hurricane hit Anguilla. Censky and her colleagues subsequently caught or sighted green iguanas on Anguilla, and on an islet half a kilometre off shore. The population still survived on Anguilla in 1998 and included at least one reproductively active female. Iguanas and related lizards, by the way, are especially good at colonising islands, all over the world. Iguanas even occur on Fiji and Tonga, which are much more remote than the West Indian islands.

I can't resist remarking how chilling this kind of 'it only had to happen once' logic becomes when you apply it to contingencies nearer home. The principle of nuclear deterrence, and the only remotely defensible justification for possessing nuclear weapons, is that nobody will dare risk a first strike, for fear of massive retaliation. What are the odds against a mistaken missile launch: a dictator who goes mad; a computer system that malfunctions; an escalation of threats that gets out of hand? What are the odds against a terrible mistake, initiating Armageddon? A hundred to one against, within any one year? I would be more pessimistic. We came awfully close in 1963. What might happen in Kashmir? Israel? Korea? Even if the odds per year are as low as one in a hundred, a century is a very short time, given the scale of the disaster we are talking about. It only has to happen once.

Let's return to a happier topic, the New World monkeys. As well as walking quadrupedally above branches, like many Old World monkeys, some New World monkeys suspend themselves like gibbons, and even brachiate. The tail is prominent in all the New World monkeys, and in the spider monkeys, woolly monkeys and howler monkeys it is prehensile, wielded like an extra arm. They can happily hang from the tail alone, or from any combination of arms, legs and tail. The tail doesn't have a hand at the end, but you almost believe it has, when you watch a spider monkey.*

New World monkeys also include some spectacularly acrobatic leapers, as well as the only nocturnal anthropoids, the owl monkeys. Like owls and cats, owl monkeys have large eyes – the largest eyes of all the monkeys or apes. Pygmy marmosets are the size of a dormouse, smaller than any other anthropoid. The largest howler monkeys, however, are only about as big as a large gibbon. Howlers resemble gibbons, too, in being good at hanging and swinging from their arms, and in being very noisy – but where gibbons sound like New York police sirens in full cry, a troop of howler monkeys, with their resonating hollow bony voice boxes, remind me more of a ghost squadron of jet planes, roaring eerily through the treetops. As it happens, howler monkeys have a particular tale to tell us Old World monkeys – about the way we see colour, for they have independently arrived at the same solution.

THE HOWLER MONKEY'S TALE
Written with Yan Wong

New genes aren't added to the genome out of thin air. They originate as duplicates of older genes. Then, over evolutionary time, they go their separate ways by mutation, selection and drift. We don't usually see this happening but, like detectives arriving on the scene after a crime, we can piece together what must have happened from the evidence that remains. The genes involved in colour vision provide a striking example. For reasons that will emerge, the howler monkey is especially well placed to tell the tale.

During their formative megayears, mammals were creatures of the night. The day belonged to the dinosaurs, who probably, if their modern relatives are any guide, had superb colour vision. So, we may plausibly imagine, did the mammals' remote ancestors, the mammal-like reptiles, who filled the days before the rise of the dinosaurs. But during the mammals' long nocturnal exile, their eyes needed to snap up whatever photons were available, regardless of colour. Not

---

* Prehensile tails are also found in several other South American groups, including kinkajous (carnivores), porcupines (rodents), tree-anteaters (xenarthrans), opossums (marsupials), and even the salamander *Bolitoglossa*. Is there something special about South America? But prehensile tails also occur in pangolins, some tree rats, some skinks and chameleons not from South America!

surprisingly, for reasons of the kind that we shall examine in the Blind Cave Fish's Tale, colour discrimination degenerated. To this day most mammals, even those who have returned to live in the daylight, have rather poor colour vision, with only a two-colour system ('dichromatic'). This refers to the number of different classes of colour-sensitive cells – 'cones' – in the retina. We catarrhine apes and Old World monkeys have three: red, green and blue, and are therefore trichromatic, but the evidence suggests that we *regained* a third class of cone, after our nocturnal ancestors lost it. Most other vertebrates, such as fish and reptiles but *not* mammals, have three-cone ('trichromatic') or four-cone ('tetrachromatic') vision, and birds and turtles can be even more sophisticated. We'll come to the very special situation in the New World monkeys, and the even more special situation in the howler monkey, in a moment.

Interestingly, there is evidence that Australian marsupials differ from most mammals in having good trichromatic colour vision. Catherine Arrese and her colleagues, who discovered this in honey possums and dunnarts (it has also been demonstrated in wallabies), suggest that Australian (but not American) marsupials kept an ancestral reptilian visual pigment that the rest of the mammals lost. But mammals in general probably have the poorest colour vision among vertebrates. Most mammals see colour, if at all, only as well as a colourblind man. The notable exceptions are to be found among primates, and it is no accident that they, more than any other group of mammals, make use of bright colours in sexual display.

Unlike the Australian marsupials who perhaps never lost it, we can tell by looking at our relatives among the mammals that we primates did not retain trichromatic vision from our reptilian ancestors but rediscovered it – not once, but twice independently: first in the Old World monkeys and apes; and second in the New World howler monkeys, although not among the New World monkeys generally. Howler monkey colour vision is like that of apes, but different enough to betray its independent origin.

Why would good colour vision be so important that trichromacy evolved independently in New and Old World monkeys? A favoured suggestion is that it has to do with eating fruit. In a predominantly green forest, fruits stand out by their colours. This, in turn, is no accident. Fruits have probably evolved bright colours to attract frugivores, such as monkeys, who play the vital role of spreading and manuring their seeds. Trichromatic vision also assists in the detection of younger, more succulent leaves (often pale green, some-times even red), against a background of darker green – but that is presumably not to the advantage of the plants.

Colour dazzles our awareness. Colour words are among the first adjectives that infants learn, and the ones they most eagerly tie to any noun that's going. It is hard to remember that the hues we perceive are labels for electromagnetic radiations of only slightly differing wavelengths. Red light has a wavelength around 700 billionths of a metre, violet around 420 billionths of a metre, but the whole gamut of visible electromagnetic radiation that lies between these bounds is an almost ludicrously narrow window, a tiny fraction of the total spectrum whose wavelengths range from kilometres (some radio waves) down to fractions of a nanometre (gamma rays).

All eyes on our planet are set up in such a way as to exploit the wavelengths of electromagnetic radiation in which our local star shines brightest, and which pass through the window of our atmo-sphere. For an eye that has committed itself to biochemical tech-niques suitable for this loosely bounded range of wavelengths, the laws of physics impose sharper bounds to the portion of the electro-magnetic spectrum that can be seen using those techniques. No animal can see far into the infrared. Those that come closest are pit vipers, who have pits in the head which, while in no sense focusing a proper image with infrared rays, allow these snakes to achieve some directional sensitivity to the heat generated by their prey. And no animal can see far into the ultraviolet although some, bees for instance, can see a bit further than we can. But on the other hand, bees can't see our red: for them it is infrared. All animals agree that 'light' is a narrow band of electromagnetic wavelengths lying some-where between ultraviolet at the short end and infrared at the long end. Bees, people and snakes differ only slightly in where they draw the lines at each end of 'light'.

An even narrower view is taken by each of the different kinds of light-sensitive cells within a retina. Some cones are slightly more sensitive towards the red end of the spectrum, others towards the blue. It is the comparison between cones that makes colour vision possible, and the quality of colour vision depends largely on how many different classes of cones there are to compare. Dichromatic animals have only two populations of cones interspersed with one another. Trichromats have three, tetrachromats four. Each cone has a graph of sensitivity, which peaks somewhere in the spectrum and fades away, not particularly symmetrically, on either side of the peak. Out beyond the edges of its sensitivity graph, the cell may be said to be blind.

Suppose a cone's sensitivity peaks in the green part of the spectrum. Does this mean, if that cell is firing impulses towards the brain, that it is looking at a green object like grass or a billiard table? Emphatically not. It is just that the cell would need more red light (say) to achieve the same firing rate as a given amount of green light. Such a cell would behave identically towards bright red light or dimmer green light.* The nervous system can tell the colour of an object only by *comparing* the simultaneous firing rates of (at least) two cells that favour different colours. Each one serves as a 'control' for the other. You can get an even better idea of the colour of an object by comparing the firing rate of three cells, all with different sensitivity graphs.

Colour television and computer screens, doubtless because they are designed for our trichromatic eyes, also work on a three-colour system. On a normal computer monitor, each 'pixel' consists of three dots placed too close together for the eye to resolve. Each dot always glows with the same colour – if you look at the screen at sufficient magnification you always see only the same three colours, usually red, green and blue although other combinations can do the job. Flesh tones, subtle shades – any hue you wish – can be achieved by manipulating the intensities with which these three primary colours glow. Tetrachromatic turtles, for example, might be disappointed by the unrealistic (to them) pictures on our television and cinema screens.

By comparing the firing rates from just three kinds of cones, our brains can perceive a huge range of hues. But most placental mammals, as already stated, are not trichromats but dichromats, with only two populations of cones in their retinas. One class peaks in the violet (or in some cases the ultraviolet), the other class peaks somewhere between green and red. In us trichromats, the short wavelength cones peak between violet and blue, and they are normally called blue cones. Our other two classes of cones can be called

---

* This raises an intriguing possibility. Imagine that a neurobiologist inserts a tiny probe into, say, a green cone and stimulates it electrically. The green cell will now report 'light' while all other cells are silent. Will the brain 'see' a 'super green' hue such as could not possibly be achieved by any real light? Real light, no matter how pure, would always stimulate all three classes of cones to differing extents.

green cones and red cones. Confusingly, even the 'red' cones peak at a wavelength that is actually yellowish. But their sensitivity curve as a whole stretches into the red end of the spectrum. Even if they peak in the yellow, they still fire strongly in response to red light. This means that, if you subtract the firing rate of a 'green' cone from that of a 'red' cone, you'll get an especially high result when looking at red light. From now on I shall forget about peak sensitivities (violet, green and yellow) and refer to the three classes of cones as blue, green and red. In addition to cones, there are also rods: light-sensitive cells of a different shape from cones, which are especially useful at night and which are not used in colour vision at all. They'll play no further part in our story.

The chemistry and the genetics of colour vision are rather well understood. The main molecular actors in the story are opsins: protein molecules which serve as visual pigments sitting in the cones (and rods). Each opsin molecule works by attaching to, and encasing, a single molecule of retinal: a chemical derived from vitamin A.* The retinal molecule has been forcibly kinked beforehand to fit it into the opsin. When hit by a single photon of light of an appropriate colour, the kink straightens out. This is the signal to the cell to fire a nervous impulse, which says to the brain 'my kind of light *here*'. The opsin molecule is then recharged with another kinked retinal molecule, from a store in the cell.

Now, the important point is that not all opsin molecules are the same. Opsins, like all proteins, are made under the influence of genes. DNA differences result in opsins that are sensitive to light of different wavelengths, and this is the genetic basis of the two-colour or three-colour systems we have been talking about. Of course, since all genes are present in all cells, the difference between a red cone and a blue cone is not which genes they possess, but which genes they turn on. And there is some kind of rule that says that any one cone *only* turns on one class of gene.

The genes that make our green and red opsins are very similar to each other, and they are on the X chromosome (the sex chromosome of which females have two copies and males only one). The gene that makes the blue opsin is a bit different, and lies not on a sex chromosome but on one of the ordinary non-sex chromosomes called

---

* Carrots are rich in beta-carotene from which vitamin A can be made: hence the rumour – rumours can be true – that carrots improve vision.

autosomes (in our case it is chromosome 7). Our green and red cells have clearly been derived from a recent gene duplication event, and much longer ago they must have diverged from the blue opsin gene in another duplication event. Whether an individual has dichromatic or trichromatic vision depends on how many distinct opsin genes it has in its genome. If it has, say, blue- and green-sensitive opsins but not red, it will be a dichromat.

That's the background to how colour vision works in general. Now, before we come to the special case of the howler monkey itself and how it became trichromatic, we need to understand the strange dichromatic system of the rest of the New World monkeys (some lemurs have it too, by the way, and not all New World monkeys do – for example, nocturnal owl monkeys have monochromatic vision). For the purposes of this discussion, 'New World monkey' temporarily excludes howler monkeys and other exceptional species. We'll come to the howler monkeys later.

First, set aside the blue gene as an unvarying fixture on an autosome, present in all individuals whether male or female. The red and green genes, on the X chromosome, are more complicated and will occupy our attention. Each X chromosome has only one locus where a red or a green* allele might sit. Since a female has two X chromosomes, she has two opportunities for a red or green gene. But a male, with only one X chromosome, has *either* a red or a green gene but *not* both. So a typical male New World monkey has to be dichromatic. He has only two kinds of cones: blue plus *either* red or green. By our standards, all males are colourblind, but they are colourblind in two different ways; some males within a population lack green opsins, others lack red opsins. All have blue.

Females are potentially more fortunate. Having two X chromosomes, they could be lucky enough to have a red gene on one and a green gene on the other (plus the blue which again goes without saying). Such a female would be a trichromat.† But an unlucky

---

* Actually, red and green are only two out of a range of possibilities at this locus, but we have enough complications to be going on with. For the purposes of this tale they will be firmly 'red' and 'green'.

† As for ensuring that, in any one cone, only the red or the green opsin gene, but not both, is turned on, this happens to be easy for females. They already have a mechanism for turning the whole of one X chromosome off in any cell. A random half of the cells deactivates one of the two X chromosomes, the other half the other one. This is important, because all the genes on an X chromosome are set up to work if only one is active – necessary because males only have one X chromosome.

female might have two reds, or two greens, and would therefore be a dichromat. By our standards such females are colourblind, and in two ways, just like males.

A population of New World monkeys such as tamarins or squirrel monkeys, therefore, is an oddly complicated mixture. All males, and some females, are dichromats: colourblind by our standards but in two alternative ways. Some females, but no males, are trichromats, with true colour vision which is presumably similar to ours. Experimental evidence with tamarins searching for food in camouflaged boxes showed that trichromatic individuals were more successful than dichromats. Perhaps foraging bands of New World monkeys rely on their lucky trichromat females to find food that most of them would otherwise miss. On the other hand, there is a possibility that the dichromats, either alone or in collusion with dichromats of the other kind, might have strange advantages. There are anecdotes of bomber crews in the Second World War deliberately recruiting one colourblind member because he could spot certain types of camouflage better than his otherwise more fortunate trichromat comrades. Experimental evidence confirms that human dichromats can indeed break certain forms of camouflage that fool trichromats. Is it possible that a troop of monkeys consisting of trichromats and two kinds of dichromats might collectively find a greater variety of fruits than a troop of pure trichromats? This might sound far-fetched, but it is not silly.

The red and the green opsin genes in New World monkeys constitute an example of a 'polymorphism'. Polymorphism is the simultaneous existence, in a population, of two or more alternative versions of a gene, where neither is rare enough to be just a recent mutant. It is a well-established principle of evolutionary genetics that visible polymorphisms like this don't just happen without good reason. Unless something very special is going on, monkeys with the red gene will be either better off, or worse off, than monkeys with the green gene. We don't know which, but it is highly unlikely that they would be exactly equally good. And the inferior kind should go extinct.

A stable polymorphism in a population, then, indicates that something special is going on. What sort of thing? Two main suggestions have been made for polymorphisms in general, and either might apply to this case: frequency-dependent selection, and heterozygous advantage. Frequency-dependent selection happens when the rarer type is at an advantage, simply by virtue of being rarer. So, as the type which we had thought was 'inferior' starts to go extinct, it ceases to be inferior and bounces back. How could this be? Well, suppose 'red' monkeys are especially good at seeing red fruits while 'green' monkeys are especially good at seeing green fruits. In a population dominated by red monkeys, most of the red fruits will be already taken, and a lone green monkey, able to see green fruits, might be at an advantage – and vice versa. Even if that is not especially plausible, it is an example of the *kind* of special circumstance that can maintain both types in a population, without one of them going extinct. It is not hard to see that something along the lines of our 'bomber crew' theory might be the kind of special circumstance that maintains a polymorphism.

Turning now to heterozygous advantage, the classic example – cliché almost – is sickle-cell anaemia in humans. The sickling gene is bad, in that individuals with two copies of it (homozygotes) have damaged blood corpuscles that look like sickles, and suffer from debilitating anaemia. But it is good in that individuals with only one copy (heterozygotes) are protected against malaria. In areas where malaria is a problem, the good outweighs the bad, and the sickling gene tends to spread through the population, in spite of the adverse effects on individuals unlucky enough to be homozygotes.* Professor John Mollon and his colleagues, whose research is mainly responsible for uncovering the polymorphic system of colour vision in New World monkeys, propose that the heterozygous advantage enjoyed by the trichromatic females is enough to favour the coexistence of the red and green genes in the population. But the howler monkey does it better, and this brings us to the teller of the tale itself.

Howler monkeys have managed to enjoy the virtues of both sides of the polymorphism, by combining them in one chromosome. They have done this by means of a lucky translocation. Translocation is a special kind of mutation. A chunk of chromosome somehow gets pasted into a different chromosome by mistake, or into a different place on the same chromosome. This seems to have happened to a lucky mutant ancestor of the howler monkeys, which consequently ended up with *both* a red gene and a green gene next door to one

---

* This sadly affects many African-Americans, who no longer live in a malarial country but inherit the genes of ancestors who did. Another example is the debilitating disease cystic fibrosis whose gene, in the heterozygous condition, seems to confer protection against cholera.

another on a single X chromosome. This monkey would have been well on its evolutionary way towards becoming a true trichromat, even if it was a male. The mutant X chromosome spread through the population until, now, all howler monkeys have it.

It was easy for howler monkeys to perform this evolutionary trick, because the three opsin genes were already knocking around the population in New World monkeys: it is just that, with the exception of a few lucky females, any one individual monkey had only two of them. When we apes and Old World monkeys independently did the same kind of thing, we did it differently. The dichromats from which we sprang were dichromats in only one way: there wasn't a polymorphism to take off from. Evidence suggests that the doubling up of the opsin gene on the X chromosome in our ancestry was a true duplication. The original mutant found itself with two tandem copies of an identical gene, say two greens next door to each other on the chromosome, and it therefore was not a near-instant trichromat like the ancestral howler monkey mutant. It was a dichromat, with a blue and two green genes. The Old World monkeys became trichromats gradually in subsequent evolution, as natural selection favoured a divergence of the colour sensitivities of the two X opsin genes, towards green and red respectively.

When a translocation happens, it isn't just the gene of interest that is seen to move. Sometimes its travelling companions – its neighbours on the original chromosome who move with it to the new chromosome – can tell us something. And so it is in this case. The gene called *Alu* is well known as a 'transposable element': a short, virus-like piece of DNA that replicates itself around the genome, as a sort of parasite, by subverting the cell's DNA replication machinery. Was *Alu* responsible for moving the opsin? It seems so. We find the 'smoking gun' when we look at the details. There are *Alu* genes at both ends of the duplicated region. Probably the duplication was an unintended by-product of parasitic reproduction. In some long-forgotten monkey of the Eocene Epoch, a genomic parasite near to the opsin gene tried to reproduce, accidentally replicated a much larger chunk of DNA than intended, and set us on the road to three-colour vision. Beware, by the way, of the temptation – it is all too common – to think that, because a genomic parasite seems, with hindsight, to have done us a favour, genomes therefore harbour parasites in the hope of future favours. That isn't how natural selection works.

Whether engineered by *Alu* or not, mistakes of this kind still sometimes happen. When two X chromosomes line up, prior to crossing over, it is possible for them to line up incorrectly. Instead of lining the red gene on one chromosome with the corresponding red on the other, the similarity of the genes can confuse the lining-up process so that a red is lined up with a green. If crossing over then happens it is 'unequal': one chromosome could end up with an extra green (say) while the other X chromosome gets no green gene at all. Even if crossing over doesn't happen, a process called 'gene conversion' can take place, where a short sequence of one chromosome is converted to the matching sequence in the other. With misaligned chromosomes, a part of the red gene may be replaced by the equivalent part of the green gene, or vice versa. Both unequal crossing over and misaligned gene conversion can lead to red–green colourblindness.

Men suffer more frequently from red–green colourblindness than women (the suffering is not great, but it is still a nuisance and they presumably are deprived of aesthetic experiences enjoyed by the rest of us) because if they inherit one faulty X chromosome they do not have another to serve as a backup. Nobody knows whether they see blood and grass in the way the rest of us see blood, or in the way the rest of us see grass, or whether they see both in some completely different way. Indeed, it may vary from person to person. All we know is that people who are red–green colourblind think grass-like things are pretty much the same colour as blood-like things. In humans, dichromatic colourblindness afflicts about two per cent of males. Don't be confused, incidentally, by the fact that other kinds of red–green colourblindness are more common (affecting about eight per cent of males). These individuals are called anomalous trichromats: genetically they are trichromats, but one of their three kinds of opsins doesn't work.*

Unequal crossing over doesn't always make things worse. Some X chromosomes end up with more than two opsin genes. The extra ones nearly always seem to be green rather than red. The record

---

* Mark Ridley, in *Mendel's Demon* (retitled *The Cooperative Gene* in America), points out that the eight per cent (or higher) figure applies to Europeans, and others with a history of good medicine. Hunter-gatherers, and other 'traditional' societies closer to the cutting edge of natural selection, show a lower percentage. Ridley suggests that a relaxation of natural selection has allowed colourblindness to increase. The whole business of colourblindness is treated, in characteristically original fashion, by Oliver Sacks in *The Island of the Colour-Blind*.

number is a staggering twelve extra green genes, arrayed in tandem. But there is no evidence that individuals with extra green genes can see any better. Nevertheless, the high mutation rate along this part of the X chromosome means that not all 'green' genes in the population are exactly the same as each other. So it is theoretically possible for a female, with her two X chromosomes, to have not trichromatic vision but vision which is tetrachromatic (or even pentachromatic, if her red genes also differ). I don't know that anybody has tested this.

It is possible that an uneasy thought has occurred to you. I have talked as though the acquisition, by mutation, of a new opsin automatically confers enhanced colour vision. But of course differences between the colour sensitivities of cones are no earthly use unless the brain has some means of knowing which kind of cone is sending it messages. If it were achieved by genetic hard wiring – this brain cell is hooked up to a red cone, that nerve cell is hooked up to a green cone – the system would work, but it couldn't cope with mutations in the retina. How could it? How could brain cells be expected to 'know' that a new opsin, sensitive to a different colour, has suddenly become available and that a particular set of cones, in the huge population of cones in the retina, have turned on the gene for making the new opsin?

It seems that the only plausible answer is that the brain learns. Presumably it compares the firing rates that originate in the population of cone cells in the retina and 'notices' that one sub-population of cells fires strongly when tomatoes and strawberries are seen; another sub-population when looking at the sky; another when looking at grass. This is a 'toy' speculation, but I suppose something like it enables the nervous system swiftly to accommodate a genetic change in the retina. My colleague Colin Blakemore, with whom I raised the matter, sees this problem as one of a family of similar problems that arise whenever the central nervous system has to adjust itself to a change in the periphery.*

The final lesson of the Howler Monkey's Tale is the importance of gene duplication. The red and the green opsin genes are clearly derived from a single ancestral gene that xeroxed itself to a different part of the X chromosome. Farther back in time, we may be sure, it

---

* I expect that some such learning must be used by birds and reptiles, who enhance their range of colour sensitivities by planting tiny coloured oil droplets over the surface of the retina.

was a similar duplication that separated the blue* autosomal gene from what was to become the red/green X-chromosomal gene. It is common for genes on completely different chromosomes to belong to the same 'gene family'. Gene families have arisen by ancient DNA duplications followed by divergence of function. Various studies have found that a typical human gene has an average probability of duplication of about 0.1 to 1 per cent per million years. DNA duplication can be a piecemeal affair, or it can happen in bursts, for example when a newly virulent DNA parasite like *Alu* spreads throughout the genome, or when a genome is duplicated wholesale. (Entire-genome duplication is common in plants, and is postulated to have happened at least twice in our ancestry, during the origination of the vertebrates.) Regardless of when or how it happens, accidental DNA duplication is one of the major sources of new genes. Over evolutionary time, it isn't only genes that change, within genomes. Genomes themselves change.

---

* Or ultraviolet or whatever it was in those days. Presumably the exact colour sensitivities of all these classes of opsin have been modified over the evolutionary years anyway.

TARSIERS

We anthropoid pilgrims have arrived at Rendezvous 7, 58 million years ago in the dense and varied forests of the Palaeocene Epoch. There we greet a little evolutionary trickle of cousins, the tarsiers. We need a name for the clade that unites anthropoids and tarsiers, and it is haplorhines. The haplorhines consist of Concestor 7, perhaps our 6-million-greats-grandparent, and all its descendants: tarsiers, 'monkeys' and apes.

The first thing you notice about a tarsier is its eyes. Looking at the skull, it is almost the only thing there is to notice: a pair of eyes on legs pretty well sums up a tarsier. Each one of its eyes is as large as its entire brain, and the pupils open very wide too. The skull seen head-on seems to be wearing a pair of fashionably outsize, not to say giant, spectacles. Their huge size makes the eyes hard to rotate in their sockets but tarsiers, like some owls, are equal to the challenge. They rotate the whole head, on an extremely flexible neck, through nearly 360 degrees. The reason for their huge eyes is the same as in owls and night monkeys – tarsiers are nocturnal. They rely on moonlight, starlight and twilight, and need to sweep up every last photon they can.

Other nocturnal mammals have a tapetum lucidum – a reflecting layer behind the retina, which turns photons back in their tracks, so giving the retinal pigments a second chance to intercept them. It is the tapetum that makes it easy to spot cats and other animals at night.* Shine a torch all around you. It will catch the attention of any animals in the vicinity, and they'll look straight at your light out of curiosity. The beam will be reflected back off the tapetum. Sometimes you can locate dozens of pairs of eyes with a single sweep of the torch. If electric light beams had been a feature of the environment in which animals evolved, they might well not have evolved a tapetum lucidum, as it is such a giveaway.

Tarsiers, surprisingly, have no tapetum lucidum. It has been suggested that their ancestors, along with other primates, passed

---

* Most nocturnal birds, too, have reflecting eyes, but not the owlet nightjars (Aegothelidae) of Australasia, nor the Galapagos swallowtailed gull *Creagrus furcatus,* the only nocturnal gull in the world.

Tarsiers join

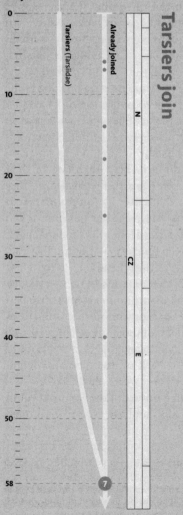

Mya

0 —

Tarsiers (Tarsiidae)

Already joined

N

CZ

E

10 —

20 —

30 —

40 —

50 —

58 —

⑦

Tarsiers join. Recent morphological and molecular studies place the five tarsier species as the sister group to the apes and monkeys, rather than allied to the lemurs as previously thought.

Image: Philippine tarsier (*Tarsius syrichta*).

through a diurnal phase and lost the tapetum. This is supported by the fact that tarsiers have the same weird system of colour vision as most of the New World monkeys. Several groups of mammals that were nocturnal in the time of the dinosaurs became diurnal when the death of the dinosaurs made it safe to do so. The suggestion is that the tarsiers sub-sequently returned to the night, but for some reason the evolutionary avenue of regrowing the tapetum was blocked to them. So they achieved the same result, of capturing as many photons as possible, by making their eyes very big indeed.*

The other descendants of Concestor 7, the 'monkeys' and apes, also lack a tapetum lucidum, not surprisingly given that they are all diurnal except the owl monkeys of South America. And the owl monkeys, like the tarsiers, have compensated by growing very large eyes – although not quite so large, in proportion to the head, as those of the tarsiers. We can make a good guess that Concestor 7 also lacked a tapetum lucidum and was prob-ably diurnal. What else can we say about it?

Apart from being diurnal, it may have been quite tarsier-like. The reason for saying this is that there are some plausible fossils called the omomyids dating from about the right period. Concestor 7 might have been something like an omomyid, and the omomyids were quite tarsier-like. Their eyes were not so big as modern tarsiers', but big enough to suggest that they were nocturnal. Perhaps Concestor 7 was a diurnal version of an omomyid, living in trees. Of its two descendant lineages, one stayed in the light and blossomed into the anthropoid monkeys and apes. The other reverted to the darkness and became the modern tarsiers.

Eyes apart, what is to be said about tarsiers? They are outstanding leapers, with long legs like frogs or grasshoppers. A tarsier can jump more then 3 metres horizontally and 1.5 metres vertically. They have

---

* On this theory, if the tarsiers had managed to regrow a tapetum they wouldn't need such huge eyes and this would have been a good thing. The absolutely largest eyes in the entire animal kingdom are those of the giant squid, nearly a foot in diameter. They too have to cope with very low light levels, this time not because they are nocturnal but because so little light penetrates to the great depths of ocean that they inhabit.

been called furry frogs. It is probably no accident that they resemble frogs too in uniting the two bones of the lower legs, the tibia and the fibula, to make a single strong bone, the tibiofibula. All anthropoids have nails instead of claws, and tarsiers do too, with the curious exception of 'grooming claws' on the second and third toes.

We can't guess with any certainty where Rendezvous 7 takes place. But we might just note that North America is rich in early omomyid fossils of the right period, and that it was in those days firmly joined to Eurasia via what is now Greenland. Perhaps Concestor 7 was a North American.

LEMURS, BUSHBABIES AND THEIR KIN

Gathering the little leaping tarsiers into our pilgrimage, we head off back towards Rendezvous 8, where we are to be joined by the rest of the primates traditionally called prosimians: the lemurs, pottos, bushbabies and lorises. We need a name for those 'prosimians' that are not tarsiers. 'Strepsirhines' has become customary. It means 'split nostril' (literally twisted nose). It is a slightly confusing name. All it means is that the nostril is shaped like a dog's. The rest of the primates, including us, are haplorhines (simple nose: our nostrils are each just a simple hole).

We haplorhine pilgrims, then, greet our strepsirhine cousins, of which the great majority are lemurs, at Rendezvous 8. Various dates have been suggested for this point. I have taken it as 63 million years into the past, a commonly accepted date and one just 'before' our passage back into the Cretaceous Period. Bear in mind, however, that a few researchers imagine this rendezvous even further back in time, during the Cretaceous itself. At 63 million years ago, the Earth's vegetation and climate had rebounded from their drastic disturbance when the Cretaceous – and the dinosaurs – came to an end (see 'The Great Cretaceous Catastrophe'). The world was largely wet and forested, with at least the northern continents covered in a relatively restricted mix of deciduous conifers, and a scattering of flowering plant species.

Perhaps in the branches of a tree, we encounter Concestor 8, seeking fruit or maybe an insect. This most recent common ancestor of all surviving primates is approximately our 7-million-greats-grandparent. Fossils that might help us reconstruct what Concestor 8 was like include the large group called plesiadapiforms. They lived about the right time, and they have many of the qualities you would expect of the grand ancestor of all the primates. Not all of them, however, which makes their supposed position close to the primate ancestor controversial.

Of the living strepsirhines, the majority are lemurs, living exclusively in Madagascar, and we'll come to them in the tale that follows. The others divide into two main groups, the leaping bushbabies and

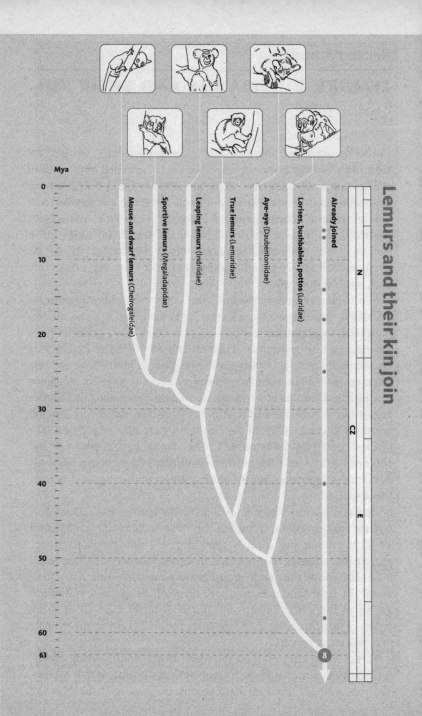

Lemurs and their kin join

Mya

Mouse and dwarf lemurs (Cheirogaleidae)

Sportive lemurs (Megaladapidae)

Leaping lemurs (Indriidae)

True lemurs (Lemuridae)

Aye-aye (Daubentoniidae)

Lorises, bushbabies, pottos (Loridae)

Already joined

N

CZ

E

8

the creeping lorises and pottos. When I was a child of three in Nyasaland (now Malawi) we had a pet bushbaby. Percy was brought in by a local African, and was probably an orphaned juvenile. He was tiny: small enough to perch on the rim of a glass of whisky, into which he would dip his hand and drink with evident enjoyment. He slept during the day, clasping the underside of a beam in the bathroom. When his 'morning' came (in the evening), if my parents failed to catch him in time (which was often, because he was extremely agile and a terrific leaper) he would race to the top of my mosquito net and urinate on me from above. When leaping, for example onto a person, he did not exhibit the common bushbaby habit of urinating on his hands first. On the theory that 'urine washing' is for scent-marking, this would make sense given that he was not an adult. On the alternative theory that the urine improves grip, it is less clear why he didn't do it.

I shall never know to which of the 17 species of bushbaby Percy belonged, but he was most certainly a leaper, not a creeper. The creepers are the pottos of Africa and the lorises of Asia. They move much more slowly – especially the 'slow loris' of the Far East, which is a stealth hunter, inching along a branch until within reach of prey, whereupon it lunges with great speed.

Bushbabies and pottos remind us that a tropical forest is a three-dimensional world like the sea. Seen from above the canopy, the green waves at its surface billow towards the horizon. Dive down into the darker green world beneath, and you pass through distinct layers, again as in the sea. The animals of the forest, like fish in the sea, find it as easy to move up and down as horizontally. But, also as in the sea, each species in practice specialises in making its living at a particular level. In the West African forests by night, the surface canopy is the province of the pygmy bushbabies hunting insects, and

Opposite: **Lemurs and their kin join.** The living primates can be divided into the lemurs and their kin, and the rest. The time of this divergence is debated – some experts place it as much as 20 million years earlier, with a consequent increase in the age of Concestors 9, 10, and 11. The five Madagascan lemur families (30 or so species) and the loris family (18 species) are known as 'strepsirhines'. The order of branching within lemurs in this strepsirhine phylogeny remains controversial.

Images, left to right: pygmy mouse lemur (*Microcebus myoxinus*); red-tailed sportive lemur (*Lepilemur ruficaudatus*); indri (*Indri indri*); white-fronted brown lemur (*Eulemur fulvus albifrons*); aye-aye (*Daubentonia madagascariensis*); slender loris (*Loris tardigradus*).

the fruit-eating pottos. Below the level of the canopy, the trunks of the trees are separated by gaps, and this is the domain of the needle-clawed bushbaby whose eponymous equipment enables it to cling to the trunks after leaping the gaps between them. Deeper still, in the understorey, the golden potto and the closely related angwantibo hunt caterpillars. At dawn, the nocturnal bushbabies and pottos give place to day-hunting monkeys, who parcel up the forest into similarly stratified layers. The same kind of stratification goes on in the South American forests, where as many as seven species of (marsupial) possum can be found, each at its own level.

The lemurs are descended from those early primates who happened to find themselves marooned in Madagascar during the time when monkeys were evolving in Africa. Madagascar is a large enough island to serve as a laboratory for natural experiments in evolution. The tale of Madagascar will be told by one of the lemurs, by no means the most typical of them, the aye-aye *Daubentonia*. I don't remember much from the discourse on lemurs that Harold Pusey – wise and learned warhorse of the lecture hall – gave to my generation of Oxford zoologists, but I do remember the haunting refrain with which he concluded almost every sentence about lemurs: 'Except *Daubentonia*.' 'EXCEPT *Daubentonia!*' Despite appearances, *Daubentonia*, the aye-aye, is a perfectly respectable lemur, and lemurs are the most famous inhabitants of the great island of Madagascar. The Aye-Aye's Tale is about Madagascar, textbook showcase of biogeographical natural experiments, a tale not just of lemurs but of all of Madagascar's peculiar – in the original sense of the word – fauna and flora.

THE AYE-AYE'S TALE

A British politician once described a rival (who later went on to become their party leader) as having 'something of the night' about him. The aye-aye conveys a similar impression, and indeed it is wholly nocturnal – the largest primate to be so. It has disconcertingly wide-set eyes in a ghostly pale face. The fingers are absurdly long: the fingers of an Arthur Rackham witch. 'Absurd' only by human standards, however, for we may be sure those fingers are long for a good reason: aye-ayes with shorter fingers would be penalised by natural selection, even if we don't know why. Natural selection is a strong enough theory to be predictive in this fashion, now that science no longer needs convincing of its truth.

One finger, the middle finger, is unique. Hugely long and thin, even by aye-aye standards, it is used specifically to make holes in dead wood and lever out grubs. Aye-ayes detect prey in wood by drumming with the same long finger, and listening for the changes in tone that betray an insect underneath.* That isn't quite all they use the long middle finger for. At Duke University, which surely has the largest collection of lemurs outside Madagascar, I have seen an aye-aye, with great delicacy and precision, insert the long middle finger up its own nostril – in quest of what, I don't know. The late Douglas Adams wrote a wonderful chapter about the aye-aye in *Last Chance to See,* his travel book about his journeys with the zoologist Mark Carwardine.

> The aye-aye is a nocturnal lemur. It is a very strange-looking creature that seems to have been assembled from bits of other animals. It looks a little like a large cat with a bat's ears, a beaver's teeth, a tail like a large ostrich feather, a middle finger like a long dead twig and enormous eyes that seem to peer past you into a totally different world which exists just over your left shoulder . . . Like virtually everything that lives on Madagascar, it does not exist anywhere else on earth.

What wonderfully pithy writing, how sadly missed its author. Adams and Carwardine's purpose in *Last Chance to See* was to call attention to the plight of endangered species. The 30 or so surviving species of lemurs are relics of a much larger fauna that survived up until Madagascar was invaded by destructive humans about 2,000 years ago.

Madagascar is a fragment of Gondwana (see page 290) which became separated from what is now Africa about 165 million years ago, and finally separated from what became India about 90 million years ago. This order of events may seem surprising but, as we shall see, once India had shaken itself free of Madagascar it moved away unusually fast by the sub-lorisoid standards of plate tectonics.

Setting aside bats (which presumably flew in) and human introductions, Madagascar's terrestrial inhabitants are descendants either of the ancient Gondwana fauna and flora, or of rare immigrants

---

* The same habit, with the same long finger (except it is the fourth finger instead of the third), has convergently evolved in a group of New Guinean marsupials, the striped possum and the trioks (*Dactylopsila*). These marsupials seem to be champion convergers, by the way. They are striped in the same pattern as skunks. And like skunks, they emit a powerful smell in defence.

rafted in with improbable good luck from elsewhere. It is a natural botanical and zoological garden, which houses about five per cent of all the plant and animal land species in the world, more than 80 per cent of them being found nowhere else. Yet, notwithstanding this astonishing richness of species, it is also remarkable for the number of major groups that are totally absent. Unlike Africa or Asia, Madagascar has no native antelopes, no horses or zebras, no giraffes, no elephants, no rabbits, no elephant shrews, no members of the cat or the dog family: none of the expected African fauna at all, although fossil remains suggest that several species of hippo survived until recent times. There are bushpigs which seem to have arrived quite recently, perhaps introduced by humans. (We shall return to the aye-aye and the other lemurs at the end of the tale.)

Madagascar has three members of the mongoose family, which are clearly related to each other and must have arrived in the form of a single founder species from Africa, and subsequently branched. Of these, the most famous is the fossa, a sort of giant mongoose the size of a beagle but with a very long tail. Its smaller relatives are the falanouc, and the fanaloka whose Latin name, confusingly, is *Fossa fossa*. The fossa's own Latin name is something quite different.

There is a group of peculiarly Madagascan rodents, nine genera in all, united in one subfamily, the Nesomyinae. These include a burrowing giant rat-like form, a tree-climber, a tufted-tail 'marsh rat' and a jumping jerboa-like form. It has long been controversial whether these peculiarly Madagascan rodents result from a single immigration event, or several. If there was a single founder, it would mean that its descendants, since arriving in Madagascar, evolved to fill all these different rodent niches: a very Madagascan story. Recent molecular evidence shows that a couple of species on the African mainland are more closely related to some Madagascan rodents than some Madagascan rodents are to each other. This might seem to indicate multiple immigration from Africa. However, a closer look at the evidence supports a more surprising hypothesis. It seems that all the Madagascan rodents are descended from a single founder who arrived, not from Africa but from *India*. If this is right, the affinities with two African rodents would indicate further rafting from Madagascar to Africa. The ancestors of the African species came from India, via Madagascar. It is as though the Indian Ocean favours rafting in a westerly direction. And once again we mustn't forget that

India would have been closer to Madagascar when the immigration happened.

Six out of the eight species of baobab tree are unique to Madagascar, and its count of 130 species of palm trees dwarfs the number found in the whole of Africa (see plate 6). Some authorities think chameleons originated there. Certainly, two-thirds of the world's species of chameleons are native to Madagascar. And there is a peculiarly Madagascan family of shrew-like animals, the tenrecs. Once classified in the Order Insectivora, they are nowadays placed with the Afrotheria whom we shall meet at Rendezvous 13. They probably arrived on Madagascar as two different founder populations from Africa, before any other mammals. They have now diversified into 27 species, including some that resemble hedgehogs, some that resemble shrews, and one that lives largely underwater like a water shrew. The resemblances are convergent – independently evolved, in typical Madagascan fashion. Madagascar being isolated, there were no 'true' hedgehogs and no 'true' water shrews. So tenrecs, who had the good fortune to be on the spot, evolved to become the local equivalents of hedgehogs and water shrews.

Madagascar has no monkeys or apes at all, and that set the scene for the lemurs themselves. By lucky chance, some time later than 63 million years ago, a founder population of early strepsirhine primates accidentally found their way to Madagascar. As usual, we have no idea how this happened. The evolutionary split (Rendezvous 8, at 63 Mya) was later than Madagascar's geographical separation from Africa (165 Mya) and India (88 Mya), so we can't say the lemurs' ancestors were Gondwanan residents sitting there all along. In several places in this book I have used 'rafting' as a kind of shortened code for 'fluke sea-crossing by some means unknown, of great statistical improbability, which only had to happen once, and which we know must have happened at least once because we see the later consequences'. I should add that 'great statistical improbability' is in there for form's sake. The evidence, as we saw at Rendezvous 6, is actually that 'rafting' in this general sense is commoner than intuition would expect. The classic example is the swift recolonisation of the remnants of Krakatoa after it was abruptly destroyed by a catastrophic volcanic event. E. O. Wilson's *The Diversity of Life* has a beautiful account.

In Madagascar, the consequences of the lucky rafting were dramatic and delightful: lemurs great and small, ranging in size from the pygmy mouse lemur, smaller than a hamster, to the recently extinct *Archaeoindris,* which was heavier than a large silverback gorilla and looked like a bear; familiar lemurs like the ringtailed, with their long, striped, hairy-caterpillar tails wafting in the air as the troop runs along the ground; or the indri, or the dancing sifaka which may be the most bipedally accomplished primate after ourselves.

And of course there is the aye-aye, teller of this tale. The world will be a sadder place when the aye-aye goes extinct, as I fear it may. But a world without Madagascar would be not just sadder – it would be impoverished. If you wiped out Madagascar, you would destroy only about a thousandth of the world's total land area, but fully four per cent of all species of animals and plants.

For a biologist, Madagascar is the Island of the Blest. Along our pilgrim voyage, it is the first of five large – in some cases very large indeed – islands, whose isolation, at crucial junctures in Earth history, radically structured the diversity of mammals. And not just mammals. Something similar happens with insects, birds, plants and fish, and when we are eventually joined by more distant pilgrims we shall find other islands playing the same role – not all of them dry land islands. The Cichlid's Tale will persuade us that each of the great African lakes is its own watery Madagascar, and cichlid fishes are its lemurs.

The islands or island continents that have shaped the evolution of mammals are, in the order we shall visit them, Madagascar, Laurasia (the great northern continent which was once isolated from its southern counterpart, Gondwana), South America, Africa, and Australia. Gondwana itself might be added to the list, for, as we shall discover at Rendezvous 15, it too bred its own unique fauna, before it broke up into all our Southern Hemisphere continents. The Aye-Aye's Tale has shown us the faunistic and floristic extravagance of Madagascar. Laurasia is the ancient home, and Darwinian proving-ground, of the huge influx of pilgrims we shall meet at Rendezvous 11, the laurasiatheres. At Rendezvous 12 we shall be joined by a strange band of pilgrims, the xenarthrans, who served their evolutionary apprenticeship on the then island continent of South America, and who will tell us the tale of the others who shared it. At Rendezvous 13 we find the afrotheres, another hugely varied group of mammals, whose diversity was honed on the island continent of Africa. Then, at Rendezvous 14 it is the turn of Australia and the marsupials. Madagascar is the microcosm which sets the pattern – large enough to follow it, small enough to display it in exemplary clarity.

THE GREAT CRETACEOUS CATASTROPHE

Rendezvous 8, where our pilgrims meet the lemurs 63 million years ago, was our last rendezvous 'before', in our backward journey, we burst through the 65-million-year barrier, the so-called K/T boundary, which separates the Age of Mammals from the much longer Age of Dinosaurs that preceded it.* The K/T was a watershed in the fortunes of the mammals. They had been small, shrew-like creatures, nocturnal insectivores, their evolutionary exuberance held down under the weight of reptilian hegemony for more than 100 million years. Suddenly the pressure was released and, in a geologically very short time, the descendants of those shrews expanded to fill the ecological spaces left by the dinosaurs.

What caused the catastrophe itself? A controversial question. At the time there was extensive volcanic activity in India, spewing out lava flows covering well over a million square kilometres (the 'Deccan Traps') which must have had a radical effect on the climate. However, a variety of evidence is building a consensus that the final deathblow was more sudden and more drastic. It seems that a projectile from space – a large meteorite or comet – hit Earth. Detectives proverbially reconstruct events from cigar ash and footprints. The ash in this case is a worldwide layer of the element iridium at just the right place in the geological strata. Iridium is normally rare in the Earth's crust but common in meteorites. The sort of impact we are talking about would have pulverised the incoming bolide, and scattered its remains as dust throughout the atmosphere, from which it would eventually have rained down all over the Earth's surface. The footprint – 100 miles wide and 30 miles deep – is a titanic impact crater, Chicxulub, at the tip of the Yucatan peninsula in Mexico.

* K/T stands for Cretaceous–Tertiary, with 'K' rather than 'C' because 'C' had already been granted by geologists to the Carboniferous Period. Cretaceous comes from *creta*, the Latin for chalk. The German for chalk is Kreide, hence the K. The 'Tertiary' was part of a now defunct system of nomenclature, and covered the first five epochs of the Cenozoic Era. The boundary is now called Cretaceous–Palaeogene (see the Geological Timescale in the General Prologue). Nevertheless, the abbreviation 'K/T' remains in common use, and I will use it here.

Space is full of moving objects, travelling in random directions and at a great variety of speeds relative to one another. There are many more ways in which objects can be travelling at high speeds relative to us than low speeds. So, most of the objects that hit our planet are travelling very fast indeed. Fortunately, most of them are small and burn up in our atmosphere as 'shooting stars'. A few are large enough to retain some solid mass all the way to the planet's surface. And, once in a few tens of millions of years, a very large one catastrophically collides with us. Because of their high velocity relative to Earth, these massive objects release an unimaginably large quantity of energy when they collide. A gunshot wound is hot because of the velocity of the bullet. A colliding meteorite or comet is likely to be travelling even faster than a high-velocity rifle bullet. And where the rifle bullet weighs only ounces, the mass of the celestial projectile that ended the Cretaceous and slew the dinosaurs was measured in gigatons. The noise of the impact, thundering round the planet at a thousand kilometres per hour, probably deafened every living creature not burned by the blast, suffocated by the wind-shock, drowned by the 150-metre tsunami that raced around the literally boiling sea, or pulverised by an earthquake a thousand times more violent than the largest ever dealt by the San Andreas fault. And that was just the immediate cataclysm. Then there was the aftermath – the global forest fires, the smoke and dust and ash which blotted out the sun in a two-year nuclear winter that killed off most of the plants and stopped dead the world's food chains.

No wonder all the dinosaurs, with the notable exception of the birds, perished – and not just the dinosaurs, but about half of all other species too, particularly the marine ones.* The wonder is that any life at all survives these cataclysmic visitations. By the way, the one that ended the Cretaceous and the dinosaurs is not the biggest – that honour falls to the mass extinction that marks the end of the Permian, about a quarter of a billion years ago, in which some 95 per cent of all species went extinct. Recent evidence suggests that an even larger comet or meteorite may have been responsible for that mother of all extinctions. We are uneasily aware that a similar catastrophe could hit us at any moment. Unlike the dinosaurs in the Cretaceous, or the pelycosaurian (mammal-like) reptiles in the Permian, astronomers

---

* It is tempting to see the catastrophe as strangely selective. The deep sea Foraminifera (protozoa in tiny shells which fossilise in enormous numbers and are therefore much used by geologists as indicator species) were almost entirely spared.

would give us several years' warning, or at least months. But this would not be a blessing for, at least with present-day technology, there is nothing we could do to prevent it. Fortunately, the odds that this will happen in any particular person's lifetime are, by normal actuarial standards, negligible. At the same time, the odds that it will happen in *some* unfortunate individuals' lifetime are near certainty. Insurance companies are just not used to thinking that far ahead. And the unfortunate individuals concerned will probably not be human, for the statistical likelihood is that we shall be extinct before then anyway.

A rational case can be mounted that humanity should start research into defensive measures now, to bring the technology up to the point where, if a credible warning were sounded, there would be time to put measures into effect. Present-day technology could only minimise the impact, by storing a suitable balance of seeds, domestic animals, machines including computers and databases full of accumulated cultural wisdom, in underground bunkers with privileged humans (now *there's* a political problem). Better would be to develop so far only dreamed-of technologies to avert the catastrophe by diverting or destroying the intruder. Politicians who invent external threats from foreign powers, in order to scare up economic or voter support for themselves, might find that a potentially colliding meteor answers their ignoble purpose just as well as an Evil Empire, an Axis of Evil, or the more nebulous abstraction 'Terror', with the added benefit of encouraging international co-operation rather than divisiveness. The technology itself is similar to the most advanced 'star wars' weapons systems, and to that of space exploration itself. The mass realisation that humanity as a whole shares common enemies could have incalculable benefits in drawing us together rather than, as at present, apart.

Evidently, since we exist, our ancestors survived the Permian extinction, and later the Cretaceous extinction. Both catastrophes, and the others that have also occurred, must have been extremely unpleasant for them, and they survived by the skin of their teeth, possibly deaf and blind but just capable of reproducing, otherwise we wouldn't be here. Perhaps they were hibernating at the time, and didn't wake up until after the nuclear winter that is thought to follow such catastrophes. And then, in the fullness of evolutionary time, they reaped the benefits. In the case of the Cretaceous survivors, there were now no dinosaurs to eat them, no dinosaurs to compete with them. You might think there was a down side: no dinosaurs for them to eat. But few mammals were large enough, and few dinosaurs small enough, to make that much of a loss. There can be no doubt that the mammals flowered massively after the K/T, but the form of the flowering and how it relates to our rendezvous points is debatable. Three 'models' have been suggested, and now is the time to discuss them. The three shade into each other, and I shall present them in their extreme forms only for simplicity. For reasons of clarity, as I believe, I shall change their usual names to the Big Bang Model, the Delayed Explosion Model, and the Non-explosive Model. There are parallels in the controversy over the so-called Cambrian Explosion, to be discussed in the Velvet Worm's Tale.

1. The Big Bang Model, in its extreme form, sees a single mammal species surviving the K/T catastrophe, a sort of Palaeocene Noah. Immediately after the catastrophe, the descendants of this Noah started proliferating and diverging. On the Big Bang Model, most of the rendezvous points occurred in a bunch, just this side of the K/T boundary – the backwards way of viewing the rapidly divergent branching of the Noah's descendants.

2. The Delayed Explosion Model acknowledges that there was a major explosion of mammal diversity after the K/T boundary. But the mammals of the explosion were not descended from a single Noah, and most of the rendezvous points between mammal pilgrims pre-date the K/T boundary. When the dinosaurs suddenly left the scene, there were lots of little shrew-like lineages who survived to step into their shoes. One 'shrew' evolved into carnivores, a second 'shrew' evolved into primates, and so on. These different 'shrews', although probably quite similar to each other, traced their separate ancestry deep into the past, eventually to unite way back in the Age of Dinosaurs. Those ancestors followed, in parallel, their long fuses into the future through the Age of Dinosaurs to the K/T boundary. Then they all exploded in diversity, more or less simultaneously, when the dinosaurs disappeared. The consequence is that the concestors of modern mammals long pre-date the K/T boundary, although they only started diverging from each other in appearance and way of life after the death of the dinosaurs.

3. The Non-explosive Model doesn't see the K/T boundary as marking any kind of sharp discontinuity in the evolution of mammalian diversity at all. Mammals just branched and branched, and

this process went on before the K/T boundary in much the same way as it went on after it. As with the Delayed Explosion Model, the concestors of modern mammals pre-date the K/T boundary. But in this model they had already diverged considerably by the time the dinosaurs disappeared.

Of the three models, the evidence, especially molecular evidence but increasingly fossil evidence too, seems to favour the Delayed Explosion Model. Most of the major splits in the mammal family tree go way back, deep into dinosaur times. But most of those mammals that coexisted with dinosaurs were pretty similar to each other, and remained so until the removal of the dinosaurs freed them to explode into the Age of Mammals. A few members of those major lineages haven't changed much since those early times, and they consequently resemble each other, even though the common ancestors that they share are extremely ancient. Eurasian shrews and tenrec shrews, for example, are very similar to each other, probably not because they have converged from different starting points but because they haven't changed much since primitive times. Their shared ancestor, Concestor 13, is thought to have lived about 105 million years ago, nearly as long before the K/T boundary as the K/T is before the present.

COLUGOS AND TREE SHREWS

Rendezvous 9 occurs 70 million years into the past, still in the time of the dinosaurs and before the flowering of mammalian diversity properly began. Actually, the flowering of flowers themselves had only just begun. Flowering plants, while diverse, had been previously restricted to disturbed habitats such as those uprooted by elephantine dinosaurs or ravaged by fire, but by now had gradually evolved to include a range of forest-canopy trees and understorey bushes. Concestor 9, which was something like our 10-million-greats-grandparent, was the common ancestor we share with a pair of squirrel-like mammal groups. Well, one of them is squirrel-like and the other more like a flying squirrel. They are the 18 species of tree shrews and the two species of colugos or 'flying lemurs', all from South East Asia.

The tree shrews are all very similar to each other, and are placed in the family Tupaiidae. Most live like squirrels, in trees, and some species resemble squirrels even down to having long, fluffy tails. The resemblance, however, is superficial. Squirrels are rodents. Tree shrews are certainly not rodents. As to what they are, well, that is partly what the next tale will be about. Are they shrews, as their common name would suggest? Are they primates, as certain authorities have long thought? Or are they something else altogether? The pragmatic solution has been to place them in their own, uncertainly placed, mammalian order, the Scandentia (Latin *scandere*, to climb). But in seeking concestor points, we cannot avoid the problem so easily. The Colugo's Tale contains my justification – or apology? – for the solution I have adopted, which is to unite the colugos and the tree shrews 'before' they join our pilgrimage.

Colugos have long been known as flying lemurs, prompting the obvious put-down: they neither fly nor are lemurs. Recent evidence suggests that they are closer to lemurs than was realised even by those responsible for the misnomer. And, while they don't have powered flight like a bat or a bird, they are adept gliders. The two species, *Cynocephalus volans*, the Philippine colugo, and *C. variegatus*, the Malayan colugo, have a whole order to themselves, the Dermoptera. It means 'wings of skin'. Like the flying squirrels of America and

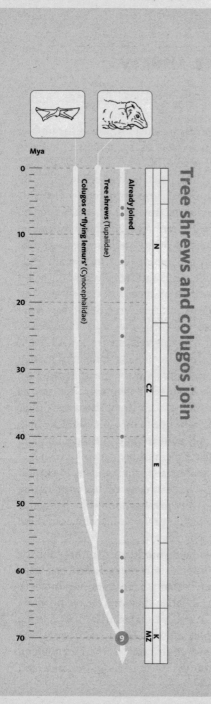

Tree shrews and colugos join

Mya

0
10
20
30
40
50
60
70

Colugos or 'flying lemurs' (Cynocephalidae)

Tree shrews (Tupaiidae)

Already joined

N

CZ

E

K

MZ

9

Tree shrews and colugos join.
This is one of the most uncertain phylogenies in the book (see the Colugo's Tale). The scheme shown here, which groups the 16 species of tree shrew with the two colugos as a sister group to the primates, is advocated by some molecular taxonomists. The dates of this and the next rendezvous are not well established.

Images, left to right: Malayan colugo (*Cynocephalus variegatus*); northern tree shrew (*Tupaia belangeri*).

Eurasia, the more distantly related flying scaly-tailed squirrels of Africa, and the marsupial gliders of Australia and New Guinea, colugos have a single large flap of skin, the patagium, which works a bit like a controlled parachute. Unlike that of the other gliders, the colugo's patagium embraces the tail as well as the limbs, and it extends right to the tips of the fingers and toes. Colugos are also, with a 'wing' span of 70 centimetres, larger than any of those other gliders. Colugos can glide more than 70 metres through the forest at night, to a distant tree, with little loss of height.

The fact that the patagium stretches right to the tip of the tail, and to the tips of the fingers and toes, suggests that the colugos are more deeply committed to the gliding way of life than other mammalian gliders. And indeed, they are pretty inept on the ground. They more than make up for it in the air, where their huge parachute gives them the run of large areas of forest at high speed. This necessitates good stereoscopic vision for steering accurately at night towards a target tree, avoiding fatal collisions, and then making a precision landing. And indeed they have large stereoscopic eyes, excellent for night vision.

Colugos and tree shrews have unusual reproductive systems, but in very different directions. Colugos resemble marsupials in that their young are born early in embryonic development. Having no marsupial pouch, the mother presses the patagium into service. The tail region of the patagium is folded forwards to form a makeshift pouch in which the (usually single) young sits. The mother often hangs upside down from a branch like a sloth, and the patagium then looks and feels like a hammock for the baby.

To be a baby colugo peeping over the edge of a warm, furry hammock sounds appealing. A baby tree shrew, on the other hand, receives perhaps less maternal care than any other baby mammal. The mother tree shrew, at least in several of the species, has two nests, one in which she herself lives, the other in which the babies are deposited. She visits them only to feed them, and then only for the briefest possible time, between five and ten minutes. And she visits them for this brief feed only once in every 48 hours. In the meantime, with no mother to keep them warm as any other baby mammal would have, the little tree shrews need to heat themselves from their food. To this end, the mother's milk is exceptionally rich.

The affinities of the tree shrews and the colugos, to each other and to the rest of the mammals, are subject to dispute and uncertainty.

There is a lesson in that very fact, and it is the lesson of the Colugo's Tale.

THE COLUGO'S TALE

The colugo could tell a tale of nocturnal gliding through the forests of South East Asia. But for the purposes of our pilgrimage it has a more down-to-earth tale to tell, whose moral is a warning. It is the warning that our apparently tidy story of concestors, rendezvous points, and the sequence in which pilgrims join us, is heavily subject to disagreement and revision as new research is done. The phylogeny diagram at Rendezvous 9 shows one recently supported view. According to this view, which I am provisionally accepting here, the pilgrims we primates greet at Rendezvous 9 are an already united band consisting of the colugos and the tree shrews. A few years ago, the colugos would not have entered into this picture. Orthodox taxonomy would have had the tree shrews alone joining the primates at this rendezvous: the colugos would have joined us further down the road, not even very close.

There is no guarantee that our present picture will stay settled. New evidence may resurrect our previous view, or it may prompt a completely different one. Some researchers even think the colugos are closer to the primates than the tree shrews are. If they are right, Rendezvous 9 is where we primates are joined by the colugos. We'd have to wait for the tree shrews at Rendezvous 10, and the numbering of concestors from then on would need to be increased by one. But that is not the view I have adopted. Doubt and uncertainty may seem rather unsatisfactory as the moral for a tale, but it is an important lesson that must be taken on board before our pilgrimage to the past proceeds much further. The lesson will apply to many other rendezvous.

I could have signalled my uncertainty by having multi-way splits ('polytomies': see the Gibbon's Tale) in my phylogenetic trees. This is the solution adopted by certain authors, notably Colin Tudge in his masterly phylogenetic summary of all life on Earth, *The Variety Of Life*. But having polytomies on some branches risks giving false confidence in the others. The revolution in mammalian systematics involving the laurasiatheres and afrotheres (Rendezvous 11 to 13) happened after Tudge's book was published, as recently as 2000, and so some areas of his classification which he considered resolved have now been transformed. Were he to bring out a new edition, it would surely be radically changed. Very possibly the same will happen with this book, and it isn't just the colugos and tree shrews. The position of tarsiers (Rendezvous 7), and the grouping of lampreys with hagfishes (Rendezvous 22) are unsure. The affinities of the afrotheres (Rendezvous 13) and the coelacanths (Rendezvous 19) are still slightly unsure. The ordering of our rendezvous with cnidarians and ctenophores (Rendezvous 28 and 29) could be the wrong way round.

Other rendezvous, such as that with the orang utans, are as near certain as it is possible to be, and there are many more in that happy category. There are also some borderline cases. So, rather than make what comes close to a subjective judgement about which groups deserve fully resolved trees and which do not, I have nailed my more-or-less uncertain colours to the mast in 2004, explaining the doubts in the text whenever possible (apart from a single rendezvous, number 37, where the order is so unsure that even the experts are not willing to hazard a guess). In the fullness of time, I fear that some (but relatively few, I hope) of my rendezvous points and their phylogenies will turn out to be wrong, in the light of new evidence.*

Earlier systems of taxonomy that were not tied to the evolution-standard might be controversial, in the way that matters of taste or judgement are controversial. A taxonomist might argue that, for reasons of convenience in exhibiting museum specimens, tree shrews should be grouped with shrews and colugos with flying squirrels. In such judgements there is no absolutely right answer. The phyletic taxonomy adopted in this book is different. There is a correct tree of life,† but we don't yet know what it is. There is still room for human judgement, but it is judgement about what will eventually turn out to be the undisputable truth. It is only because we haven't looked at enough details yet, especially molecular details, that we are still unsure what that truth is. The truth really is hanging up there waiting to be discovered. The same cannot be said for judgements of taste or of museum convenience.

---

* Creationist misquotation alert: Creationists, please do not quote this as indicating that 'the evolutionists can't agree about anything' with the implication that the whole massive underlying theory can therefore be thrown out.
† With the slight reservation that this tree will actually be a majority consensus among gene trees, as explained in the closing paragraphs of the Gibbon's Tale.

RODENTS AND RABBITKIND

Rendezvous 10 occurs 75 million years into our journey. It is here that our pilgrims are joined – overwhelmed, rather – by a teeming, scurrying, gnawing, whisker-quivering plague of rodents. For good measure, we also greet at this point the rabbits, including the very similar hares and jack-rabbits, and the rather more distant pikas. Rabbits were once classified as rodents, because they also have very prominent gnawing teeth at the front – indeed they outpoint the rodents, with an extra pair. They were then separated off, and are still placed in their own order, Lagomorpha, as opposed to Rodentia. But modern authorities group the lagomorphs together with the rodents in a 'cohort' called Glires. In the terms of this book, the lagomorph pilgrims and the rodent pilgrims joined up with each other 'before' the whole lot of them joined our pilgrimage. Concestor 10 is approximately our 15-million-greats-grandparent. It is the latest ancestor we share with a mouse, but the mouse is connected to it through a very much larger number of greats, because of short generation times.

Rodents are one of the great success stories of mammaldom. More than 40 per cent of all mammal species are rodents, and there are said to be more individual rodents in the world than all other mammals combined. Rats and mice have been the hidden beneficiaries of our own Agricultural Revolution, and they have travelled with us across the seas to every land in the world. They devastate our granaries and our health. Rats and their cargo of fleas were responsible for the

Opposite: **Rodents and rabbits join.** Experts generally accept that the 70 or so species of rabbit relatives and the approximately 2,000 rodents (two-thirds of which are in the mouse family) group together. Recent genetic studies place this group as the sister to the primates, colugos, and tree shrews. Parts of the branching order within the rodents are not entirely established, but a phylogeny similar to this is supported by most molecular data.

Images, left to right: capybara (*Hydrochaeris hydrochaeris*); Cape mole rat (*Georychus capensis*); Cape porcupine (*Hystrix africaeaustralis*); red squirrel (*Sciurus vulgaris*); common dormouse (*Muscardinus avellanarius*); springhare (*Pedetes capensis*); European beaver (*Castor fiber*); bank vole (*Clethrionomys glareolus*); northern birch mouse (*Sicista betulina*); Arctic hare (*Lepus arcticus*); American pika (*Ochotona princeps*).

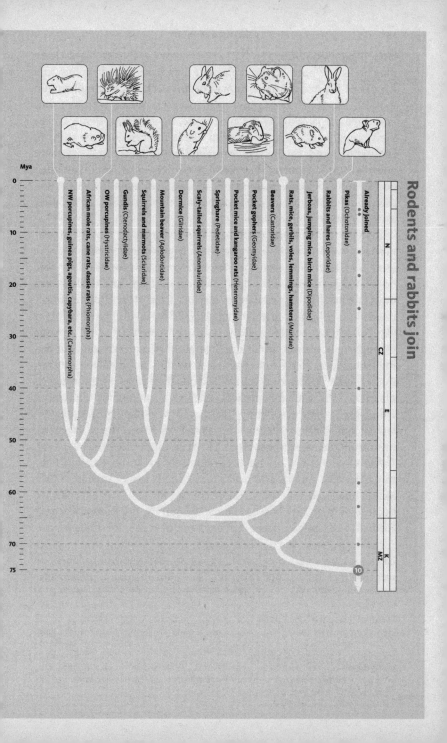

Rodents and rabbits join

Mya

NW porcupines, guinea pigs, agoutis, capybara, etc. (Caviomorpha)

African mole rats, cane rats, dassie rats (Phiomorpha)

OW porcupines (Hystricidae)

Gundis (Ctenodactylidae)

Squirrels and marmots (Sciuridae)

Mountain beaver (Aplodontidae)

Dormice (Gliridae)

Scaly-tailed squirrels (Anomaluridae)

Springhare (Pedetidae)

Pocket mice and kangaroo rats (Heteromyidae)

Pocket gophers (Geomyidae)

Beavers (Castoridae)

Rats, mice, gerbils, voles, lemmings, hamsters (Muridae)

Jerboas, jumping mice, birch mice (Dipodidae)

Rabbits and hares (Leporidae)

Pikas (Ochotonidae)

Already joined

10

Great Plague (traditionally, but now controversially, the Black Death may also have been bubonic plague), they have spread typhus, and have been blamed for more human deaths in the second millennium than all wars and revolutions put together. When even the four horsemen are laid low by the apocalypse, it will be rats that scavenge their remains, rats that will swarm like lemmings over the ruins of civilisation. And, by the way, lemmings are rodents, too – northern voles who, for reasons that are not entirely clear, build up their populations to plague proportions in so-called 'lemming years', and then indulge in frantic – though not wantonly suicidal as is falsely alleged – mass migrations.

Rodents are gnawing machines. They have a pair of very prominent incisor teeth at the front, perpetually growing to replace massive wear and tear. The gnawing masseter muscles are especially well developed in rodents. They don't have canine teeth, and the large gap or diastema that separates their incisors from their back teeth improves the efficiency of their gnawing. Rodents can gnaw their way through almost anything. Beavers fell substantial trees by gnawing through their trunks. Mole rats live entirely underground, tunnelling, not with their front paws like moles, but purely with their incisor teeth.* Different species of rodents have penetrated the deserts of the world (gundis, gerbils), the high mountains (marmots, chinchillas), the forest canopy (squirrels, including flying squirrels), rivers (water voles, beavers, capybaras), rainforest floor (agoutis), savannah (maras, springhares), and Arctic tundra (lemmings).

Most rodents are mouse-sized, but they range up through marmots, beavers, agoutis and maras to the sheep-sized capybaras of the South American waterways. Capybaras are prized for meat, not just because of their large size but because, bizarrely, the Roman Catholic Church traditionally deemed them honorary fish for Fridays, presumably because they live in water. Large as they are, modern capybaras are dwarfed by various giant South American rodents that went extinct only quite recently. The giant capybara, *Protohydrochoerus*, was the size of a donkey. *Telicomys* was an even larger rodent

---

* With the exception of one of the 15 species, they gnaw their way through the earth. The extreme troglodytes among mole rats, the naked mole rats, form tandem trains to mass-produce burrows, kicking back up the line the soil gnawed from the raw earth face by the lead worker. I use 'worker' advisedly, for the naked mole rats are further remarkable in being the nearest approach to social insects that the mammal world has to offer. They even *look* a bit like oversized termites – extraordinarily ugly by our standards, but they themselves are blind so they presumably don't care.

the size of a small rhinoceros which, like the giant capybara, went extinct at the time of the Great American Interchange, when the Isthmus of Panama ended South America's island status. These two groups of giant rodents were not particularly closely related to each other, and seem to have evolved their gigantism independently.

A world without rodents would be a very different world. It is less likely to come to pass than a world dominated by rodents and free of people. If nuclear war destroys humanity and most of the rest of life, a good bet for survival in the short term, and for evolutionary ancestry in the long term, is rats. I have a post-Armageddon vision. We and all other large animals are gone. Rodents emerge as the ultimate post-human scavengers. They gnaw their way through New York, London and Tokyo, digesting spilled larders, ghost supermarkets and human corpses and turning them into new generations of rats and mice, whose racing populations explode out of the cities and into the countryside. When all the relics of human profligacy are eaten, populations crash again, and the rodents turn on each other, and on the cockroaches scavenging with them. In a period of intense competition, short generations perhaps with radioactively enhanced mutation-rates boost rapid evolution. With human ships and planes gone, islands become islands again, with local populations isolated save for occasional lucky raftings: ideal conditions for evolutionary divergence. Within 5 million years, a whole range of new species replace the ones we know. Herds of giant grazing rats are stalked by sabretoothed predatory rats.* Given enough time, will a species of intelligent, cultivated rats emerge? Will rodent historians and scientists eventually organise careful archaeological digs (gnaws?) through the strata of our long-compacted cities, and reconstruct the peculiar and temporarily tragic circumstances that gave ratkind its big break?

THE MOUSE'S TALE

Of all the thousands of rodents, the house mouse, *Mus musculus*, has a special tale to tell because it has become the second most intensively studied mammal species after our own. Much more than the proverbial guinea pig, the mouse is a main staple of medical, physiological and genetic laboratories the world over. In particular,

---

* Dougal Dixon long ago foresaw this, and he had the talent to paint it, in his imaginative book *After Man: A Zoology of the Future*.

the mouse is one of very few mammals apart from ourselves whose genome has so far been completely sequenced.

Two things about these recently sequenced genomes have sparked unwarranted surprise. The first is that mammal genomes seem rather small: of the order of 30,000 genes or maybe even less. And the second is that they are so similar to each other. Human dignity seemed to demand that our genome should be much larger than that of a tiny mouse. And shouldn't it be absolutely larger than 30,000 genes anyway?

This last expectation has led people, including some who should know better, to deduce that the 'environment' must be more important than we thought, because there aren't enough genes to specify a body. That really is a breathtakingly naive piece of logic. By what standard do we *decide* how many genes you need to specify a body? This kind of thinking is based on a subconscious assumption which is wrong: the assumption that the genome is a kind of blueprint, with each gene specifying its own little piece of body. As the Fruit Fly's Tale will tell us, it is not a blueprint, but something more like a recipe, a computer program, or a manual of instructions for assembly.

If you think of the genome as a blueprint, you might expect a big, complicated animal like yourself to have more genes than a little mouse, with fewer cells and a less sophisticated brain. But, as I said, that isn't the way genes work. Even the recipe or instruction-book model can be misleading unless it is properly understood. My colleague Matt Ridley develops a different analogy which I find beautifully clear, in his book *Nature via Nurture*. Most of the genome that we sequence is not the book of instructions, or master computer program, for building a human or a mouse, although parts of it are. If it were, we might indeed expect our program to be larger than the mouse's. But most of the genome is more like the dictionary of words available for writing the book of instructions – or, we shall soon see, the set of subroutines that are called by the master program. As Ridley says, the list of words in *David Copperfield* is almost the same as the list of words in *The Catcher in the Rye*. Both draw upon the vocabulary of an educated native speaker of English. What is completely different about the two books is the order in which those words are strung together.

When a person is made, or when a mouse is made, both embryologies draw upon the same dictionary of genes: the normal vocabulary of mammal embryologies. The difference between a person and a mouse comes out of the different orders with which the genes, drawn from that shared mammalian vocabulary, are deployed, the different places in the body where this happens, and its timing. All this is under the control of particular genes whose business it is to turn other genes on, in complicated and exquisitely timed cascades. But such controlling genes constitute only a minority of the genes in the genome.

Don't misunderstand 'order' as meaning the order in which the genes are strung out along the chromosomes. With notable exceptions, which we shall meet in the Fruit Fly's Tale, the order of genes along a chromosome is as arbitrary as the order in which words are listed in a vocabulary – usually alphabetical but, especially in phrase books for foreign travel, sometimes an order of convenience: words useful in airports; words useful when visiting the doctor; words useful for shopping, and so on. The order in which genes are stored on chromosomes is unimportant. What matters is that the cellular machinery finds the right gene when it needs it, and it does this using methods that are becoming increasingly understood. In the Fruit Fly's Tale, we'll return to those few cases, very interesting ones, where the order of genes arranged on the chromosome is non-arbitrary in something like the foreign phrase-book sense. For now, the important point is that what distinguishes a mouse from a man is mostly not the genes themselves, nor the order in which they are stored in the chromosomal 'phrase-book', but the order in which they are turned on: the equivalent of Dickens or Salinger choosing words from the vocabulary of English and arranging them in sentences.

In one respect the analogy of words is misleading. Words are shorter than genes, and some writers have likened each gene to a sentence. But sentences aren't a good analogy, for a different reason. Different books are not put together by permuting a fixed repertoire of sentences. Most sentences are unique. Genes, like words but unlike sentences, are used over and over again in different contexts. A better analogy for a gene than either a word or a sentence is a toolbox subroutine in a computer.

The computer I happen to be familiar with is the Macintosh, and it is some years since I did any programming so I am certainly out of date with the details. Never mind – the principle remains, and it is true of other computers too. The Mac has a toolbox of routines stored in ROM (Read Only Memory) or in System files permanently loaded at start-up time. There are thousands of these toolbox routines, each one doing a particular operation, which is likely to be needed, over and over again, in slightly different ways, in different programs. For example the toolbox routine called ObscureCursor hides the cursor from the screen until the next time the mouse is moved. Unseen to you, the ObscureCursor 'gene' is called every time you start typing and the mouse cursor vanishes. Toolbox routines lie behind the familiar features shared by all programs on the Mac (and their imitated equivalents on Windows machines): pulldown menus, scrollbars, shrinkable windows that you can drag around the screen with the mouse, and many others.

The reason all Mac programs have the same 'look and feel' (that very similarity famously became the subject of litigation) is precisely that all Mac programs, whether written by Apple, or by Microsoft, or by anybody else, call the same toolbox routines. If you are a programmer who wishes to move a whole region of the screen in some direction, say following a mouse drag, you would be wasting your time if you didn't invoke the ScrollRect toolbox routine. Or if you want to place a check mark by a pulldown menu item, you would be mad to write your own code to do it. Just write a call of CheckItem into your program, and the job is done for you. If you look at the text of a Mac program, whoever wrote it, in whatever programming language and for whatever purpose, the main thing you'll notice is that it consists largely of invocations of familiar, built-in toolbox routines. The same repertoire of routines is available to all programmers. Different programs string calls of these routines together in different combinations and sequences.

The genome, sitting in the nucleus of every cell, is the toolbox of DNA routines available for performing standard biochemical functions. The nucleus of a cell is like the ROM of a Mac. Different cells, for example liver cells, bone cells and muscle cells, string 'calls' of these routines together in different orders and combinations when performing particular cell functions including growing, dividing, or secreting hormones. Mouse bone cells are more similar to human bone cells than they are to mouse liver cells – they perform very similar operations and need to call the same repertoire of toolbox routines in order to do so. This is the kind of reason why all mammal genomes are approximately the same size as each other – they all need the same toolbox.

Nevertheless, mouse bone cells do behave differently from human bone cells; and this too will be reflected in different calls to the toolbox in the nucleus. The toolbox itself is not identical in mouse and man, but it might as well be identical without in principle jeopardising the main differences between the two species. For the purpose of building mice differently from humans, what matters is differences in the calling of toolbox routines, more than differences in the toolbox routines themselves.

THE BEAVER'S TALE

A 'phenotype' is that which is influenced by genes. That pretty much means everything about a body. But there is a subtlety of emphasis which flows from the word's etymology. *Phaino* is Greek for 'show', 'bring to light', 'make appear', 'exhibit', 'uncover', 'disclose', 'manifest'. The phenotype is the external and visible manifestation of the hidden genotype. The *Oxford English Dictionary* defines it as 'the sum total of the observable features of an individual, regarded as the consequence of the interaction of its genotype with its environment' but it precedes this definition by a subtler one: 'A type of organism distinguishable from others by observable features.'

Darwin saw natural selection as the survival and reproduction of certain types of organism at the expense of rival types of organism. 'Types' here doesn't mean groups or races or species. In the subtitle of *The Origin of Species*, the much misunderstood phrase 'preservation of favoured races' most emphatically does not mean races in the normal sense. Darwin was writing before genes were named or properly understood, but in modern terms what he meant by 'favoured races' was 'possessors of favoured genes'.

Selection drives evolution only to the extent that the alternative types owe their differences to genes: if the differences are not inherited, differential survival has no impact on future generations. For a Darwinian, phenotypes are the manifestations by which genes are judged by selection. When we say that a beaver's tail is flattened to serve as a paddle, we mean that genes whose phenotypic expression included a flattening of the tail survived by virtue of that phenotype. Individual beavers with the flat-tailed phenotype survived as a consequence of being better swimmers; the responsible genes survived inside them, and were passed on to new generations of flat-tailed beavers.

At the same time, genes that expressed themselves in huge, sharp incisor teeth capable of gnawing through wood also survived.

Individual beavers are built by permutations of genes in the beaver gene pool. Genes have survived through generations of ancestral beavers because they have proved good at collaborating with other genes in the beaver gene pool, to produce phenotypes that flourish in the beaver way of life.

At the same time again, alternative co-operatives of genes are surviving in other gene pools, making bodies that survive by prosecuting other life trades: the tiger co-operative, the camel co-operative, the cockroach co-operative, the carrot co-operative. My first book, *The Selfish Gene*, could equally have been called *The Co-operative Gene* without a word of the book itself needing to be changed. Indeed, this might have saved some misunderstanding (some of a book's most vocal critics are content to read the book by title only). Selfishness and co-operation are two sides of a Darwinian coin. Each gene promotes its own selfish welfare, by co-operating with the other genes in the sexually stirred gene pool which is that gene's environment, to build shared bodies.

But beaver genes have special phenotypes quite unlike those of tigers, camels or carrots. Beavers have lake phenotypes, caused by dam phenotypes. A lake is an *extended phenotype*. The extended phenotype is a special kind of phenotype, and it is the subject of the rest of this tale, which is a brief summary of my book of that title. It is interesting not only in its own right but because it helps us to understand how conventional phenotypes develop. It will turn out that there is no great difference of principle between an extended phenotype like a beaver lake, and a conventional phenotype like a flattened beaver tail.

How can it possibly be right to use the same word, phenotype, on the one hand for a tail of flesh, bone and blood, and on the other hand for a body of still water, stemmed in a valley by a dam? The answer is that both are manifestations of beaver genes; both have evolved to become better and better at preserving those genes; both are linked to the genes they express by a similar chain of embryological causal links. Let me explain.

The embryological processes by which beaver genes shape beaver tails are not known in detail, but we know the kind of thing that goes on. Genes in every cell of a beaver behave as if they 'know' what kind of cell they are in. Skin cells have the same genes as bone cells, but different genes are switched on in the two tissues. We saw this in the Mouse's Tale. Genes, in each of the different kinds of cells in a beaver's tail, behave as if they 'know' where they are. They cause their respective cells to interact with each other in such a way that the whole tail assumes its characteristically hairless flattened form. There are formidable difficulties in working out how they 'know' which part of the tail they are in, but we understand in principle how these difficulties are overcome; and the solutions, like the difficulties themselves, will be of the same general kind when we turn to the development of tiger feet, camel humps and carrot leaves.

They are also of the same general kind in the development of the neuronal and neurochemical mechanisms that drive behaviour. Copulatory behaviour in beavers is instinctive. A male beaver's brain orchestrates, via hormonal secretions into the blood, and via nerves controlling muscles tugging on artfully hinged bones, a symphony of movements. The result is precise co-ordination with a female, who herself is moving harmoniously in her own symphony of movements, equally carefully orchestrated to facilitate the union. You may be sure that such exquisite neuromuscular music has been honed and perfected by generations of natural selection. And that means selection of genes. In beaver gene pools, genes survived whose phenotypic effects on the brains, the nerves, the muscles, the glands, the bones, and the sense organs of generations of ancestral beavers improved the chances of those very genes passing through those very generations to arrive in the present.

Genes 'for' behaviour survive in the same kind of way as genes 'for' bones, and skin. Do you protest that there aren't 'really' any genes for behaviour; only genes for the nerves and muscles that make the behaviour? You are still wrecked among heathen dreams. Anatomical structures have no special status over behavioural ones, where 'direct' effects of genes are concerned. Genes are 'really' or 'directly' responsible only for proteins or other immediate biochemical effects. All other effects, whether on anatomical or behavioural phenotypes, are indirect. But the distinction between direct and indirect is vacuous. What matters in the Darwinian sense is that *differences* between genes are rendered as *differences* in phenotypes. It is only differences that natural selection cares about. And, in very much the same way, it is differences that geneticists care about.

Remember the 'subtler' definition of phenotype in the *Oxford English Dictionary*: 'A type of organism distinguishable from others by observable features.' The key word is distinguishable. A gene 'for' brown eyes is not a gene that directly codes the synthesis of a brown pigment. Well, it might happen to be, but that is not the point. The point about a gene 'for' brown eyes is that its possession makes a *difference* to eye colour *when compared* with some alternative version of the gene – an 'allele'. The chains of causation that culminate in the difference between one phenotype and another, say between brown and blue eyes, are usually long and tortuous. The gene makes a protein which is different from the protein made by the alternative gene. The protein has an enzymatic effect on cellular chemistry, which affects X which affects Y which affects Z which affects . . . a long chain of intermediate causes which affects . . . the phenotype of interest. The allele makes the *difference* when its phenotype is compared with the corresponding phenotype, at the end of the correspondingly long chain of causation that proceeds from the alternative allele. Gene differences cause phenotypic differences. Gene changes cause phenotypic changes. In Darwinian evolution alleles are selected, vis à vis alternative alleles, by virtue of the differences in their effects on phenotypes.

The beaver's point is that this comparison between phenotypes can happen anywhere along the chain of causation. All intermediate links along the chain are true phenotypes, and any one of them could constitute the phenotypic effect by which a gene is selected: it only has to be 'visible' to natural selection, nobody cares whether it is visible to us. There is no such thing as the 'ultimate' link in the chain: no final, definitive phenotype. Any consequence of a change in alleles, anywhere in the world, however indirect and however long the chain of causation, is fair game for natural selection, so long as it impinges on the survival of the responsible allele, relative to its rivals.

Now, let's look at the embryological chain of causation leading to dam-building in beavers. Dam-building behaviour is a complicated stereotypy, built into the brain like a fine-tuned clockwork mechanism. Or, as if to follow the history of clocks into the electronic age, dam-building is hard wired in the brain. I have seen a remarkable film of captive beavers imprisoned in a bare, unfurnished cage, with no water and no wood. The beavers enacted, 'in a vacuum', all the stereotyped movements normally seen in natural building behaviour when there is real wood and real water. They seem to be placing virtual wood into a virtual dam wall, pathetically trying to build a ghost wall with ghost sticks, all on the hard, dry, flat floor of their prison. One feels sorry for them:

it is as if they are desperate to exercise their frustrated dam-building clockwork.

Only beavers have this kind of brain clockwork. Other species have clockwork for copulation, scratching and fighting, and so do beavers. But only beavers have brain clockwork for dam-building, and it must have evolved by slow degrees in ancestral beavers. It evolved because the lakes produced by dams are useful. It is not totally clear what they are useful for, but they must have been useful for the beavers who built them, not just any old beavers. The best guess seems to be that a lake provides a beaver with a safe place to build its lodge, out of reach for most predators, and a safe conduit for transporting food. Whatever the advantage it must be a substantial one, or beavers would not devote so much time and effort to building dams. Once again, note that natural selection is a predictive theory. The Darwinian can make the confident prediction that, if dams were a useless waste of time, rival beavers who refrained from building them would survive better and pass on genetic tendencies not to build. The fact that beavers are so anxious to build dams is very strong evidence that it benefited their ancestors to do so.

Like any other useful adaptation, the dam-building clockwork in the brain must have evolved by Darwinian selection of genes. There must have been genetic variations in the wiring of the brain which affected dam-building. Those genetic variants that resulted in improved dams were more likely to survive in beaver gene pools. It is the same story as for all Darwinian adaptations. But which is the phenotype? At which link in the chain of causal links shall we say the genetic difference exerts its effect? The answer, to repeat it, is all links where a difference is seen. In the wiring diagram of the brain? Yes, almost certainly. In the cellular chemistry that, in embryonic development, leads to that wiring? Of course. But also *behaviour* – the symphony of muscular contractions that is behaviour – this too is a perfectly respectable phenotype. Differences in building behaviour are without doubt manifestations of differences in genes. And, by the same token, the *consequences* of that behaviour are also entirely allowable as phenotypes of genes. What consequences? Dams, of course. And lakes, for these are consequences of dams. Differences between lakes are influenced by differences between dams, just as differences between dams are influenced by differences between behaviour patterns, which in turn are consequences of differences between genes. We may say that the characteristics of a dam, or of a lake, are true phenotypic effects of genes, using exactly the logic we use to say that the characteristics of a tail are phenotypic effects of genes.

Conventionally, biologists see the phenotypic effects of a gene as confined within the skin of the individual bearing that gene. The Beaver's Tale shows that this is unnecessary. The phenotype of a gene, in the true sense of the word, may extend outside the skin of the individual. Birds' nests are extended phenotypes. Their shape and size, their complicated funnels and tubes where these exist, all are Darwinian adaptations, and so must have evolved by the differential survival of alternative genes. Genes for building behaviour? Yes. Genes for wiring up the brain so it is good at building nests of the right shape and size? Yes. Genes for nests of the right shape and size? Yes, by the same token, yes. Nests are made of grass or sticks or mud, not bird cells. But the point is irrelevant to the question of whether differences between nests are influenced by differences between genes. If they are, nests are proper phenotypes of genes. And nest differences surely must be influenced by gene differences, for how else could they have been improved by natural selection?

Artefacts like nests and dams (and lakes) are easily understood examples of extended phenotypes (see plate 7). There are others where the logic is a little more . . . well, extended. For example, parasite genes can be said to have phenotypic expression in the bodies of their hosts. This can be true even where, as in the case of cuckoos, they don't live inside their hosts. And many examples of animal communication – as when a male canary sings to a female and her ovaries grow – can be rewritten in the language of the extended phenotype. But that would take us too far from the beaver, whose tale will conclude with one final observation. Under favourable conditions the lake of a beaver can span several miles, which may make it the largest phenotype of any gene in the world.

LAURASIATHERES

Eighty-five million years ago, in the hot-house world of the Upper Cretaceous, we greet Concestor 11, approximately our 25-million-greats-grandparent. Here we are joined by a much more diverse band of pilgrims than the rodents and rabbits who swelled our party at Rendezvous 10. Zealous taxonomists recognise their shared ancestry by giving them a name, Laurasiatheria, but it is seldom used because, in truth, this is a miscellaneous bunch. The rodents are all built to the same toothy design and have proliferated and diversified, presumably because it works so well. 'Rodents' therefore really means something strong; it unites animals that have much in common. 'Laurasiatheria' is as awkward as it sounds. It unites highly disparate mammals which have only one thing in common: their pilgrims all joined up with each other 'before' they join us. They all hail, originally, from the old northern continent of Laurasia.

And what a diverse crew these laurasian pilgrims are, some of them flying, some of them swimming, many of them galloping, half of them nervously looking over their shoulder for fear of being eaten by the other half. They belong to seven different orders, the Pholidota (pangolins), Carnivora (dogs, cats, hyenas, bears, weasels, seals, etc.), the Perissodactyla (horses, tapirs and rhinos), Cetartiodactyla (antelopes, deer, cattle, camels, pigs, hippos and . . . well, we'll come to the surprise member of this group later), Microchiroptera and Megachiroptera (respectively small and big bats) and Insectivora (moles, hedgehogs and shrews, but NOT elephant shrews or tenrecs: we have to wait for Rendezvous 13 to meet them).

Carnivora is an irritating name because, after all, it simply means meat-eater, and meat-eating has been invented literally hundreds of times independently in the animal kingdom. Not all carnivores are Carnivora (spiders are carnivores and so was the hoofed *Andrews-archus*, the largest meat-eater since the end of the dinosaurs) and not all Carnivora are carnivores (think of the gentle giant panda, eating almost nothing but bamboo). Within the mammals the order Carnivora does appear to be a genuinely monophyletic clade: that is, a group of animals, all descended from a single concestor who would

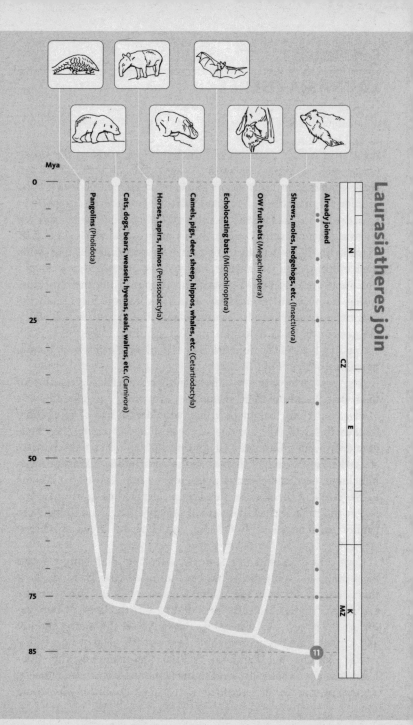

Laurasiatheres join

Mya

0

25

50

75

85

Pangolins (Pholidota)

Cats, dogs, bears, weasels, seals, walrus, etc. (Carnivora)

Horses, tapirs, rhinos (Perissodactyla)

Camels, pigs, deer, sheep, hippos, whales, etc. (Cetartiodactyla)

Echolocating bats (Microchiroptera)

OW fruit bats (Megachiroptera)

Shrews, moles, hedgehogs, etc. (Insectivora)

Already joined

N

CZ

E

K

MZ

11

have been classified as one of them. Cats (including lions, cheetahs and sabretooths), dogs (including wolves, jackals and Cape hunting dogs), weasels and their kind, mongooses and their kind, bears (including pandas), hyenas, wolverines, seals, sea lions and walruses, all are members of the laurasiatherian order Carnivora, and all are descended from a concestor which would have been placed in the same order.

Carnivores and their prey need to outrun each other, and it is not surprising that the demands of fleetness have pushed them in similar evolutionary directions. You need long legs for running, and the great laurasiatherian herbivores and carnivores have, independently and in different ways, added extra length to their legs by commandeering bones which, in us, are inconspicuously buried within the hands (metacarpals) or feet (metatarsals). The 'cannon-bone' of a horse is the enlarged third metacarpal (or metatarsal) fused together with two tiny 'splint' bones that are vestiges of the second and fourth metacarpal (metatarsal). In antelopes and other even-toed ungulates, the cannon-bone is a fusion of the third and fourth metacarpal (metatarsal). Carnivores, too, have elongated their metacarpals and metatarsals, but these five bones have stayed separate instead of fusing together or disappearing altogether, as in horses, cattle and the rest of the so-called ungulates.

Unguis is Latin for nail, and ungulates are animals that walk on their nails – hooves. But the ungulate way of walking has been invented several times and ungulate is a descriptive term rather than a respectable taxonomic name. Horses, rhinos and tapirs are odd-toed ungulates. Horses walk on a single toe, the middle one. Rhinos and tapirs walk on the middle three, as did early horses and some atavistically mutant horses today. Even-toed or cloven-hoofed ungulates walk on two toes, the third and fourth. The convergent

Opposite: **Laurasiatheres join.** In the early 2000s, genetic studies led to a revolution in mammalian taxonomy. According to this new view, there are four major groups of placental mammal. One is our current band (mostly consisting of rodents and primates). Consistently found to be its closest relative is another major group, the 2,000 or so species of laurasiathere. The laurasiathere phylogeny drawn here is considered reasonably certain by proponents of this new classification.

Images, left to right: Cape pangolin (*Manis temminckii*); polar bear (*Ursus maritimus*); Malayan tapir (*Tapirus indicus*); hippopotamus (*Hippopotamus amphibius*); ghost bat (*Macroderma gigas*); Indian flying fox (*Pteropus giganteus*); European hedgehog (*Erinaceus europaeus*).

resemblances between the two-toed cattle family and the one-toed horse family are modest compared to the convergent resemblances of both, separately, to certain extinct South American herbivores. A group called the litopterns independently, and earlier, 'discovered' the horse habit of walking on a single middle toe. Their leg skeletons are almost identical to those of horses. Other South American herbivores, among the so-called notoungulates, independently discovered the cattle/antelope habit of walking on toes three and four. Such stunning resemblances really did fool a senior Argentinian zoologist in the nineteenth century, who thought that South America was the evolutionary nursery of many of our great groups of mammals. In particular, he believed that litopterns were early relatives of the true horse (perhaps with a little national pride that his country might have been the cradle of that noble animal).

The laurasiathere pilgrims now joining us include small animals as well as the large ungulates and carnivores. Bats are remarkable for all sorts of reasons. They are the only surviving vertebrates to put up any sort of competition to birds in flight, and very impressive aerobats they are. With nearly a thousand species, they far outnumber all other orders of mammals except rodents. And bats have perfected sonar (the sound equivalent of radar) to a higher degree than any other group of animals, including human submarine designers.[*]

The other main group of small laurasiatheres are the so-called insectivores. The order Insectivora includes shrews, moles, hedgehogs and other small, snouty creatures which eat insects and small terrestrial invertebrates like worms, slugs and centipedes. As with Carnivora, I shall use a capital letter to denote the taxonomic group, Insectivora, as opposed to insectivore with a small i, which means just anything that eats insects. So, a pangolin (or scaly anteater) is an insectivore but not an Insectivore. A mole is an Insectivore, which actually eats insects. As I have already remarked, it is a pity the early taxonomists used names like Insectivora and Carnivora, which are only loosely correlated to the descriptions of preferred diet with which they are so readily confused.

Related to carnivores like dogs, cats and bears are the seals, sea lions and walruses. We shall soon hear the Seal's Tale, which is about mating systems. I find seals interesting for another reason, too: they

---

[*] I would have liked, at this point, to insert the Bat's Tale, but it would be pretty much the same as a chapter in another of my books, so I won't. Incidentally, I had to exercise similar restraint over 'the Spider's Tale', 'the Fig Tree's Tale' and half a dozen others.

have moved into the water, and have modified themselves in that direction to about half the extent that dugongs have, or whales have. And that reminds me – there is one other major group of laurasiatheres that we haven't dealt with. On to the Hippo's Tale, for a real surprise.

THE HIPPO'S TALE

When I was a schoolboy studying Greek, I learned that *hippos* meant 'horse' and *potamos* 'river'. Hippopotamuses were river horses. Later, when I gave up Greek and read Zoology, I was not too disconcerted to learn that hippopotamuses weren't close to horses after all. Instead, they were classified firmly with pigs, in the middle of the even-toed ungulates or artiodactyls. I have now learned something so shocking that I am still reluctant to believe it, but it looks as though I am going to have to. Hippos' closest living relatives are whales (see plate 8). The even-toed ungulates include whales! Whales, needless to say, don't have hooves at all, whether odd- or even-toed. Indeed, they don't have toes, so it might be less confusing if we adopt the scientific name, artiodactyls (which is actually just the Greek for even-toed, so the change doesn't help much). For completeness, I should add that the equivalent name for the horse order is Perissodactyla (Greek for uneven-toed). Whales, it would now seem from strong molecular evidence, are artiodactyls. But since they previously had been placed in the order Cetacea, and since Artiodactyla was also a well-established name, a new composite has been coined: Cetartiodactyla.

Whales are wonders of the world. They include the largest organisms that have ever moved. They swim with up-and-down movements of the spine derived from the mammalian gallop, as opposed to the side to side wave motion of the spine of a swimming fish or a running lizard. Or presumably a swimming ichthyosaur which, in other respects, looked rather like a dolphin, except for the tell-tale vertical tail, where the dolphin's is horizontal for galloping through the sea. A whale's front limbs are used for steering and stabilising. There are no externally visible hind limbs at all, but some whales have small vestigial pelvic and leg bones buried deep in their bodies.

It would not be too hard to believe that whales are closer cousins to even-toed ungulates than they are to any other mammals. A bit strange, perhaps, but not shocking to accept that some remote ancestor branched to the left and went to sea to give rise to the whales, while it branched to the right to give all the even-toed ungulates. What is shocking is that, according to the molecular evidence, whales are deeply embedded *within* the even-toed ungulates. Hippos are closer cousins to whales than hippos are to anything else including other even-toed ungulates such as pigs.* On their backward journey, the hippo pilgrims and the whale pilgrims unite with each other 'before' the two of them join the ruminants, and then the other even-toed ungulates such as pigs. Whales are the surprise inclusion that I coyly referred to when I introduced the cetartiodactyls at Rendezvous 11. It is known as the Whippo Hypothesis.

All this supposes that we believe the testimony of the molecules.† What do the fossils say? To my initial surprise, the new theory fits quite nicely. Most of the great orders of mammals (though not the subdivisions within them) go a long way back into the age of dinosaurs, as we saw in connection with the Great Cretaceous Catastrophe. Rendezvous 10 (with the rodents and rabbits) and Rendezvous 11 (the one we have just reached) both took place during the Cretaceous Period at the height of the dinosaur regime. But mammals in those days were all rather small, shrew-like creatures, whether their respective descendants were destined to become mice or hippos. The real growth of mammal diversity started suddenly after the dinosaurs went extinct 65.5 million years ago. It was then that the mammals were able to blossom into all economic trades vacated by the dinosaurs. Large body size was just one thing that became possible for mammals only when the dinosaurs were gone. The process of divergent evolution was swift, and a huge range of mammals, of all sizes and shapes, roamed the land within 5 million years of 'liberation'. Five to ten million years later, in the late Palaeocene to early Eocene Epoch, there are abundant fossils of even-toed ungulates.

Another 5 million years later, in the early to middle Eocene, we find a group called the archaeocetes. The name means 'old whales',

---

* By the way, we even had it wrong when we classified hippos most closely with pigs within the artiodactyls. The molecules suggest that the sister group to the hippo–whale clade is the ruminants: the cows, sheep and antelopes. Pigs lie outside them all.

† The molecular evidence for this radical view is what I have referred to in the Gibbon's Tale as a Rare Genomic Change (RGC). Transposable element genes that are highly recognisable are found in particular places in the genome, and are presumably inherited from the hippo–whale ancestor. While this is very strong testimony, it is nevertheless prudent to look at the fossils too.

and most authorities accept that among these animals are to be found the ancestors of modern whales. An early one of these, *Pakicetus* from Pakistan, seems to have spent at least some of its time on land. Later ones include the unfortunately named *Basilosaurus* (unfortunate not because of Basil but because saurus means lizard: when first discovered, *Basilosaurus* was thought to be a marine reptile, and the rules of naming rigidly enforce priority, even though we now know better*). It had an immensely long body, and would have been a good candidate for the giant sea serpent of legend, if only it were not long extinct. Around the time that whales were represented by the likes of *Basilosaurus*, the contemporary hippo ancestors may have been members of a group called the anthracotheres, some reconstructions of which make them look quite like hippos.

Returning to the whales, what of the antecedents of the archaeocetes, before they re-invaded the water? If the molecules are right that whales' closest affinities are with hippos, we might be tempted to seek their ancestors among fossils which show some evidence of herbivory. On the other hand, no modern whale or dolphin is herbivorous. The completely unrelated dugongs and manatees, by the way, show that it is perfectly possible for a purely marine mammal to have a purely herbivorous diet. But whales eat either planktonic crustacea (baleen whales), fish or squid (dolphins and most toothed whales), or large prey such as seals (killer whales). This has led people to look for whale ancestors among carnivorous land mammals, beginning with Darwin's own speculation, sometimes ridiculed though I have never understood why:

> In North America the black bear was seen by Hearne swimming for hours with widely open mouth, thus catching, like a whale, insects in the water. Even in so extreme a case as this, if the supply of insects were constant, and if better adapted competitors did not already exist in the country, I can see no difficulty in a race of bears being rendered, by natural selection, more and more aquatic in their structure and habits, with larger and larger mouths, till a creature was produced as monstrous as a whale (*Origin of Species*, 1859, p. 184).

As an aside, this suggestion of Darwin illustrates an important general point about evolution. The bear seen by Hearne was evidently an

---

* The celebrated Victorian anatomist Richard Owen tried to get the name changed to *Zeuglodon*, and Haeckel followed suit in his phylogeny, reproduced in plate 9. But we are stuck with *Basilosaurus*.

enterprising individual, feeding in an unusual way for its species. I suspect that major new departures in evolution often start in just such a way, with a piece of lateral thinking by an individual who discovers a new and useful trick, and learns to perfect it. If the habit is then imitated by others, including perhaps the individual's own children, there will be a new selection pressure set up. Natural selection will favour genetic predispositions to be good at learning the new trick, and much will follow. I suspect that something like this is how 'instinctive' feeding habits such as tree-hammering in woodpeckers, and mollusc-smashing in thrushes and sea otters, got their start.[*]

For a long time, people looking over the available fossils for a plausible antecedent to the archaeocetes have favoured the mesonychids, a large group of land mammals that flourished in the Palaeocene Epoch, just after the extinction of the dinosaurs. The mesonychids seem to have been largely carnivorous, or omnivorous like Darwin's bear, and they fit with what we all – before the coming of the hippo theory – thought a whale ancestor ought to be. An additionally nice thing about the mesonychids is that they had hooves. They were hoofed carnivores, perhaps a bit like wolves but running on hooves![†] Could they, then, have given rise to the even-toed ungulates, as well as to the whales? Unfortunately, the idea doesn't fit with the hippo theory specifically. Even though the mesonychids seem to be cousins of today's even-toed ungulates (and there are reasons for believing this over and above their hooves) they are no closer to hippos than they are to all the rest of the cloven-hoofed animals. We keep coming back to the molecular shocker: whales are not just cousins of all the artiodactyls, they are buried within the artiodactyls, closer to hippos than hippos are to cows and pigs.

Gathering all this together, we can sketch a forward chronology as follows. Molecular evidence puts the split between camels (plus llamas) and the rest of the artiodactyls at 65 million years, more or less exactly when the last dinosaurs died. Don't imagine, by the way, that the shared ancestor looked anything like a camel. In those days,

---

[*] This idea has a name, the Baldwin Effect, although the idea was independently proposed by Lloyd Morgan in the same year, and by Douglas Spalding earlier. I follow Alister Hardy's way of developing it in *The Living Stream*. For some reason it is a favourite of mystics and obscurantists.
[†] The fearsome *Andrewsarchus* was one of them.

all mammals looked more or less like shrews. But 65 million years ago, the 'shrews' that were going to give rise to camels split from the 'shrews' that were going to give rise to all the rest of the artiodactyls. The split between pigs and the rest (mostly ruminants) took place 60 million years ago. The split between ruminants and hippos took place about 55 million years ago. Then the whale lineage split off from the hippo lineage not long afterwards, say about 54 million years ago, which gives time for primitive whales such as the semi-aquatic *Pakicetus* to have evolved by 50 million years ago. Toothed whales and baleen whales parted company much later, around 34 million years ago, around the time when the earliest baleen whale fossils are found.

Perhaps I was exaggerating a little when I implied that a traditional zoologist like me should be positively upset at the discovery of the hippo–whale connection. But let me try to explain why I was genuinely disconcerted when I first read about it a few years ago. It wasn't just that it was different from what I had learned as a student. That wouldn't have worried me at all, in fact I would have found it positively exhilarating. What worried me, and still does to some extent, was that it seemed to undermine all generalisations that one might wish to make about groupings of animals. The life of a molecular taxonomist is too short to allow a pairwise comparison of every species with every other species. Instead what one does is take two or three whale species, say, and assume that they are representative of whales as a group. It is tantamount to the assumption that the whales are a clade, sharing a common ancestor which is not shared by the other animals with which one is making the comparison. It is assumed not to matter, in other words, which whale you take to stand for all. Similarly, lacking the time to test every species of rodent, say, or artiodactyl, we might take blood* from a rat, and from a cow. It doesn't matter which artiodactyl you take to compare with the representative whale because, yet again, we assume that the artiodactyls are a good clade, so it doesn't make any difference whether we take a cow, a pig, a camel or a hippo.

But now we are told that it does matter. Camel blood and hippo blood really will give a different comparison with whale blood because hippos are closer cousins to whales than they are to camels.

---

* Actually, blood is not the best source of DNA in mammals because, unusually among vertebrates, their red blood corpuscles don't have nuclei.

See where this lands us. If we can't trust the artiodactyls to hang together as a group, represented by any one of their number, how can we be sure that any group will hang together? Can we even assume that hippos hang together, such that it doesn't matter whether we choose a pygmy hippo or a common hippo for comparison with whales? What if whales are closer to pygmy hippos than to common hippos? Actually we probably can rule that out, because fossil evidence suggests that the two hippo genera split apart about as recently as our split from chimpanzees, and that really does leave too little time to evolve all the different kinds of whales and dolphins.

It is more problematical whether all the whales hang together. On the face of it, the toothed whales and the baleen whales might well represent two entirely separate returns to the sea from the land. Indeed, that very possibility has often been advocated. The molecular taxonomists who demonstrated the hippo connection very wisely did take DNA from both a toothed whale and a baleen whale. They found that the two whales are indeed much closer cousins to each other than they are to a hippo. But again, how do we know that 'the toothed whales' hang together as a group? And the same for 'the baleen whales'? Maybe all the baleen whales are related to a hippo except the minke whale, which is related to a hamster. No, I don't believe that, and I really do think the baleen whales are a united clade, sharing a common ancestor which is not shared by anything that is not a baleen whale. But can you see how the hippo–whale discovery shakes the confidence?

We could regain our confidence if we could think of a good reason why whales might be special in this respect. If whales are glorified artiodactyls, they are artiodactyls that suddenly took off, evolutionarily speaking, leaving the rest of the artiodactyls behind. Their closest cousins, the hippos, remained relatively static, as normal, respectable artiodactyls. Something happened in the history of the whales that made them flip into evolutionary overdrive. They evolved so much faster than all the rest of the artiodactyls that their origin within that group was obscured, until molecular taxonomists came along and uncovered it. So, what is special about the history of the whales?

When you write it down like that, the solution leaps off the page. Leaving the land and becoming wholly aquatic was a bit like going into outer space. When we go into space we are weightless (not, by the way, because we are a long way from the Earth's gravity, as many people think, but because we are in free fall like a parachutist before he pulls the ripcord). A whale floats. Unlike a seal or a turtle, which still comes on land to breed, a whale never stops floating. It never has to contend with gravity. A hippo spends time in the water, but it still needs stout, treetrunk-like legs and strong leg muscles for the land. A whale doesn't need legs at all, and indeed it doesn't have any. Think of a whale as what a hippo would like to be if only it could be freed from the tyranny of gravity. And of course there are so many other odd things about living the whole time in the sea that it comes to seem far less surprising that whale evolution should have spurted as it did, leaving hippos behind, stranded on land and stranded in the middle of the artiodactyls. This suggests that I was unduly alarmist a few paragraphs back.

Much the same thing happened in the other direction, 300 million years earlier, when our fish ancestors emerged from the water onto the land. If whales are glorified hippos, we are glorified lungfish. The emergence of legless whales from within the middle of the artiodactyls, leaving the rest of the artiodactyls 'behind', should not seem more surprising than the emergence of four-legged land animals from one particular group of fish, leaving those fish 'behind'. That, at any rate, is how I rationalise the hippo–whale connection, and recover my lost zoological composure.

EPILOGUE TO THE HIPPO'S TALE

Zoological composure be blowed. My attention was drawn to the following while this book was in its final stages of preparation. In 1866, the great German zoologist Ernst Haeckel drew up a schematic evolutionary tree of mammals (see plate 9). I had often seen the full tree reproduced in histories of zoology, but I had never before noticed the position of the whales and hippos in Haeckel's scheme. Whales are 'Cetacea', as today, and Haeckel presciently placed them close to the artiodactyls. But the real stunner is where he put the hippos. He called them by the unflattering name 'Obesa' and he classified them not in the artiodactyls but as a tiny twig on the branch leading to Cetacea.* Haeckel classified hippos as the sister group to

---

* Haeckel didn't get everything right, however. He put the sirenians (dugongs and manatees) in with whales.

the whales: hippos, in his vision, were more closely related to whales than they were to pigs, and all three were more closely related to each other than to cows.

> . . . there is no new thing under the sun. Is there anything whereof it may be said, See, this is new? It hath been already of old time, which was before us.
>
> ECCLESIASTES 1: 9–10

THE SEAL'S TALE

Most wild animal populations have approximately equal numbers of males and females. There's a good Darwinian reason for this, which was clearly seen by the great statistician and evolutionary geneticist R. A. Fisher. Imagine a population in which the numbers were unequal. Now, individuals of the rarer sex will on average have a reproductive advantage over individuals of the commoner sex. This is not because they are in demand and have an easier time finding a mate (although that might be an additional reason). Fisher's reason is a deeper one, with a subtle economic slant. Suppose there are twice as many males as females in the population. Now, since every child born has exactly one father and one mother, the average female must, all other things being equal, have twice as many children as the average male. And vice versa if the population sex ratio is reversed. It is simply a question of allocating the available posterity among the available parents. So, any general tendency for parents to favour sons rather than daughters, or daughters rather than sons, will immediately be counteracted by natural selection for the opposite tendency. The only evolutionarily stable sex ratio is 50/50.

But it isn't quite that simple. Fisher spotted an economic subtlety in the logic. What if it costs twice as much to rear a son, say, as to rear a daughter, presumably because males are twice as big? Well, now, the reasoning changes. The choice that faces a parent is no longer, 'Shall I have a son or a daughter?' It is now, 'Shall I have a son or – for the same price – two daughters?' The balanced sex ratio in the population is now twice as many females as males. Parents who favour sons on the grounds that males are rare, will see their advantage precisely undermined by the extra cost of making males. Fisher divined that the true sex ratio equalised by natural selection is not the ratio of numbers of males to numbers of females. It is the ratio of economic spending on rearing sons to economic spending on rearing daughters. And what does economic spending mean? Food? Time? Risk? Yes, in practice all these are likely to be important, and for Fisher the agent doing the spending was always parents. But economists use a more general expression of cost, which they call opportunity cost. The true cost to a parent of making a child is measured in lost opportunities to make other children. This opportunity cost was named Parental Expenditure by Fisher. Under the name Parental Investment, Robert L. Trivers, a brilliant intellectual successor to Fisher, used the same idea to elucidate sexual selection. Trivers was also the first to understand clearly the fascinating phenomenon of parent–offspring conflict, in a theory that has been carried further in startling directions by the equally brilliant David Haig.

As ever, and at the risk of boring those of my readers not handicapped by a little learning in philosophy, I once again must stress that the purposeful language I have used is not to be treated literally. Parents do not sit down and discuss whether to have a son or a daughter. Natural selection favours, or disfavours, genetic tendencies to invest food or other resources in such a way as to lead eventually to equal or unequal parental expenditure on sons and daughters, over the whole of a breeding population. In practice this will often amount to equal numbers of males and females in the population.

But what about those cases where a minority of males holds the majority of females in harems? Does this violate Fisher's expectations? Or those cases where males parade in front of females in a 'lek', and the females look them over and choose their favourite? Most females have the same favourite, so the end result is the same as for a harem: polygyny – disproportionate access to a majority of females by a privileged minority of males. That minority of males ends up fathering most of the next generation, with the rest of the males hanging about as bachelors. Does polygyny violate Fisher's expectations? Surprisingly, no. Fisher still expects equal investment in sons and daughters, and he is right. Males may have a lower expectation of reproducing at all, but if they do reproduce they reproduce in spades. Females are unlikely to have no children but they are also unlikely to have very many. Even under conditions of extreme polygyny it evens out and Fisher's principle holds.

Some of the most extreme examples of polygyny are to be found among the seals. Seals haul themselves out onto beaches to breed, often in huge 'rookeries', heaving with intense sexual and aggressive activity. In a famous study of elephant seals by the California zoologist Burney LeBoeuf, four per cent of the males accounted for 88 per cent of all copulations seen. No wonder the rest of the males are dissatisfied, and no wonder elephant-seal fights are among the fiercest in the animal kingdom.

Elephant seals are named for their trunks (short, by elephant standards, and used for social purposes only), but it could equally be for their size. Southern elephant seals can weigh 3.7 tonnes, more than some cow elephants. Only the bull seals reach this weight, however, and that is one of the central points of the tale. Cow elephant seals are typically less than a quarter the weight of bulls, by whom they, and the calves, are regularly flattened as the bulls charge about fighting each other.*

Why are males so much bigger than females? Because large size helps them to win harems. Most young seals, of whichever sex, are born to a giant father who won a harem, rather than a smaller male who failed to win a harem. Most young seals, of whichever sex, are born to a relatively small mother whose size was optimised to the business of giving birth and rearing babies, rather than the business of winning fights.

The separate optimisation of male and female characteristics comes about through selection of genes. People are sometimes surprised to learn that the genes concerned are present in both sexes. Natural selection has favoured so-called sex-limited genes. Sex-limited genes are present in both sexes but turned on in only one sex. For example, genes that tell the developing seal: 'If you are male grow very big and fight' are favoured at the same time as genes that say, 'If you are female, grow small and don't fight.' Both classes of genes are passed on to sons and to daughters, but each is expressed in one sex and not the other.

If we look at mammals overall, we notice a generalisation. Sexual dimorphism – meaning a big difference between males and females – tends to be most marked in polygynous species, especially those with a harem-style society. As we've seen, there are good theoretical reasons why this should be so, and we've also seen that the seals and sea lions go farthest out along this particular limb.

---

* Don't be surprised that bull seals flatten calves of their own species. Any calf squashed is no more likely to be the bull's own child than the child of any rival bull. Therefore there is no Darwinian selection against squashing.

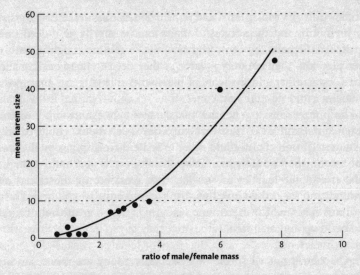

Relationship between sexual dimorphism and harem size. Each point represents a species of seal or sea lion. Adapted from Alexander *et al.* [5].

The graph above comes from a study made by the eminent zoologist Richard D. Alexander of the University of Michigan and his colleagues. Each blob in the graph represents one species of seal or sea lion, and you can see that there is a strong relationship between sexual dimorphism and harem size. In extreme cases, for example southern elephant seals and northern fur seals which are the two blobs at the top of the graph, males may be more than six times the weight of females. And, sure enough, in these species the successful males – a minority, to put it mildly – have huge harems. Two extreme species can't be used to draw general conclusions. But a statistical analysis of the known data for seals and sea lions confirms that the trend we think we see is real (the odds against its being a chance effect are more than 5,000 to 1). There is weaker evidence in the same direction from ungulates and from monkeys and apes.

To repeat the evolutionary rationale for this, males have a great deal to gain, and also a great deal to lose, from fighting other males. Most individuals born, of either sex, are descended from a long line of male ancestors who succeeded in winning harems, and a long line of female ancestors who were members of them. Therefore most individuals, whether male or female, and whether they themselves go on to be winners or losers, inherit genetic equipment for helping male bodies to win harems and female bodies to join them. Size is at a premium, and the successful males can be very large indeed (see plate 10). Females, by contrast, have little to gain from fighting other females, and they are only as large as they need to be to survive and be good mothers. Individuals of both sexes inherit genes that make females avoid fighting and concentrate on child-rearing. Individuals of both sexes inherit genes that make males fight against other males, even at the expense of time that could have been spent helping to rear children. If only males could agree to settle their disputes by the toss of a coin, they would presumably shrink over evolutionary time to the size of the females or smaller, with great economic savings all round, and they could give their time to looking after children. Their surplus mass, which in extreme cases must cost a great deal of food to build up and maintain, is the price they pay to be competitive with other males.

Of course, not all species are like seals. Many are monogamous and the sexes are much more alike. Species in which the sexes are the same size tend, with some exceptions such as horses, not to have harems. Species in which males are markedly bigger than females tend to have harems, or to practise some other form of polygyny. Most species are either polygynous or monogamous, presumably depending on their different economic circumstances. Polyandry (females mated to more than one male) is rare. Among our close relatives, gorillas have a harem-based polygynous breeding system and gibbons are faithfully monogamous. We could have guessed this from their sexual dimorphism, and lack of it respectively. A large male gorilla weighs twice as much as a typical female, while gibbon males and females are approximately equal in size. Chimpanzees are more indiscriminately promiscuous.

Can the Seal's Tale tell us something about our own natural breeding system, before civilisation and custom obliterated the traces? Our sexual dimorphism is moderate but undeniable. Lots of women are taller than lots of men, but the tallest men are taller than the tallest women. Lots of women can run faster, lift heavier weights, throw javelins further, play better tennis, than lots of men. But for humans, unlike for racehorses, the underlying sexual dimorphism precludes sex-blind open competition at the top level in almost any sport you care to name. In most physical sports, every single one of the world's top hundred men would beat every single one of the world's top hundred women.

Even so, by the standards of seals and many other animals, we are only slightly dimorphic. Less so than gorillas, but more than gibbons. Perhaps our slight dimorphism means our female ancestors lived sometimes monogamously, sometimes in small harems. Modern societies vary so much that you can find examples to support almost any preconception. The *Ethnographic Atlas* of G. P. Murdock, published in 1967, is a brave compilation. It lists particulars of 849 human societies, surveyed all over the world. From it we might hope to count numbers of societies that permit harems versus numbers that enforce monogamy. The problem with counting societies is that it is seldom obvious where to draw lines, or what to count as independent. This makes it hard to do proper statistics. Nevertheless, the atlas does its best. Of those 849 societies, 137 (about 16 per cent) are monogamous, four (less than one per cent) are polyandrous, and a massive 83 per cent (708) are polygynous (males can have more than one wife). The 708 polygynous societies are divided about equally into those where polygyny is permitted by the rules of the society but rare in practice, and those where it is the norm. To be brutally precise, of course, 'norm' refers to harem membership for females and harem aspiration for males. By definition, given equal numbers of men and women, the majority of men miss out. The harems of some Chinese Emperors and Ottoman Sultans broke the most extravagant records of the elephant seals and fur seals. Yet our physical dimorphism is small when compared with the seals, and also probably – although this evidence is disputed – when compared with the australopithecines. Would this mean that australopithecine chiefs had harems even larger than Chinese Emperors?

No. We mustn't apply the theory in a naive way. The correlation between sexual dimorphism and harem size is only a loose one. And physical size is only one indicator of competitive strength. For elephant seals, male size is presumably important, because they win their harems by physically fighting other males, biting them or overpowering them by sheer weight of blubber. Size is probably not negligible in hominids. But any sort of differential power, which enables some males to control a disproportionate number of females, can take the place of physical size. In many societies, political clout plays this role. Being a friend of the chief – or, better, being the chief – empowers an individual: enables him to intimidate rivals in a way that is equivalent to the physical intimidation of a large bull seal over a smaller one. Or there may be massive inequalities in economic wealth. You don't fight for wives, you buy them. Or you pay soldiers to fight for them on your behalf. The Sultan or the Emperor may be a physical wimp, yet he may still secure a harem larger than any bull seal. The point I am moving towards is that even if australopithecines were much more dimorphic in size than us, our evolution from them may not, after all, have been a move away from polygyny itself. It may just have been a shift in the weapons used for male competition: from sheer size and brute force to economic power and political intimidation. Or, of course, we also may have shifted towards more genuine sexual equality.

For those of us with a distaste for sexual inequality, it is a consoling hope that cultural polygyny, as distinct from brute-force polygyny, might be rather easy to get rid of. On the face of it, this seems to have happened in those societies, such as (non-Mormon) Christian societies, which became officially monogamous. I say 'on the face of it' and 'officially', because there is also some evidence that apparently monogamous societies are not quite what they seem. Laura Betzig is a historian with a Darwinian turn of mind, and she has uncovered intriguing evidence that overtly monogamous societies like ancient Rome and medieval Europe were really polygynous under the surface. A rich nobleman, or Lord of the Manor, may have had only one legal wife but he had a *de facto* harem of female slaves, or housemaids and tenants' wives and daughters. Betzig cites other evidence that the same was true of priests, even those who were notionally celibate.

These historical and anthropological facts have been seen by some scientists as suggesting, together with our moderate sexual dimorphism, that we evolved under a polygynous breeding regime. But sexual dimorphism is not the only clue we can get from biology. Another interesting signal from the past is testis size.

Our closest relatives, the chimpanzees and bonobos, have extremely large testes. They are not polygynous like gorillas, nor are they monogamous like gibbons. Female chimpanzees in oestrus normally copulate with more than one male. This promiscuous mating pattern is not polyandry, which means the stable bonding of one female with more than one male. It does not predict any simple pattern of sexual dimorphism. But it did suggest to the British biologist Roger Short an explanation for the large testes: chimpanzee genes have been passed down the generations via spermatozoa that had to battle it out in competition with rival sperms from several

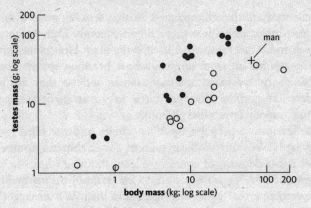

Relationship between testis mass and body mass. Each point represents a species of primate. Adapted from Harvey and Pagel [132].

males inside the same female. In such a world, sheer numbers of spermatozoa matter, and this demands big testes. Male gorillas, on the other hand, have small testes but powerful shoulders and huge resonating chests. Gorilla genes do their competing via male fights and chest-thumping threats to win females, which pre-empts subsequent sperm competition inside females. Chimpanzees compete via sperm proxies inside vaginas. This is why gorillas have pronounced sexual dimorphism and small testes, while chimpanzees have large testes and weak sexual dimorphism.

My colleague Paul Harvey, with various collaborators including Roger Short, tested the idea using comparative evidence from monkeys and apes. They took twenty genera of primates and weighed their testes. Well, actually, they went into the library and gathered published information on testis masses. Large animals obviously tend to have larger testes than small animals, so they had to correct for that. Their method was the one explained in the Handyman's Tale for brains. They placed each monkey or ape genus as a point on a graph (above) of testis mass against body mass and, for the same reasons as we saw in the Handyman's Tale, they took logarithms of both. The points fell around a straight line, from marmosets at the bottom to gorillas at the top. As with the brains, the interesting question was which species have relatively big testes for their size, and which smaller. Of all the points scattered about the line, which ones fall above the line and which below?

The results are suggestive. The filled black symbols all represent animals that are like chimpanzees in that females mate with more than one male, and where there is consequently likely to be sperm competition. The chimpanzee itself is the black blob at the top. The open circles are all from animals whose breeding system does not involve much sperm competition, either because they are harem breeders like gorillas (the open circle to the far right) or they are faithfully monogamous like gibbons.

The separation between the open circles and the filled blobs is satisfying.* We seem to have support for the sperm competition hypothesis. And now, of course, we want to know where we fall on the graph. How big are our testes? Our position on the graph (see small vertical cross) is close to the orang utan. We seem to cluster with the open circles rather than with the black blobs. We are not like chimpanzees, and probably have not had to contend with much sperm competition in our evolutionary history. But this graph says nothing about whether the breeding system of our evolutionary past was like a gorilla's (harem) or like a gibbon's (faithful monogamy). That sends us back to the evidence of sexual dimorphism and anthropology, both of which suggest mild polygyny: a small tendency in the direction of harems.

If there is indeed evidence that our recent evolutionary ancestors were weakly polygynous, I hope it needs no saying that this should not be used to justify a moral or political stance, one way or the other. 'You can't get an ought from an is' has been said so often it is in danger of becoming tedious. It is none the less true for that. Let's hasten on to our next rendezvous.

---

* In plots of this kind, it is important to include only data that are independent of each other, otherwise you can unfairly inflate the result. Harvey and colleagues sought to avoid this danger by counting genera instead of species. It is a step in the right direction, but the ideal solution is that urged by Mark Ridley in *The Explanation of Organic Diversity*, and fully endorsed by Harvey: look at the family tree itself and count neither species nor genera but independent evolutions of the characteristics of interest.

XENARTHRANS

Rendezvous 12, about 95 million years ago in the time of our 35-million-greats-grandparent, is where we meet the xenarthran pilgrims from South America, which at that time had fairly recently torn itself away from Africa and was a very large island – just the thing for fostering the evolution of a unique fauna. The xenarthrans are a rather odd group of mammals, consisting of the armadillos, sloths and anteaters and their extinct relatives. Their name means 'alien joints', referring to the peculiar way their vertebrae join onto each other: they have extra articulations between their lumbar vertebrae, which strengthen the backbone for the digging that so many of them go in for. Among anteaters, only South American ones are xenarthrans. Other mammals such as pangolins and aardvarks also eat ants and are called, respectively, scaly anteaters and ant bears. All 'anteaters', by the way, might just as well be called termite-eaters – they are very fond of termites.

The xenarthrans have a tale to tell of South America, and it falls to the armadillo to tell it. We shall cover the diversity of the Xenarthra themselves during the course of the tale.

THE ARMADILLO'S TALE

Zoologically speaking, South America is a sort of giant Madagascar. Like Madagascar, it split off from Africa, but from the west rather than the east side, around the same time, or a bit later than Madagascar. Like Madagascar, South America was cut off from the rest of the world during most of the period of mammal evolution. Its long purdah, which ended only about 3 million years ago, led to South America becoming a gigantic natural experiment culminating in a unique and fascinating mammal fauna. Like Australia but unlike Madagascar, South America's fauna was rich in marsupials. In South America's case, marsupials filled most of the carnivorous niches. Unlike Australia, South America also had plenty of placental (non-marsupial) mammals, including armadillos and other xenarthrans, and various uniquely South American 'ungulates', now all extinct,

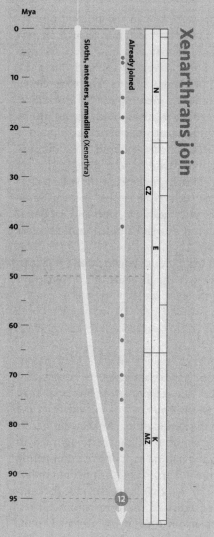

Mya

Xenarthrans join

Sloths, anteaters, armadillos (Xenarthra)

Already joined

N

CZ

E

K
MZ

12

Xenarthrans join. Of the four major placental mammal groups identified by molecular taxonomists, the two earliest branching are the afrotheres (see Rendezvous 13) and the South American xenarthrans (approximately 30 species of sloths, anteaters, and armadillos). Conceivably, further data may reverse the order of Rendezvous 12 and 13, but here we draw the current consensus.

Image: six-banded armadillo (*Euphractus sexcinctus*).

which evolved entirely independently of the even-toed and odd-toed ungulates of the rest of the world.

We have already seen that monkeys and rodents entered South America, probably in separate rafting incidents long after the continent wrenched itself free from Africa. When the monkeys and the rodents arrived, they found a continent already thickly populated with unique mammals. These 'old timers', to borrow from the book *Splendid Isolation* by the great American zoologist G. G. Simpson, belong to three main groups. The Xenarthra are one, and there were some marsupials, which we'll come to later. The remaining old timers can all be loosely called ungulates. As we saw at Rendezvous 11, 'ungulate' is not a taxonomically precise word. These South American old timers did the same herbivorous job as horses, rhinos and camels, but they evolved independently of them.

Unlike Madagascar and Australia, South America's isolation came to an end naturally, before human travel brought all zoological isolation more or less to an end. The rise of the Isthmus of Panama, as recently as 3 million years ago, led to the Great American Interchange. The separate faunas of North and South America were free to travel along the narrow corridor formed by the Isthmus, to each other's continents. This enriched the two faunas, but then some extinctions occurred on both sides, presumably at least partly as a result of competition.

Because of the Great American Interchange, there are now tapirs (odd-toed ungulates) and peccaries (even-toed ungulates) in South America – animals that entered from North America even though tapirs are now extinct there and peccaries greatly reduced. Because of the Interchange, there are now jaguars in South America. Before, there were no cats, or any members of the order Carnivora. Instead, there were carnivorous marsupials, some bearing a fearsome resemblance to the sabretooths (true cats) which were their contemporaries in North America. Since the Interchange there have been armadillos in North America, including glyptodonts – giant armadillos with what look comically like tweed caps on their heads, and formidably spiked clubs on the end of their tails, perhaps wielded against sabretooths, both marsupial and placental. Alas, the glyptodonts went extinct, surprisingly recently, as did the giant ground sloths, lumbering ground-dwelling cousins of today's tree sloths. Ground sloths are often depicted rearing up on their hind legs to feed on trees, and they may also have knocked them down, as elephants do today. The largest of them were, indeed, comparable in size to elephants, six metres long and between three and four tonnes in weight. Ground sloths (though not the very largest of them) penetrated North America as far north as Alaska.

Coming the other way, llamas, alpacas, guanacos and vicuñas, all members of the camel family, are now confined to South America, but camels originally evolved in North America. They spread into Asia and then Arabia and Africa quite recently, presumably via Alaska, where they gave rise to the Bactrian camels of the Mongolian steppe, and the dromedaries of the hot deserts. The horse family, too, did most of their evolving in North America but then went extinct there, which makes poignant the baffled surprise with which the Native Americans responded to the horses, reintroduced from Eurasia under the infamous conquistadores.

The anteaters don't seem to have made it into North America, but three genera survive in South America, and very unusual mammals they are. They have no teeth at all and the skull, especially in the case of *Myrmecophaga*, the large ground-dwelling anteater, has become little more than a long, curved tube, a kind of straw for imbibing ants and termites which are chivvied out of their nests by means of a long sticky tongue. And let me tell you something amazing about them. Most mammals, like us, secrete hydrochloric acid into our stomachs to aid digestion, but South American anteaters don't. Instead, they rely upon the formic acid from the ants that they eat. This is typical of the opportunism of natural selection.

Of the other 'old timers' of South America, the marsupials survive only in the form of the opossums (which are also now common in North America), the very different 'shrew opossums' (confined to the Andes), and the single mouse-like monito del monte (which, strangely enough, seems to have emigrated back to South America from Australia). We shall meet them properly when we get to Rendezvous 14.

The old South American 'ungulates' are all extinct, and more's the pity because they were amazing creatures. Simpson's name 'old timers' only means that their ancestors have been in South America for a very long time, probably since that continent broke free of Africa. They evolved and diversified during the same long period as our more familiar mammals were evolving and diversifying in the Old World. Many of them flourished up to the time of the Great American Interchange and, in some cases, beyond. The litopterns split early into horse-like and camel-like forms, which probably (from the position of the nose bones) had a trunk like an elephant. Another group, the pyrotheres, also probably had a trunk, and may have been quite elephant-like in other respects. They were certainly very large. The South American mammal fauna rather ran to massive rhino-like forms, some of whose fossil bones were first found by Darwin. The notoungulates included huge rhinoceros-like toxodons, and smaller rabbit-like and rodent-like forms.

The Armadillo's Tale is the tale of South America in the Age of Mammals. It is the tale of a gigantic raft, like Madagascar, Australia and India cut adrift by the breakup of Gondwana. Madagascar we have already dealt with in the Aye-Aye's Tale. Australia will be the subject of the Marsupial Mole's Tale. India would have been a fourth raft experiment, except that it travelled north so fast that it reached Asia rather early, and so its fauna became integrated with that of Asia during the latter half of the Age of Mammals. Africa, too, was a gigantic island during the rise of the mammals, not so isolated as South America and not for so long. But long enough for a large and very diverse group of mammals to go their own way in isolation, closer cousins to each other than to the rest of the mammals, though you'd never guess it to look at them. These are the Afrotheria, and we are about to meet them at Rendezvous 13.

AFROTHERES

The Afrotheria are the last placental mammals to join our pilgrimage. They originated in Africa as their name suggests, and they include the elephants, the elephant shrews, the dugongs and manatees (also known as sea elephants or sea cows), the hyraxes, the aardvarks or ant bears, and probably the tenrecs of Madagascar and the golden moles of southern Africa. The next pilgrims we shall greet will be our far more distant cousins the marsupials, so the Afrotheria – all of them equally – are our most distant non-marsupial cousins. Concestor 13 lived 105 million years ago, and was our 45-million-greats-grandparent, or thereabouts. Once again, it looked similar to Concestor 12 and Concestor 11, all pretty much like shrews.

I never saw an elephant shrew until I revisited the beautiful country of Malawi which, as Nyasaland, had been my childhood home. My wife and I spent some time in the Mvuu Game Reserve, just south of the great Rift Valley lake which gave the country its name and on whose sandy beaches I had spent my first bucket-and-spade holidays long ago. In the game reserve, we benefited from our African guide's encyclopaedic knowledge of the animals, his sharp eye for spotting them, and his engaging turns of phrase in calling them to our attention. Elephant shrews always elicited from him the same joke, which seemed to improve with each repeating: 'One of the small five.' (See plate 11.)

Elephant shrews, which are named for their long trunk-like noses, are larger than European shrews, and they run higher, on longer legs – a little suggestive of miniature antelopes. The smaller of the 15 species jump. Elephant shrews used to be more numerous and diverse, and included some plant-eating species as well as the

Opposite: **Afrotheres join.** The new phylogeny of placental mammals recognises the split between the 70 or so afrothere species and all other placentals as the earliest division within the group. However, the order of Rendezvous 12 and 13 is not entirely resolved. Within the Afrotheria, there is still debate about the order of branching between the elephants, sirenians, and hyraxes; the position of the aardvark; and the tenrecs and golden moles.

Images, left to right: Cape elephant shrew (*Elephantulus edwardii*); Grant's golden mole (*Eremitalpa granti*); aardvark (*Orycteropus afer*); West Indian manatee (*Trichechus manatus*); African elephant (*Loxodonta africana*); rock hyrax (*Procavia capensis*).

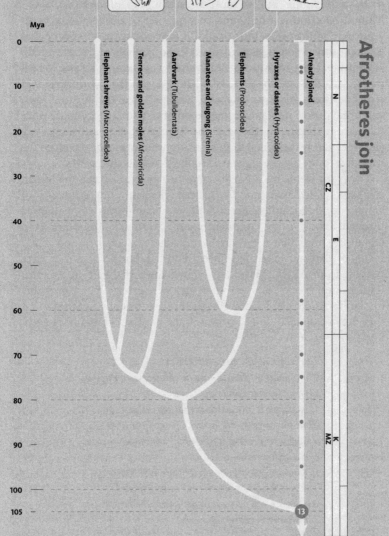

Mya

Afrotheres join

Elephant shrews (Macroscelidea)

Tenrecs and golden moles (Afrosoricida)

Aardvark (Tubulidentata)

Manatees and dugong (Sirenia)

Elephants (Proboscidea)

Hyraxes or dassies (Hyracoidea)

Already joined

N

CZ

E

K

MZ

13

insectivorous ones that survive today. Elephant shrews have the prudent habit of devoting time and attention to making runways for themselves to use later when escaping predators. This sounds like foresight and so, in a way, it is. But it should not be taken to imply deliberate intention (although, as always, that cannot be ruled out). Animals often behave as if they know what is good for them in the future, but we must be careful not to forget that 'as if'. Natural selection is a beguiling counterfeiter of deliberate purpose.

In spite of their dear little trunks, it never occurred to anyone that elephant shrews might be particularly close to elephants. It was always assumed that they were just African versions of Eurasian shrews. Recent molecular evidence, however, astonishes us with the information that elephant shrews are closer cousins to elephants than they are to shrews, and some people now prefer their alternative name, sengi, to distance them from shrews. By the way, the 'trunks' of elephant shrews are almost certainly incidental to their relationship to elephants. From the small five to the big, we come next to the elephants themselves.

Today the elephants are reduced to two genera: *Elephas* the Indian elephant, and *Loxodonta* the African elephant, but elephants of various kinds including mastodons and mammoths once roamed through almost every continent except Australia. There are even tantalising hints that they may have made it to Australia too. Fragments of elephant fossils have been reported, but perhaps they were flotsam from Africa. Mastodons and mammoths were in America until around 12,000 years ago when they were exterminated, probably by the Clovis people. Mammoths died out so recently in Siberia that they are occasionally found frozen in the permafrost and have even, poets sing, been made into soup:

> THE FROZEN MAMMOTH
> *This Creature, though rare, is still found to the East*
> *Of the Northern Siberian Zone.*
> *It is known to the whole of that primitive group*
> *That the carcass will furnish an excellent soup,*
> *Though the cooking it offers one drawback at least*
> *(Of a serious nature I own):*
> *If the skin be but punctured before it is boiled*
> *Your confection is wholly and utterly spoiled.*
> *And hence (on account of the size of the beast)*
> *The dainty is nearly unknown.*

HILAIRE BELLOC

As with all of the Afrotheria, Africa is the ancient home of elephants, mastodons and mammoths, the root of their evolution and the site of most of their diversification. Africa has also become the home of lots of other mammals such as antelopes and zebras, and the carnivores that prey on them, but those are laurasiatheres, who came into Africa later, from the great northern continent of Laurasia. The Afrotheria are the African 'old timers'.

The elephant order is called Proboscidea, after their long proboscis or trunk, which is an enlarged nose. The many purposes for which the trunk is used include drinking, which may have been its primitive use. Drinking, when you are a very tall animal like an elephant or a giraffe, is a problem. Food, for elephants and giraffes, mostly grows on trees which may be partly why they are so tall in the first place. But water finds its own level, which tends to be uncomfortably low. Kneeling down to the water is a possibility. Camels do it. But it is hard work getting up again, even more so for elephants or giraffes. Both solve the problem by sucking the water up through a long siphon. Giraffes stick their head on the end of the siphon – the neck. Giraffe heads therefore have to be rather small. Elephants keep their head – which can therefore be bigger and brainier – at the base of the siphon. Their siphon is, of course, the trunk, and it comes in handy for lots of other things as well. I have elsewhere quoted Oria Douglas-Hamilton on the elephant trunk. Much of her life has been devoted, with her husband Iain, to the study and preservation of wild elephants. It is an angry passage, prompted by the horrifying sight of a mass 'cull' of elephants in Zimbabwe.

> I looked at one of the discarded trunks and wondered how many millions of years it must have taken to create such a miracle of evolution. Equipped with fifty thousand muscles and controlled by a brain to match such complexity, it can wrench and push with tonnes of force. Yet, at the same time, it is capable of performing the most delicate operations such as plucking a small seed-pod to pop in the mouth. This versatile organ is a siphon capable of holding four litres of water to be drunk or sprayed over the body, as an extended finger and as a trumpet or loud speaker. The trunk has social functions, too; caresses, sexual advances, reassurances, greetings and mutually inter-twining hugs . . . And yet there it lay, amputated like so many elephant trunks I had seen all over Africa.

Proboscideans also run to tusks, which are greatly enlarged incisor teeth. Modern elephants have tusks only in the upper jaw, but some extinct proboscideans had tusks in the lower jaw as well, or instead. *Deinotherium* had large down-curving tusks in the lower jaw and no tusks at all in the upper jaw. *Amebelodon*, a North American member of the large group of early proboscideans called gomphotheres, had elephant-like tusks in the upper jaw and flat, spade-like tusks in the lower jaw. Perhaps they were indeed used as spades, to dig up tubers (see plate 12). This speculation doesn't, by the way, conflict with the one about the evolution of the trunk as a siphon to obviate the need to kneel for a drink. The lower jaw, with its two flat spades on the end, was so long that a standing gomphothere could easily have used it to dig in the ground.

In *The Water Babies*, Charles Kingsley wrote that the elephant 'is first cousin to the little hairy coney of Scripture . . .' The primary meaning of coney in the English dictionary is rabbit, and two of the word's four occurrences in the Bible are explanations of why the rabbit is not kosher: 'And the coney, because he cheweth the cud, but divideth not the hoof; he is unclean unto you' (Leviticus 11:5 and the very similar passage in Deuteronomy 14:7). But Kingsley can't have meant rabbit, because he goes on to say that the elephant is 13th or 14th cousin to the rabbit. The other two biblical references refer to an animal that lives among rocks: Psalm 104 ('The high hills are a refuge for the wild goats; and the rocks for the conies') and Proverbs 30:26 ('The conies are but a feeble folk, yet make they their houses in the rocks'). Here, it is generally agreed, coney means hyrax, dassie or rock badger, and Kingsley, that admirably Darwinian clergyman, was right.

Well, he was right at least until those tiresome modern taxonomists burst in. Textbooks say that the elephants' closest living cousins were the hyraxes, which agrees with Kingsley. But recent analysis shows that we must also include dugongs and manatees in the mix, perhaps even as the closest living relatives of the elephants, with hyraxes the sister group. Dugongs and manatees are purely marine mammals who never come ashore even to breed, and it looks as though we were misled in the same way as over the hippos and the whales. Purely marine mammals are freed from the constraints of terrestrial gravity and can evolve rapidly in their own special direction. Hyraxes and elephants, left behind on the land, have remained more similar to each other, just as hippos and pigs did. With hindsight, dugongs' and manatees' slightly trunk-like nose and small eyes in a wrinkled face give them a faintly elephantine appearance, but it is probably an accident.

Dugongs and manatees belong to the order Sirenia. The name comes from their supposed resemblance to the sirens of myth, although this is not, it has to be said, very convincing. Their slow, sleepy-lagoon style of swimming could perhaps have been thought mermaid-like, and they suckle their young with a pair of breasts under the flippers. But one can't help feeling that the sailors who first spotted the likeness must have been at sea for a very long time. Sirenians are, with whales, the only mammals that never come on land at any time. One species, the Amazon manatee, lives in freshwater; the other two manatees are found in the sea as well. Dugongs are exclusively marine, and all four species are vulnerable to extinction, which inspired my wife to design a T-shirt: Dugoing Dugong Dugone. A heart-rending story concerns the fifth species, the enormous Steller's sea cow, which lived in the Bering Straits and weighed over 5 tonnes. It was hunted to extinction a mere 27 years after its discovery by Bering's ill-fated crew in 1741, showing how vulnerable sirenians can be.

As with whales and dolphins, the front limbs of sirenians have become flippers and they have no hind limbs at all. Sirenians are also known as sea cows, but they are not related to cows and they don't ruminate. Their vegetarian diet requires an immensely long gut and a low energy budget. The high-speed aquabatics of a carnivorous dolphin contrast dramatically with the lazy drifting of the vegetarian dugong: guided missile to dirigible balloon.

There are small afrotheres, too. Golden moles and tenrecs seem to be related to each other, and most modern authorities place them within the afrotheres. Golden moles live in southern Africa where they do the same job as moles do in Eurasia, and do it beautifully, swimming through the sand as if it were water. Tenrecs live mostly in Madagascar. There are some semi-aquatic 'otter shrews' that are actually tenrecs in western Africa. As we saw in the Aye-Aye's Tale, Madagascan tenrecs include shrew-like forms, hedgehog-like forms and an aquatic species which probably returned to the water independently of the African ones.

MARSUPIALS

Here we are, 140 million years ago, at the base of the Cretaceous when Concestor 14, in round figures our 80-million-greats-grandparent, lived in the shadow of the dinosaurs. As the Elephant Bird's Tale will recount, South America, Antarctica, Australia, Africa, and India, which had been part of the great southern supercontinent of Gondwana, were just starting to break apart (a map of approximately this period is shown in plate 19). Consequently, changes in climate had plunged the world into a (geologically) short-lived cold period, with snow and ice blanching the poles during the winter months. Only a few flowering plants grew in the temperate forests of coniferous trees and the plains of ferns that covered the northern and southern parts of the globe, and there were correspondingly few of the pollinating insects that we know today. It is in such a world that the entire massed pilgrims of the placental mammals – horses and cats, sloths and whales, bats and armadillos, camels and hyenas, rhinoceroses and dugongs, mice and men – all now represented by a small insectivore, greet the other great group of mammals, the marsupials.

Marsupium means pouch in Latin. Anatomists use it as a technical term for any pouch, such as the human scrotum. But the most famous pouches in the animal kingdom are those in which kangaroos and other marsupials keep their young. Marsupials are born as tiny embryos equipped only to crawl – crawl for their tiny lives

Opposite: **Marsupials join.** Three major lines of living mammals are recognised, based upon their method of reproduction. These are the egg-laying mammals (monotremes), the pouched mammals (marsupials) and the placental mammals (including ourselves). Morphology and most DNA studies agree in grouping the marsupials and placentals together, making Rendezvous 14 the divergence of the 270 or so species of marsupial from the 4,500 or so placental mammals. It is generally accepted that marsupials fall into the seven orders shown here. Their interrelationships are not firmly established: particularly problematic is the position of the South American 'monito del monte'.

Images, left to right: red kangaroo (*Macropus rufus*); Tasmanian devil (*Sarcophilus harrisii*); southern marsupial mole (*Notoryctes typhlops*); bilby (*Macrotis lagotis*); Virginia opossum (*Didelphis virginiana*).

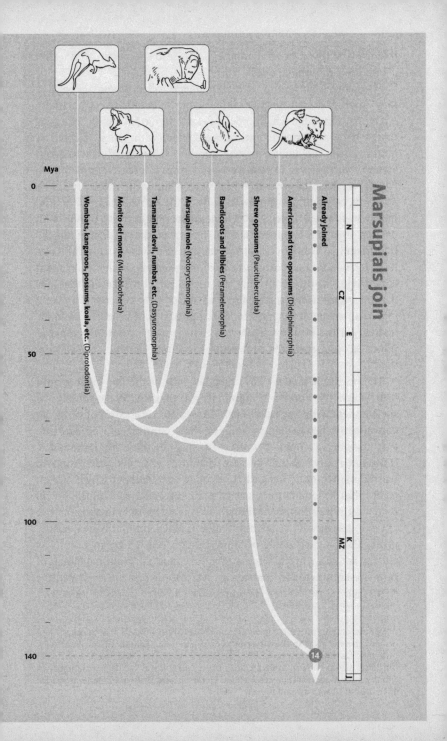

Marsupials join

Mya

0

50

100

140

Wombats, kangaroos, possums, koala, etc. (Diprotodontia)

Monito del monte (Microbiotheria)

Tasmanian devil, numbat, etc. (Dasyuromorphia)

Marsupial mole (Notoryctemorphia)

Bandicoots and bilbies (Peramelemorphia)

Shrew opossums (Paucituberculata)

American and true opossums (Didelphimorphia)

Already joined

14

N

CZ

E

K

MZ

through the forest of their mother's fur, into the pouch where they clamp their mouths to a teat.

The other main group of mammals are called placentals because they nourish their embryos with various versions of a placenta: a large organ through which miles of capillary blood vessels belonging to the baby are brought into close contact with miles of capillary blood vessels belonging to the mother. This excellent exchange system (for it serves to remove wastes from the foetus as well as to feed it) enables the baby to be born very late in its career. It enjoys the protection of its mother's body until, in the case of hoofed herbivores for example, it is capable of keeping up with the herd on its own legs, and even running away from predators. Marsupials do it differently. The pouch is like an external womb, and the large teat, to which the baby becomes attached as a semipermanent appendage, works a bit like an umbilical cord. Later, the joey detaches itself from the nipple and sucks only occasionally like a placental infant. It emerges from the pouch as if in a second birth, and uses it less and less frequently as a temporary refuge. Kangaroo pouches open forwards but many marsupial pouches open backwards.

The marsupials, as we have seen, are one of the two great groups into which the surviving mammals are divided. We normally associate them with Australia which, from a faunistic point of view, can conveniently be deemed to embrace New Guinea. It is unfortunate that no widely recognised word exists to unite these two landmasses. 'Meganesia' and 'Sahul' are not memorable or evocative enough. Australasia won't do because it includes New Zealand which, zoologically, has little in common with Australia and New Guinea. I shall coin Australinea for my purpose.* An Australinean animal hails from mainland Australia, Tasmania or New Guinea, but not New Zealand. From a zoological point of view, though not a human one, New Guinea is like a tropical wing of Australia, and the mammal faunas of both are dominated by marsupials. Marsupials also, as we saw in the Armadillo's Tale, have a long and older history of association with

---

* The Australinean fauna extends a bit beyond New Guinea towards Asia. Wallace's Line, named after the great co-discoverer of natural selection, separates the predominantly Australian fauna from that of Asia. Surprising as it might seem, the line passes between two small islands of the Indonesian archipelago, Lombok and Bali, which are separated only by a fairly narrow (but deep) strait. Further north, Wallace's Line separates the larger islands of Sulawesi and Borneo.

South America, where they still occur, mostly in the shape of some dozens of species of opossums.

Although present-day American marsupials are nearly all opossums, it hasn't always been so. If we take fossils into account, most of the range of marsupial diversity is in South America. Older fossils have been found in North America, but the oldest of all marsupial fossils is from China. They went extinct in Laurasia but survived in two of the main relics of Gondwana, namely South America and Australinea. And it is Australinea which is the main stage for modern marsupial diversity. It is generally agreed that marsupials came to Australinea from South America via Antarctica. Those fossil marsupials that have been found in Antarctica are not in themselves plausible ancestors of the Australinean forms, but this is probably just because so few Antarctic fossils have been found anyway.

It happens that Australinea, for much of its history since it split off from Gondwana, has had no placental mammals. It is not unlikely that all Australia's marsupials stem from a single introduction of an opossum-like founder animal from South America, via Antarctica. We don't know exactly when, but it can't have been much later than 55 million years ago, which is approximately when Australia (more especially Tasmania) pulled far enough away from Antarctica to be inaccessible to island-hopping mammals. It could have been much earlier, depending upon how inhospitable Antarctica was to mammals. American opossums are no more closely related to the animals that Australians call possums than they are to any other Australian marsupials. Other American marsupials, mostly fossils, seem to be more distantly related. Most of the major branches in the marsupial family tree, in other words, are American, which is one reason why we think the marsupials originated in America and migrated to Australinea, rather than the other way around. But the Australinean branch of the family diversified mightily after their homeland became isolated. The isolation came to an end around 15 million years ago when Australinea (specifically New Guinea) reached close enough to Asia to allow the arrival of bats and (presumably island-hopping) rodents. Then, much more recently, dingos arrived (in trading canoes, we must suppose) and finally a whole host of other animals, such as rabbits, camels and horses, introduced by European immigrants. Most ridiculously of all, these included foxes in order that they should be hunted – an eloquent comment on the claim that this pursuit can be justified as pest control.

Together with the monotremes who join us next, the evolving Australian marsupials were carried, on the great raft that Australia became, out into South Pacific isolation. There, for the next 40 million years the marsupials (and monotremes) had Australia to themselves. If there were other mammals at the beginning* they died out early. Dinosaur shoes were waiting to be filled, in Australia as well as in the rest of the world. From our point of view the exciting thing about Australia is that it was isolated for a very long time, and it had a very small founder population of marsupial mammals, conceivably even a single species.

And the results? They were dazzling. Of approximately 270 surviving species of marsupial in the world, about three-quarters are Australinean (the rest are all American, mostly opossums plus a few other species such as the enigmatic *Dromiciops*, the monito del monte). The 200 Australinean species (give or take a few depending on whether we are lumpers or splitters)† have branched to fill the whole range of 'trades' formerly occupied by the dinosaurs, and independently occupied by other mammals in the rest of the world. The Marsupial Mole's Tale goes through some of these trades, one by one.

THE MARSUPIAL MOLE'S TALE

There is a living to be made underground, a living which moles (family Talpidae) have made familiar to us in Eurasia and North America. Moles are dedicated burrowing machines, their hands modified into spades, their eyes, which would be useless underground, all but completely degenerate. In Africa, the mole niche is filled by golden moles (family Chrysochloridae). These are superficially very similar to Eurasian moles and for years they have been placed in the same order: Insectivora. In Australia, as we would expect, the niche is filled by a marsupial, *Notoryctes*, the marsupial mole.‡

---

* A couple of teeth which seem to belong to condylarths (a group of extinct placental mammals) have been found, but nothing younger than 55 million years.
† These more-or-less self-evident terms have become technical terms for taxonomists who habitually lump animals (or plants) into a few large groups, or who habitually split them into lots of small groups. Splitters proliferate names, in extreme cases where fossils are concerned, elevating almost every specimen they discover to species status.
‡ *Necrolestes*, a South American marsupial of the Miocene Epoch, also appears to have been a 'mole'. Its name, rather inappropriately, translates as 'grave robber'.

Marsupial moles look like true (talpid) moles and golden moles, they feed on worms and insect larvae like true moles and golden moles, and they burrow like true moles and even more like golden moles. True moles leave an empty tunnel behind them as they dig looking for prey. Golden moles, at least the ones that live in deserts, 'swim' through sand, which collapses behind them, and marsupial moles do the same. Evolution has fashioned the 'spades' of talpid moles out of all five fingers of the hand. Marsupial moles and golden moles use two (or in some golden moles three) claws. The tail is short in talpid moles and marsupial moles, and completely invisible in golden moles. All three are blind and have no visible ears. Marsupial moles (it's what marsupial MEANS) have a pouch, in which the prematurely born (by placental standards) young are housed.

The similarities of these three 'moles' are convergent: independently evolved for their digging habit, from different beginnings, from non-digging ancestors. And it is a three-way convergence. Although golden moles and Eurasian moles are more closely related to each other than either is to marsupial moles, their common ancestor was surely not a specialised burrower. All three resemble each other because they all dig. Incidentally, we are so used to the idea that mammals stepped into dinosaur shoes, it is surprising to reflect that no dinosaurian 'mole' has so far been found. Both fossilised burrows and special organs adapted for digging have been described for the 'mammal-like reptiles' that preceded the dinosaurs, but never convincingly for dinosaurs themselves.

Australinea is home not just to marsupial moles but to a dramatic cast list of marsupials, each of which plays more or less the same role as a placental mammal on another continent. There are marsupial 'mice' (better called marsupial shrews because they eat insects), marsupial 'cats', 'dogs', 'flying squirrels', and a gallery of counterparts to animals familiar in the rest of the world. In some cases the resemblance is very striking. Flying squirrels such as *Glaucomys volans* of the American forests look and behave very like such denizens of the Australian eucalypt forests as the sugar glider (*Petaurus breviceps*) or the mahogany glider (*Petaurus gracilis*). These are also called 'flying phalangers', although actually they are not members of the family Phalangeridae (cuscuses and brushtail possums). The American flying squirrels are true squirrels, related to our familiar tree squirrels. In Africa, interestingly, the flying squirrel trade is plied by the so-called scaly-tailed squirrels or Anomaluridae which, though also rodents, are not true squirrels. The marsupials of Australia, too, have produced three lineages of gliders, which evolved the habit independently. Returning to placental gliders, we have already met, at Rendezvous 9, the mysterious 'flying lemurs' or colugos which differ from the flying squirrels and the marsupial gliders in that their tail is included in the gliding membrane, as well as all four limbs.

Thylacinus, the Tasmanian wolf, is one of the most famous examples of convergent evolution. Thylacines are sometimes called Tasmanian tigers because of their striped backs, but it is an unfortunate name. They are much more like wolves or dogs. They were once common all over Australia and New Guinea, and they survived in Tasmania until living memory. There was a bounty on their scalps until 1909, the last authenticated specimen sighted in the wild was shot in 1930, and the last captive thylacine died in Hobart Zoo in 1936. Most museums have a stuffed specimen. They are easy to tell from a true dog because of the stripes on the back but the skeleton is harder to distinguish. Zoology students of my generation at Oxford had to identify 100 zoological specimens as part of the final exam. Word soon got around that, if ever a 'dog' skull was given, it was safe to identify it as *Thylacinus* on the grounds that anything as obvious as a dog skull had to be a catch. Then one year the examiners, to their credit, double bluffed and put in a real dog skull. In case you are interested, the easiest way to tell the difference is by the two prominent holes in the palate bone, which are characteristic of marsupials generally. Dingos, of course, are not marsupials but real dogs, probably introduced by aboriginal man. It may have partly been competition from dingos that drove the thylacines extinct on mainland Australia. Dingos never reached Tasmania, which may be why thylacines survived there until European settlers drove them extinct. But fossils show that there were other species of thylacine in Australia that went extinct too early for humans or dingos to bear the blame.

The 'natural experiment' of Australinean 'alternative mammals' is often demonstrated in a series of pictures, each matching an Australinean marsupial with its more familiar placental counterpart. But not all ecological counterparts resemble each other. There doesn't seem to be any placental equivalent of the honey possum. It is easier to see why there is no marsupial equivalent of whales: quite apart from the difficulty of managing a pouch under water, 'whales' would not be subject to the isolation that allowed the Australian marsupials to evolve separately. Similar reasoning explains why there are not marsupial bats. And although kangaroos could be described as the Australinean equivalent of antelopes they look very different, because so much of their body is built around their unusual hind-legged hopping gait with massive counterbalancing tail. Yet the range of 68 species of Australinean kangaroos and wallabies matches the range of 72 species of antelopes and gazelles in diet and way of life. The overlap is not perfect. Some kangaroos will take insects if they get a chance, and fossils tell of a large carnivorous kangaroo which must have been terrifying. There are placental mammals outside Australia which leap kangaroo-fashion, but these are mostly small rodents, like the jumping jerboas. The springhare of Africa is also a rodent, not a true hare, and is the only placental mammal that one might actually mistake for a kangaroo (or rather a small wallaby). Indeed my colleague Dr Stephen Cobb, when teaching zoology at the University of Nairobi, was amused to find himself being excitedly contradicted by his pupils when he told them that kangaroos are confined to Australia and New Guinea.

The lessons of the Marsupial Mole's Tale on the importance of convergence in evolution – real convergence in the forward direction, not the backwards coalescence of this book's central metaphor – will be taken up in the final chapter, The Host's Return.

MONOTREMES

Rendezvous 15 takes place approximately 180 million years ago in the half-monsoonal, half-arid world of the Lower Jurassic. The southern continent of Gondwana was still just about connected to the great northern continent of Laurasia – the first time on our backwards journey that we find all major land-masses collected into a contiguous 'Pangaea'. In forward time, the split of Pangaea would have momentous consequences for the descendants of Concestor 15, perhaps our 120-million-greats-grandparent. Our rendezvous is a rather one-sided affair. The new pilgrims that join all the rest of the mammals here represent only three genera: *Ornithorhynchus anatinus*, the duckbilled platypus which lives in Eastern Australia and Tasmania; *Tachyglossus aculeatus*, the short-beaked echidna which lives all over Australia and New Guinea; and *Zaglossus*, the long-beaked echidna, which is confined to the highlands of New Guinea.[*] Collectively the three genera are known as monotremes.

Several tales have developed the theme of island continents as nurseries of major animal groups: Africa for the afrotheres, Laurasia for the laurasiatheres, South America for the xenarthrans, Madagascar for the lemurs, Australia for most of the surviving marsupials. But it is looking increasingly as though there was a much earlier continental separation among the mammals. According to one supported theory, long before the demise of the dinosaurs, the mammals were split into two major groups called the australosphenidans and the boreosphenidans. Australo, once again, doesn't mean Australian, it means southern. And boreo means northern, as in the northern aurora borealis. The australosphenidans were those early mammals that evolved in the great southern continent of Gondwana. And the boreosphenidans evolved in the northern continent of Laurasia, in a sort of earlier incarnation long before the evolution of the laurasiatheres we know today. The monotremes are the only surviving representatives of the australosphenidans. All the

---

[*] Three species of *Zaglossus* have been distinguished, one of them called, I am delighted to say, *Z. attenboroughi*.

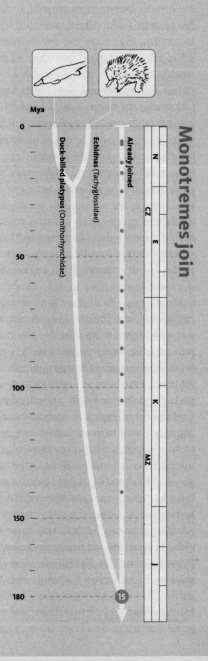

Monotremes join

Mya

0

—

—

N

CZ

E

50

—

—

—

100

—

K

—

MZ

—

150

—

J

—

180

15

Echidnas (Tachyglossidae)

Already joined

Duck-billed platypus (Ornithorhynchidae)

Monotremes join. Living mammals, numbering fewer than 5,000 species, all have fur and suckle their young. Those we have met so far – the placental and marsupial mammals – are thought to share a common northern hemisphere origin in the Jurassic Period. The five monotreme species are the sole survivors of a once diverse lineage of southern hemisphere mammals which retained the habit of laying eggs.

Images, left to right: duck-billed platypus (*Ornithorhynchus anatinus*); short-beaked echidna (*Tachyglossus aculeatus*) rest of the mammals, the therians, including the marsupials that we now associate with Australia, are descended from the northern boreosphenidans. Those therians who later became associated with the south, and with the breakup of Gondwana – for instance the afrotheres of Africa and the marsupials of South America and Australia – were boreosphenidans who had migrated south into Gondwana long after their northern origins.

Let's now turn to the monotremes themselves. The echidnas live on dry land and eat ants and termites. The platypus lives mostly in water where it feeds on small invertebrates in the mud. Its 'bill' really does look like that of a duck. The echidnas' bill is more tubular. Somewhat surprisingly, by the way, molecular evidence suggests that the concestor of echidnas with platypuses lived more recently than the fossil platypus *Obdurodon*, which lived and looked essentially like a modern platypus except that it had teeth inside its duckbill. This would mean that echidnas are modified platypuses who left the water within the last 20 million years, lost the webbing between their toes, narrowed the duckbill to make an anteater's probing tube, and developed protective spines.

One respect in which the monotremes resemble reptiles and birds has given them their name. Monotreme means single hole in Greek. As with reptiles and birds, the anus, the urinary tract and the reproductive tract empty into a single shared opening, the cloaca. Even more reptilian is that eggs, not babies, emerge from that cloaca. And not microscopic eggs like all other mammals, but two-centimetre eggs with a tough white leathery shell, containing nutriment to feed the baby until it is ready to hatch, which it eventually does like a reptile or bird with the aid of an egg-tooth on the end of its 'bill'.

Monotremes have some other typically reptilian features too, such as the interclavicle bone near the shoulder, which reptiles, but no therian mammals, possess. On the other hand the monotreme skeleton also has a number of standard mammal traits. Their lower jaw consists of a single bone, the dentary. Reptile lower jaws have three additional bones, around the hinge with the main skull. During the evolution of the mammals, these three bones migrated away from the lower jaw into the middle ear, where, renamed the hammer, the anvil and the stirrup, they transmit sound from the eardrum to the inner ear in a cunning way that physicists call impedance-matching. Monotremes are firmly with the mammals on this point. Their inner ear itself, however, is more reptilian or bird-like, in that the

Could your ancestor have looked like this? Drawing of *Henkelotherium,* a eupantothere, by Elke Gröning. (The leaf form shown is that of modern ginkgos; the leaves of Jurassic ginkgos would have been more finely divided.)

cochlea, the tube in the inner ear that detects sounds of different pitch, is more nearly straight than the snail-shaped coil which all other mammals have, and which gives the organ its name.

Monotremes are again with the mammals in secreting milk for their young: that most proverbially mammalian of substances. But again, they slightly spoil the effect by lacking discrete nipples. Instead, the milk oozes out from pores over a wide area of skin on the ventral surface, where it is licked up by the baby clinging to the hairs on the mother's belly. Our ancestors probably did the same. Monotreme limbs sprawl sideways a little more than those of a typical mammal. You can see this in the weird rolling gait of echidnas: not quite lizard-like, but not entirely mammal-like either. It adds to the impression that the monotremes are sort of inter-mediate between reptiles and mammals.

What did Concestor 15 look like? There is of course no reason to think it was like either an echidna or a platypus. It was, after all, *our* ancestor, as well as theirs, and we've all had a very long time to evolve since. Fossils of the right vintage in the Jurassic Period belong to various types of small shrew-like or rodent-like animals such as *Morganucodon* and the large group known as multituberculates. The charming picture on page 241 is of another of these early mammals, a eupantothere, up a ginkgo tree.

THE DUCKBILL'S TALE

An early Latin name of the platypus was *Ornithorhynchus paradoxus*. It seemed so weird when first discovered that a specimen sent to a museum was thought to be a hoax: bits of mammal and bits of bird stitched together. Others have wondered whether God was having a bad day when he created the platypus. Finding some spare parts left over on the workshop floor, he decided to unite rather than waste them. More insidiously (because they aren't joking) some zoologists write the monotremes off as 'primitive', as though sitting around being primitive was a full-time way of life. To question this is a purpose of the Duckbill's Tale.

Since Concestor 15, platypuses have had exactly the same time to evolve as the rest of the mammals. There is no reason why either group should be more primitive than the other (primitive, remember, precisely means 'resembling the ancestor'). Monotremes might be more primitive than us in some respects, such as laying eggs. But there is no reason at all why primitiveness in one respect should dictate primitiveness in another. There is no substance called Essence of Antiquity that pervades the blood and soaks into the bones. A primitive bone is a bone that has not changed much for a long time. There is no rule that says the neighbouring bone has to be primitive too, not even a faint presumption in that direction – at least unless a further case is made. There's no better illustration than the eponymous duck bill itself. It has evolved far, even if other parts of the platypus have not.

The platypus bill seems comical, its resemblance to that of a duck made the more incongruous by its relatively large size, and also because a duck's bill has a certain intrinsic laughableness, perhaps borrowed from Donald. But humour does an injustice to this wondrous apparatus. If you want to think in terms of an incongruous graft, forget all about ducks. A more telling comparison is the extra nose grafted onto a Nimrod reconnaissance aircraft. The American equivalent is AWACS, more familiar but less appropriate for my comparison in that the AWACS 'graft' is on top of the fuselage rather than at the front like a bill.

The point is that the platypus bill is not just a pair of jaws for dabbling and feeding, as in a duck. It is that too, though it is rubbery rather than horny like a duck's bill. But far more interestingly, the platypus bill is a reconnaissance device, an AWACS organ. Platypuses hunt crustaceans, insect larvae and other small creatures in the mud at the bottom of streams. Eyes aren't much use in mud, and the platypus keeps them tight shut while hunting. Not only that, it closes its nostrils and its ears as well. See no prey, hear no prey, smell no prey: yet it finds prey with great efficiency, catching half its own weight in a day.

If you were a sceptical investigator of somebody claiming a 'sixth sense', what would you do? You'd blindfold him, stop his ears and his nostrils, and then set him some task of sensory perception. Platypuses go out of their way to do the experiment for you. They switch off three senses which are important to us (and perhaps to them on land), as if to concentrate all their attention on some other sense. And the clue is given by one further feature of their hunting behaviour. They swing the bill in movements called saccades, side to side, as they swim. It looks like a radar dish scanning . . .

One of the first scientific descriptions of the platypus, Sir Everard Home's publication in the *Philosophical Transactions of the Royal Society* for 1802, was farsighted. He noticed that the branch of the trigeminal nerve that innervates the face is

> uncommonly large. We should be led by this circumstance to believe that the sensibility of the different parts of the bill is very great, and therefore that it answers the purpose of a hand, and is capable of nice discrimination in its feeling.

Sir Everard didn't know the half of it. It's the reference to a hand that tells. The great Canadian neurologist Wilder Penfield published a famous picture of a human brain, together with a diagram showing the proportions given over to different parts of the body. The map of a part of the brain given over to controlling muscles in different parts of the body, on one side, is shown on page 244. Penfield made a similar map of parts of the brain concerned with the sense of touch in different parts of the body. The striking thing about both maps is the huge prominence given to the hand. The face, too, is prominent, especially the parts controlling jaw movements, in chewing and speaking. But it is the hand that you really notice when you see a Penfield 'homunculus'. The image reproduced in plate 13 is another way of representing the same thing. This grotesque has his body

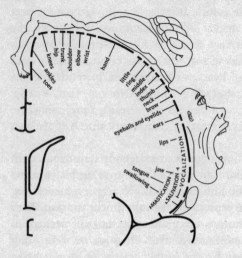

Penfield brain map.
Adapted from Penfield
and Rasmussen [222].

distorted in proportion to the amount of brain given over to different parts. Again it shows that the human brain is hand-heavy.

Where is all this leading? My account of the Duckbill's Tale is indebted to the distinguished Australian neurobiologist Jack Pettigrew and his colleagues, including Paul Manger, and one of the fascinating things they did was to prepare a 'platypunculus', the platypus equivalent of a Penfield homunculus. The first thing to say is that it is far more accurate than the Penfield homunculus, which was based on very scanty data. The platypunculus is a very thorough piece of work. You can see three little platypus maps on the upper part of the brain: separate representations, in different parts of the brain, of sensory information from the body surface. What matters to the animal is that there is an orderly spatial mapping between each part of the body and the corresponding part of the brain.

Notice that the hands and feet, coloured black on the three maps, are approximately in proportion to the body itself, unlike the case of the Penfield homunculus with its vast hands. What is not in proportion in the platypunculus is the bill. The bill's maps are the huge areas reaching down from the maps of the rest of the body. Where the human brain is hand-heavy, the platypus brain is bill-heavy (see plate 14). Sir Everard Home's guess is looking good. But, as we shall see, in one respect the bill is even better than a hand: it can reach out and 'feel' things that it is not touching. It can feel at a distance. It does it by electricity.

When any animal, such as a freshwater shrimp which is a typical platypus prey, uses its muscles, weak electric fields are inevitably generated. With sufficiently sensitive apparatus these can be detected, especially in water. Given dedicated computer power to handle data from a large array of such sensors, the source of the electric fields can be calculated. Platypuses don't, of course, calculate as a mathematician or a computer would. But at some level in their brain the equivalent of a calculation is done, and the result is that they catch their prey.

Platypuses have about 40,000 electrical sensors distributed in longitudinal stripes over both surfaces of the bill. As the platypunculus shows, a large proportion of the brain is given over to processing the data from these 40,000 sensors. But the plot thickens. In addition to the 40,000 electrical sensors, there are about 60,000 mechanical sensors called push rods, scattered over the surface of the bill. Pettigrew and his co-workers have found nerve cells in the brain that receive inputs from mechanical sensors. And they have found other brain cells that respond to both electrical and mechanical sensors (so far they have found no brain cells that respond to electrical sensors only). Both kinds of cell occupy their correct position on the spatial map of the bill, and they are layered in a way that is reminiscent of the human visual brain, where layering assists binocular vision. Just as our layered brain combines information from the two eyes to construct a stereo percept, the Pettigrew group suggests that the platypus might be combining the information from electrical and mechanical sensors in some similarly useful way. How might this be done?

They propose the analogy of thunder and lightning. The flash of lightning and the crack of thunder happen at the same moment. We see the lightning instantaneously, but the thunder takes longer to reach us, travelling at the relatively slow speed of sound (and incidentally the bang becomes a rumble because of echoes). By timing the lag between lightning and thunder, we can calculate how far away the storm is. Perhaps the electrical discharges from the prey's muscles are the platypus's lightning, while the thunder is the waves of disturbance in the water caused by the prey animal's movements. Is the platypus brain set up to compute the time lag between the two, and hence calculate how far away the prey is? It seems likely.

As for pinpointing the prey's direction, this must be done by

Remote pins and needles. The electric sensory world of the platypus. From Manger and Pettigrew [181].

comparing the inputs from different receptors all over the map, presumably aided by the scanning side-to-side movements of the bill, just as a man-made radar uses the rotation of the dish. With such a huge array of sensors projecting to mapped arrays of brain cells, the platypus very likely forms a detailed three-dimensional image of any electrical disturbances in its vicinity.

Pettigrew and his colleagues prepared this contour map of lines of equal electrical sensitivity around the bill of the platypus. When you think of a platypus, forget duck, think Nimrod, think AWACS; think huge hand feeling its way, by remote pins and needles; think lightning flashing and thunder rumbling, through the watery mud of Australia.

The platypus is not the only animal to use this kind of electrical sense. Various fish do it, including paddlefish such as *Polyodon spathula*. Technically 'bony' fish, paddlefish have secondarily, with their relatives the sturgeons, evolved a cartilaginous skeleton like a shark. Unlike sharks, however, paddlefish live in freshwater, often turbid rivers where again eyes are not much use. The 'paddle' is shaped pretty much like the upper jaw of a platypus's bill, though it is not a jaw at all but an extension of the cranium. It can be extremely long, often as much as one-third of the body length. It reminds me of a Nimrod aircraft even more than the platypus does.

The paddle is obviously doing something important in the life of the fish, and it has in fact been clearly demonstrated that it is doing the same job as the platypus bill – detecting electric fields from prey animals. As with the platypus, the electrical sensors are set into pores deployed in longitudinal lines. The two systems are independently evolved, however. Platypus electrical pores are modified mucus glands. Paddlefish electric pores are so similar to the pores used by sharks for electrical sensing, called ampullae of Lorenzini, that they have been given the same name. But where the platypus arranges its sensory pores in a dozen or so narrow stripes along the length of the bill, the paddlefish has two broad stripes, on either side of the midline of the paddle. Like the platypus, the paddlefish has an enormous number of sensory pores – actually even more than the platypus. Both the paddlefish and the platypus are far more sensitive to electricity than any one of their sensors by itself. They must be doing some sort of sophisticated signal summation from different sensors.

There is evidence that the electrical sense is more important for juvenile paddlefish than for adults. Adults who have accidentally lost their paddle have been found alive and apparently healthy, but no juveniles have been found to survive up any creek without a paddle. This may be because juvenile paddlefish, like adult platypuses, target and catch individual prey. Adult paddlefish feed more like planktivorous baleen whales, sieving their way through the mud, catching prey *en masse*. They grow big on this diet, too – not as big as whales, but as long and as heavy as a man, larger than most animals that swim in freshwater. Presumably if you are sieving plankton as an adult, you have less need of an accurate prey-locator than if you are darting after individual prey as a juvenile.

Platypus and paddlefish, then, have independently hit upon the same ingenious trick (see plate 15). Has any other animal discovered it? Whilst doing his D.Phil. work in China, my research assistant Sam Turvey encountered an extremely unusual trilobite called *Reedocalymene*. Otherwise a 'bog-standard' trilobite (similar to the Dudley Bug, *Calymene*, which features on the coat of arms of the town of Dudley), *Reedocalymene* has one unique and remarkable feature: a huge flattened rostrum, like that of a paddlefish, sticking out a whole body length in front. It can't have been for streamlining, since this trilobite, unlike many others, was obviously unfitted for swimming above the sea bed. A defensive purpose is also unlikely for various reasons. Like a paddlefish, sturgeon or platypus bill, the trilobite's rostrum is studded with what look like sensory receptors, probably used for detecting prey. Turvey knows of no modern arthropods with an electrical sense (interesting in itself, given the versatility of the arthropods), but he would put money on

Reedocalymene being yet another 'paddlefish' or 'platypus'. He is hoping to start work on it soon.

Other fish, though lacking the Nimrod-like 'antenna' of the platypus and the paddlefish, have an even more sophisticated electrical sense. Not content with picking up electrical signals inadvertently given off by prey, these fish generate their own electric fields. They navigate and detect prey by reading the distortions in these self-generated fields. Along with various cartilaginous rays, two groups of bony fish, the gymnotid family of South America and the mormyrid family of Africa, have independently developed this to a high art.

How do these fish make their own electricity? The same way the shrimps and insect larvae and other prey of the platypus inadvertently do it: with their muscles. But whereas the shrimps can't help making a little electricity because that is what muscles just do, the electric fish gang their blocks of muscle together just like batteries in series.[*] A gymnotid or mormyrid electric fish has a battery of muscle blocks arranged in series along its tail, each generating a low voltage and adding up to a higher voltage. The electric eel (not a true eel but another South American freshwater gymnotid) takes it to an extreme. It has a very long tail into which it can pack a much larger battery of electrical cells than a fish of normal length. It stuns its prey with electric shocks which may exceed 600 volts and can be fatal to people. Other freshwater fish, such as the African electric catfish *Malapterurus* and the marine electric ray *Torpedo* also generate enough volts to kill, or at least knock out, their prey.

These high-voltage fish seem to have pushed, to a literally stunning extreme, a capacity which was originally a kind of radar used by the fish to find its way around and detect prey. Weakly electric fish such as the South American *Gymnotus* and the unrelated African *Gymnarchus* have an electrical organ like the electric eel's but much shorter – their battery consists of fewer modified muscle plates in series – and a weakly electric fish typically generates less than one volt. The fish holds itself like a rigid stick in the water, for a very good reason as we shall see, and electric current flows along curved lines that would have delighted Michael Faraday. All along the sides of the body are pores containing electrical sensors – tiny voltmeters. Obstacles or prey items distort the field in various ways, which are

[*] Of course, the word battery in its original electrical sense means a battery of cells in series, as opposed to a single cell. If your transistor radio takes six 'batteries', a pedant would insist that it takes one battery of six cells.

detected by these little voltmeters. By comparing the readings of the different voltmeters and correlating them with the fluctuations of the field itself (sinusoidal in some species, pulsed in others) the fish can calculate the location of obstacles and prey. They also use their electric organs and sensors to communicate with one another.

A South American electric fish such as *Gymnotus* is remarkably similar to *Gymnarchus*, its African opposite number, but there is one revealing difference. Both have a single long fin running the length of the midline, and both use it for the same purpose. They can't throw the body into the normal sinuous waves of a swimming fish because it would distort their electrical sense. Both are obliged to keep the body rigid, so they swim by means of the longitudinal fin, which waves sinuously just like a normal fish should. It means they swim slowly, but presumably it is worth it to get the benefits of a good clear signal. The beautiful fact is that *Gymnarchus* has its longitudinal fin on its back, while *Gymnotus* and the other South American electric fish, including the electric 'eel', keep their longitudinal fin on their belly. It is for such cases that 'the exception that proves the rule' was coined.

Returning to the platypus, the sting in the tale is actually in the hind claws of the male platypus. True venomous stings, with hypodermic injection, are found in various invertebrate phyla, and in fish and reptiles among vertebrates – but never in birds or mammals other than the platypus (unless you count the toxic saliva of solenodons and some shrews that makes their bites slightly venomous). Among mammals, the male platypus is in a class of its own, and it may be in a class of its own among venomous animals too. The fact that the sting is found only in males suggests, rather surprisingly, that it is aimed not at predators (as in bees) nor at prey (as in snakes) but at rivals. It is not dangerous but is extremely painful, and is unresponsive to morphine. It looks as though platypus venom works directly on pain receptors themselves. If scientists could understand how this is done, there is a hope that it might give a clue to how to resist the pain caused by cancer.

This tale began by chiding those zoologists who call the platypus 'primitive' as though that were any kind of explanation for the way it is. At best it is a description. Primitive means 'resembling the ancestor' and there are many respects in which this is a fair description of a platypus. The bill and the sting are interesting exceptions. But the more important moral of the tale is that even an animal that is genuinely primitive in all respects is primitive for a reason. The ancestral characteristics are good for its way of life, so there is no reason to change. As Professor Arthur Cain of Liverpool University liked to say, an animal is the way it is because it needs to be.

WHAT THE STAR-NOSED MOLE SAID TO THE DUCKBILLED PLATYPUS

The star-nosed mole, who had joined the pilgrimage along with the other laurasiatheres at Rendezvous 11, listened to the Duckbill's Tale with close attention, and with growing recognition in what was left of his vestigial, pin-prick eyes. 'Yes!' he squeaked, too high for some of the larger pilgrims to hear, and he clapped his spades with excitement. 'That's just the way it is for me . . . well, sort of.'

No, it won't do, I wanted to follow Chaucer in having at least one section devoted to what one pilgrim said to another, but I'll limit it to the heading and first paragraph, and now revert to my practice of telling the tale itself in my own words. Bruce Fogle (*101 Questions Your Dog Would Ask Its Vet*) or Olivia Judson (*Dr Tatiana's Sex Advice to All Creation*) might get away with it, but not me.

Pushes the envelope of touch beyond our dreams. An in-your-face view of a star-nosed mole, *Condylura cristata*.

The star-nosed mole, *Condylura cristata*, is a North American mole which, in addition to burrowing and hunting for worms like other moles, is a good swimmer too, hunting for underwater prey – it often tunnels deep into river banks. It is also more at home above ground than other moles, where it still prefers damp, soggy places. It has large spade hands like other moles.

What sets it apart is the remarkable nose that gives it its name. Surrounding the two forwards-pointing nostrils, there is an extraordinary ring of fleshy tentacles, like a baby sea anemone with 22 arms. The tentacles are not used to grasp things. Nor are they an aid to smelling, which is the next hypothesis that might occur to us. Nor, despite the beginning of this section, are they an electrical radar like that of the platypus. Their true nature has been beautifully worked out by Kenneth Catania and Jon Kaas of Vanderbilt University, Tennessee. The star is a touch-sensitive organ, like a super-sensitive human hand, but lacking the grasping function of the hand and emphasising its sensitivity instead. But it isn't just any ordinary touch-sensitive organ. The star-nosed mole pushes the envelope of touch beyond our dreams. The skin of its nose is more sensitive than any other area of skin anywhere among the mammals, not excluding the human hand.

There are 11 tentacles arcing round each nostril, labelled 1 to 11 in order. Tentacle 11, which lies close to the midline and just below the level of the nostril, is special, as we shall see in a moment. Although they are not used for grasping, the tentacles are moved, independently or in particular groupings. The surface of each tentacle is carpeted with a regular array of little round bumps called Eimer's Organs, each one a unit of touch sensitivity, and each one wired up by between seven nerve fibres (for tentacle 11) and four nerve fibres (most of the other tentacles).

The density of Eimer's organs is the same for all tentacles. Tentacle 11, being smaller, has fewer of them, but it has more nerves supplying each one. Catania and Kaas were able to map the tentacles to the brain. They found (at least) two independent maps of the nose star in the cerebral cortex. In each of these two brain areas, the parts of the brain corresponding to each tentacle are laid out in order. And tentacle 11 again is special. It is more sensitive than the rest. Once an object has been first detected by any of the tentacles, the animal then moves the star so that tentacle 11 can examine it carefully. Only then is the decision taken whether to eat

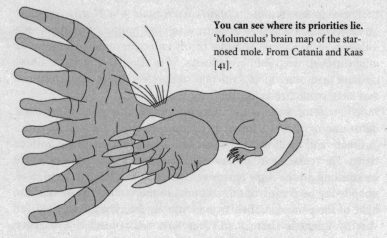

You can see where its priorities lie.
'Molunculus' brain map of the star-nosed mole. From Catania and Kaas [41].

it or not. Catania and Kaas refer to tentacle 11 as the 'fovea' of the star.* More generally, they say:

> Although the nose of the star-nosed mole acts as a tactile sensory surface, there are anatomical and behavioral similarities between the mole's sensory system and the visual system of other mammals.

If the star is not an electrical sensor, whence the empathy with the platypus with which I opened this section? Catania and Kaas constructed a schematic model of the relative amount of brain tissue given over to different parts of the body surface. It is a molunculus, by analogy with Penfield's homunculus and Pettigrew's platypunculus. And just look at it!†

You can see where the star-nosed mole's priorities lie. You can get a feel for the world of the star-nosed mole. And *feel* is the right word. This animal lives in a tactile world, dominated by the tentacles of the nose, with a subsidiary interest in the large spade hands and the whiskers.

What is it like to be a star-nosed mole? I am tempted to propose the star-nosed counterpart to an idea I once offered for bats. Bats live in a world of sound, but what they do with their ears is pretty much the same as what, say, insect-hawking birds like swallows do with

---

* The fovea is the small area in the middle of the human retina where cone cells are concentrated so that acuity, and colour vision, are both maximal. We read with our fovea, recognise each other's faces, and do everything that needs fine visual discrimination.

† Note that parts of the 'molunculus' are hidden behind parts that we can see.

their eyes. In both cases the brain needs to construct a mental model of a three-dimensional world, to be navigated at high speed, with obstacles to be avoided and small moving targets to catch. The model of the world needs to be the same, whether it is constructed and updated with the aid of light rays or sound echoes. My conjecture was that a bat probably 'sees' the world (using echoes) in pretty much the same way as a swallow, or a person, sees the world using light.

I even went so far as to speculate that bats hear in colour. The hues that we perceive have no necessary link with the particular wavelengths of light that they represent. The sensation that I call red (and nobody knows if my red is the same as yours) is an arbitrary label for light of long wavelengths. It could equally well have been used for short wavelengths (blue), and the sensation that I call blue used for long wavelengths. Those hue sensations are available in the brain for tying to whatever, in the outside world, is most convenient. In bat brains those vivid qualia would be wasted on light. They are more likely to be used as labels tied to particular qualities of echo, perhaps textures of surfaces on obstacles or prey.

My conjecture now is that a star-nosed mole 'sees' with its nose. And my speculation is that it uses those same qualia that we call colour, as labels for tactile sensations. Similarly, I want to guess that duckbilled platypuses 'see' with the bill, and use the qualia we call colour as internal labels for electrical sensations. Could this be why platypuses close their eyes tight shut when they are hunting electrically with the bill? Could it be because the eyes and the bill are competing, in the brain, for internal qualia labels, and to use both senses at once would lead to confusion?

MAMMAL-LIKE REPTILES

The monotremes having joined us, the entire company of mammal pilgrims now walks back 130 million unbroken years, the longest gap yet between any two milestones, to Rendezvous 16 where we are to meet an even larger band of pilgrims than our own, the sauropsids: reptiles and birds. That pretty much means all vertebrates that lay large eggs with a waterproof shell on land. I have to say 'pretty much', partly because monotremes, who have already joined us, also lay that kind of egg. Even turtles that are otherwise wholly marine haul themselves up the beach to lay their eggs. Plesiosaurs may have done the same. Ichthyosaurs, however, were so specialised for swimming that, like the dolphins who later resembled them, they presumably couldn't come on shore at all. They independently discovered how to give birth to live young – as we know from mothers fossilised in the act.[*]

I said that our pilgrims walked through 130 million years without milestones, but of course 'without milestones' is true only within the conventions of this book: we are recognising as milestones only rendezvous with living pilgrims. Our ancestral lineage indulged in fertile evolutionary branching during that time, as we know from the rich fossil record of 'mammal-like reptiles', but none of the branches among these counts as a 'rendezvous' because, as it turned out, none of them survived. There are therefore no modern representatives to set off as pilgrims from the present. When we met a similar problem with the hominids, we decided to give certain fossils honorary status as 'shadow pilgrims'. Since we are pilgrims seeking our ancestors, pilgrims who actually want to know what our 100-million-greats-grandparent looked like, we cannot ignore the mammal-like reptiles and jump straight to Concestor 16. Concestor 16, as we shall see, looked like a lizard. The gap from Concestor 15, which looked like a shrew, is too great to leave unbridged. We have to examine the mammal-like reptiles as shadow pilgrims, as though they were living pilgrims joining our march – although they shall not actually tell

[*] Some extant lizards have also discovered live birth.

tales. But first, some background information on the timespan involved, because it is very long.

The intervening years without rendezvous milestones span half the Jurassic, the whole of the Triassic, the whole of the Permian and the final 10 million years of the Carboniferous. As the pilgrimage moves from the Jurassic back into the hotter and drier world of the Triassic – one of the hottest periods in the planet's history, when all the landmasses were joined together, forming Pangaea – we pass the late Triassic mass extinction, when three-quarters of all species went extinct. But this is nothing compared to the next transition, from the Triassic Period back into the Permian. At the Permo–Triassic boundary, a staggering 90 per cent of all species perished without descendants, including all the trilobites and several other major groups of animals. The trilobites, to be fair, had already been declining over a long period. But the end-Permian mass extinction was the most devastating of all time. There is some evidence from Australia that this extinction, like the Cretaceous one, was caused by a massive bolide collision. Even the insects took a severe knock, the only one in their history. At sea, bottom-dwelling communities were almost wiped out. On land, the Noah among the mammal-like reptiles was *Lystrosaurus*. Immediately after the catastrophe, the squat, short-tailed *Lystrosaurus* became extremely abundant over the whole world, rapidly occupying vacant niches.

The natural association with apocalyptic carnage needs to be tempered. Extinction is the eventual fate of nearly all species. Perhaps 99 per cent of all species that have ever existed have gone extinct. Nevertheless, the rate of extinctions per million years is not fixed and only occasionally rises above 75 per cent, the threshold arbitrarily recognised for a 'mass' extinction. Mass extinctions are spikes in the rate of extinction, rising above the background rate.

The diagram on the next page shows rates of extinction per million years.[*] Something happened at the time of those spikes. Something bad. Perhaps a single catastrophic event, such as the collision with a massive celestial rock that killed the dinosaurs 65 million years ago in the Cretaceous–Palaeogene extinction. Or, in other cases among the five spikes, the agony may have been drawn out. What Richard Leakey and Roger Lewin have called the Sixth Extinction is the one

---

[*] The absolute figures are lower than 75 per cent, because they refer to genera not species. Absolute figures for species are higher than for genera, because each genus contains lots of species so it's harder to extinguish a genus than a species.

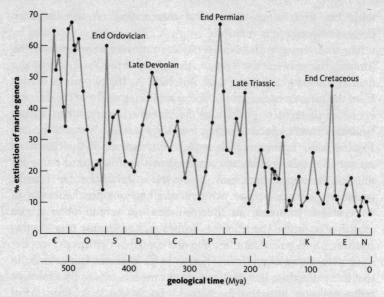

Percent extinction of marine genera throughout the Phanerozoic Eon. Adapted from Sepkoski [260].

now being perpetrated by *Homo sapiens* – or *Homo insipiens* as my old German teacher William Cartwright preferred to say.[*]

Before we get to the mammal-like reptiles, we face a somewhat tiresome point of terminology. Terms like reptile and mammal can refer to 'clades' or 'grades' – the two are not exclusive. A clade is a set of animals consisting of an ancestor and all its descendants. The 'birds' constitute a good clade. 'Reptile', as traditionally understood, is not a good clade because it excludes birds. Biologists consequently refer to the reptiles as 'paraphyletic'. Some reptiles (e.g. crocodiles) are closer cousins of some non-reptiles (birds) than they are of other reptiles (turtles). To the extent that reptiles all have something in common, they are members of a *grade*, not a clade. A grade is a set of animals that have reached a similar stage in a recognisably progressive evolutionary trend.

Yet another informal grade name, favoured by American zoologists, is 'herp'. Herpetology is the study of reptiles (except birds) and

---

[*] A remarkable, bushy-browed, slow-spoken man who called a spade a spade and was seldom seen without one, Mr Cartwright discovered environmental activism long before its vogue, and filled his lessons with ecology, to the detriment of our German but the advantage of our humane education.

amphibians. 'Herp' is a rare kind of word: an abbreviation for which there is no long form. A herp is simply the kind of animal studied by a herpetologist, and that is a pretty lame way to define an animal. The only other name that comes close is the biblical 'creeping thing'.

Another grade name is fish. 'Fish' include sharks, various extinct fossil groups, teleosts (bony fish such as trout and pike) and coelacanths. But trout are closer cousins to humans than they are to sharks (and coelacanths are even closer cousins to humans than trout are). So 'fish' is not a clade because it excludes humans (and all mammals, birds, reptiles and amphibians). Fish is a grade name for animals that sort of look fishy. It is more or less impossible to make grade terminology precise. Ichthyosaurs and dolphins look sort of fishy, and very possibly would taste fishy if we were to eat them, but they don't count as members of the fish 'grade', because they *reverted* to fishiness via ancestors that were non-fishy.

Grade terminology works well for you if you have a strong belief in evolution marching progressively in one direction, in parallel lines from a shared starting point. If, say, you think that a whole lot of related lineages were all independently evolving in parallel from amphibianhood through reptilehood towards mammalhood, you could speak of passing through the reptile grade on the way to the mammal grade. Something like that parallel march may have happened. It was the view that I was brought up with, by my own respected teacher of vertebrate palaeontology Harold Pusey. I have a lot of time for it, but it is not something to be taken for granted in general, nor necessarily enshrined in terminology.

If we swing to the other extreme and adopt strict cladistic terminology, the word reptile can be rescued only if it is deemed to include birds. This is the course favoured by the authoritative 'Tree of Life' project founded by the Maddison brothers.* There's a lot to be said for following them, and also for the very different tactic of replacing 'mammal-like reptile' by 'reptile-like mammal'. But the word reptile has become so ingrained in its traditional sense that I fear it would confuse to change it now. Also, there are times when strict cladistic purism can give ludicrous results. Here's a *reductio ad absurdum*. Concestor 16 must have had an immediate descendant on the mammal side and an immediate descendant on the lizard/crocodile/

---

* This excellent resource is continually updated at *http://tolweb.org/tree*. The website has a delightful disclaimer: 'The Tree is under construction. Please have patience: the real Tree took over 3,000,000,000 years to grow.'

dinosaur/bird, or 'sauropsid', side. These two must have been all but identical to each other. In fact there must have been a time when they could hybridise with each other. Yet the strict cladist would insist on calling one of them a sauropsid and the other one a mammal. Fortunately we don't often reach such a *reductio* in practice, but such hypothetical cases are good to quote when cladistic purists start getting above themselves.

We are so used to the idea of mammals as successors to the dinosaurs, that we may find it surprising that the mammal-like reptiles flourished before the rise of the dinosaurs. They filled the same range of niches as the dinosaurs were later to fill, and as the mammals themselves were to fill even later still. Actually they filled those niches not once but several times in succession, separated by large-scale extinctions. In the absence of milestones supplied by rendezvous with living pilgrims, I shall recognise three shadowy milestones to bridge the gap between the shrew-like Concestor 15 (which unites us to the monotremes) and the lizard-like Concestor 16 (which unites us to birds and dinosaurs).

Your 150-million-greats-grandmother might have been something a bit like a *Thrinaxodon*, which lived in the Middle Triassic and whose fossils have been found in Africa and Antarctica, then joined to each other within Gondwana. It is too much to hope that it was *Thrinaxodon* itself, or any other particular fossil that we happen to have found. *Thrinaxodon*, like any fossil, should be thought of as a cousin of our ancestor, not the ancestor itself. It was a member of a group of mammal-like reptiles called the cynodonts. The cynodonts were so mammal-like, it is tempting to call them mammals. But who cares what we call them? They are almost perfect intermediates. Given that evolution has happened, it would be weird if there were not intermediates like the cynodonts.

The cynodonts were among several groups that radiated from an earlier group of mammal-like reptiles called the therapsids. Your 160-million-greats-grandfather was probably a therapsid, living in the Permian Period, but it is hard to pick out a particular fossil to represent it. The therapsids dominated the land trades before the dinosaurs arrived in the Triassic Period, and even in the Triassic itself they gave the dinosaurs a run for their money. They included some huge animals: herbivores three metres long, with large and probably ferocious carnivores to prey on them. But our therapsid ancestor was probably a smaller and more insignificant creature. It seems to be a rule that large or specialised animals, such as the fearsome fanged gorgonopsids or the tusk-bearing herbivorous dicynodonts (see plate 16), don't have a long-term evolutionary future but belong to the 99 per cent of species destined for extinction. The Noah species, the one per cent from which we later animals are all descended – whether we ourselves are large and spectacular in our own time or not – tend to be smaller and more retiring.

The early therapsids were a bit less mammal-like than their successors, the cynodonts, but more mammal-like than their predecessors, the pelycosaurs, who constituted the early radiation of mammal-like reptiles. Before the therapsids, your 165-million-greats-grandmother was almost certainly a pelycosaur although, once again, it would be foolhardy to attempt to single out a particular fossil for that honour. The pelycosaurs were the earliest wave of mammal-like reptiles. They flourished in the Carboniferous Period, when the great coalfields were being laid down. The best-known pelycosaur is *Dimetrodon*, the one with the great sail on its back. Nobody knows how *Dimetrodon* used its sail. It may have been a solar panel to help the animal warm up to a temperature where it could use its muscles, and/or perhaps it was a radiator to cool down in the shade, when things got too hot. Or it could have been a sexual advertisement, a bony equivalent of a peacock's fan. The pelycosaurs mostly went extinct during the Permian – all except for the Noah-pelycosaurs who sprouted the second wave of mammal-like reptiles, the therapsids. The therapsids then spent the early part of the Triassic Period 'reinventing many of the lost body forms of the Late Permian'.[*]

The pelycosaurs were considerably less mammal-like than the therapsids, which in turn were less mammal-like than the cynodonts. For example, the pelycosaurs sprawled on their bellies like lizards, with legs splayed out sideways. They probably had a fish-like wiggle to their gait. The therapsids, and then the cynodonts and finally the mammals, raised their bellies progressively higher off the ground – their legs became more vertical and their gait less reminiscent of a fish on land. Other 'mammalisation' trends – perhaps recognised as progressive only with the hindsight of the mammals that we are – include the following. The lower jaw became reduced to a single bone, the dentary, as its other bones were commandeered by the ear

[*] This happy turn of phrase is from *Mass Extinctions and Their Aftermath* by A. Hallam and P. B. Wignall.

(as discussed at Rendezvous 15). At some point, though fossils aren't much help in pinning it down, our ancestors developed hair and a thermostat, milk and advanced parental care, and complex teeth specialised for different purposes.

I have dealt with the evolution of our mammal-like reptile anchors – 'shadow pilgrims' – as three successive waves: pelycosaurs, therapsids and cynodonts. The mammals themselves are the fourth wave, but their evolutionary invasion into the familiar range of ecotypes was postponed 150 million years. First, the dinosaurs had to have their go, which lasted twice as long as all three waves of mammal-like reptiles put together.

On our backwards march, the earliest of our three groups of 'shadow pilgrims' have brought us to a rather lizard-like pelycosaur 'Noah', our 165-million-greats-grandparent, who lived in the Triassic Period, about 300 million years ago. We have almost penetrated back to Rendezvous 16.

SAUROPSIDS

Concestor 16, our approximately 170-million-greats-grandparent, lived some 310 million years ago in the second half of the Carboniferous, a time of vast swamps of giant club moss trees in the tropics (the origin of most coal) and an extensive ice cap at the South Pole. This rendezvous point is where a huge throng of new pilgrims joins us: the sauropsids. The sauropsids are by far the largest contingent of newcomers we have yet had to deal with along our Pilgrims' Way. For most of the years since Concestor 16 lived, sauropsids, in the form of dinosaurs, dominated the planet. Even today, with the dinosaurs gone, there still are more than three times as many sauropsid species as mammals. At Rendezvous 16, approximately 4,600 mammal pilgrims greet 9,600 bird pilgrims and 7,770 pilgrims from the rest of the reptiles: crocodiles, snakes, lizards, tuataras, turtles. They are the main group of land vertebrate pilgrims. The only reason I am regarding them as joining us, rather than we joining them, is that we arbitrarily chose to see the journey through human eyes.

Seen through sauropsid eyes, the last to join their pilgrimage 'before' the rendezvous with us were the turtles (using the word in its American sense to include tortoises as well as aquatic turtles and terrapins). The sauropsid contingent, therefore, consists of the turtles and the rest. 'The rest' are a union of two major groups: the lizard-like reptiles which include snakes, chameleons, iguanas, Komodo dragons and tuataras; and the dinosaur-like reptiles or archosaurs, which include pterodactyls, crocodiles and birds. The great aquatic reptile groups such as ichthyosaurs and plesiosaurs are not dinosaurs and seem to be, if anything, closer to the lizard-like reptiles. Pterodactyls have less claim to be called dinosaurs than birds do. Birds are an offshoot of one particular order of dinosaurs, the saurischians. The saurischian dinosaurs, such as *Tyrannosaurus* and the gigantic sauropods, are closer to birds than they are to the other main group of dinosaurs, the unfortunately-named ornithischians such as *Iguanodon, Triceratops*, and the duckbilled hadrosaurs. Ornithischian means 'bird-hipped', but the resemblance is superficial and confusing.

The relationship of birds to saurischian dinosaurs is made secure

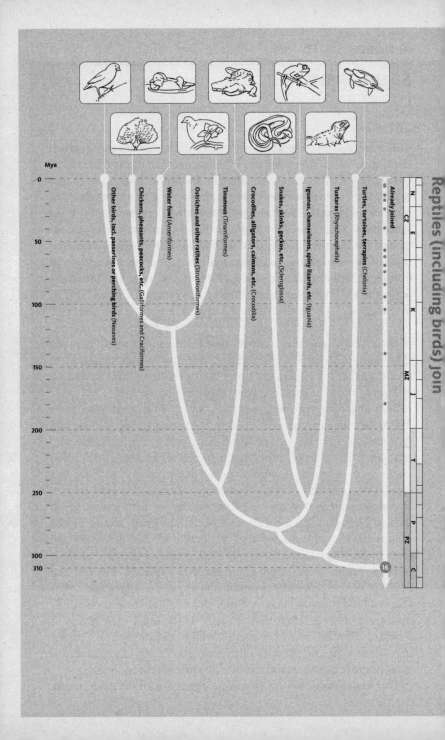

Mya

Other birds, incl. passerines or perching birds (Neoaves)

Chickens, pheasants, peacocks, etc. (Galliformes and Craciformes)

Water fowl (Anseriformes)

Ostriches and other ratites (Struthioniformes)

Tinamous (Tinamiformes)

Crocodiles, alligators, caimans, etc. (Crocodilia)

Snakes, skinks, geckos, etc. (Scleroglossa)

Iguanas, chamaeleons, spiny lizards, etc. (Iguania)

Tuataras (Rhynchocephalia)

Turtles, tortoises, terrapins (Chelonia)

Already joined

16

by recent spectacular finds of feathered dinosaurs in China. Tyrannosaurs are closer cousins to birds than they are even to other saurischians such as the large plant-eating sauropods *Diplodocus* and *Brachiosaurus*.

These, then, are the sauropsid pilgrims, the turtles, lizards and snakes, crocodiles, and birds, together with the huge concourse of shadowpilgrims – the pterosaurs in the air, the ichthyosaurs, plesiosaurs and mosasaurs in water, and above all the dinosaurs on land. Focused as this book is on pilgrims from the present, it is not appropriate to expatiate on the dinosaurs, who dominated the planet for so long, and who would dominate it yet, but for the cruel – no, indifferent – bolide that laid them low. It seems added cruelty to treat them now so indifferently.* They do survive after a fashion – the special and beautiful fashion of birds – and we shall do them homage by listening to four tales of birds. But first, *in memoriam*, Shelley's well-known *Ode to a Dinosaur*:

> *I met a traveller in an antique land*
> *Who said: 'Two vast and trunkless legs of stone*
> *Stand in the desert . . . Near them, on the sand,*
> *Half-sunk, a shattered visage lies, whose frown,*
> *And wrinkled lip, and sneer of cold command,*
> *Tell that its sculptor well those passions read*
> *Which yet survive, stamped on these lifeless things,*
> *The hand that mocked them and the heart that fed:*
> *And on the pedestal these words appear:*

* Other books have done them proud, for example David Norman's *Dinosaur!* and Robert Bakker's *The Dinosaur Heresies*, not forgetting Robert Mash's delightfully affectionate *How to Keep Dinosaurs*.

Opposite: **Reptiles (including birds) join.** A breakthrough in the evolution of terrestrial vertebrates was the *amnion*, a waterproof yet breathable egg membrane. Two early-diverging lineages of these 'amniotes' survive today: the synapsids (represented by the mammals), and the sauropsids (17,000 living species of 'reptiles' and birds) who join us here. The phylogeny shown here is reasonably secure.

Images, left to right: medium ground finch (*Geospiza fortis*); Indian peafowl (*Pavo cristatus*); mandarin duck (*Aix galericulata*); solitary tinamou (*Tinamus solitarius*); Nile crocodile (*Crocodylus niloticus*); red-sided garter snake (*Thamnophis sirtalis parietalis*); Mediterranean chameleon (*Chamaeleo chamaeleon*); tuatara (*Sphenodon punctatus*); green turtle (*Chelonia mydas*).

"My name is Ozymandias, king of kings:
Look on my works, ye Mighty, and despair!"
Nothing beside remains. Round the decay
Of that colossal wreck, boundless and bare
The lone and level sands stretch far away.'

PROLOGUE TO THE GALAPAGOS FINCH'S TALE

The human imagination is cowed by antiquity, and the magnitude of geological time is so far beyond the ken of poets and archaeologists it can be frightening. But geological time is large not only in comparison to the familiar timescales of human life and human history. It is large on the timescale of evolution itself. This would surprise those, from Darwin's own critics on, who have complained of insufficient time for natural selection to wreak the changes the theory requires of it. We now realise that the problem is, if anything, opposite. There has been too much time! If we measure evolutionary rates over a short time, and then extrapolate, say, to a million years, the potential amount of evolutionary change turns out to be hugely greater than the actual amount. It is as though evolution must have been marking time for much of the period. Or, if not marking time, wandering around this way and that, with meandering fluctuations drowning out, in the short term, whatever trends there might be in the long.

Evidence of various kinds, and theoretical calculations, all point towards this conclusion. Darwinian selection, if we impose it artificially as hard as we are able, can drive evolutionary change at a rate far faster than we ever see in nature. To see this, we cash in on the lucky fact that our forebears, whether they fully understood what they were doing or not, have for centuries been selectively breeding domestic animals and plants (see the Farmer's Tale). In all cases these spectacular evolutionary changes have been achieved in no more than a few centuries or, at most, millennia: far faster than even the fastest evolutionary changes that we can measure in the fossil record. No wonder Charles Darwin made much of domestication in his books.

We can do the same thing under more controlled experimental conditions. The most direct test of a hypothesis about nature is an *experiment*, in which we deliberately and artificially mimic the crucial element of nature in the hypothesis. If you have a hypothesis that, say, plants grow better in soil that contains nitrates, you don't just analyse soils to see if there are nitrates there. You experimentally *add* nitrate to some soils but not others. So it is with Darwinian selection. The hypothesis about nature is that non-random survival over generations leads to a systematic shift in average form. The experimental test is to engineer just such non-random survival, in an attempt to steer evolution in some desired direction. That is what artificial selection is. The neatest experiments simultaneously select two lines in opposite directions from the same starting point: say one line making larger animals and the other making smaller ones. Obviously if you want to get decent results before dying of old age you must choose a creature with a faster life cycle than your own.

Fruit flies and mice measure their generations in weeks and months, not decades as we do. In one experiment, *Drosophila* fruit flies were split into two 'lines'. One line was bred, over several generations, for a positive tendency to approach light. In each generation, the most strongly light-seeking individuals were allowed to breed. The other line was bred systematically in the opposite direction, over the same number of generations, for a tendency to shun light. In a mere 20 generations, dramatic evolutionary change was achieved, in both directions. Would the divergence go on for ever at the same rate? No, if only because the available genetic variation would eventually run out and we'd have to wait for new mutations. But before this happens, a great deal of change can be achieved.

Maize has a longer generation time than *Drosophila*. But in 1896 the Illinois State Agricultural Laboratory started breeding for oil content in maize seeds. A 'high line' was selected for increased oil content, and a low line simultaneously selected for decreased oil (see plate 17). Fortunately this experiment has been continued far longer than the research career of any normal scientist, and it is possible to see, over 90 or so generations, an approximately linear increase in oil content in the high line. The low line has decreased its oil content less rapidly, but that is presumably because it is hitting the floor of the graph: you can't have less oil than zero.

This experiment, like the *Drosophila* one and like many others of the same type, brings home the potential power of selection to drive evolutionary change very fast indeed. Translate 90 generations of maize, or 20 generations of *Drosophila*, even 20 elephant generations, into real time, and you have something that is still negligible on the geological scale. One million years, which is too short to notice in most parts of the fossil record, is 20,000 times as long as it takes to triple the oil content of maize seeds. Of course this doesn't mean a million years of selection could multiply the oil content by 60,000. Quite apart from running out of genetic variation, there's a limit to how much oil a maize seed can pack in. But these experiments serve to warn against looking at apparent trends spread over millions of fossil years, and naively interpreting them as responses to steadily sustained selection pressures.

Darwinian selection pressures are out there, for sure. And they are immensely important, as we shall see throughout this book. But selection pressures are not sustained and uniform over the sort of timescales that can normally be resolved by fossils, especially in older parts of the fossil record. The lesson of the maize and the fruit flies is that Darwinian selection could meander hither and yon, back and forth, ten thousand times, all within the shortest time we can measure in the record of the rocks. My bet is that this happens.

Yet there are major trends over longer timescales, and we have to be aware of them too. To repeat an analogy I have used before, think of a cork, bobbing about off the Atlantic coast of America. The Gulf Stream imposes an overall eastward drift in the average position of the cork, which will eventually be washed up on some European shore. But if you measure its direction of movement during any one minute, buffeted by waves and eddies and whirlpools, it will seem to move west as often as east. You won't notice any eastward bias unless you sample its position over much longer periods. Yet the eastward bias is real, it is there, and it too deserves an explanation.

The waves and eddies of natural evolution are usually too slow for us to see in our little lifetimes, or at least within the short compass of a typical research grant. There are a few notable exceptions. The school of E. B. Ford, the eccentric and fastidious scholar from whom my generation of Oxford zoologists learned our genetics, devoted decades of research to tracking the year-by-year fortunes of particular genes in wild populations of butterflies, moths and snails. Their results in some cases seem to have straightforward Darwinian explanations. In other cases the noise of buffeting waves drowns out the signal of whatever Gulf Streams may have been tugging the undertow, and the results are enigmatic. The point I am now making is that such enigma is to be expected by any mortal Darwinian – even a Darwinian with a research career as long as Ford's. One of the main messages Ford himself drew from his life's work was that the selection pressures actually to be found in nature, even if they don't always pull in the same direction, are orders of magnitude stronger than anything dreamed of by the most optimistic founders of the neo-Darwinian revival. And this again underlines the point: why doesn't evolution go much faster than it does?

THE GALAPAGOS FINCH'S TALE

The Galapagos archipelago is volcanic, and no more than 5 million years old. During that brief existence, a spectacular quantity of diversity has evolved – most famously among the 14 species of finches widely, though perhaps wrongly, believed to have been Darwin's principal inspiration.* The Galapagos finches are among the most thoroughly studied wild animals in existence. Peter and Rosemary Grant have devoted their professional lives to following the year-by-year fortunes of these small island birds. And in the years between Charles Darwin and Peter Grant (who himself bears a pleasing facial resemblance to Darwin) the great (but clean-shaven) ornithologist David Lack also paid them a perceptive and productive visit.†

The Grants and their colleagues and students have been returning yearly to the Galapagos Islands for more than a quarter of a century, trapping finches, individually marking them, measuring their beaks and wings, and more recently taking blood samples for DNA analysis to establish paternities and other relationships. There has probably never been a more complete study of the individuals and genes of any wild population. The Grants know in minute detail exactly what is happening to the bobbing corks which are the finch populations, as they are tossed this way and that in the sea of evolution by selection pressures that change every year.

In 1977 there was a severe drought, and the food supply plummeted. The total number of individual finches of all species on the

* Stephen Gould discusses the matter in 'Darwin at sea – and the virtues of port', one of the essays collected in *The Flamingo's Smile*.
† See his 1947 book *Darwin's Finches*. In 1994, the work of the Grants was the basis of another excellent book, *The Beak of the Finch*, by Jonathan Weiner. Peter Grant's own classic monograph of 1986, *Ecology and Evolution of Darwin's Finches*, was republished in 1999.

small island of Daphne Major dropped from 1,300 in January to less than 300 by December. The population of the dominant species, *Geospiza fortis*, the medium ground finch, dropped from 1,200 to 180. The cactus finch, *G. scandens*, fell from 280 to 110. Figures for other species confirmed that 1977 was a finch *annus horribilis*. But the Grant team didn't just count the numbers of each species dying and living. Being Darwinians, they looked at the *selective* mortality figures within each species. Were individuals with certain characteristics more likely to survive the catastrophe than others? Did the drought selectively change the relative composition of a population?

Yes, it did. Within the *G. fortis* population, the survivors were on average more than five per cent larger than those that succumbed. And the average beak after the drought was 11.07 mm long compared with 10.68 mm before. The mean depth of beak had similarly gone up from 9.42 mm to 9.96 mm. These differences may seem tiny but, within the sceptical conventions of statistical science, they were too consistent to be due to chance. But why would a drought year favour such changes? The team already had evidence that larger birds with larger beaks are more efficient than average birds at dealing with the big, tough, spiky seeds such as those of the weed *Tribulus*, which were just about the only seeds to be found during the worst of the drought. A different species, the large ground finch *G. magnirostris*, is the professional when it comes to handling *Tribulus* seeds. But Darwinian survival of the fittest is all about the relative survival of individuals within a species, not the relative survival of one species compared to another. And within the population of medium ground finches, the largest individuals with the largest beaks survived best. The average *G. fortis* individual became a tiny bit more like *G. magnirostris*. The Grant team had observed a small episode of natural selection in action, during a single year.

They witnessed another episode after the drought ended, which pushed the finch populations in the same evolutionary direction, but for a different reason. As with many species of bird, *G. fortis* males are larger than females, and they have larger beaks, which presumably equipped them to survive the drought better. Before the drought there were about 600 males and 600 females. Of the 180 individuals who survived, 150 were male. The rains, when they finally returned in January 1978, unleashed boom conditions which were ideal for breeding. But now there were five males for every female. Understandably, there was fierce competition among the males for the scarce females. And the males who won these sexual competitions, the new winners among the already larger-than-normal surviving males, again tended to be the largest males with the largest beaks. Once again, natural selection was driving the population to evolve larger body size and larger beaks, but for a different reason. As to why females prefer large males, the Seal's Tale has primed us to see significance in the fact that male *Geospiza* – the more competitive sex – are larger than females anyway.

If large size is such an advantage, why weren't the birds just larger in the first place? Because in other years, non-drought years, natural selection favours smaller individuals with smaller beaks. The Grants actually witnessed this in the years following 1982–83 when there happened to be an El Niño flood. After the flood, the balance of seeds changed. The large tough seeds of plants such as *Tribulus* became rare in comparison with the smaller, softer seeds of plants like *Cacabus*. Now smaller finches with smaller beaks came into their own. It wasn't that large birds couldn't eat small, soft seeds. But they needed more of them to maintain their larger bodies. So smaller birds now had a slight edge. And, within the population of medium ground finches, the tables were turned. The evolutionary trend of the drought years was reversed.

The differences in beak size between the successful and the unsuccessful birds in the drought year seem awfully small, don't they? Jonathan Weiner quotes a telling anecdote about this, from Peter Grant:

> Once, just as I was beginning a lecture, a biologist in the audience interrupted me: 'How much difference do you claim to see,' he asked me, 'between the beak of a finch that survives and the beak of a finch that dies?'
>
> 'One half of a millimetre, on average,' I told him.
>
> 'I don't believe it!' the man said. 'I don't believe a half of a millimetre really matters so much.'
>
> 'Well, that's the fact,' I said. 'Watch my data and then ask questions.'
> And he asked no questions.

Peter Grant calculated that it would take only 23 bouts of 1977-style drought on Daphne Major to turn *Geospiza fortis* into *G. magnirostris*. It wouldn't literally be *magnirostris*, of course. But it is a vivid way to visualise the origin of species, and how rapidly it can happen. Darwin little knew, when he met them and failed to label them properly, what powerful allies 'his' finches would eventually turn out to be.[*]

THE PEACOCK'S TALE

The peacock's 'tail' is not its true morphological tail (the true tail of a bird is the diminutive 'parson's nose'), but a 'fan' made of long back feathers. The Peacock's Tale is exemplary for this book because, in true Chaucerian style, it carries a message or moral from one pilgrim, which helps other pilgrims to understand themselves. In particular, when I was discussing two of the major transitions in human evolution, I looked forward to when the peacock would join our pilgrimage and give us the benefit of his (and I mean his in this case, not her) tale. It is, of course, a tale of sexual selection. Those two hominid transitions were our shift from four legs to two, and the subsequent enlargement of our brain. Let's add a third, perhaps less important but very characteristically human feature: our loss of body hair. Why did we become the Naked Ape?

There were lots of ape species in Africa in the late Miocene. Why did one of them suddenly and rapidly start evolving in a very different direction from the rest – indeed from the rest of the mammals? What picked out this one species and sent it hurtling at high speed in new and strange evolutionary directions: first to become bipedal, then to become brainy, and at some point to lose most of its body hair?

Rapid, apparently arbitrary spurts of evolution in quirky directions say one thing to me: sexual selection. This is where we have to start listening to the peacock. Why does the peacock have a train that dwarfs the rest of its body, quivering and shimmering in the sun with glorious eye-spot motifs of royal purple and green? Because generations of peahens have chosen peacocks who flaunted ancestral equivalents of these extravagant advertisements. Why does

[*] Hawaii is an even more remote volcanic archipelago, and about as young as the Galapagos. Hawaii's Robinson Crusoe bird was a honey-creeper, whose descendants rapidly evolved to 'do a Galapagos' – even evolving a 'woodpecker'. Similarly, about 400 original insect immigrant species spawned all 10,000 endemic Hawaiian species, including a unique carnivorous caterpillar and a semi-marine cricket. Apart from a bat and a seal, there are no native Hawaiian mammals. Alas, to quote E. O. Wilson's beautiful book, *The Diversity of Life*, 'Most of the honeycreepers are gone now. They retreated and vanished under pressure from overhunting, deforestation, rats, carnivorous ants, and malaria and dropsy carried in by exotic birds introduced to "enrich" the Hawaiian landscape.'

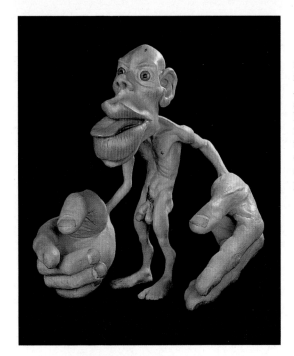

13. The human brain is hand-heavy
A Penfield homunculus showing parts of the human body blown up in proportion to the area of brain cortex dedicated to their sensory perception (see page 243).

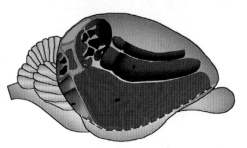

14. The platypus brain is bill-heavy
'Platypunculus', from Pettigrew *et al* [225] (see page 244).

15. The same ingenious trick?
The paddlefish (*Polyodon spathula*) (see page 247).

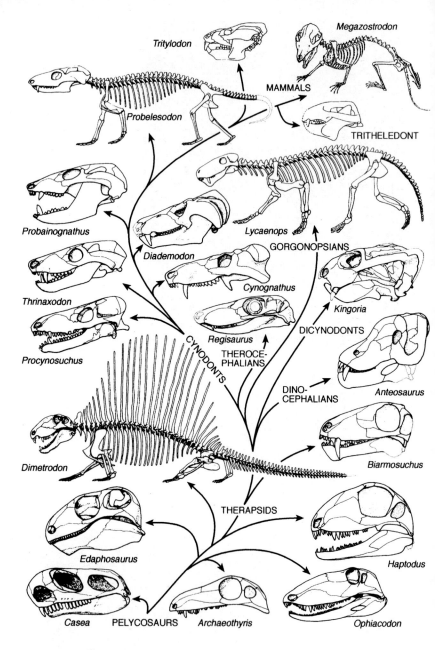

16. Before the dinosaurs

The phylogenetic relationships of the mammal-like reptiles. Adapted from Tom Kemp [1515] (see page 259).

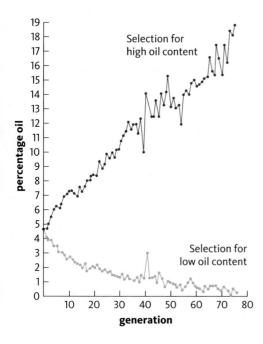

17. **The power of selection**
Impact of selection for high and low oil content in maize seeds over 90 generations. Adapted from Dudley and Lambert [85] (see page 265).

18. **Quirky, whimsical evolution**
Wilson's bird of paradise (*Diphyllodes respublica*) (see page 271).

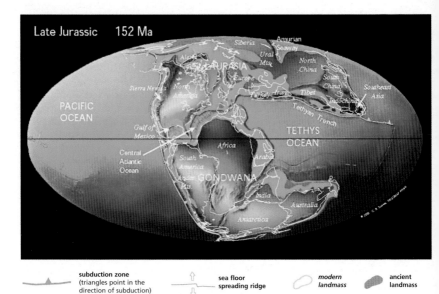

Late Jurassic   152 Ma

| | subduction zone (triangles point in the direction of subduction) | | sea floor spreading ridge | | *modern landmass* | | ancient landmass |

19. The Earth in the Upper Jurassic, about 150 million years ago [257]
The supercontinent of Pangaea had separated into Laurasia (in the north) and Gondwana (in the south), and the Atlantic Ocean was beginning to form. Gondwana itself was about to fragment. The climate was very warm (see page 291).

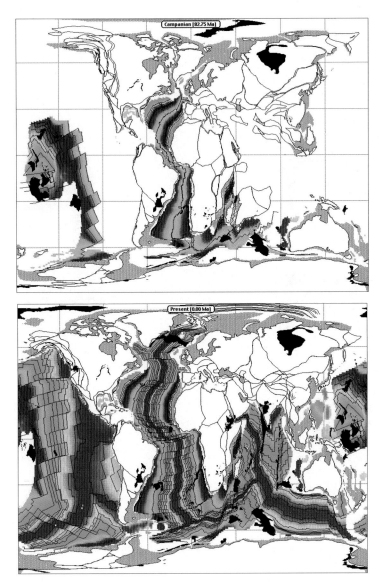

20. These two maps represent the ages of rocks on the seafloor, based on their remnant magnetism. The map at the top shows the Earth at 68 million years ago. The map below shows the Earth today. The false-colour bands show seafloor rocks from the Cretaceous period pushed back as new seafloor formed, widening the Atlantic (see page 298).

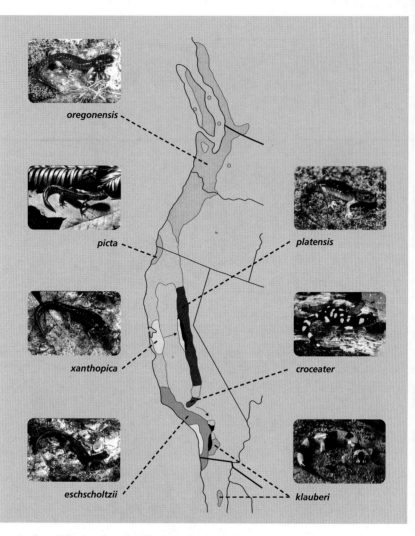

21. Strikes a blow against the discontinuous mind
Ensatina populations around the Central Valley, California. Stippled
areas indicate zones of transition. Map adapted from Stebbins (2003)
(see page 309).

Opposite
22. Concestor 18
The land vertebrates evolved from lobe-finned fish, such as the one reconstructed
here. The name comes from the prominent lobes on all fins, except the dorsal fin
and the heterocercal (asymmetrical) tail. Artistic reconstruction by Malcolm
Godwin (see page 330).

23. 'I would not have been more surprised if I had seen a dinosaur walking down the street'
Coelacanth (*Latimeria chalumnae*) photographed off the Comoros Islands in the Indian Ocean (see page 335).

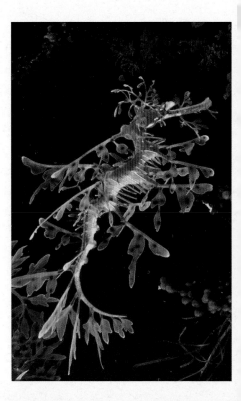

24. NOT seaweed
Leafy sea dragon (*Phycodurus equus*) (see page 340).

the male twelve-wired bird of paradise have red eyes and a black ruff with an iridescent green fringe, while Wilson's bird of paradise catches the eye with a scarlet back, yellow neck and blue head? Not because something in their respective diets or habitats predisposes these two species to their different colour schemes. No, these differences, and those that so conspicuously mark out all the other species of bird of paradise, are arbitrary, whimsical, unimportant to anybody – except female birds of paradise (see plate 18). Sexual selection does this kind of thing. Sexual selection produces quirky, whimsical evolution that runs away in apparently arbitrary directions, feeding on itself to produce wild flights of evolutionary fancy.

On the other hand, sexual selection also tends to magnify differences between the sexes – sexual dimorphism (see the Seal's Tale). Any theory that attributes human brains, bipedality or nakedness to sexual selection has got to face up to a major difficulty. There is no evidence that one sex is brainier than the other, nor that one sex is more bipedal than the other. It is true that one sex tends to be more naked than the other, and Darwin made use of this in his own sexual selection theory of the loss of human hair. He supposed that ancestral males chose females rather than the other way around as is normal in the animal kingdom, and that they preferred hairless females. When one sex evolves ahead of the other (in this case the female sex towards hairlessness) the other sex can be thought of as 'dragged in its wake'. It's the kind of explanation we more or less have to offer for that old chestnut, male nipples. It is not implausible to invoke it for the evolution of partial nakedness in man, dragged in the wake of more total nakedness in woman. The 'dragged in its wake' theory works less well for bipedality and braininess. The mind boggles – recoils even – when trying to imagine a bipedal member of one sex walking out with a quadrupedal member of the other. Nevertheless, the 'dragged in its wake' theory has a role to play.

There are circumstances in which sexual selection can favour monomorphism. My own suspicion, and Geoffrey Miller's in The Mating Mind, is that human mate choice, unlike perhaps that of peafowl, goes both ways. Moreover, our criteria for choice may be different when we look for a long-term partner than when we seek a one-night stand.

For the moment, we return to the simpler world of peacocks and peahens where females do the choosing and males strut around and aspire to be chosen. One version of the idea assumes that choice of mate (in this case choice by peahens) is arbitrary and whimsical compared with, for example, choice of food or choice of habitat. But you could reasonably ask why this should be so. According to at least one influential theory of sexual selection, that of the great geneticist and statistician R. A. Fisher, there is a very good reason. I have expounded the theory in detail in another book (*The Blind Watchmaker*, Chapter 8) and will not do so again here. The essential point is that male appearance and female taste evolve together in a kind of explosive chain reaction. Innovations in the consensus of female taste within a species, and corresponding changes in male appearance, are amplified in a runaway process which drives both of them in lockstep, further and further in one direction. There is no overweening reason for this one direction to be chosen: it just happens to be the direction in which the evolutionary trend started. The ancestors of peahens happened to take a step in the direction of preferring a larger fan. That was enough for the explosive engine of sexual selection. It kicked in and, within a very short time by evolutionary standards, peacocks were sprouting larger and more iridescent fans, and females couldn't get enough of them.

Every species of bird of paradise, many other birds, and fish and frogs, beetles and lizards, zoomed off in their own evolutionary directions, all bright colours or weird shapes – but different bright colours, different weird shapes. What matters for our purpose is that sexual selection, according to a sound mathematical theory, is apt to drive evolution to take off in arbitrary directions and push things to non-utilitarian excess. The suggestion arose in the chapters on human evolution that this is just what the sudden inflation of the brain looks like. So does the sudden loss of body hair, and even the sudden take-off into bipedality.

Darwin's *Descent of Man* is largely devoted to sexual selection. His lengthy review of sexual selection in non-human animals prefaces his advocacy of sexual selection as the dominant force in the recent evolution of our species. His treatment of human nakedness begins by dismissing – more glibly than his modern followers find comfortable – the possibility that we lost our hair for utilitarian reasons. His faith in sexual selection is reinforced by the observation that in all races, however hairy or however hairless, the women tend to be less hairy than the men. Darwin believed that ancestral men found hairy women unattractive. Generations of men chose the most naked women as mates.* Nakedness in men was dragged along in the evolutionary wake of nakedness in women, but never quite caught up, which is why men remain hairier than women.

For Darwin, the preferences that drove sexual selection were taken for granted – given. Men just prefer smooth women, and that's that. Alfred Russel Wallace, the co-discoverer of natural selection, hated the arbitrariness of Darwinian sexual selection. He wanted females to choose males not by whim but on merit. He wanted the bright feathers of peacocks and birds of paradise to be tokens of underlying fitness. For Darwin, peahens choose peacocks simply because, in their eyes, they are pretty. Fisher's later mathematics put that Darwinian theory on a sounder mathematical footing. For Wallaceans,† peahens choose peacocks not because they are pretty but because their bright feathers are a token of their underlying health and fitness.

In post-Wallace language, a Wallacean female is, in effect, reading a male's genes by their external manifestations from which she judges their quality. And it is a startling consequence of some sophisticated neo-Wallacean theorising that males are expected to go out of their way to make it easy for females to read their quality, even if their quality is poor. This piece of theory – progression of theories, rather – which we owe to A. Zahavi, W. D. Hamilton and A. Grafen, would take us too far afield, interesting as it is. My best attempt at expounding it is in the endnotes to the second edition of *The Selfish Gene*.

This brings us to the first of our three questions about human evolution. Why did we lose our hair? Mark Pagel and Walter Bodmer have made the intriguing suggestion that hairlessness evolved to reduce ectoparasites such as lice and, in keeping with the theme of this tale, as a sexually selected advertisement of freedom from parasites. Pagel and Bodmer followed Darwin's invocation of sexual selection, but in the neo-Wallacean version of W. D. Hamilton.

Darwin did not try to explain female preference, but was content to postulate it to explain male appearance. Wallaceans seek evolutionary explanations for sexual preferences themselves. Hamilton's favoured explanation is all about advertising health. When individuals choose their mates, they are looking for health, freedom from

---

* Of course, like my colleague Desmond Morris, I am using 'naked' to mean hairless rather than unclothed.
† Helena Cronin's term, in her wonderful book *The Ant and the Peacock*.

parasites, or signs that the mate is likely to be good at evading or combating parasites. And individuals seeking to be chosen advertise their health: make it easy for the choosers to read their health, whether it is good or bad. Patches of bare skin in turkeys and monkeys are conspicuous screens on which the health of their possessors is displayed. You can actually see the colour of the blood through the skin.

Humans don't just have bare skin on their rumps like monkeys. They have bare skin all over, except on the top of the head, under the arms and in the pubic region. When we get ectoparasites such as lice, they are often confined to these very regions. The crab louse, *Phthirus pubis*, is mainly found in the pubic region, but also infests the armpits, beard and even eyebrows. The head louse, *Pediculus humanus capitus*, infests only the hairs of the head. The body louse, *P. h. humanus*, is a subspecies in the same species as the head louse which, interestingly, is believed to have evolved from it only after we began to wear clothes. Some workers in Germany have looked at the DNA of head lice and body lice to see when they diverged, with a view to dating the invention of clothes. They put it at 72,000 years, plus or minus 42,000.

Lice need hair, and Pagel and Bodmer's first suggestion is that the benefit of losing our body hair was that it reduced the real estate available to lice. Two questions arise. Why, if losing hair is such a good idea, have other mammals who also suffer from ectoparasites kept theirs? Those, such as elephants and rhinos, that could afford to lose their hair because they are large enough to keep warm without it, have indeed lost it. Pagel and Bodmer suggest that it was the invention of fire and clothes that enabled us to dispense with our hair. This immediately leads to the second question. Why have we retained hair on our heads, under our arms and in the pubic region? There must have been some overriding advantages. It is entirely plausible that hair on the top of the head protects against sunstroke, which can be very dangerous in Africa where we evolved. As for armpit and pubic hair, it probably helps disseminate the powerful pheromones (airborne scent signals) that our ancestors certainly used in their sex lives, and which we still use more than many of us realise.

So, the straightforward portion of the Pagel/Bodmer theory is that ectoparasites such as lice are dangerous (lice carry typhus and other serious diseases), and ectoparasites prefer hair to bare skin. Getting rid of hair is a good way to make life difficult for these unpleasant and dangerous parasites. It is also much easier for us to see and pick off ectoparasites like ticks if we have no hair. Primates spend a substantial amount of time doing this, to themselves and to each other. It has become, indeed, a major social activity and, as a by-product, a vehicle for bonding.

But I find the most interesting angle on the Pagel/Bodmer theory is one that they treat rather briefly in their paper: sexual selection, which is why it belongs in the Peacock's Tale. Nakedness is not only bad news for lice and ticks. It is good news for choosers trying to discover whether a would-be sexual partner *has* lice or ticks. The Hamilton/Zahavi/Grafen theory predicts that sexual selection will enhance whatever it takes to help choosers tell whether would-be mates have parasites. Hairlessness is a beautiful example. On closing the Pagel/Bodmer paper I thought of T. H. Huxley's famous words: how extremely stupid not to have thought of that.

But nakedness is a small matter. As promised, let's turn now to bipedality and brains. Can the peacock help us to understand those two larger events in human evolution – the rise onto our hind legs, and the inflation of our brain? Bipedality came first, and I shall discuss it first. In Little Foot's Tale, I mentioned various theories of bipedality, including the recent squat-feeding theory of Jonathan Kingdon, which I find very convincing. I said that I was postponing my own suggestion to the Peacock's Tale.

Sexual selection, and its power to drive evolution in non-utilitarian arbitrary directions, is the first ingredient in my theory of the evolution of bipedality. The second is a tendency to imitate. The English language even has a verb, to ape, meaning to copy, although I am not sure how apt it is. Among all the apes, humans are the champion copyists, but chimpanzees do it too, and there is no reason to think australopithecines did not. The third ingredient is the widespread habit among apes generally of rising temporarily onto the hind legs, including during sexual and aggressive displays. Gorillas do it to drum on their chests with their fists. Male chimpanzees also thump their chests, and they have a remarkable display called the rain dance which involves leaping about on the hind legs. A captive chimpanzee called Oliver habitually and for preference walks on his hind legs. I have seen a film of him walking, and his stance is surprisingly erect – not a shambling totter, almost a military gait. So un-chimp-like is Oliver's walk that he has been the subject of bizarre speculations. Until DNA tests showed him to be a chimpanzee, *Pan troglodytes*, people have thought he might be a chimp/human hybrid, a chimp/bonobo hybrid, even a relict australopithecine. Unfortunately Oliver's biography is hard to piece together, and nobody seems to know whether he was taught to walk as a trick for a circus or fair-ground sideshow, or whether it is an odd idiosyncrasy: he might even be a genetic mutant. Oliver aside, orang utans are slightly better on their hind legs than chimpanzees; and wild gibbons actually run across clearings bipedally, in a style which is not very different from the way they run along branches in trees – when they are not brachiating under them.

Putting all these ingredients together, my suggestion for the origin of human bipedality is this. Our ancestors, like other apes, walked on all fours when not up in trees, but reared up on their hind legs from time to time, perhaps in something like a rain dance, or to pick fruits off low branches, or to move from one squat-feeding position to another, or to wade across rivers, or to show off their penises, or for any combination of reasons, just as modern apes and monkeys do. Then – this is the crucial additional suggestion I am adding – something unusual happened in one of those ape species, the one from which we are descended. A *fashion* for walking bipedally arose, and it arose as suddenly and capriciously as fashions do. It was a gimmick. An analogy might be found in the legend (probably false, alas) that the Spanish lisp sprang from the fashionable imitation of an admired courtier, or, in another version of the legend, a king of the Habsburg dynasty, or an infanta, who had a speech defect.

It'll be easiest if I tell the story in a sex-biased way, with females choosing males, but remember that it could have been the other way around. In my vision, an admired or dominant ape, a Pliocene Oliver perhaps, gained sexual attractiveness and social status through his unusual virtuosity in maintaining the bipedal posture, perhaps in some ancient equivalent of a rain dance. Others imitated his gimmicky habit and it became 'cool', 'it', 'the thing to do' in a local area, just as local bands of chimpanzees have habits of nut-cracking or termite-fishing which spread by fashionable imitation. In my teenage years, a more than usually inane popular song had the refrain,

> *Everybody's talking*
> *'Bout a new way of walking!*

And, while this particular line was probably chosen in the service of a lazy rhyme, it is undoubtedly true that styles of walking have a kind of contagiousness and are imitated because they are admired. The boarding school that I attended, Oundle in central England, had a ritual whereby the senior boys paraded into the chapel after the rest of us were in our places. Their mutually imitated style of walking, a mixture of swagger and lumbering roll (which I now, as a student of animal behaviour and a colleague of Desmond Morris, recognise as a dominance display) was so characteristic and idiosyncratic that my father, who saw it once a term on Parents' Day, gave it a name, 'the Oundle Roll'. The socially observant writer Tom Wolfe has named a particular loose-limbed gait of American dudes, fashionable in a certain social sector, the Pimp Roll. At the time of writing, the abject sycophancy of the British Prime Minister to the US President has earned him the title 'Bush's Poodle'. Several commentators have noticed that, especially when in his company, he imitates Bush's macho 'cowboy swagger', with arms held out to the sides as though ready to reach for two pistols.

Returning to our imagined sequence of events among human ancestors, females in the local area of the fashion preferred to mate with males who adopted the new way of walking. They preferred them for the same reason as individuals wanted to join the fashion: because it was admired in their social group. And now the next step in the argument is crucial. Those who were especially good at the fashionable new walk would be most likely to attract mates and sire children. But this would be of evolutionary significance only if there was a genetic component to the variation in ability to do 'the walk'. And this is entirely plausible. We are talking, remember, about a quantitative shift in the amount of time spent doing an existing activity. It is unusual for a quantitative shift in an existing variable *not* to have a genetic component.

The next step in the argument follows standard sexual selection theory. Those choosers whose taste conforms to the majority taste will tend to have children who inherit, from their mothers' choice of mate, skill in walking according to the bipedal fashion. They will also have daughters who inherit their mothers' taste in males. This dual selection – on males for possessing some quality and on females for admiring the self-same quality – is the ingredient for explosive, runaway selection, according to the Fisher theory. The key point is that the precise direction of the runaway evolution is arbitrary and unpredictable. It could have been opposite. Indeed, in another local population perhaps it was in the opposite direction. An explosive evolutionary excursion, in an arbitrary and unpredictable direction, is just the kind of thing we need, if we are to explain why one group of apes (who became our ancestors) suddenly evolved in the direction of bipedality while another group of apes (the ancestors of the chimpanzees) did not. An additional virtue of the theory is that this evolutionary spurt would have been exceptionally fast: just what we need in order to explain the otherwise puzzling closeness in time of Concestor 1 and the supposedly bipedal Toumai and *Orrorin*.

Let's turn now to the other great advance in human evolution, the enlargement of the brain. The Handyman's Tale discussed various theories, and again we left sexual selection till last, postponing it until the Peacock's Tale. In *The Mating Mind*, Geoffrey Miller argues that some very high percentage of human genes, perhaps up to 50 per cent, express themselves in the brain. Yet again, for the sake of clarity, it's convenient to tell the story from one point of view only – females choosing males – but it could go the other way: or both ways simultaneously. A female who seeks a penetrating and thorough reading of the quality of a male's genes would do well to concentrate on his brain. She can't literally look at the brain, so she looks at its works. And, following the theory that males should make it easy by advertising their quality, males will not hide their mental light under a bony bushel but bring it out into the open. They will dance, sing, sweet-talk, tell jokes, compose music or poetry, play it or recite it, paint cave walls or Sistine chapel ceilings. Yes, yes, I know Michelangelo might not, as it happens, have been interested in impressing females. It is still entirely plausible that his brain was 'designed' by natural selection for impressing females, just as – whatever his personal preferences – his penis was designed for impregnating them. The human mind, on this view, is a mental peacock's tail. And the brain expanded under the same kind of sexual selection as drove the enlargement of the peacock's tail. Miller himself favours the Wallacean rather than the Fisherian version of sexual selection, but the consequence is essentially the same. The brain gets bigger, and it does so swiftly and explosively.

The psychologist Susan Blackmore, in her audacious book *The Meme Machine*, has a more radical sexual selection theory of the human mind. She makes use of what have been called 'memes', units of cultural inheritance. Memes are not genes, and they have nothing to do with DNA except by analogy. Whereas genes are transmitted via fertilised eggs (or via viruses), memes are transmitted via imitation. If I teach you how to make an origami model of a Chinese junk, a meme passes from my brain to yours. You may then teach two other people the same skill, each of whom teaches two more, and so on. The meme is spreading exponentially, like a virus. Assuming we have all done our teaching work properly, later 'generations' of the meme will not be detectably different from earlier ones. All will produce the same origami 'phenotype'.* Some junks may be more perfect than others, as some paper-folders take more trouble, say. But quality will not deteriorate gradually and progressively over the 'generations'. The meme is passed on, whole and intact like a gene, even if its detailed phenotypic expression varies. This particular example of a meme is a good analogue for a gene, specifically a gene in a virus. A manner of speaking, or a skill in carpentry, might be more dubious candidates for memes because – I am guessing – progressively later 'generations' in a lineage of imitation will probably become progressively more different from the original generation.

Blackmore, like the philosopher Daniel Dennett, believes that memes played a decisive role in the process that made us human. In Dennett's words:

> The haven all memes depend on reaching is the human mind, but a human mind is itself an artifact created when memes restructure a human brain in order to make it a better habitat for memes. The avenues for entry and departure are modified to suit local conditions, and strengthened by various artificial devices that enhance fidelity and prolixity of replication: native Chinese minds differ dramatically from native French minds, and literate minds differ from illiterate minds.†

It would be Dennett's view that the main difference between anatomically modern brains before the cultural Great Leap Forward and after it is that the latter are swarming with memes. Blackmore goes further. She invokes memes to explain the evolution of the large

---

* As we saw in the Beaver's Tale, phenotype normally means the external appearance by which a gene manifests itself, for example eye colour. Obviously I am here using it in an analogous sense: the visible phenotype of a meme otherwise buried in the brain, as opposed to the phenotype of a gene buried in a chromosome. This is also a good analogy for the 'self-normalising' that I mentioned in the General Prologue, under 'Renewed Relics'. See also my foreword to Blackmore's book.

† Dennett makes constructive use of the theory of memes in various places, including *Consciousness Explained* (from which the quotation is taken) and *Darwin's Dangerous Idea*.

human brain. It can't be only memes, of course, because we are talking about major anatomical change here. Memes may manifest themselves in the circumcised penis phenotype (which sometimes passes, in quasi-genetic fashion, from father to son), and they might even manifest themselves in body shape (think of a transmitted fashion for slimming, or elongating the neck with rings). But a doubling in brain size is another matter. This has got to come about through changes in the gene pool. So what role does Blackmore see for memes in the evolutionary expansion of the human brain? This is, again, where sexual selection comes in.

People are most apt to copy their memes from admired models. This is a fact that advertisers bet money on: they pay footballers, film stars and supermodels to recommend products – people who have no expertise to judge them. Attractive, admired, talented or otherwise celebrated people are potent meme donors. The same people also tend to be sexually attractive and therefore, at least in the sort of polygamous society in which our ancestors probably lived, potent gene donors. In every generation, the same attractive individuals contribute more than their fair share of both genes and memes to the next generation. Now Blackmore assumes that part of what makes people attractive is their meme-generating minds: creative, artistic, loquacious, eloquent minds. And genes help to make the kind of brains that are good at generating attractive memes. So, quasi-Darwinian selection of memes in the meme pool goes hand in hand with genuinely Darwinian sexual selection of genes in the gene pool. It is yet another recipe for runaway evolution.

What, on this view, is the exact role of memes in the evolutionary swelling of the human brain? I think the most helpful way to look at it is this. There are genetic variations in brains which would remain unnoticed without memes to bring them out into the open. For example, the evidence is good that there is a genetic component to variation in musical ability. The musical talent of members of the Bach family probably owed much to their genes. In a world full of musical memes, genetic differences in musical ability shine through and are potentially available for sexual selection. In a world before musical memes entered human brains, genetic differences in musical ability would still have been there, but would not have manifested themselves, at least not in the same way. They would have been unavailable for sexual, or natural, selection. Memetic selection cannot change brain size by itself, but it can bring into the open genetic variation that would otherwise have remained under cover. This could be seen as a version of the Baldwin Effect, which we met in the Hippo's Tale.

The Peacock's Tale has used Darwin's beautiful theory of sexual selection to pick up a number of questions about human evolution. Why are we naked? Why do we walk on two legs? And why do we have big brains? I do not want to go out on a limb for sexual selection as the universal answer to all outstanding questions about human evolution. In the particular case of bipedalism, I am at least as persuaded by Jonathan Kingdon's 'squat feeding' theory. But I applaud the current vogue for giving sexual selection another serious look, after its long neglect since Darwin first proposed it. And it does provide a ready answer to the supplementary question that so often lurks behind the main questions: why, if bipedalism (or braininess or nakedness) was such a good idea for us, do we not see it in other apes? Sexual selection is good at that, because it predicts sudden evolutionary spurts in arbitrary directions. On the other hand, the lack of sexual dimorphism in braininess and in bipedality demands some special pleading. Let's leave the matter there. It needs more thought.

THE DODO'S TALE

Land animals, for obvious reasons, have a hard time reaching remote oceanic islands such as the Galapagos archipelago, or Mauritius. If, through the much invoked freak accident of rafting inadvertently on a detached mangrove, they do happen to find themselves on an island like Mauritius, an easy life is likely to open up. This is precisely because it is hard to get to the island in the first place, so the competition and the predation are usually not so fierce as on the mainland left behind. As we have seen, this is probably how monkeys and rodents arrived in South America.

If I say it is 'hard' to colonise an island, I must hasten to forestall the usual misunderstanding. A drowning individual may try desperately to reach land, but no species ever *tries* to colonise an island. A species is not the kind of entity that tries to do anything. Individuals of a species may happen, by luck, to find themselves in a position to colonise an island previously uninhabited by their kind. The individuals concerned can then be expected to take advantage of the vacuum, and the consequence may be that their species, with hindsight, is said to have colonised the island. The descendants of the species may subsequently change their ways, over evolutionary time, to accommodate the unfamiliar island conditions.

And now here is the point of the Dodo's Tale. It is hard for land animals to reach an island, but it is a lot easier if they have wings. Like the ancestors of the Galapagos finches . . . or the ancestors of the dodo, whoever they were. Flying animals are in a special situation. They don't need the proverbial mangrove raft. Their wings carry them, perhaps as a freak accident, blown on a gale, to a distant island. Having arrived on wings, they find that they no longer need them. Especially because islands often lack predators. This is why island animals, as Darwin noted on the Galapagos, are often remarkably tame. And this is what makes them easy meat for sailors. The most famous example is the dodo, *Raphus cucullatus*, cruelly named *Didus ineptus* by Linnaeus, the father of taxonomy.

The very name dodo comes from the Portuguese for stupid. Stupid is unfair. When Portuguese sailors arrived on Mauritius in 1507, the abundant dodos were completely tame, and approached the sailors in a manner which cannot have been far from 'trusting'. Why would they not trust, for their ancestors had not encountered a predator for thousands of years? Alas for trust. The unfortunate dodos were clubbed to death by Portuguese, and later Dutch, sailors – even though they were deemed 'unpalatable'. Presumably it was 'sport'. Extinction took less than two centuries. As so often, it came about through a combination of killing and more indirect effects. Humans introduced dogs, pigs, rats, and religious refugees. The first three ate dodo eggs, and the last planted sugar cane and destroyed habitats.

Conservation is a very modern idea. I doubt that extinction, and what it means, entered anybody's head in the seventeenth century. I can hardly bear to tell the story of the Oxford Dodo, the last dodo stuffed in England. Its owner and taxidermist, John Tradescant, was induced to bequeath his large collection of curios and treasures to the infamous (some say) Elias Ashmole, which is why the Ashmolean Museum in Oxford is not called the Tradescantian as (some say) it should be. Ashmole's curators (some say, probably falsely) later decided to burn, as rubbish, all of Tradescant's dodo except the beak and one foot. These are now in my place of work, the University Museum of Natural History, where they memorably inspired Lewis Carroll. Also Hilaire Belloc:

The Dodo used to walk around
And take the sun and air.
The sun yet warms his native ground –
The Dodo is not there!

The voice, which used to squawk and squeak
Is now for ever dumb –
Yet you may see his bones and beak
All in the Mu-se-um.

The white dodo, *Raphus solitarius*, was alleged to have met the same fate on the neighbouring island of Réunion.* And Rodriguez, the third island of the Mascarene archipelago, housed, and lost for the same reason, a slightly more distant relative, the Rodriguez solitaire, *Pezophaps solitaria*.

The ancestors of the dodos had wings. Their forebears were flying pigeons who arrived on the Mascarene Islands under their own muscle power, perhaps aided by a freak wind. Once there, they had no need to fly any more – nothing to flee – and so lost their wings. Like Galapagos and Hawaii, these islands are recent volcanic creations, none of them more than seven million years old. Molecular evidence suggests that the dodo and solitaire probably arrived on the Mascarene Islands from the East, not from Africa or Madagascar as we might otherwise have supposed. Perhaps the solitaire did the bulk of its evolutionary divergence before it finally arrived on Rodriguez, retaining enough wing power to get there from Mauritius.

Why bother to lose the wings? They took a long time to evolve, why not hang on to them in case one day they might come in useful again? Alas (for the dodo) that is not the way evolution thinks. Evolution doesn't think at all, and certainly not ahead. If it did, the

---

* However, my astoundingly knowledgeable research assistant Sam Turvey informs me that the white dodo almost certainly never existed: 'White dodos are figured in a few seventeenth-century paintings, and contemporary travellers made reference to large, white birds on Réunion, but the accounts are vague and possibly confused, and no raphid skeletal material is known from the island. Although the species has been given the scientific name *Raphus solitarius*, and the eccentric Japanese naturalist Masauji Hachisuka defended the occurrence of two dodo species on Réunion (which he named *Victoriornis imperialis* and *Ornithaptera solitaria*), it is more likely that the early accounts either refer to an extinct Réunion ibis (*Threskiornis solitarius*), for which skeletal material is known and which was apparently similar to the living white sacred ibis, or to immature specimens of the grey brown dodo of Mauritius. Alternatively, they may merely be the product of artistic licence.'

dodos would have kept their wings, and the Portuguese and Dutch sailors would not have had sitting targets for their vandalism.

The late Douglas Adams was moved by the sad case of the dodo. In one of the episodes of *Doctor Who* that he wrote in the 1970s, the aged Professor Chronotis's college room in Cambridge serves as a time machine, but he uses it for one purpose only, his secret vice: he obsessively and repeatedly visits seventeenth-century Mauritius in order to *weep for the dodo.* Because of a strike at the BBC, this episode of *Doctor Who* was never broadcast, and Douglas Adams later recycled the haunting dodo *motif* in his novel *Dirk Gently's Holistic Detective Agency.* Call me sentimental, but I must pause for a moment – for Douglas, and for Professor Chronotis and what he wept for.

Evolution, or its driving engine natural selection, has no foresight. In every generation within every species, the individuals best equipped to survive and reproduce contribute more than their fair share of genes to the next generation. The consequence, blind as it is, is the nearest approach to foresight that nature admits. Wings might be useful a million years hence when sailors arrive with clubs. But wings will not help a bird contribute offspring and genes to the next generation, in the immediate here and now. On the contrary wings, and especially the massive breast muscles needed to power them, are an expensive luxury. Shrink them, and the resources saved can now be spent on something more immediately useful such as eggs: immediately useful for surviving and reproducing the very genes that programmed the shrinkage.

That's the kind of thing natural selection does all the time. It is always tinkering: here shrinking a bit, there expanding a bit, constantly adjusting, putting on and taking off, optimising immediate reproductive success. Survival in future centuries doesn't enter into the calculation, for the good reason that it isn't really a calculation at all. It all happens automatically, as some genes survive in the gene pool and others don't.

The sad end of the Oxford Dodo (Alice's Dodo, Belloc's Dodo) is mitigated by a happier sequel. A group of Oxford scientists in the laboratory of my colleague Alan Cooper obtained permission to take a tiny sample from inside one of the foot bones. They also obtained a thigh bone of a solitaire found in a cave on Rodriguez. These bones yielded enough mitochondrial DNA to allow detailed, letter-by-letter sequence comparisons between the two extinct birds and a wide range of living birds. The results confirm that, as long suspected, dodos were modified pigeons. It is also no surprise that, within the pigeon family, the closest relative of the dodo is the solitaire, and vice versa. What is less expected is that these two extinct flightless giants are nested deep inside the pigeon family tree. In other words, dodos are more closely related to some flying pigeons than those flying pigeons are to other flying pigeons; despite the fact that, to look at them, you'd expect all the flying pigeons to be more closely related to each other, with the dodos out on a limb. Among pigeons, the dodos are most close to the Nicobar, *Caloenus nicobarica,* a beautiful pigeon from South-East Asia. In turn, the group consisting of the Nicobar pigeon and the dodos is most closely related to the Victoria crowned pigeon, a splendid bird from New Guinea, and *Didunculus,* a rare Samoan toothbilled pigeon which looks quite like a dodo and whose name even means 'little dodo'.*

The Oxford scientists comment that the nomadic lifestyle of the Nicobar pigeon makes it ideally suited to invade remote islands, and Nicobar-type fossils are known from Pacific islands as far east as the Pitcairns. These crowned and toothbilled pigeons, they go on to point out, are large, ground-dwelling birds who rarely fly. It looks as though this whole subgroup of pigeons habitually colonise islands and then lose their power of flight and become larger and more dodo-like. The dodo itself and the solitaire have pushed the trend to extremes.

Something like the Dodo's Tale has been repeated on islands all over the world. Many different families of birds, most of which are dominated numerically by flying species, have evolved flightless forms on islands. Mauritius itself had a large flightless rail, *Aphanapteryx bonasia,* also now extinct, which may on occasion have been confused with the dodo. Rodriguez had a related species, *A. leguati.* Rails seem to lend themselves to Dodo's Tale island-hopping followed by flightlessness. In addition to the Indian Ocean forms, there is a flightless rail in the Tristan da Cunha group in the South Atlantic; and most of the Pacific islands have – or had – their own species of flightless rail. Before man ruined the Hawaiian avifauna, there were more than twelve species of flightless rail in that archipelago. More than a quarter of all the world's 60-odd living species of rail are

---

* A fossilised giant flightless pigeon, *Natunaornis,* approaching the dodo in size, has recently been found in Fiji.

flightless, and all flightless rails live on islands (if you count large islands like New Guinea and New Zealand). Perhaps as many as 200 species have gone extinct on tropical Pacific islands since human contact.

Again on Mauritius, and also now extinct, was a large parrot *Lophopsittacus mauritianus*. This crested parrot was a poor flyer and may have occupied a niche similar to the still (just) surviving kakapo of New Zealand.* New Zealand is, or was, home to a large number of flightless birds belonging to many different families. One of the more striking was the so-called adzebill – a stout, chunky bird, distantly related to cranes and rails. There were different species of adzebill on the North and South Islands, but neither island had any mammals except (for the obvious reason which underlies the Dodo's Tale) bats, and it is easy to imagine that the adzebills made their living in a rather mammal-like way, filling a gap in the market.

In all these cases, the evolutionary story is almost certainly a version of the Dodo's Tale. Ancestral flying birds are carried by their wings to a remote island where an absence of mammals opens up opportunities for making a living on the ground. Their wings are no longer useful in the way that they were on the mainland, so the birds give up flying, and their wings and costly wing muscles degenerate. There is one notable exception, one of the oldest and the most famous of all the groups of flightless birds: the ratites, the ostrich order. The evolutionary story of the ratites is very different from all the rest of the flightless birds, and they have a tale of their own, the Elephant Bird's Tale.

THE ELEPHANT BIRD'S TALE

From the tales of the *Arabian Nights*, the image that most stirred my childish imagination was the roc encountered by Sinbad the Sailor, who at first thought this monstrous bird was a cloud, come over the sun:

> I had heard aforetime of pilgrims and travellers, how in a certain island dwelleth a huge bird, called the 'roc', which feedeth its young on elephants.

The legend of the roc (rucke or rukh) surfaces in several stories of the

* Also memorably celebrated by Douglas Adams, in *Last Chance to See.*

Arabian Nights – two involving Sinbad and two about Abd-al-Rahman. It is mentioned by Marco Polo as living in Madagascar, and envoys from the King of Madagascar were said to have presented the Khan of Cathay with a roc feather. Michael Drayton (1563–1631) invoked the monstrous bird's name to contrast it with the proverbially tiny wren:

> All feathered things yet ever knowne to men,
> From the huge Rucke, unto the little Wren . . .

What is the origin of the roc legend? And if it is pure fantasy, whence the recurrent connection with Madagascar?

Fossils from Madagascar tell us that a gigantic bird, the elephant bird *Aepyornis maximus*, lived there, perhaps until as late as the seventeenth century,* although more probably around 1000 AD. The elephant bird finally succumbed, perhaps partly through people stealing its eggs which were up to a metre in circumference† and would have provided as much food as 200 chicken eggs. The elephant bird was three metres tall and weighed nearly half a tonne – as much as five ostriches. Unlike the legendary roc (which used its 16-metre wingspan to carry Sinbad aloft as well as elephants) the real elephant bird could not fly, and its wings were (relatively) small like an ostrich's. But, though a cousin, it would be wrong to imagine it as a scaled-up ostrich: it was a more robust, heavy-set bird, a kind of feathered tank with a big head and neck, unlike the ostrich's slender periscope. Given how legends readily grow and inflate, *Aepyornis* is a plausible progenitor of the roc.

The elephant bird was probably vegetarian, unlike the fabulously jumbo-phagous roc, and unlike earlier groups of giant carnivorous birds such as the phorusrhachoid family of the New World. These could grow to the same height as *Aepyornis*, with a fearsomely hooked beak which, as if in justification of their nickname of 'feathered tyrannosaurs', looks capable of swallowing a medium-sized lawyer whole. These monstrous cranes seem at first sight better casting for the role of the terrifying roc than *Aepyornis*, but they went extinct too long ago to have started the legend, and in any case Sinbad (or his real-life Arab counterparts) never visited the Americas.

---

* Actually there were several related species, in two genera, *Aepyornis* and *Mullerornis*. But *A. maximus*, as its name suggests, most deserves to be called the elephant bird.
† Not diameter – it is not quite as surprising as it sounds.

They've gone and there ain't no moa.
Sir Richard Owen with the skeleton of
Dinornis, the giant moa. Owen, to
whom we owe the term dinosaur, was
the first to describe the moa.

The elephant bird of Madagascar is the heaviest bird known to have lived, but it was not the longest. Some species of moa could reach a height of 3.5 metres, but only if the neck was raised, as in Richard Owen's mounting (see photo). In life, it seems, they normally carried the head only a little way above the back. But the moa cannot have generated the roc legend, for New Zealand, too, was well beyond Sinbad's ken. About ten moa species existed in New Zealand, ranging in size from turkey to double-sized ostrich.* Moas are extreme among flightless birds in that they have no trace of wings at all, not even buried vestiges of wing bones. They thrived in both the North and South Islands of New Zealand until the recent invasion by the Maori people, about 1250 AD. They were easy prey, no doubt for the same reason as the dodo. Except for the (extinct) Haast's eagle, the largest eagle ever to have lived, they had known no predators for tens of millions of years, and the Maoris slaughtered them all, eating the choicer parts and discarding the rest, belying, not for the first time, the wishful myth of the noble savage living in respectful harmony with his environment. By the time the Europeans arrived, only a few centuries after the Maoris, the last moa was gone. Legends

---

* Kiwis are smaller than turkeys but are no longer regarded as dwarf moas. As we shall see, they are closer cousins of the emus and cassowaries and arrived later from Australia.

and tall stories of sightings persist to this day, but the hope is forlorn. In the words of a plaintive song, to be sung in a mournful New Zealand accent:

> No moa, no moa
> In old Ao-tea-roa.*
> Can't get 'em.
> They've et 'em;
> They've gone and there ain't no moa!

Elephant birds and moas (but not the carnivorous phorusrhachoids nor various other extinct flightless giants) were ratites, an ancient family of birds, which now includes the rheas of South America, the emus of Australia, the cassowaries of New Guinea and Australia, the kiwis of New Zealand, and the ostrich, now confined to Africa and Arabia but previously common in Asia and even Europe.

I take delight in the power of natural selection, and it would have given me satisfaction to report that the ratites evolved their flightlessness separately in different parts of the world, conforming to the message of the Dodo's Tale. In other words, I would have liked the ratites to have been an artificial assemblage, driven to superficial resemblance by parallel pressures in different places. Alas, this is not so. The true tale of the ratites, which I am attributing to the elephant bird, is very different. And I must say that it eventually turns out to be, in its way, even more fascinating. The Elephant Bird's Tale, taken together with its Epilogue, is a tale of Gondwana, and of continental drift or, as it is now called, plate tectonics.

The ratites are a truly natural group. Ostriches, emus, cassowaries, rheas, kiwis, moas and elephant birds really are more closely related to each other than they are to any other birds. And their shared ancestor was flightless too. Probably it originally lost its wings, following good 'Dodo's Tale' reasons, after flying to some long-forgotten island off Gondwana. But that was before the ratites divided into the separate forms whose descendants we now find on the different southern continents and islands. Moreover, the split of the ratites from the rest of the birds is extremely old. The ratites are a genuinely ancient group in the following sense. Surviving birds fall into two groups. On the one hand are the ratites and the tinamous (a group of South American birds which can fly). On the other hand are

---

* The Maori name for New Zealand.

all the rest of the surviving birds put together. So if you are a bird, *either* you are a ratite/tinamou *or* you are in with the rest, and the division between these two categories is the oldest split among surviving birds. I have to say surviving birds because there are several groups of extinct birds (including flightless as well as flying forms) which lie outside the cousinship of all modern birds.

The ratites, then, are a natural group, with a common ratite ancestor that was also flightless. This is not to deny that an earlier ancestor of all the ratites flew. Of course it did, for why else would they (most of them) have vestigial wings? There is even fossil evidence in the form of *Lithornis,* a flying relative of the ratites, which lived in North America in the Palaeocene and Eocene Epochs. But the last common ancestor of all the surviving ratites had already reduced its wings to stubby vestiges long before its descendants branched into the various groups of ratites we see today. This deprives us of our accustomed Dodo's Tale of ancestors flying across the sea to distant lands and then each independently losing its wings. The ratites reached their present separated homelands without benefit of flight. How did they get there?

They walked. All the way.* How is this possible? It is the whole point of the Elephant Bird's Tale. The sea wasn't there – there was nothing to cross. What we now know as separate continents were joined together, and the great flightless birds walked dry shod.

When I was a small child in Africa my father regaled my little sister and me with bedtime stories, as we lay under our mosquito nets and marvelled at his luminous wristwatch, about a 'Broncosaurus' who lived faaaaaaaaaaaar away in a place called Gonwonky-land. I forgot all about this until much later when I learned about the great southern continent of Gondwanaland.

One hundred and fifty million years ago, Gondwanaland, or Gondwana,† consisted of everything that we now know as South America, Africa, Arabia, Antarctica, Australasia, Madagascar and India. The southern tip of Africa was touching Antarctica, and tilted to the 'right'. There was therefore a triangular gap between the east coast of Africa and the north coast of Antarctica – but it wasn't really

---

* With the probable exception of the kiwi, as we shall see.
† 'Gondwanaland' is criticised as a tautology, because *vana* in Sanskrit means land (actually forest). I shall not use it. But it has the virtue of distinguishing the giant continent from the region of central Madhya Pradesh where the Gonds live, which is still called Gondwana and which gave its name to the Gondwana geological series.

a gap because it was filled by India. India was in those days separated from the rest of Asia (Laurasia) by an ocean, the Tethys, whose centre roughly corresponds today to the modern Indian Ocean and whose westernmost reaches turned into today's Mediterranean Sea. Madagascar nestled between India and Africa, joined on both sides. Australia with New Guinea, and the embryonic New Zealand, were also joined to Antarctica, further round the coast from India (see plate 19).

But Gondwana was about to break up. You can see where the tale is going. When the ratite birds first roamed Gondwana, they could walk from any of the places where they later dwelt, to any other. Ratite fossils have even been found in Antarctica, which we know from plant fossils to have been covered with warm, subtropical forest at the time. Ancestral ratites wandered freely over the whole continent of Gondwana with no inkling that their homeland was destined to be broken up into chunks separated by thousands of miles of ocean. When it did break up, the ratites went too. They rafted all right. But their raft was not the proverbial fragment of mangrove. It was the very ground beneath their feet. And there was plenty of time for them to evolve their separate ways, on their separately retreating landmasses.

The breakup happened rather suddenly and explosively, by the standards of geological time. About 150 million years ago, India (still with Madagascar attached) started to break away from Africa. As the gap widened between Africa and India/Madagascar, by about 140 million years ago clear water began to open up between the other side of India and Antarctica, and between Australia and Antarctica. A little later, South America began to draw away from the west side of Africa, and by 120 million years ago an immensely long and narrow, angled channel separated the two. The last place you could walk across was where West Africa just hung on by a thread to what is now Brazil. By then, a similarly long, narrow channel had opened up between Antarctica and the new south coast of Australia. Around 80 million years ago, Madagascar split off from India, and remained approximately in its present position while India began a spectacularly fast migration north, eventually to crunch into the south coast of Asia and raise the Himalayas. During the same period, the other fragments of Gondwana had continued to drift apart, each bearing its manifest of ratite passengers – ancestral rheas on the new South American continent, ancestral elephant birds on India/Madagascar, ancestral emus on Australia, ancestral ostriches on . . . but no, let's postpone that matter.

Plant fossils tell us that Cretaceous Antarctica was sub-tropical, lush with vegetation, and a fine place for animals to live.* The dearth of fossils that have actually been found cannot reflect a corresponding dearth of animals; a rich vegetation like that must have supported an equally rich fauna. As I've already mentioned, among the few fossil animals that *have* been found are large ratites, some as large as moas, and it seems likely that these birds were abundant in Cretaceous Antarctica. If not necessarily Ratite Grand Central Station, Antarctica provided a clement and ratite-friendly land bridge linking Africa and South America on one side of the world to Australia and New Zealand on the other, and India/Madagascar too.

From the point of view of a wandering ratite ancestor, what matters is not when the great bulk of its particular continent separated from the rest of Gondwana. What matters is the last moment when it could still have walked across the gap. For example, by 100 million years ago, Africa was widely separated from Antarctica in the south and from India/Madagascar in the east. From these points of view, Africa was already an island. It was also widely separated from South America, almost all the way along its west coast. But there was still that lingering bridge, between the southern margin of the West African bulge and a part of what is now Brazil. This was the last moment of contact between the ancestors of the rhea and the rest of what had once been Gondwana. We have other dates of last contact between the various elements of the Gondwanan continental diaspora.

Is there a coincidence between the times at which the various continents and islands split apart geographically, and the times at which, on molecular genetic evidence, the corresponding ratite bird lineages split apart in evolution? Or, if that is too much to ask, are the two sets of timings at least compatible with each other? Yes, they are. And (with the exception of the kiwi and, in an interesting sense which I shall come to, the ostrich) they are incompatible with the alternative hypothesis that the ratites distributed themselves among

---

* Although, strange-seeming to us, the southernmost parts would still have spent a substantial fraction of the year in darkness. Presumably this led to all sorts of behavioural adaptations for which there are no modern counterparts, for today extreme latitude goes with extreme cold.

their present landmasses *after* those landmasses separated from one another.

Alan Cooper and his colleagues at Oxford, whom we met in the Dodo's Tale, have compared the molecular genetics of all the ratite birds. Doing this for surviving birds is easy. You just take blood from zoo specimens of ostriches, emus and the rest. Indeed, lots of sequences have already been published in the technical literature. But the Cooper team achieved the additional coup of sequencing mitochondrial DNA from two genera of moas and an elephant bird, for which they had only old bones borrowed from museums. Remarkably, the team managed to piece together the entire mitochondrial genome of both genera of moas, although they were at least 700 years dead. The material from elephant birds was less well-preserved, but still they managed to sequence some elephant bird DNA. These ancient DNA sequences could then be compared with each other and with the sequences of the surviving ratites. The molecular clock technique allowed them to put approximate dates on the evolutionary divergences among the ratites.

It would be nice to be able to say that the molecular splits among the ratites do indeed coincide precisely with the geographical splits between their homelands. Unfortunately, the dating is not precise enough to be sure. Remember, too, that island-hopping remains an option, even for flightless animals, for quite a while after their continents have drawn apart, so the exact moment at which the various parts of Gondwana broke away is not too significant. Flightless birds, after all, are no *more* flightless than mammals such as the monkeys and rodents that somehow crossed from Africa to South America, or the iguanas blown by the hurricane to Anguilla. To make it harder, Gondwana fractured into most of its parts pretty much simultaneously (again using the word in its geological sense of 'give or take a few million years'). What the molecular sequences now let us say with confidence is that the ancestral splits among the ratites are very old – old enough to be fully compatible with the view that their ancestors were already in their separate southern hemisphere homelands when they pulled apart.

Here's the best guess at what happened. Think of Antarctica as the unit from which the other continents split. Of course the continents split from each other as well, but it helps to have a reference point and Antarctica is conveniently central, which makes it easy to visualise. What is more, as we have seen, during the period that matters for our tale, in the Cretaceous Period spanning roughly the 40 million years on either side of the 100-million-year mark, Antarctica was by no means the frozen waste it is now. Is this because Antarctica was at a more clement latitude? No, it was only somewhat north of its present position. It was warm because coastal shapes in those days happened to direct warm currents from the tropics to far southern latitudes, in a more dramatic version of the way the Gulf Stream fosters palm trees in western Scotland today. One of the consequences of the break-up of Gondwana was that the warm current was no longer directed south. Antarctica reverted to the icy climate appropriate to its latitude, and it has been cold ever since.

So, there were plenty of ratites in Antarctica, at just the right time. The rest of the tale is straightforward. South America was already well populated with the ancestors of the rheas. New Zealand broke away from Antarctica about 70 million years ago, carrying the ancestral moas as cargo. The molecular data suggest that the moas had already diverged from the other ratites, about 80 million years ago. Australia lost contact with Antarctica around 56 million years ago. This fits with the molecular evidence that the moas split off from other ratites earlier (82 million years) than the Australian ratites, the emu and cassowary, who diverged from one another around 30 million years ago. The kiwis are probably the one exception to the rule that the ratite birds walked everywhere. They are not closely related to the moas. Their affinities are with the Australian ratites, and they presumably island-hopped from Australia to New Zealand via New Caledonia. As for the elephant bird, it stayed in Madagascar after India broke away 75 million years ago, and remained there until the arrival of man.

I said I would return to the ostrich. From about 90 million years ago, it was no longer possible to cross by land between Africa and any other part of the former Gondwana. This therefore constitutes the last moment at which, one might have thought, the ostrich, being an African bird, could have diverged from the rest of the ratites. In fact, however, the molecular evidence suggests that the ostrich line diverged later, around 75 million years ago. How can this be?

The argument is a little intricate, so let me repeat the problem. The geographic evidence suggests that Africa was already separated from the rest of the former Gondwana by about 90 million years ago, yet the molecular evidence suggests that the ostrich split off from other Gondwana birds around 75 million years ago. Where were the ancestors of the ostrich during the intervening 15 million years? Presumably not in Africa, for the reason we have just seen. They could have been anywhere in the rest of Gondwana, because all the other bits – South America, Australia, New Zealand and Indo-Madagascar, remained connected to each other, if only via Antarctica, and if only by lingering land bridges.

How, then, do modern ostriches manage to end up in Africa? Alan Cooper has an ingenious theory. India/Madagascar remained connected to Antarctica, via a large land bridge called the Kerguelen Plateau (now submerged) until 75 million years ago, when what is now Sri Lanka pulled away. Up till that moment, the ancestors of the ostrich and the elephant bird were still in contact with Antarctica – and hence with the rest of Gondwana except Africa which had earlier separated. Cooper believes that the ancestors of the ostrich and the elephant bird were in India/Madagascar at this time of separation. On this hypothesis, we can regard 75 million years as the last moment the ostrich and elephant bird lineages could have split away from the other ratites, and this fits well with the molecular data. Then, about 5 million years later, India detached itself from Madagascar, taking what were to become ostriches along for the ride, and leaving what were to become the elephant birds behind.

So far, so good. But we are still left with the riddle which began this part of our tale. If the ancestors of the ostrich were ensconced in the then-island of India, how did they finally get to Africa? Now we come to the final part of Cooper's theory. India, you will remember, after parting from Madagascar, sped off north to its present position as part of Asia. Cooper believes that it carried with it the ostrich ancestors, who took advantage of the collision to spill out into Asia. Then, once in Asia, the ostrich line fanned out in a great loop to the north. There are still ostriches in Arabia, and there are ostrich fossils in Asia, including India, and even in Europe. At that time, as now, Africa was connected to Asia via Arabia, and by this route the ostriches finally arrived, perhaps about 20 million years ago, in Africa where we now find them. According to Cooper, the ancestral ostriches were by no means the only animals to catch the India ferry to Asia. He suggests that India's cargo of Gondwanan animals played a major part in the recolonisation of Asia after the catastrophe that killed the dinosaurs.

The legend of the roc, the fabulous great bird with the strength to shift elephants, is a wonder of childhood. But isn't the true story of how the very continents themselves are shifted, through thousands of miles, an even greater wonder, more worthy of the adult imagination? We look at the details in the epilogue to this tale.

EPILOGUE TO THE ELEPHANT BIRD'S TALE

The theory of plate tectonics, as it is now called, is one of the success stories of modern science. When my father was at Oxford in the 1930s, what was then called the theory of continental drift was widely, though not universally, ridiculed. It was associated with the German meteorologist Alfred Wegener (1880–1930) but others had suggested something similar before him. Several people had noticed the snug fit of the east coast of South America and the west coast of Africa, but it was generally written off as coincidence. There were some even more striking coincidences in distributions of animals and plants, which had to be explained by postulating land bridges between the continents. But scientists mostly thought that the map had altered through the sea level fluctuating up and down, rather than the continents themselves drifting sideways. The name Gondwana was originally coined for a continent consisting of Africa and South America in their present positions but with the South Atlantic drained. Wegener's idea that the continents themselves had drifted was far more revolutionary – and controversial.

Even when I was an undergraduate, in the 1960s instead of the 1930s, it was not an open-and-shut case. Charles Elton, the veteran Oxford ecologist, lectured us on the subject. At the end of his lecture he took a vote (I am sorry to say, because democracy is no way to establish a truth) and I think we were fairly evenly split. That all changed shortly after I graduated. Wegener turns out to have been much closer to the truth than most of his contemporaries who ridiculed him. The main thing he got wrong was that he thought the existing continental masses floated in the semi-liquid mantle, and ploughed their way through it like rafts through the sea. The modern theory of plate tectonics sees the whole surface of the Earth – the seafloor as well as the visible continents – as a set of *plates*. Continents are thick, less dense parts of plates that bulge up into the atmosphere forming mountains, and down into the mantle. Plate boundaries are, as often as not, under the sea. Indeed, we shall understand the theory best if we forget all about the sea: pretend it isn't there. We'll bring it back to flood the low ground later.

Plates do not plough through a sea, either of water or of molten rock. Instead, the entire surface of the Earth is armoured, covered by plates, sliding over the surface, sometimes diving under another plate in the process known as subduction. As a plate moves, it does not leave a gap behind it as Wegener imagined. Instead, the 'gap' is continuously filled by new material welling up from the deep layers of Earth's mantle and contributing to the substance of the plate, in the process called seafloor spreading. In some ways, plate seems too rigid an image: a better metaphor is a conveyor belt, or a roll-top desk. I'll describe it using the neatest, clearest example – the Mid-Atlantic Ridge.

The Mid-Atlantic Ridge is an underwater canyon 16,000 kilometres long, which snakes its way in a huge S-bend down the middle of the North and South Atlantic. The ridge is a zone of volcanic upwelling. Molten rock pushes up from the deeps of the mantle. It then fountains out sideways to east and west, like two desk roll-tops. The east-going roll-top pushes Africa away from the middle of the Atlantic. The west-going roll-top pushes South America in the other direction. That is why these two continents are moving away from one another, at a rate of about one centimetre per year which is, as somebody has imaginatively pointed out, about the rate at which fingernails grow, although rates of movement, from plate to plate, are quite variable. It is the same force that originally pushed them apart when Gondwana split. There are similar zones of volcanic upwelling in the floor of the Pacific Ocean, of the Indian Ocean, and in various other places (though they are sometimes called rises rather than ridges). These spreading ocean ridges are the driving engines of plate movement.

The language of 'pushing', however, is grossly misleading, if it suggests that the upwelling from the seafloor pushes the plate from behind. Indeed, how could an object as massive as a continental plate be moved by shoving from behind? It isn't. Rather, the crust and top part of the mantle are moved by the circulating currents in the molten rock beneath. A plate is not so much pushed from behind as dragged by the current in the fluid on which it floats, tugging on the underside of the whole expanse of the plate.

The evidence for plate tectonics is elegantly compelling and the theory is now proved beyond reasonable doubt. If you measure the age of the rocks on either side of a ridge such as the Mid-Atlantic Ridge, you notice a truly remarkable thing. The rocks that are closest to the ridge are the youngest. The further you travel, sideways from the ridge, the older are the rocks. The result is that if you plot 'isochrons' (that is, contour lines of equal age) they run parallel to the ridge itself, snaking with it down the North Atlantic and then the South Atlantic. This is true on both sides of the ridge. The isochrons on one side of the ridge are almost perfectly mirrored on the other side (see plate 20).

Imagine that we set off to cross the bottom of the Atlantic in a submersible tractor, due east at the tenth parallel, from the Brazilian port of Maceio towards the cape of Barra do Cuanza in Angola, just missing Ascension Island on the way. As we go, we sample the rocks beneath our caterpillar tracks (tyres couldn't stand the pressure). For reasons that follow from the theory of volcanic seafloor spreading, we shall be interested only in the igneous basalt (solidified lava) lying at the base of whatever sedimentary rocks may have been deposited above it. It is these igneous rocks which, according to the theory, constitute the roll-top or conveyor belt, as South America moves westward and Africa east. We shall drill down through the sediments – which may be quite thick in places, having been laid down over millions of years – and take samples from the hard volcanic rocks beneath.

For the first 50 kilometres of our eastward journey, we are on the continental shelf. This doesn't count as sea bottom at all, for our purposes. We haven't left the continent of South America, it is just that there is some shallow water above our heads. In any case, for the purpose of explaining plate tectonics, we are ignoring water. But now we descend rapidly to the sea bottom proper, take our first sample from the true seafloor, and analyse radiometrically the date of the basalt under the sediments. Here on the western edge of the Atlantic, it turns out to be of Lower Cretaceous age, some 140 million years old. We continue our eastward journey, taking samples at regular intervals from the volcanic rocks at the base of the sediments, and we find a remarkable fact: they become steadily younger. Five hundred kilometres from our starting point, we are well into the Upper Cretaceous, younger than 100 million years. About 730 kilometres into our journey, although we shall see no discrete border because we are looking only at volcanic rock, we cross the 65-million-year boundary between the Cretaceous and the Palaeogene period, the geological instant when, on land, the dinosaurs suddenly disappeared. The sequence of decreasing age continues. As we drive due east, the volcanic rocks under the sea get steadily younger and younger. Sixteen hundred kilometres from our starting point we are in the Pliocene, looking at young rocks, contemporary with woolly mammoths in Europe and Lucy in Africa.

When we reach the Mid-Atlantic Ridge itself, about 1,620 kilometres from South America and slightly further (at this latitude) from Africa, we notice that the rocks of our sample are now so young, they are of our own time. They have only just erupted out of the depths of the sea bottom. Indeed, if we are very lucky we may see an eruption in the particular part of the Mid-Atlantic Ridge that we are crossing. But we would have to be lucky because, notwithstanding the image of a continuously moving roll-top conveyor belt, it isn't literally continuous. How could it be, given that the roll-top moves on average at one centimetre per year? When there is an eruption, the rocks are shifted more than one centimetre. But correspondingly, eruptions occur less often than once per year in any one place along the ridge.

Having crossed the Mid-Atlantic Ridge, we continue our eastward journey in the direction of Africa, again drawing up samples of volcanic rock from beneath the sediments. And what we now notice is that the ages of the rocks are a mirror image of what we measured before. The rocks are now getting progressively older as we move away from the central ridge, and this continues all the way to Africa and the eastern margin of the Atlantic. Our last sample, just short of the African continental shelf, shows rocks of Lower Cretaceous age, just like their mirror images on the western side, hard by South America. Indeed, the whole sequence is reflected about the Mid-Atlantic Ridge, and the mirroring is even more precise than you could know from radiometric dating alone. What follows is extremely elegant.

In the Redwood's Tale, we shall meet the ingenious dating technique known as dendrochronology. Tree rings result from the fact that trees have an annual growing season, and not all years are equally favourable, so a signature pattern of thick and thin rings develops. Such fingerprint signatures, when they occasionally arise in nature, are a natural gift to science, to be seized eagerly whenever we encounter them. It is a particularly fortunate fact that something like tree rings, although on a larger timescale, is imprinted into volcanic lava as it cools and solidifies. It works like this. While lava is still liquid, molecules within it behave like tiny compass needles, and

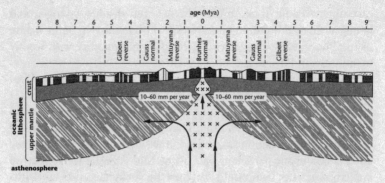

Magnetic stripes on either side of an oceanic ridge. Dark stripes represent normal polarity; white, reverse polarity. Geologists group these into magnectic intervals dominated by normal or reverse polarity. The symmetry of the stripes was first identified as evidence for seafloor spreading by Fred Vine and Drummond Matthews in a classic 1963 paper in *Nature* [296]. The crust and rigid top layer of the mantle, together known as the lithosphere, is pushed apart by convection currents in magma in the semi-rigid layer of the mantle below (the asthenosphere). The distinctive pattern of stripes allows us to identify the age of rocks on the sea bed back to about 150 million years ago. Older seafloor has been destroyed by subduction.

become aligned to the magnetism of the Earth. When the lava solidifies as rock, the compass needles are petrified in their current position. Igneous rock therefore acts as a weak magnet, whose polarity is a frozen record of the Earth's magnetic field at the moment of solidification. This polarity, which is easy to measure, tells us the direction of the magnetic North Pole at the moment when the rock solidified.

Now comes the fortunate fact. The polarity of the Earth's magnetic field reverses at irregular but, by geological standards, quite frequent intervals, on a timescale of tens, or hundreds, of thousands of years. You can immediately see the exciting consequence of this. As the two conveyor belts stream out west and east from the Mid-Atlantic Ridge, their measured magnetic polarity will exhibit stripes, reflecting the flips of the Earth's field, frozen at the moment of solidification of the rock. The pattern of stripes on the west side will mirror the precise pattern of stripes on the east side, because both sets of rocks shared the same magnetic field when they spewed together, as liquid, out of the mid-ocean ridge. It is possible to match up exactly which stripe on the east side of the ridge belongs with which stripe on the west side, and the two stripes can be dated (they have the same date as each other, of course, because they were liquid at the same time when they gushed together out of the ridge). The same pattern of stripes will be found on either side of the spreading zones on all the other ocean floors, although the distances between the mirrored stripes will vary because not all conveyor belts move at the same speed. You could not ask for more compelling evidence.

There are complications. The pattern of parallel stripes does not snake down the seafloor in a simple unbroken way. It is subject to numerous fractures – 'faults'. I chose the tenth parallel south of the equator deliberately for our caterpillar tractor journey, because it happens not to be complicated by any fault lines. At another latitude, our sequence of gradually changing age would have been interrupted by occasional hiccups as we crossed a fault line. But the general picture of parallel isochrons is entirely clear from the geological map of the whole Atlantic floor.

The evidence for the seafloor spreading theory of plate tectonics is, then, very solid, and the dating of the various tectonic events such as the parting of particular continents, is, by geological standards, accurate. The plate tectonics revolution has been one of the swiftest, yet at the same time most decisive, in the whole history of science.

AMPHIBIANS

Three hundred and forty million years ago, in the early Carboniferous Period, only about 30 million years beyond the great milestone of Rendezvous 16, we amniotes (the name that unites mammals with reptiles and birds) meet our amphibian cousins at Rendezvous 17. Pangaea had not yet come together, and northern and southern landmasses surrounded a pre-Tethys ocean. A south polar ice cap was beginning to form, there were tropical forests of club mosses around the equator, and the climate was probably something like that of today, although the flora and fauna were of course very different.

Concestor 17, in the vicinity of our 175-million-greats-grandparent, is the ancestor of all surviving tetrapods. Tetrapod means four feet. We who don't walk on four feet are lapsed tetrapods, recently lapsed in our case, much less recently in the case of birds, but we are all called tetrapods. More to the point, Concestor 17 is the grand ancestor of the huge throng of land vertebrates. Despite my earlier strictures on the conceit of hindsight, the emergence of fish onto the land was a major transition in our evolutionary history.

Three main bands of modern amphibian pilgrims have joined forces long 'before' they meet up with us amniotes. They are the frogs (and toads: the distinction is not a zoologically helpful one), the salamanders (and newts, which are those species that return to the water to breed), and the caecilians (moist, legless burrowers or swimmers, with a superficial resemblance to earthworms or snakes). The frogs have no tail as adults but a vigorously swimming tail as larvae. The salamanders have a long tail in the adult as well as the larval stage, and their body proportions most resemble ancestral amphibians, as judged by fossils. The caecilians have no limbs – not even internal traces of the pectoral and pelvic girdles that supported the limbs of their ancestors. The great length of the caecilian body is achieved by multiplying up the vertebrae in the trunk region (up to 250, compared to 12 in frogs), and their ribs, which provide useful support and protection. The tail, oddly, is very short or even absent: if caecilians had legs, their hind legs would be right at the posterior

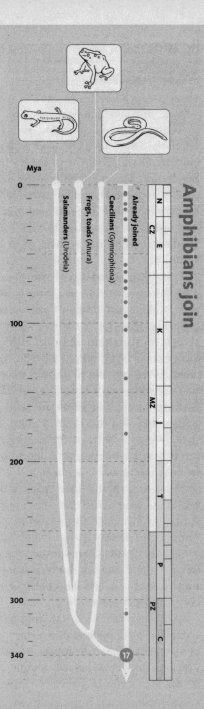

Amphibians join

Mya

Salamanders (Urodela)

Frogs, toads (Anura)

Caecilians (Gymnophiona)

Already joined

N | CZ | E | K | MZ | J | T | P | PZ | C

0

100

200

300

340

17

Amphibians join. At odds with several fossil studies, genetic studies consistently unite the 5,000 or so described amphibian species in a single group, sister to the amniotes. Molecular taxonomy has been followed here, but there are disagreements on the order of branching between the three amphibian groups.

Images, left to right: Monterey ensatina (*Ensatina eschscholtzii eschscholtzii*); blue poison frog (*Dendrobates azureus*); caecilian (*Ichthyophis* sp.).

tip of the body, which is where some extinct amphibians actually kept them.

Even if they live on land as adults, most amphibians reproduce in water, while amniotes (except in secondarily evolved cases such as whales, dugongs and ichthyosaurs) reproduce on land. Amniotes reproduce either viviparously, giving birth to live young, or with a relatively large, tough-shelled waterproof egg. In both cases the embryo floats in its own 'private pond'. Amphibian embryos are much more likely to float in a real pond, or something equivalent to one. The amphibian pilgrims who join us at Rendezvous 17 may spend part of their time on land, but they are seldom far from water and, at least at some stage in their life cycle, they usually return to it. Those that reproduce on land go to some lengths to contrive watery conditions.

Trees provide relatively safe havens, and frogs have discovered ways to reproduce in them without losing the vital tie to water. Some exploit the small pools of rainwater that form in the rosettes of bromeliad plants. Male African grey tree frogs, *Chiromantis xerampelina*, co-operate to whip up a thick white foam, with their back legs, from a liquid secreted by the females. This foam hardens to a crust on the outside, protecting the moist interior which serves as a nest for the group's eggs. The tadpoles develop inside the wet foam nest, up a tree. When they are ready, in the next rainy season, they wriggle free and drop into puddles of water below the tree, where they develop into frogs. Other species use the foam-nest technique, but they don't co-operate to do so. Instead one male beats up a foam from the secretion of one female.

Some frog species have made interesting transitions in the direction of true viviparity – live birth. The female of the South American marsupial frog (various species of the genus *Gastrotheca*) transfers her fertilised eggs to her back, where they become covered by a layer of skin. There the tadpoles develop and can clearly be seen wriggling under the skin of their mother's back until they eventually burst out. Again, several other species do something similar, probably independently evolved.

Another South American frog species, named *Rhinoderma darwinii* after its illustrious discoverer, practises a most unusual version of viviparity. The male appears to eat the eggs that he has fertilised. The eggs don't travel down his gut, however. Like many male frogs he has a commodious vocal sac, used as a resonator to amplify the voice, and it is in this moist chamber that the eggs lodge. There they develop, until they are finally vomited out as fully formed froglets, forgoing the freedom to swim as tadpoles.

The key difference between amphibians and amniotes is that amniote skins and eggshells are waterproof. Amphibian skin typically lets water evaporate through it, at the same rate you'd expect from a body of standing water of the same area. As far as the water under the skin is concerned, there might almost as well not be any skin. This is very different from reptiles, birds and mammals, where one of the main roles of the skin is to serve as a barrier to water. There are exceptions among amphibians – most notably among various species of desert frogs in Australia. These exploit the fact that even deserts can have flood times, though brief and well spaced out. During such rare and intermittent times of high rainfall, each frog makes a water-filled cocoon in which it buries itself in a state of torpor, for two years or, by some accounts, even as long as seven years. Some species of frogs can withstand temperatures well below the normal freezing point of water, by manufacturing glycerol as an antifreeze.

Almost no amphibians live in sea water, and it is therefore not surprising that, unlike lizards, they are seldom found on remote islands.* Darwin noted this in more than one of his books, and also the fact that frogs that have been artificially introduced to just such islands thrive there. He presumed that lizard eggs are protected from sea water by their hard shells, whereas sea water promptly kills frogspawn. Frogs are, however, found on all continents except Antarctica, and have probably been there continuously since before the continents broke up. They are a very successful group.

Frogs remind me of birds in one respect. Both have a body plan which is a somewhat bizarre modification of the ancestral one. That is not particularly remarkable, but birds and frogs have taken this bizarre body plan and made it the basis for a whole new range of variation. There are not quite so many frog species as birds, but the more than 4,000 species of frogs, in every part of the world, are impressive enough. Just as the bird body plan is obviously a design

---

* Sam Turvey tells me that the two frog species with the remotest island distribution, the Fijian frogs *Platymantis vitiensis* and *P. vitianus* (closely related and presumably descended from a single colonising ancestor), develop completely in the egg rather than having a free-swimming tadpole. They appear more salt-tolerant than most frogs, with *P. vitianus* sometimes found on beaches. These unusual characteristics, if present in their colonising ancestor as seems likely, would have pre-adapted them for island-hopping.

for flying, even in birds such as ostriches that don't fly, the best way to understand the body plan of an adult frog is as a highly specialised jumping machine. Some species can jump spectacular distances, up to 50 body lengths in the well-named rocket frog of Australia (*Litoria nasuta*). The largest frog in the world, the goliath frog (*Conraua goliath*) of West Africa, which is the size of a small dog, is said to jump three metres. Not all frogs jump, but all are descended from jumping ancestors. They are at least lapsed jumpers, just as ostriches are lapsed fliers. Some tree-dwelling species, such as Wallace's flying frog, *Rhacophorus nigropalmatus*, prolong the jump by spreading their long fingers and toes, whose webbing acts as a parachute. Indeed, they glide a bit like flying squirrels.

Salamanders and newts swim like fish when they are in water. Even on land, their legs are too small and feeble to walk or run in the sense we would recognise, and the salamander uses a sinuous swimming motion like a fish, with the legs just helping it along. Most salamanders today are quite small. The largest reach a respectable 1.5 metres, but this is still far smaller than the giant amphibians of past times, which dominated the land before the rise of the reptiles.

But what did Concestor 17 look like: the ancestor that amphibians share with reptiles and ourselves? Certainly more like an amphibian than an amniote, and more like a salamander than a frog – but probably not much like either. The best fossils are in Greenland which, during the Devonian Period, was on the equator. These possibly transitional fossils have been much studied,[*] among them *Acanthostega*, which seems to have been wholly aquatic (showing that 'legs' originally evolved for movement in water, not on land), and *Ichthyostega*.

Concestor 17 might have been something like *Ichthyostega* or *Acanthostega*, although both were larger than we normally expect grand ancestors to be. There are some other surprises for zoologists preconditioned by acquaintance with modern animals. We tend to think the possession of five digits is deeply stamped in the hands and feet of tetrapods – the 'pentadactyl' limb is a classic zoological totem. Yet recent evidence shows that *Ichthyostega* had seven toes, *Acanthostega* had eight, and *Tulerpeton*, a third genus of Devonian tetrapod, had six. It is tempting to say the number of digits doesn't matter, is

---

[*] Notably in recent years by Dr Jennifer Clack of Cambridge University and her colleagues. See her book, *Gaining Ground: The Origin and Evolution of Tetrapods*.

functionally neutral. I doubt that. My tentative guess is that in those early times the different species really did benefit from their respective numbers of toes. They really were more efficient than other numbers would have been, for swimming or walking. Later, the tetrapod limb design hardened at five digits, probably because some internal embryological process came to rely upon that number. In the adult, the number is frequently reduced from the embryonic number – in extreme cases such as modern horses, to just one, the middle toe.

The fish group from which the amphibians sprang is the one known as the lobefins. The only surviving lobefins are the lungfish and the coelacanths,* and we shall meet them at Rendezvous 18 and 19 respectively. In Devonian times, lobefins were much more prominent in both the marine and freshwater faunas. The tetrapods probably evolved from an otherwise extinct group of lobefins called the osteolepiforms. Among osteolepiforms are *Eusthenopteron* and *Panderichthys,* both dating from the late Devonian, about the time when the first tetrapods were starting to emerge onto the land.

Why did fish first develop the changes that permitted the move out of water onto the land? Lungs, for example? And fins that you could walk on rather than, or as well as, swim with? It wasn't that they were trying to initiate the next big chapter in evolution! For years, the favoured answer to the question was one that the eminent American palaeontologist Alfred Sherwood Romer derived from the geologist Joseph Barrell. The idea was that if these fish were trying to do anything it was to get back to water. In times of drought, fish can easily become stranded in drying pools. Individuals capable of walking and of breathing air have the enormous advantage that they can forsake a doomed, drying pond and set out for a deeper one elsewhere.

This admirable theory has become unfashionable but not, I think, for uniformly good reasons. Unfortunately, Romer quoted the prevailing belief of his day that the Devonian was a time of drought, a belief that has more recently been called into question. But I don't think Romer needed his Devonian desiccated. Even at times of no particular drought, there will always be some ponds shallow enough to be in danger of becoming too shallow for some particular kind of

---

* The name lobefin is not used with universal agreement. Some authors exclude the lungfish and say that coelacanths are the only surviving lobefins. I follow the terminology of Professor Robert Carroll's *Vertebrate Palaeontology and Evolution* and include lungfish as lobefins.

fish. If ponds three feet deep would have been at risk under severe drought conditions, mild drought conditions will render ponds one foot deep at risk. It is sufficient for the Romer hypothesis that there are some ponds that dry up, and therefore some fish that could save their lives by migrating. Even if the world of the late Devonian was positively waterlogged, one could say this simply increases the number of ponds available to dry up, thereby increasing opportunities for saving the life of walking fish and the Romer theory. Nevertheless, it is my duty to record that the theory is now unfashionable. A further point against the theory is that modern fish that venture onto land do so in humid, wet areas – that is, when conditions on land are 'good' for water animals, not poor as in the Romer hypothesis.

And, to be sure, there are plenty of other good reasons for a fish to emerge, temporarily or permanently, onto land. Streams and ponds can become unusable for reasons other than drying up. They can become choked with weeds, in which case, again, a fish that can migrate over land to deeper water might benefit. If, as has been suggested *contra* Romer, we are talking Devonian swamps rather than Devonian droughts, swamps provide plenty of opportunities for a fish to benefit by walking, or slithering or flip-flopping or otherwise travelling through the marshy vegetation, in search of deep water or, indeed, food. This still retains the essential Romer idea that our ancestors left the water, not at first to colonise land, but to return to water.

The group of lobefins from which we tetrapods are derived, are today reduced to a pitiful four genera, but they once dominated the seas almost as the teleost fish do today. We are not due to meet the teleosts until Rendezvous 20, but they will help our discussion because some of them breathe air, at least occasionally, and a few even come out of the water and walk on land. A little further into our pilgrimage, we shall hear from one of them, the mudskipper, whose tale is a tale of independent, more recent encroachment onto the land.

THE SALAMANDER'S TALE

Names are a menace in evolutionary history. It is no secret that palaeontology is a controversial subject in which there are even some personal enmities. At least eight books called *Bones of Contention* are in print. And if you look at what two palaeontologists are quarrelling about, as often as not it turns out to be a name. Is this fossil *Homo erectus,* or is it an archaic *Homo sapiens?* Is this one an early *Homo habilis* or a late *Australopithecus?* People evidently feel strongly about such questions, but they often turn out to be splitting hairs. Indeed, they resemble theological questions, which I suppose gives a clue to why they arouse such passionate disagreements. The obsession with discrete names is an example of what I call the tyranny of the discontinuous mind. The Salamander's Tale strikes a blow against the discontinuous mind (see plate 21).

The Central Valley runs much of the length of California, bounded by the Coastal Range to the west and by the Sierra Nevada to the east. These long mountain ranges link up at the north and the south ends of the valley, which is therefore surrounded by high ground. Throughout this high ground lives a genus of salamanders called *Ensatina.* The Central Valley itself, about 40 miles wide, is not friendly to salamanders, and they are not found there. They can move all round the valley but normally not across it, in an elongated ring of more or less continuous population. In practice any one salamander's short legs in its short lifetime don't carry it far from its birthplace. But genes, persisting through a longer timescale, are another matter. Individual salamanders can interbreed with neighbours whose parents may have interbred with neighbours further round the ring, and so on. There is therefore potentially gene flow all around the ring. Potentially. What happens in practice has been elegantly worked out by the research of my old colleagues at the University of California at Berkeley, initiated by Robert Stebbins and continued by David Wake.

In a study area called Camp Wolahi, in the mountains to the south of the valley, there are two clearly distinct species of *Ensatina* which do not interbreed. One is conspicuously marked with yellow and black blotches. The other is a uniform light brown with no blotches. Camp Wolahi is in a zone of overlap, but wider sampling shows that the blotched species is typical of the eastern side of the Central Valley which, here in Southern California, is known as the San Joaquin Valley. The light brown species, on the contrary, is typically found on the western side of the San Joaquin.

Non-interbreeding is the recognised criterion for whether two populations deserve distinct species names. It therefore should be straightforward to use the name *Ensatina eschscholtzii* for the plain western species, and *Ensatina klauberi* for the blotched eastern species – straightforward but for one remarkable circumstance, which is the nub of the tale.

If you go up to the mountains that bound the north end of the Central Valley, which up there is called the Sacramento Valley, you'll find only one species of *Ensatina*. Its appearance is intermediate between the blotched and the plain species: mostly brown, with rather indistinct blotches. It is not a hybrid between the two: that is the wrong way to look at it. To discover the right way, make two expeditions south, sampling the salamander populations as they fork to west and east on either side of the Central Valley. On the east side, they become progressively more blotched until they reach the extreme of *klauberi* in the far south. On the west side, the salamanders become progressively more like the plain *eschscholtzii* that we met in the zone of overlap at Camp Wolahi.

This is why it is hard to treat *Ensatina eschscholtzii* and *Ensatina klauberi* with confidence as separate species. They constitute a 'ring species'. You'll recognise them as separate species if you only sample in the south. Move north, however, and they gradually turn into each other. Zoologists normally follow Stebbins's lead and place them all in the same species, *Ensatina eschscholtzii*, but give them a range of subspecies names. Starting in the far south with *Ensatina eschscholtzii eschscholtzii*, the plain brown form, we move up the west side of the valley through *Ensatina eschscholtzii xanthoptica* and *Ensatina eschscholtzii oregonensis* which, as its name suggests, is also found further north in Oregon and Washington. At the north end of California's Central Valley is *Ensatina eschscholtzii picta*, the semi-blotched form mentioned before. Moving on round the ring and down the east side of the valley, we pass through *Ensatina eschscholtzii platensis* which is a bit more blotched than *picta*, then *Ensatina eschscholtzii croceater* until we reach *Ensatina eschscholtzii klauberi* (which is the very blotched one that we previously called *Ensatina klauberi* when we were considering it to be a separate species).

Stebbins believes that the ancestors of *Ensatina* arrived at the north end of the Central Valley and evolved gradually down the two sides of the valley, diverging as they went. An alternative possibility is that they started in the south as, say, *Ensatina eschscholtzii eschscholtzii*, then evolved their way up the west side of the valley, round the top and down the other side, ending up as *Ensatina eschscholtzii klauberi* at the other end of the ring. Whatever the history, what happens today is that there is hybridization all round the ring, except where the two ends of the line meet, in the far south of California.

As a complication, it seems that the Central Valley is not a total barrier to gene flow. Occasionally, salamanders seem to have made it across, for there are populations of, for example, *xanthoptica,* one of the western subspecies, on the eastern side of the valley, where they hybridise with the eastern subspecies, *platensis.* Yet another complication is that there is a small break near the south end of the ring, where there seem to be no salamanders at all. Presumably they used to be there, but have died out. Or maybe they are still there but have not been found: I am told that the mountains in this area are rugged and hard to search. The ring is complicated, but a ring of continuous gene flow is, nevertheless, the predominant pattern in this genus, as it is with the better-known case of herring gulls and lesser black-backed gulls around the Arctic Circle.

In Britain the herring gull and the lesser black-backed gull are clearly distinct species. Anybody can tell the difference, most easily by the colour of the wing backs. Herring gulls have silver-grey wing backs, lesser black-backs, dark grey, almost black. More to the point, the birds themselves can tell the difference too, for they don't hybridise although they often meet and sometimes even breed alongside one another in mixed colonies. Zoologists therefore feel fully justified in giving them different names, *Larus argentatus* and *Larus fuscus.*

But now here's the interesting observation, and the point of resemblance to the salamanders. If you follow the population of herring gulls westward to North America, then on around the world across Siberia and back to Europe again, you notice a curious fact. The 'herring gulls', as you move round the pole, gradually become less and less like herring gulls and more and more like lesser black-backed gulls until it turns out that our Western European lesser black-backed gulls actually are the other end of a ring-shaped continuum which started with herring gulls. At every stage around the ring, the birds are sufficiently similar to their immediate neighbours in the ring to interbreed with them. Until, that is, the ends of the continuum are reached, and the ring bites itself in the tail. The herring gull and the lesser black-backed gull in Europe never interbreed, although they are linked by a continuous series of interbreeding colleagues all the way round the other side of the world.

Ring species like the salamanders and the gulls are only showing us in the spatial dimension something that must always happen in the time dimension. Suppose we humans, and the chimpanzees, were a ring species. It could have happened: a ring perhaps moving up one side of the Rift Valley, and down the other side, with two completely separate species co-existing at the southern end of the ring, but an unbroken continuum of interbreeding all the way up and back round the other side. If this were true, what would it do to our attitudes to other species? To apparent discontinuities generally?

Many of our legal and ethical principles depend on the separation between *Homo sapiens* and all other species. Of the people who regard abortion as a sin, including the minority who go to the lengths of assassinating doctors and blowing up abortion clinics, many are unthinking meat-eaters, and have no worries about chimpanzees being imprisoned in zoos and sacrificed in laboratories. Would they think again, if we could lay out a living continuum of intermediates between ourselves and chimpanzees, linked in an unbroken chain of interbreeders like the Californian salamanders? Surely they would. Yet it is the merest accident that the intermediates all happen to be dead. It is only because of this accident that we can comfortably and easily imagine a huge gulf between our two species – or between any two species, for that matter.

I have previously recounted the case of the puzzled lawyer who questioned me after a public lecture. He brought the full weight of his legal acumen to bear on the following nice point. If species A evolves into species B, he reasoned closely, there must come a point when a child belongs to the new species B but his parents still belong to the old species A. Members of different species cannot, by definition, interbreed with one another, yet surely a child would not be so different from its parents as to be incapable of interbreeding with their kind. Doesn't this, he wound up, wagging his metaphorical finger in the special way that lawyers, at least in courtroom dramas, have perfected as their own, undermine the whole idea of evolution?

That is like saying, 'When you heat a kettle of cold water, there is no particular moment when the water ceases to be cold and becomes hot, therefore it is impossible to make a cup of tea.' Since I always try to turn questions in a constructive direction, I told my lawyer about the herring gulls, and I think he was interested. He had insisted on placing individuals firmly in this species or that. He didn't allow for the possibility that an individual might lie half way between two species, or a tenth of the way from species A to species B. Exactly the same limitation of thought hamstrings the endless debates about exactly when in the development of an embryo it becomes human (and when, by implication, abortion should be regarded as tantamount to murder). It is no use saying to these people that, depending upon the human characteristic that interests you, a foetus can be 'half human' or 'a hundredth human'. 'Human', to the qualitative, absolutist mind, is like 'diamond'. There are no halfway houses. Absolutist minds can be a menace. They cause real misery, human misery. This is what I call the tyranny of the discontinuous mind, and it leads me to develop the moral of the Salamander's Tale.

For certain purposes names, and discontinuous categories, are exactly what we need. Indeed, lawyers need them all the time. Children are not allowed to drive; adults are. The law needs to impose a threshold, for example the seventeenth birthday. Revealingly, insurance companies take a very different view of the proper threshold age.

Some discontinuities are real, by any standards. You are a person and I am another person and our names are discontinuous labels that correctly signal our separateness. Carbon monoxide really is distinct from carbon dioxide. There is no overlap. A molecule consists of a carbon and one oxygen, or a carbon and two oxygens. None has a carbon and 1.5 oxygens. One gas is deadly poisonous, the other is needed by plants to make the organic substances that we all depend upon. Gold really is distinct from silver. Diamond crystals really are different from graphite crystals. Both are made of carbon, but the carbon atoms naturally arrange themselves in two quite distinct ways. There are no intermediates.

But discontinuities are often far from so clear. My newspaper carried the following item during a recent flu epidemic. Or was it an epidemic? That question was the burden of the article.

Official statistics show there are 144 people in every 100,000 suffering from flu, said a spokeswoman for the Department of Health. As the usual gauge of an epidemic is 400 in every 100,000, it is not being officially treated as an epidemic by the Government. But the spokeswoman added: 'Professor Donaldson is happy to stick by his version that this is an epidemic. He believes it is many more than 144 per 100,000. It is very confusing and it depends on which definition you choose. Professor Donaldson has looked at his graph and said it is a serious epidemic.'

What we know is that some particular number of people are suffering from flu. Doesn't that, in itself, tell us what we want to know? Yet for the spokeswoman, the important question is whether this counts as an 'epidemic'. Has the proportion of sufferers crossed the rubicon of 400 per 100,000? This is the great decision which Professor Donaldson had to make, as he pored over his graph. You'd think he might have been better employed trying to do something about it, whether or not it counted officially as an epidemic.

As it happens, in the case of epidemics, for once there really is a natural rubicon: a critical mass of infections above which the virus, or bacterium, suddenly 'takes off' and dramatically increases its rate of spreading. This is why public health officials try so hard to vaccinate more than a threshold proportion of the population against, say, whooping cough. The purpose is not just to protect the individuals vaccinated. It is also to deprive the pathogens of the opportunity to reach their own critical mass for 'take-off'. In the case of our flu epidemic, what should really worry the spokeswoman for the Ministry of Health is whether the flu virus has yet crossed its rubicon for take-off, and leapt abruptly into high gear in its spread through the population. This should be decided by some means other than reference to magic numbers like 400 per 100,000. Concern with magic numbers is a mark of the discontinuous mind, or qualitative mind. The funny thing is that, in this case, the discontinuous mind overlooks a genuine discontinuity, the take-off point for an epidemic. Usually there isn't a genuine discontinuity to overlook.

Many Western countries at present are suffering what is described as an epidemic of obesity. I seem to see evidence of this all around me, but I am not impressed by the preferred way of turning it into numbers. A percentage of the population is described as 'clinically obese'. Once again, the discontinuous mind insists on separating people out into the obese on one side of a line, the non-obese on the other. That is not the way real life works. Obesity is continuously distributed. You can measure how obese each individual is, and you can compute group statistics from such measurements. Counts of numbers of people who lie above some arbitrarily defined threshold of obesity are not illuminating, if only because they immediately prompt a demand for the threshold to be specified and maybe redefined.

The same discontinuous mind also lurks behind all those official figures detailing the numbers of people 'below the poverty line'. You can meaningfully express a family's poverty by telling us their income, preferably expressed in real terms of what they can buy. Or you can say 'X is as poor as a church mouse' or 'Y is as rich as Croesus' and everybody will know what you mean. But spuriously precise counts or percentages of people said to fall above or below some arbitrarily defined poverty *line* are pernicious. They are pernicious because the precision implied by the percentage is instantly belied by the meaningless artificiality of the 'line'. Lines are impositions of the discontinuous mind. Even more politically sensitive is the label 'black', as opposed to 'white', in the context of modern society – especially American society. This is the central issue in the Grasshopper's Tale, and I'll leave it for now, except to say that I believe race is yet another of the many cases where we don't need discontinuous categories, and where we should do without them unless an extremely strong case in their favour is made.

Here's another example. Universities in Britain award degrees that are classified into three distinct classes, First, Second and Third Class. Universities in other countries do something equivalent, if under different names, like A, B, C etc. Now, my point is this. Students do not really separate neatly into good, middling and poor. There are not discrete and distinct classes of ability or diligence. Examiners go to some trouble to assess students on a finely continuous numerical scale, awarding marks or points that are designed to be added to other such marks, or otherwise manipulated in mathematically continuous ways. The score on such a continuous numerical scale conveys far more information than classification into one of three categories. Nevertheless, only the discontinuous categories are published.

In a very large sample of students, the distribution of ability and prowess would normally be a bell curve with few doing very well, few doing very badly and many in between. It might not actually be a symmetrical bell like the picture on page 316, but it would certainly be smoothly continuous, and it would become smoother as more and more students are added in.

A few examiners (especially, I hope I'll be forgiven for adding, in non-scientific subjects) seem actually to believe that there really is a discrete entity called the First-Class Mind, or the 'alpha' mind, and a student either definitely has it or definitely hasn't. The task of the examiner is to sort out the Firsts from the Seconds and the Seconds from the Thirds, just as one might sort sheep from goats. The likelihood that in reality there is a smooth continuum, sliding from pure sheepiness through all intermediates to pure goatiness, is a difficult one for some kinds of mind to grasp.

If, against all my expectations, it should turn out that the more students you add in, the more the distribution of exam marks approximates to a discontinuous distribution with three peaks (see lower picture), it would be a fascinating result. The awarding of First, Second and Third Class degrees might then actually be justifiable.

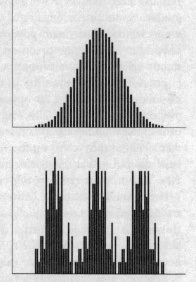

But there is certainly no evidence for this, and it would be very surprising given everything we know about human variation. As things are, it is clearly unfair: there is far more difference between the top of one class and the bottom of the same class, than there is between the bottom of one class and the top of the next class. It would be fairer to publish the actual marks obtained, or a rank order based upon those marks. But the discontinuous or qualitative mind insists on forcing people into one or other discrete category.

Returning to our topic of evolution, what about sheep and goats themselves? Are there sharp discontinuities between species, or do they merge into each other like first-class and second-class exam performances? If we look only at surviving animals, the answer is normally yes, there are sharp discontinuities. Exceptions like the gulls and the Californian salamanders are rare, but revealing because they translate into the spatial domain the continuity which is normally found only in the temporal domain. People and chimpanzees are certainly linked via a continuous chain of intermediates and a shared ancestor, but the intermediates are extinct: what remains is a discontinuous distribution. The same is true of people and monkeys, and of people and kangaroos, except that the extinct intermediates lived longer ago. Because the intermediates are nearly always extinct, we can usually get away with assuming that there is a sharp discontinuity between every species and every other. But in this book we are concerned with evolutionary history, with the dead as well as the living. When we are talking about all the animals that have ever lived, not just those that are living now, evolution tells us there are lines of gradual continuity linking literally every species to every other. When we are talking history, even apparently discontinuous modern species like sheep and dogs are linked, via their common ancestor, in unbroken lines of smooth continuity.

Ernst Mayr, distinguished elder statesman of twentieth-century evolution, has blamed the delusion of discontinuity – under its philosophical name of Essentialism – as the main reason why evolutionary understanding came so late in human history. Plato, whose philosophy can be seen as the inspiration for Essentialism, believed that actual things are imperfect versions of an ideal archetype of their kind. Hanging somewhere in ideal space is an essential, perfect rabbit, which bears the same relation to a real rabbit as a mathematician's perfect circle bears to a circle drawn in the dust. To this day many people are deeply imbued with the idea that sheep are sheep and goats are goats, and no species can ever give rise to another because to do so they'd have to change their 'essence'.

There is no such thing as essence.

No evolutionist thinks that modern species change into other modern species. Cats don't turn into dogs or vice versa. Rather, cats and dogs have evolved from a common ancestor, who lived tens of millions of years ago. If only all the intermediates were still alive, attempting to separate cats from dogs would be a doomed enterprise, as it is with the salamanders and the gulls. Far from being a question of ideal essences, separating cats from dogs turns out to be possible only because of the lucky (from the point of view of the essentialist) fact that the intermediates happen to be dead. Plato might find it ironic to learn that it is actually an imperfection – the sporadic ill-fortune of death – that makes the separation of any one species from another possible. This of course applies to the separation of human beings from our nearest relatives – and, indeed, from our more distant relatives too. In a world of perfect and complete information, fossil information as well as recent, discrete names for animals would become impossible. Instead of discrete names we would need sliding scales, just as the words hot, warm, cool and cold are better replaced by a sliding scale such as Celsius or Fahrenheit.

Evolution is now universally accepted as a fact by thinking people, so one might have hoped that essentialist intuitions in biology would have been finally overcome. Alas, this hasn't happened. Essentialism refuses to lie down. In practice, it is usually not a problem. Everyone agrees that *Homo sapiens* is a different species (and most would say a different genus) from *Pan troglodytes,* the chimpanzee. But everyone also agrees that if you follow human ancestry backward to the shared ancestor and then forward to chimpanzees, the intermediates all along the way will form a gradual continuum in which every generation would have been capable of mating with its parent or child of the opposite sex.

By the interbreeding criterion every individual is a member of the same species as its parents. This is an unsurprising, not to say platitudinously obvious conclusion, until you realise that it raises an intolerable paradox in the essentialist mind. Most of our ancestors throughout evolutionary history have belonged to different species from us by any criterion, and we certainly couldn't have interbred with them. In the Devonian Period our direct ancestors were fish. Yet, although we couldn't interbreed with them, we are linked by an unbroken chain of ancestral generations, every one of which could have interbred with their immediate predecessors and immediate successors in the chain.

In the light of this, see how empty are most of those passionate arguments about the naming of particular hominid fossils. *Homo ergaster* is widely recognised as the predecessor species that gave rise to *Homo sapiens,* so I'll play along with that for what follows. To call *Homo ergaster* a separate species from *Homo sapiens* could have a precise meaning in principle, even if it is impossible to test in practice. It means that if we could go back in our time machine and meet our *Homo ergaster* ancestors, we could not interbreed with them.* But suppose that, instead of zooming directly to the time of *Homo ergaster,* or indeed any other extinct species in our ancestral lineage, we stopped our time machine every thousand years along the way and picked up a young and fertile passenger. We transport this passenger back to the next thousand-year stop and release her (or him: let's take a female and a male at alternate stops). Provided our one-stop time traveller could accommodate to local social and linguistic customs (quite a tall order) there would be no biological

---

* I am not asserting that as a fact. I don't know if it is a fact, although I suspect that it is. It is an implication of our plausibly agreeing to give *Homo ergaster* a different species name.

barrier to her interbreeding with a member of the opposite sex from 1,000 years earlier. Now we pick up a new passenger, say a male this time, and transport him back another 1,000 years. Once again, he too would be biologically capable of fertilising a female from 1,000 years before his native time. The daisy chain would continue on back to when our ancestors were swimming in the sea. It could go back without a break, to the fishes, and it would still be true that each and every passenger transported 1,000 years before its own time would be able to interbreed with its predecessors. Yet at some point, which might be a million years back but might be longer or shorter, there would come a time when we moderns could not interbreed with an ancestor, even though our latest one-stop passenger could. At this point we could say that we have travelled back to a different species.

The barrier would not come suddenly. There would never be a generation in which it made sense to say of an individual that he is *Homo sapiens* but his parents are *Homo ergaster*. You can think of it as a paradox if you like, but there is no reason to think that any child was ever a member of a different species from its parents, even though the daisy chain of parents and children stretches back from humans to fish and beyond. Actually it isn't paradoxical to anybody but a dyed-in-the-wool essentialist. It is no more paradoxical than the statement that there is never a moment when a growing child ceases to be short and becomes tall. Or a kettle ceases to be cold and becomes hot. The legal mind may find it necessary to impose a barrier between childhood and majority – the stroke of midnight on the eighteenth birthday, or whenever it is. But anyone can see that it is a (necessary for some purposes) fiction. If only more people could see that the same applies to when, say, a developing embryo becomes 'human'.

Creationists love 'gaps' in the fossil record. Little do they know, biologists have good reason to love them too. Without gaps in the fossil record, our whole system for naming species would break down. Fossils could not be given names, they'd have to be given numbers, or positions on a graph. Or instead of arguing heatedly over whether a fossil is 'really', say, an early *Homo ergaster* or a late *Homo habilis,* we might call it *habigaster*. There's a lot to be said for this. Nevertheless, perhaps because our brains evolved in a world where most things do fall into discrete categories, and in particular where most of the intermediates between living species are dead, we often feel more comfortable if we can use separate names for things when we talk about them. I am no exception and neither are you, so I shall not bend over backwards to avoid using discontinuous names for species in this book. But the Salamander's Tale explains why this is a human imposition rather than something deeply built into the natural world. Let us use names as if they really reflected a discontinuous reality, but let's privately remember that, at least in the world of evolution, it is no more than a convenient fiction, a pandering to our own limitations.

THE NARROWMOUTH'S TALE

Microhyla (sometimes confused with *Gastrophryne*) is a genus of small frogs, the narrowmouthed frogs. There are several species, including two in North America: the eastern narrowmouth *Microhyla carolinensis,* and the Great Plains narrowmouth *Microhyla olivacea.* These two are so closely related that they occasionally hybridise in nature. The eastern narrowmouth's range extends down the east coast from the Carolinas to Florida, and west until half way across Texas and Oklahoma. The Great Plains narrowmouth extends from Baja California in the west, as far as eastern Texas and eastern Oklahoma, and as far north as northern Missouri. Its range is therefore a western mirror of the eastern narrowmouth's, and it might as well be called the western narrowmouth. The important point is that their ranges meet in the middle: there is an overlap zone running up the eastern half of Texas and into Oklahoma. As I said, hybrids are occasionally found in this overlap zone, but mostly the frogs distinguish just as well as herpetologists do. This is what justifies our calling them two different species.

As with any two species, there must have been a time when they were one. Something separated them: to use the technical term, the single ancestral species 'speciated' and became two. It is a model for what happens at every branch point in evolution. Every speciation begins with some sort of initial separation between two populations of the same species. It isn't always a geographical separation, but, as we shall see in the Cichlid's Tale, an initial separation of some kind makes it possible for the statistical distribution of genes in the two populations to move apart. This usually results in an evolutionary divergence with respect to something visible: shape or colour or behaviour. In the case of these two populations of American frogs, the western species became adapted to life in drier climates than the eastern, but the most conspicuous difference lies in their mating calls. Both are squeaky buzzes, but each buzz of the western species lasts about twice as long (2 seconds) as the eastern species, and its predominant pitch is noticeably higher: 4,000 cycles per second as against 3,000. That is to say, the predominant pitch of the western narrowmouth is about top C, the highest key on a piano, and the eastern predominant pitch is around the F# below that. These sounds are not musical, however. Both calls contain a mixture of frequencies, ranging from far below the predominant to far above. Both are buzzes, but the eastern buzz is lower. The western call, as well as being longer, begins with a distinct peep, rising in pitch before the buzz takes over. The eastern frog goes straight into its shorter buzz.

Why go into so much detail about these calls? Because what I have described is true only in the zone of overlap where the comparison between them is clearest, and that is the whole point of the tale. W. F. Blair tape-recorded frogs from a good spread of sampling locations across the United States, with fascinating results. In areas where the two species of frogs never meet one another – Florida for the eastern species and Arizona for the western – their songs are much more similar to each other in pitch: the predominant pitch of both is around 3,500 cycles per second: top A on the piano. In areas close to the zone of overlap but not quite in it, the two species are more different, but not as different as they are in the zone of overlap itself.

The conclusion is intriguing. Something is pushing the calls of these two species apart in the zone where they overlap. Blair's interpretation, which not everybody accepts, is that hybrids are penalised. Anything that helps potential miscegenators to distinguish the species and avoid the wrong one is favoured by natural selection. Such small differences as there may be are exaggerated in just that part of the country where it matters. The great evolutionary geneticist Theodosius Dobzhansky called this 'reinforcement' of reproductive isolation. Not everybody accepts Dobzhansky's reinforcement theory, but the Narrowmouth's Tale, at least, seems to offer support.

There is another good reason why closely related species might be pushed apart when they overlap. They are likely to be competing for similar resources. In the Galapagos Finch's Tale, we saw how different species of finch have partitioned the available seeds. Species with larger bills take larger seeds. Where they don't overlap, both species can take a broader range of resources – large seeds and small seeds. Where they do overlap, each species is forced, by competition from the other, to become more different from the other. The large-billed species might evolve an even larger bill, the small-billed species an even smaller bill. As usual, by the way, don't be misled by the metaphorical idea of being forced to evolve. What actually happens is that within each species, when the other species is present, individuals who happen to be more different from the competing species thrive.

This phenomenon, where two species differ from each other more when they overlap than when they don't, is called 'character displacement' or 'reverse cline'. It is easy to generalise from biological species to cases where any class of entities differ more when they encounter one another than when they are alone. The human parallels are tempting, but I shall resist. As authors used to say, this is left as an exercise for the reader.

THE AXOLOTL'S TALE

We think of young animals as small versions of the adults they are to become, but this is far from the rule. Probably a majority of animal species manage their life stories very differently. The young make their own living, as specialists in a totally different way of life from their parents. A substantial fraction of plankton consists of swimming larvae whose adulthood – if they survive, which is statistically unlikely – will be very different. In many insects the larval stage is the one that does the bulk of the feeding, building up a body that will eventually metamorphose into an adult whose only roles are dispersal and reproduction. In extreme cases such as mayflies, the adult doesn't feed at all, and – for nature is ever niggardly* – lacks a gut and other costly feeding apparatus.

A caterpillar is a feeding machine which, when it has grown to a good size on plant food, in effect recycles its own body and reconstitutes itself as an adult butterfly that flies, sucks nectar as aviation fuel, and reproduces. Adult bees, too, power their flight muscles on

---

* I use this word advisedly. In 1999 the Mayor of Washington DC accepted the resignation of an official whose description of a budget proposal as niggardly gave offence. Julian Bond, distinguished Chairman of the NAACP, correctly described the Mayor's judgement as niggardly. Inspired by the case, a nasty little student at the University of Wisconsin brought an official complaint against her professor, who had used 'niggardly' in a lecture on Chaucer. Such ignorant witch-hunting is not peculiar to the USA. In 2001, a mob of British vigilantes stoned the house of a consultant paediatrician, mistaking her for a paedophile.

nectar while they gather pollen (a very different kind of food) for the worm-like larvae. Many insect larvae live underwater before hatching into adults, who fly through the air and disperse their genes to other bodies of water. A huge diversity of marine invertebrates have adult stages that live on the sea bottom, sometimes permanently moored to one spot, but very different larval stages that disperse the genes by swimming in the plankton. These include molluscs, echinoderms (sea urchins, starfish, sea cucumbers, brittle stars), sea squirts, worms of many kinds, crabs and lobsters, and barnacles. Parasites typically have a series of distinct larval stages, each with its own characteristic way of life and diet. Often the different life stages are also parasitic, but parasitic on very different hosts. Some parasitic worms have as many as five completely distinct juvenile stages, each of which makes its living in a different way from all the others.

All this means that a single individual must carry within it the full genetic instruction set for each of the larval stages, with their different ways of making a living. A caterpillar's genes 'know' how to make a butterfly, and a butterfly's genes know how to make a caterpillar. Doubtless some of the very same genes are involved, in different ways, in making both these radically different bodies. Other genes lie dormant in the caterpillar and are turned on in the butterfly. Yet others are active in the caterpillar and are turned off and forgotten when it becomes a butterfly. But the whole set of genes is there, in both bodies, and is passed on to the next generation. The lesson is that we shouldn't be too surprised if animals as different from each other as caterpillars and butterflies occasionally evolve directly one into the other. Let me explain what I mean.

Fairy stories are filled with frogs turning into princes, or pumpkins turning into coaches drawn by white horses metamorphosed from white mice. Such fantasies are profoundly unevolutionary. They couldn't happen, not for biological reasons but mathematical ones. Such transitions would have an inherent improbability value to rival, say, a perfect deal at bridge, which means that for practical purposes we can rule them out. But for a caterpillar to turn into a butterfly is not a problem: it happens all the time, the rules having been built up over the ages by natural selection. And although no butterfly has ever been seen to turn into a caterpillar, it should not surprise us in the same way as, say, a frog turning into a prince. Frogs don't contain genes for making princes. But they do contain genes for making tadpoles.

My former Oxford colleague John Gurdon dramatically demonstrated this in 1962 when he transformed an adult frog (well, an adult frog cell!) into a tadpole (it has been suggested that this first-ever experimental cloning of a vertebrate deserves a Nobel prize). Similarly, butterflies contain genes for turning into caterpillars. I don't know what embryological hurdles would need to be surmounted in order to persuade a butterfly to metamorphose into a caterpillar. No doubt it would be very difficult. But the possibility is not completely ludicrous in the same way as the frog/prince transformation. If a biologist claimed to have induced a butterfly to turn into a caterpillar, I would study his report with interest. But if he claimed to have persuaded a pumpkin to turn into a glass coach, or a frog into a prince, I'd know he was a fraud without even looking at the evidence. The difference between the two cases is important.

Tadpoles are larvae of frogs or salamanders. Aquatic tadpoles change radically, in the process called 'metamorphosis', into a terrestrial adult frog or salamander. A tadpole may not be quite as different from a frog as a caterpillar is from a butterfly, but there's not a lot in it. A typical tadpole makes its living as a small fish, swimming with its tail, breathing underwater with gills, and eating vegetable matter. A typical frog makes its living on land, hopping rather than swimming, breathing air rather than water, and hunting live animal prey. Yet, different as they seem, we could easily imagine a frog-like adult ancestor evolving into a tadpole-like adult descendant, because all frogs contain the genes for making a tadpole. A frog 'knows' genetically how to be a tadpole, and a tadpole how to be a frog. The same is true of salamanders and they are rather more like their larvae than frogs are like theirs. Salamanders don't lose their tadpole tails, although the tails tend to lose their vertical keel shape and become rounder in cross section. Salamander larvae are often carnivorous like the adults. And, like the adults, they have legs. The most conspicuous difference is that the larvae have long, feathery external gills, but there are lots of less obvious differences too. Actually, to turn a salamander species into a species whose adult stage was a tadpole would be easy – all it would take is for the reproductive organs to mature early, with metamorphosis suppressed. Yet, if it were only the adult stages that fossilised, it would look like a major, and apparently 'improbable' evolutionary transformation.

And so we come to the axolotl, whose tale this is. It is a strange creature, native to a mountain lake in Mexico. It is of the essence of its tale that it is hard to say exactly what an axolotl is. Is it a salamander? Well, sort of. Its name is *Ambystoma mexicanum*, and it is a close relative of the tiger salamander *Ambystoma tigrinum*, which is found in the same area and more widely in North America as well. The tiger salamander, named for obvious reasons, is an ordinary salamander with a cylindrical tail and dry skin, which walks around on land. The axolotl is not at all like an adult salamander. It is like a larval salamander. In fact it is a larval salamander except for one thing. It never turns into a proper salamander and never leaves the water, but mates and reproduces while still looking and behaving like a juvenile. I nearly said the axolotl mates and reproduces while still *being* a juvenile, but this might violate the definition of a juvenile.

Definitions apart, there seems little doubt about what happened in the evolution of the modern axolotl. A recent ancestor was just an ordinary land salamander, probably very like the tiger salamander. It had a swimming larva, with external gills and a deep-keeled tail. At the end of larval life it would metamorphose, as expected, into a dry-land salamander. But then a remarkable evolutionary alteration occurred. Probably under the control of hormones, something shifted in the embryological calendar such that the sex organs and sexual behaviour matured earlier and earlier (or it may even have been a sudden change). This evolutionary regression continued until sexual maturity was arriving in what was, in other respects, clearly the larval stage. And the adult stage was chopped off the end of the life history. Alternatively, you may prefer to see the change not as an acceleration of sexual maturity relative to the rest of the body ('progenesis'), but as a slowing down of everything else, relative to sexual maturity ('neoteny').[*]

Whether the means is neoteny or progenesis, the evolutionary consequence is called paedomorphosis. It is not difficult to see its plausibility. Slowing-down or speeding-up of developmental processes, relative to other developmental processes, happens all the time in evolution. It is called heterochrony and it presumably, if you think about it, must underlie many, if not all, evolutionary changes in anatomical shape. When reproductive development varies

---

[*] Stephen Jay Gould helpfully sorts out the terminology, in his classic *Ontogeny and Phylogeny*.

heterochronically relative to the rest of development, what may evolve is a new species that lacks the old adult stage. This seems to be what happened with the axolotl.

The axolotl is just an extreme among salamanders. Many species seem to be, at least to some extent, paedomorphic. And others do other heterochronically interesting things. The various species of salamander colloquially called 'newts' have an especially revealing life history.* A newt first lives as a gilled larva in water. Then it emerges from the water and lives for two or three years as a kind of salamander on dry land, having lost its gills and the keel on its tail. But, unlike other salamanders, newts don't reproduce on land. Instead, they return to water, regaining some, but not all, of their larval characteristics. Unlike axolotls, newts don't have gills, and their need to come to the surface to breathe air is an important and competitive constraint on their underwater courtship. Unlike the larval gills, they do regain the keel of the larval tail, and in other respects they resemble a larva. But unlike a typical larva, their reproductive organs develop and they court and mate underwater. The dry land phase never reproduces and, in this sense, one might prefer not to call it the 'adult'.

You might ask why newts bother to turn into a dry land form at all, given that they are going to return to water to breed. Why not just do what axolotls do: start in water and stay in water? The answer seems to be that there is an advantage to breeding in temporary ponds which form in the wet season and are destined to dry up, and you have to be good on dry land in order to reach them (shades of Romer). Having reached a pond, how do you then reinvent your aquatic equipment? Heterochrony comes to the rescue: but heterochrony of a peculiar kind, involving going into reverse after the 'dry adult' has served its purpose of dispersing to a new, temporary pond.

Newts serve to emphasise the flexibility of heterochrony. They remind us of the point I made about how genes in one part of the life cycle 'know' how to make other parts. Genes in dry land salamanders know how to make an aquatic form because that is what they once were; and, to prove it, that is precisely what newts do.

Axolotls are, in one respect, more straightforward. They have lopped the dry land phase off the end of the ancestral life cycle. But

* A. Fink-Nottle, *in litt.*

the genes for making a dry land salamander still lurk in every axolotl. It has long been known, from the classic work of Laufberger and of Julian Huxley mentioned in the Epilogue to Little Foot's Tale, that they can be activated by a suitable dose of hormones in the laboratory. Axolotls treated with thyroxine lose their gills and become dry land salamanders, just as their ancestors once did naturally. Perhaps the same feat could be achieved by natural evolution, should selection favour it. One way might be a genetically mediated raising of the natural production of thyroxine (or an increase in sensitivity to the existing thyroxine). Maybe axolotls have undergone paedomorphic and reverse-paedomorphic evolutions repeatedly during their history. Maybe evolving animals in general are continually, though less dramatically than the axolotl, moving one way or another along an axis of paedomorphosis/reverse-paedomorphosis.

Paedomorphosis is one of those ideas of which, once you get the hang of it, you start seeing examples everywhere you look. What does an ostrich remind you of? During the Second World War my father was an officer in the King's African Rifles. His batman Ali, like many Africans of the time, had never seen most of the large wild animals for which their homeland is famous, and his first glimpse of an ostrich sprinting across the savannah elicited a shriek of astonishment: 'Big chicken, BIG CHICKEN!' Ali had it nearly right, but more penetrating would have been 'Big baby chicken!' The wings of an ostrich are silly little stubs, just like the wings of a newly hatched chick. Instead of the stout quills of a flying bird, ostrich feathers are coarse versions of the fluffy down of a baby chick. Paedomorphosis illuminates our understanding of the evolution of flightless birds such as the ostrich and the dodo. Yes, the economy of natural selection favoured downy feathers and stubby wings in a bird that did not need to fly (see the Elephant Bird's Tale and the Dodo's Tale). But the evolutionary route that natural selection employed to achieve its advantageous outcome was paedomorphosis. An ostrich is an overgrown chick.

Pekinese dogs are overgrown puppies.* Pekinese adults have the domed forehead and the juvenile gait, even the juvenile appeal, of a puppy. Konrad Lorenz has wickedly suggested that Pekineses and other babyfaced breeds like King Charles spaniels appeal to the

---

* The pompous meddling with the English language that has given us 'Beijing', 'Mumbai' and 'cosmonaut' has so far spared us 'Beijinese dog'.

maternal instincts of frustrated mothers. The breeders may or may not have known what they were trying to achieve, but they surely didn't know that they were doing it through an artificial version of paedomorphosis.

Walter Garstang, a well-known English zoologist of a century ago, was the first to emphasise the importance of paedomorphosis in evolution. Garstang's case was later taken up by his son-in-law Alister Hardy, who was my professor when I was an undergraduate. Sir Alister delighted in reciting the comic verses which were Garstang's preferred medium for communicating his ideas. They were slightly funny at the time but not, I think, quite funny enough to justify the elaborate zoological glossary which would have to accompany a reprinting here.* Garstang's idea of paedomorphosis, however, is today as interesting as ever – which doesn't necessarily mean it is right.

We can think of paedomorphosis as a kind of evolutionary gambit: Garstang's Gambit. It can in theory herald a whole new direction in evolution: can even, Garstang and Hardy believed, permit a dramatic and, by geological standards, sudden breakout from an evolutionary dead end. This seems especially promising if the life cycle sports a distinct larval phase like a tadpole. A larva that is already adapted to a different way of life from the old adult is primed to swerve evolution into a whole new direction by the simple trick of accelerating sexual maturity relative to everything else.

Among the cousins of the vertebrates are the tunicates or sea squirts. This seems surprising because adult sea squirts are sedentary filter-feeders anchored to rocks or seaweeds. How can these soft bags of water be cousins to vigorously swimming fishes? Well, the adult sea squirt may look like a bag, but the larva looks like a tadpole. It is even called a 'tadpole larva'. You can imagine what Garstang made of this, and we shall revisit the point, and unfortunately cast doubt on Garstang's theory, at Rendezvous 24 when we meet the sea squirts.

Bearing in mind the adult Pekinese as an overgrown puppy, think of the heads of juvenile apes. What do they remind you of? Wouldn't you agree that a juvenile chimpanzee or orang utan is more humanoid than an adult chimpanzee or orang utan? Admittedly it is controversial, but some biologists regard a human as a juvenile ape.

---

* A fragment of one heads the Lancelet's Tale.

An ape that never grew up. An ape axolotl. We have already met the idea in the Epilogue to Little Foot's Tale, and I shall not spell it out again here.

LUNGFISH

At Rendezvous 18, around 417 million years ago, we are joined in the warm and shallow seas of the Devonian–Silurian boundary by a tiny trickle of pilgrims who have plodded a lonely course from the present. They are the lungfish, and they join us to look at the common ancestor we share with them – an experience that may seem less strange to them than to us, for they find they have much in common with Concestor 18. Approximately our 185-million-greats-grandparent, it was a sarcopterygian, a lobefin fish, certainly much more like a lungfish than like a tetrapod (see plate 22).

There are only six species of lungfish today: *Neoceratodus forsteri* from Australia, *Lepidosiren paradoxa* from South America, and four species of *Protopterus* from Africa. The Australian lungfish looks really quite excitingly like an ancient sarcopterygian, with fleshy lobe fins like a coelacanth. The African and South American species, which are closely related to each other, have their fins reduced to long trailing tassels, and they therefore look less like the lobe-finned fish from whom they are descended. All the lungfish breathe air using lungs. The Australian lungfish has a single lung, the others have two. The African and South American species use their lungs to withstand a dry season. They burrow into the mud and stay dormant, breathing air through a little breathing hole in the mud. The Australian species, by contrast, lives in permanent bodies of water filled with weed. It takes air into its lung to supplement its gills in oxygen-poor water.

When first discovered in 1870, modern lungfish living in Queensland were united with fossil fish more than 200 million years old under the same name, *Ceratodus*. This gives an indication of how little they have changed during that time. Let's not get carried away, however. A classic study published in 1949 by the British palaeontologist T. S. Westoll showed that, although the lungfish have indeed stagnated for the last 200 million years or so, they evolved much more rapidly before that. In the Carboniferous Period, from around 350 million years ago, they were really racing along, before they slowed down almost to a stop about 250 million years ago, towards the end of the Permian Period.

The Lungfish's Tale is a tale of 'living fossils'.

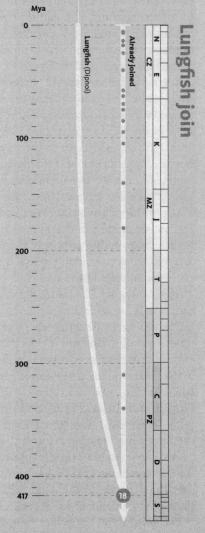

Mya

0

100

200

300

400
417

Lungfish (Dipnoi)

Already joined

18

Lungfish join

N		
CZ	E	
	K	
MZ		
	T	
	P	
PZ	C	
	D	
	S	

Lungfish join. Humans and the other 'tetrapods' could be said to be lobe-finned fish, whose arms, wings, or legs are modified lobed fins. The two other living lineages of lobefins are the coelacanths and the lungfish. The division of these three lineages at the end of the Silurian is thought to have happened in a very short space of time. This makes the order of branching difficult to sort out, even using genetic data. Nevertheless, genetic and fossil studies are starting to agree that the three lungfish species are the closest living relatives of the tetrapods, as shown here.

Image: Australian lungfish (*Neoceratodus forsteri*).

THE LUNGFISH'S TALE
Written with Yan Wong

A living fossil is an animal that, while being as alive as you or me, strongly resembles its ancient ancestors. Not much evolutionary change has occurred down the line leading to the living fossil. It is one of those random, pointless facts that the four most famous living fossils all begin with L: Lungfish, *Limulus*, *Latimeria* (the coelacanth) and *Lingula*. *Limulus*, the so-called 'horseshoe crab' (not a crab at all, but its own thing, superficially resembling a large trilobite) is placed in the same genus as *Limulus walchi* of the Jurassic, 200 million years ago. *Lingula* belongs to the phylum Brachiopoda, sometimes called lamp-shells. The kind of lamp they resemble, if any, is the Aladdin variety with its wick coming out of a kind of teapot spout, but what *Lingula* spectacularly resembles is its own ancestors of 400 million years ago. Its assignment to the very same genus has been disputed, but the fossil forms are still remarkably similar to their modern representatives. Although the anatomies, and presumably the ways of life, of these living fossils have changed rather little, their DNA texts have not stopped evolving. We cousins of lungfish have been changing massively during the hundreds of millions of years since we branched apart. But although lungfish bodies stagnated during the same time, you wouldn't guess it if you looked at the speed of evolution of their DNA.

The ray-finned fish (familiar fish, such as trout or perch) during this time have produced an amazing variety of forms. So, more familiarly, have the tetrapods – we glorified lobe-finned fish who moved out onto the land. The bodies of the lobefins themselves have evolved extremely slowly. Yet at the same time – here is the point this whole tale is leading up to – their genetic molecules seem not to have stuck to this same slow pace. If they had, the DNA sequences of lungfish and coelacanths would be much more similar to each other (and presumably to ancient ancestors) than they are to us, and to ray-finned fish. Yet they are not.

We know from fossils the approximate timings of the ancestral splits between lungfish, coelacanths, ourselves and the ray-finned fish. The first split, at about 440 million years ago, is that between the ray-finned fish and all the rest of us. The next to split off were the coelacanths, about 425 million years ago. That left the lungfish and all

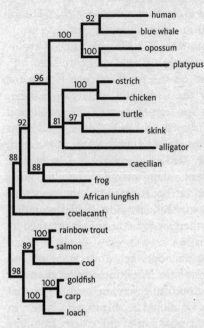

Evolutionary tree of various species from maximum likelihood analysis of DNA (see the Gibbon's Tale). Adapted from one of several trees put together by Zardoya and Meyer [324].

the rest of us. About 5 or 10 million years later still, the lungfish split off, leaving the rest of us, now called tetrapods, to make our own evolutionary way. As evolutionary time goes, all three of these splits occurred at pretty nearly the same time, at least compared to the long time over which all four lineages have been evolving ever since.

While working on a different problem, Rafael Zardoya of Spain and Axel Meyer of Germany drew the evolutionary tree above for the DNA of various species. The length of each branch is drawn to reflect the amount of evolutionary change, in mitochondrial DNA, along it.

If the DNA evolved at a constant rate, regardless of the species, then we would expect all the branches to finish lined up at the right hand edge. This clearly isn't the case. But neither do the organisms that show the least morphological change have the shortest branches. The DNA seems to have evolved at about the same rate in the lungfish and coelacanth as in the ray-finned fish. The vertebrates that colonised the land experienced a faster rate of DNA evolution, but even this is not obviously linked to morphological change. The winner and the runner-up of this molecular caucus race are the platypus and the alligator, neither of which have evolved morphologically as fast as, say, the blue whale or (vanity cannot help whispering) us.

The diagram illustrates an important fact. The rate of DNA evolution is not always constant, but neither is it obviously correlated with morphological change. The tree on the previous page is just one example. Lindell Bromham of the University of Sussex and her colleagues compared evolutionary trees based on morphological change against equivalent trees based on DNA change. And what they found confirmed the message of the Lungfish's Tale. The overall rate of genetic change is independent of morphological evolution.* This is not to say that it is constant – that would have been too good to be true. Certain lineages, such as the rodents and the nematode worms, seem to have a rather fast overall rate of molecular evolution compared to close relatives. In others, such as the cnidarians, the rate is much slower than related lineages.

The Lungfish's Tale encourages a hope that, a few years ago, no zoologist would have dared to entertain. With due caution in choosing genes, and with available methods of correcting for lineages that show variable rates of evolution, we should be able to put a figure, in millions of years, on the time of separation of any species from any other species. This bright hope is called the 'molecular clock', and it is the technique responsible for most of the quoted dates on our rendezvous points in this book. The principle of the molecular clock, and the controversies that still bedevil it, will be explained in the Epilogue to the Velvet Worm's Tale.

But now, on to Rendezvous 19 and the mysterious coelacanth.

---

* An earlier study had obtained a different result. But Bromham and her colleagues convincingly showed that the previous study had failed to allow for non-independence of data – the multiple counting problem that we met in the Seal's Tale.

COELACANTHS

Concestor 19, perhaps our 190-million-greats-grandparent, lived around 425 million years ago, just as plants were colonising the land and coral reefs expanding in the sea. At this rendezvous we meet one of the sparsest, most tenuous bands of pilgrims in this story. We know of only one genus of coelacanth alive today, and its discovery was a huge surprise when it happened. The episode is well described by Keith Thomson in his *Living Fossil: the Story of the Coelacanth*.

The coelacanths were well known in the fossil record, but thought to have gone extinct before the dinosaurs. Then, astoundingly, a living coelacanth turned up in the catch of a South African trawler in 1938. By good fortune Captain Harry Goosen, skipper of the *Nerita*, was friendly with Marjorie Courtenay-Latimer, the enthusiastic young curator of the East London Museum. It was Goosen's habit to put aside interesting finds for her, and on 22 December 1938 he telephoned to tell her he had something. She went down to the quay, and an old Scotsman of the crew showed her a motley collection of discarded fish, which at first didn't seem of any interest. She was about to leave when:

> I saw a blue fin and pushing off the fish, the most beautiful fish I had ever seen was revealed. It was 5 feet long and a pale mauve blue with iridescent silver markings.

She made a sketch of the fish, which she sent to South Africa's leading ichthyologist, Dr J. L. B. Smith, and it knocked him out. 'I would not have been more surprised if I had seen a dinosaur walking down the street.' (See plate 23.) Unfortunately, Smith took his time going to the scene, for reasons that are hard to fathom. By Keith Thomson's account, Smith didn't trust his judgement until he had sent off for a particular reference book, from Dr Barnard, a colleague in Cape Town. Smith hesitantly confessed his secret to Barnard, who was immediately sceptical. It seems that it was weeks before Smith could bring himself to go to East London and actually see the fish. Meanwhile, poor Miss Courtenay-Latimer was coping with its noisome decay. Too large to fit in a formalin jar, she wrapped it in

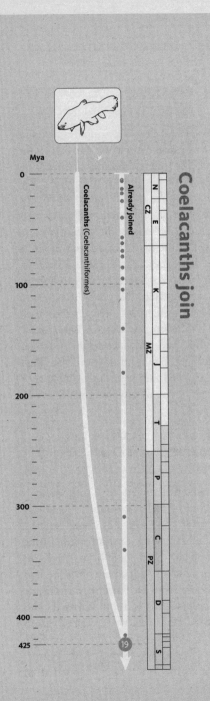

Coelacanths join. Growing consensus places coelacanths (of which there are two living species known) as the earliest diverging of the three extant lineages of lobefins.

Image: Comorean coelacanth (*Latimeria chalumnae*).

formalin-soaked cloths. These were inadequate to stave off decay, and eventually she had to have it stuffed. It was in this form that Smith finally saw it:

> Coelacanth, yes, by God! Although I had come prepared, that first sight hit me like a white-hot blast and made me feel shaky and queer, my body tingled. I stood as if stricken to stone . . . I forgot everything else and then almost fearfully went close up and touched and stroked, while my wife watched in silence . . . It was only then that speech came back, the exact words I have forgotten, but it was to tell them that it was true, it really was true, it was unquestionably a coelacanth. Not even I could doubt any more.

Smith named it *Latimeria* after Marjorie. Since then, many more have been found in deep waters around the Comoros Islands near Madagascar, and a second species turned up on the other side of the Indian Ocean, off Sulawesi. The genus has now been studied in detail, although not without the acrimony and accusations of fakery that seem – regrettably but I suppose understandably – to go with rare and very important discoveries.

RAY-FINNED FISH

Rendezvous 20 is a big one, 440 million years ago in the earliest Silurian, still with a southern ice cap left over from the cold Ordovician. Concestor 20, which I am estimating to be our 195-million-greats-grandparent, is the one that unites us to the actinopterygian or ray-finned fish, most of whom belong to the large and successful group known as teleosts. The teleost fish are the great success story among modern vertebrates – there are some 23,500 species of them. They are prominent at many levels of underwater food chains, in both salt and freshwater. They have managed to invade hot springs at one extreme, and the icy waters of the Arctic seas and high mountain lakes at the other. They thrive in acid streams, stinking marshes and saline lakes.

'Ray' refers to the fact that their fins have a skeleton similar to a Victorian lady's fan. Ray-fins lack the fleshy lobe at the base of each fin – eponym for the lobefin fish like coelacanths and Concestor 18. Unlike our arms and legs, which have relatively few bones, and muscles that can move them relative to one another within the limb, actinopterygian fins are moved mostly by muscles in the main body wall. In this respect, we are more like lobefin fish – as well we should be, for we are lobefins adjusted for life on land. Lobefin fish have muscles in the fleshy fins themselves, just as we have biceps and triceps muscles in our upper arms and Popeye muscles in our lower arms.

The ray-finned fish are mostly teleosts, plus a few odds and ends,

Opposite: **Ray-finned fish join.** The ray-finned fish are the closest relatives of we lobefins, and contain roughly the same number of described species – about 25,000. Their phylogeny is not well resolved, although it is clear that the sturgeons and paddlefish, the bichirs, the gars and the bowfin all branched off early. The phylogeny displayed here is particularly uncertain. For this reason, a few of the especially obscure groups have been omitted from this tree.

Images, left to right: plaice (*Pleuronectes platessa*); snaggletooth (*Astronesthes niger*); pike (*Esox lucius*); red-bellied piranha (*Serrasalmus nattereri*); northern anchovy (*Engraulis mordax*); green moray (*Gymnothorax prasinus*); Florida gar (*Lepisosteus platyrhincus*); Siberian sturgeon (*Acipenser baeri*).

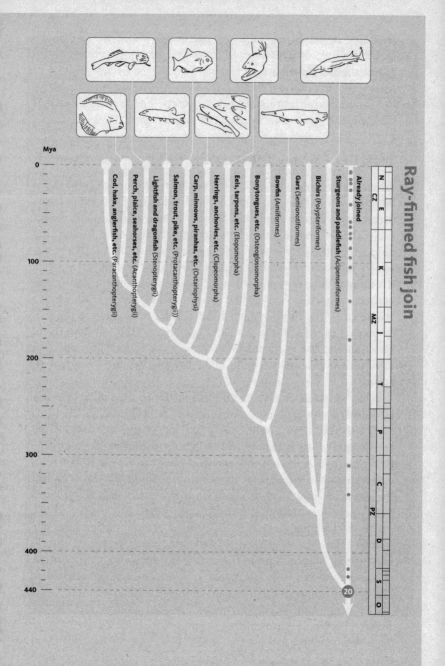

Ray-finned fish join

Mya

Cod, hake, anglerfish, etc. (Paracanthopterygii)

Perch, plaice, seahorses, etc. (Acanthopterygii)

Lightfish and dragonfish (Stenopterygii)

Salmon, trout, pike, etc. (Protacanthopterygii)

Carp, minnows, piranhas, etc. (Ostariophysi)

Herrings, anchovies, etc. (Clupeomorpha)

Eels, tarpons, etc. (Elopomorpha)

Bonytongues, etc. (Osteoglossomorpha)

Bowfin (Amiiformes)

Gars (Semionotiformes)

Bichirs (Polypteriformes)

Sturgeons and paddlefish (Acipenseriformes)

Already joined

20

N	CZ	E		K		MZ	J		T		P		C	PZ		D		S		O

including the sturgeons, and the paddlefish whom we met in the Duckbill's Tale. It is right and proper that such a hugely successful group should contribute several tales and I shall relegate most of what I have to say about them to the tales. The teleost pilgrims arrive in a jostling crowd, brilliant in their variety. The magnitude of that variety is the inspiration for the Leafy Sea Dragon's Tale.

THE LEAFY SEA DRAGON'S TALE

When my daughter was tiny, she loved to ask adults to draw fish for her. She would rush up to me when I was trying to write a book, thrust a pencil in my hand and clamour, 'Draw a fish. Daddy, draw a fish!' The cartoon fish that I would immediately draw to keep her quiet – and the only kind of fish she ever wanted me to draw – was always the same: a regulation-issue fish like a herring or a perch, streamlined side view, pointed at the front end, triangular fin top and bottom, triangular tail at the back, finally dotted with an eye bracketed by the curve of a gill cover. I don't think I ran to pectoral or pelvic fins, which was remiss of me because they all have them. The standard fish is indeed an extremely common shape, one that obviously works well over the full range of sizes from minnow to tarpon.

What would Juliet have said if I had possessed the skill to draw for her a leafy sea dragon, *Phycodurus equus*? (See plate 24.) 'NO, Daddy. NOT seaweed. Draw a fish. Draw a FISH.' The message of the Leafy Sea Dragon's Tale is that animal shapes are malleable like plasticine (see plate 25). A fish can change in evolutionary time to whatever unfishy shape is required for its way of life. Those fish that look like the standard-issue Juliet fish do so only because it suits them. It is a good shape for swimming through open water. But if survival is a matter of hanging motionless in beds of gently swaying kelp, the standard fish shape can be twisted and kneaded, pulled out in fantastically branched projections whose resemblance to the fronds of brown seaweed is so great that a botanist might be tempted to narrow it down to species (perhaps of the genus *Fucus*).

The shrimpfish, *Aeoliscus strigatus*, which lives on reefs in the western Pacific, is again much too cunningly disguised for Juliet to have been satisfied, had I drawn it as a 'fish'. Its extremely elongated body is further prolonged by a long snout, and the effect is enhanced by a dark stripe running right through the eye and straight to the very untail-like tail. The fish looks like a long shrimp, or a little like a cut-throat razor – which accounts for its other name of razorfish. It is covered with a transparent armour which, my colleague George Barlow who has watched them in the wild tells me, even feels like that of a shrimp. The resemblance to a shrimp is probably, however, no part of their camouflage. Like many teleosts, shrimpfish swim around in coordinated groups, and with military synchrony. But unlike any other teleost you might think of, shrimpfish swim with the body pointing straight down. I don't mean they swim in a vertical direction. They swim in a horizontal direction, but with the body vertical. The whole effect of this synchronised swimming is a resemblance to a stand of weeds, or, even more strikingly, to the tall spines of a giant sea urchin, among which they often take refuge. Swimming head down is a deliberate decision. When alarmed, they are perfectly capable of flipping into more conventional, horizontal mode and they then flee with surprising speed.

Or, what would Juliet have said if I had drawn for her a snipe eel (Nemichthyidae) or a gulper eel (*Eurypharynx pelecanoides*), two deep-sea eels with birds in their names? The snipe eel looks like a joke, ludicrously long and thin, with bird-like jaws that curve away from each other like a megaphone. So dysfunctional do these diverging jaws look, I can't help wondering how many of the fish have been seen alive. Could the megaphone jaws be a distortion in a dried-up museum specimen?

The gulper looks like a nightmare. With jaws ludicrously too large for its body, or so it would seem, it is capable of swallowing whole prey larger than itself – one of several deep-sea fish with this remarkable talent. It is not unusual, of course, for predators to *kill* prey larger than themselves, and then eat them in bits. Lions do it; so do spiders.* But to *swallow* a larger animal than yourself whole is hard to imagine. The gulper eel, and other deep-sea fish – such as the closely related swallower eel, and the unrelated black swallower, which is not an eel – achieve the trick. They do it by a combination of disproportionately oversized jaws and a slack distensible stomach that hangs down only when full, looking rather like some gross external tumour. After the long digestion period, the stomach shrinks again. Why the trick of prodigious swallowing should be

---

* Spiders eat the large prey not so much in bits as in liquid form. They inject it with digestive juices and then suck it in as if through a straw.

peculiar to snakes* and deep-sea fish is not obvious to me. The gulper and the swallower eel lure prey into the vicinity of their mouth with a luminous lure at the tip of the tail.

The teleost body plan seems almost indefinitely malleable over evolutionary time, tolerant of being pulled or squashed into any shape, however distantly removed from the 'standard' fish shape. The oceanic sunfish's Latin name, *Mola mola*, means millstone, and it is easy to see why. Seen from the side, it looks like a huge disc, up to an astonishing four metres in diameter and weighing up to two tonnes. The circularity of its outline is broken only by two gigantic fins on top and underneath, each one up to two metres long.

The Hippo's Tale invoked, in explanation of its dramatic difference from its whale cousins, the liberation from gravity that whales must have enjoyed, as soon as they severed all ties with the land. No doubt something similar explains the great variety of shapes that the teleost fish display. But in exploiting that liberation, teleosts have one other advantage over, for example, sharks. Teleosts cope with buoyancy in a very special way, and the pike will tell the tale.

THE PIKE'S TALE

In the sad province of Ulster, where 'the Mountains of Mourne sweep down to the sea', I know a beautiful lake. A party of children were swimming naked there one day, when somebody shouted that they had seen a large pike. Instantly all the boys – but not the girls – fled to dry land. The northern pike, *Esox lucius*, is a formidable predator of small fish. It is beautifully camouflaged, not against predators but to help it steal up on its prey. A stealth predator, and not particularly fast over a distance, it hangs almost motionless in the water, creeping imperceptibly forwards until within striking distance. During the deadly creep, it propels itself with imperceptible movements of the rear-mounted dorsal fin.

This whole hunting technique depends upon the ability to hang in the water at the desired level, like a drifting dirigible, without any effort, in perfect hydrostatic equilibrium. All locomotor work is concentrated on the clandestine business of creeping forwards. If a pike needed to swim in order to maintain its level, as many sharks

---

* Snakes do it by disarticulating their skulls. For a snake, eating a meal must be an ordeal comparable to giving birth for a woman.

do, its ambush technique would not work. Effortless maintenance, and adjustment, of hydrostatic equilibrium is what teleost fish are supremely good at, and it may be the single most significant key to their success. How do they do it? By means of the swim bladder: a modified lung filled with gas, which provides sensitive dynamic control of the animal's buoyancy. Except for some bottom-dwellers who have secondarily lost the swim bladder, all teleosts have it – not just pike and not just their prey.

The swim bladder is often explained as working like a Cartesian Diver, but I think that is not quite correct. A Cartesian Diver is a miniature diving bell containing a bubble of air, which hangs at hydrostatic equilibrium in a bottle of water. When the pressure is increased (usually by squeezing down the cork in the neck of the bottle), the bubble is compressed and less water is displaced by the diver as a whole. Therefore, by Archimedes' Principle, the diver sinks. If the cork is eased slightly upwards so that the pressure in the bottle decreases, the bubble in the diver expands, more water is displaced, and the diver floats a little higher. So, with your thumb on the cork, you can exert fine control over the level at which the diver finds its equilibrium.

The key point about a Cartesian Diver is that the number of air molecules in the bubble remains fixed, while the volume and the pressure are changed (in inverse proportions, following Boyle's Law). If fish worked like Cartesian Divers, they would use muscle power to squeeze, or relax, the swim bladder, thereby changing the pressure and volume but leaving the number of molecules the same. That would work in theory, but it isn't what happens. Instead of keeping the number of molecules fixed and adjusting the pressure, the fish adjust the number of molecules. To sink, the fish absorbs some molecules of gas from its swim bladder into the blood, thereby reducing the volume. To rise, it does the reverse, releasing molecules of gas into the swim bladder.

In some teleosts, the swim bladder is also used to assist in hearing. The fish's body being mostly water, sound waves propagate through it pretty much as they did through the water before they hit the fish. When they strike the wall of the swim bladder, however, they suddenly reach a different medium, gas. The swim bladder therefore acts as a kind of eardrum. In some species it lies right against the inner ear. In others it is connected to the inner ear by a series of small bones called the Weberian ossicles. These do a similar job to our own hammer, anvil and stirrup, but are completely different bones.

The swim bladder seems to have evolved – been 'co-opted' – from a primitive lung, and some surviving teleosts, such as bowfins, gars and bichirs, still use it for breathing. This perhaps comes as a little surprise to us, for whom breathing air seems like a significant 'advance' that went with leaving the water for the land. One might have supposed the lung to be a modified swim bladder. On the contrary, it seems that the primitive breathing lung forked in evolution and went two ways. On the one hand, it carried its old breathing function out onto the land, and we use it still. The other branch of the fork was the new and exciting one: the old lung became modified to form a genuine innovation – the swim bladder.

THE MUDSKIPPER'S TALE

On an evolutionary pilgrimage it is fitting that some of the tales, though told by surviving pilgrims, should deal with recent re-enactments of ancient evolutionary events. Teleost fish are so variable and so versatile, it is only to be expected that some of them might replay parts of the lobefins' history, and come out onto the land. The mudskipper is just such a fish out of water, and it lives to tell the tale.

A number of teleost fish species live in swampy water, poor in oxygen. Their gills cannot extract enough, and they need help from the air. Familiar aquarium fish from the swamps of South-East Asia, such as the Siamese fighting fish *Betta splendens*, frequently come to the surface to gulp air, but they still use their gills to extract the oxygen. I suppose, since the gills are wet, you could say the gulping is equivalent to locally oxygenating their gill water, as you might bubble air through your aquarium. It goes further than that, however, because the gill chamber is furnished with an auxiliary air space, richly supplied with blood vessels. This cavity is not a true lung. The true homologue of the lung in teleost fish is the swim bladder which, as the Pike's Tale has shown, they use for keeping their buoyancy neutral.

Those fish that breathe air through their gill chamber have rediscovered air breathing by a completely different route. Perhaps the most advanced exponents of the air-breathing gill chamber are the climbing perches *Anabas*. These fish also live in poorly oxygenated water and they have the habit of walking over land looking for water when their previous home has dried up. They can survive out of water for days at a time. *Anabas* is, indeed, a living, breathing example of what Romer was talking about in his (now less fashionable) theory of how fish came out onto the land.

Another group of walking teleost fish are the mudskippers, for example *Periophthalmus*, whose tale this is. Some mudskippers actually spend more time out of water than in it. They eat insects and spiders, which are not normally found in the sea. It is possible that our Devonian ancestors enjoyed similar benefits when they first left the water, for they were preceded onto the land by both insects and spiders. A mudskipper flaps its body across the mudflats, and it can also crawl using its pectoral (arm) fins, whose muscles are so well developed that they can support the fish's weight. Indeed, mudskipper courtship takes place partly on land, and a male may do push-ups, as some male lizards do, to show off his golden chin and throat to females. The fin skeleton, too, has evolved convergently to resemble that of a tetrapod such as a salamander.

Mudskippers can jump more than half a metre by bending the body to one side and suddenly straightening it – hence some of their many vernacular names, including 'mud-hopper', 'johnny jumper', 'frogfish' and 'kangaroo fish'. Another common name, 'climbing fish', comes from their habit of climbing mangrove trees looking for prey. They cling to the trees with the pectoral fins, aided by a kind of sucker which is made by bringing the pelvic fins together under the body.

Like the swamp fish already mentioned, mudskippers breathe by taking air into their moist gill chambers. They also take in oxygen through the skin, which has to be kept moist. If a mudskipper is in danger of drying out, it will roll about in a puddle. Their eyes are especially vulnerable to dryness, and they sometimes wipe them with a wet fin. The eyes bulge close together near the top of the head, where, as with frogs and crocodiles, they can be used as periscopes to see above the surface when the fish is underwater. When out on land, a mudskipper will frequently withdraw its bulging eyes into their sockets to moisten them. Before leaving the water on a land sortie, the fish will fill its gill cavities with water.

In a popular book on the conquest of the land, the author mentions an account by an eighteenth-century artist living in Indonesia who kept a 'frogfish' alive for three days in his house:

It followed me everywhere with great familiarity, much like a little dog.

The book has a cartoon of a 'frogfish' walking like a little dog, but what it actually depicts is clearly an *angler* fish: a deep-sea fish with a lure on the end of a spine sticking up above the head, used to catch smaller fish. I suspect that the cartoonist has been the victim of a misunderstanding: an instructive one because it shows what can happen if we rely on colloquial common names for animals rather than the scientific names that, whatever their faults, are designed to be unique. It is true that some people call angler fish frogfish. But it is highly implausible that the fish that followed the artist around like a dog could have been a deep-sea angler fish. It could easily have been a mudskipper, however. They do live in Indonesia, and frogfish is one of their colloquial names. A mudskipper looks, to my eyes at least, far more like a frog than an angler fish ever could, and it leaps like a frog. I conjecture that the artist's pet 'frogfish', which followed him around like a little dog, was a mudskipper.

I like the idea that we are descended from some creature which, even if it was different from a modern mudskipper in many other respects, was as adventurous and enterprising as a little dog: the nearest thing, perhaps, to a dog that the Devonian had to offer? A girlfriend of mine from long ago explained why she loved dogs: 'Dogs are such good sports.' I think the first fish to venture out onto the land must have been an archetypal good sport, whom it would be a pleasure to call ancestor.

THE CICHLID'S TALE

Lake Victoria is the third largest lake in the world, but it is also one of the youngest. Geological evidence indicates that it is only about 100,000 years old. It is home to a huge number of endemic cichlid (pronounced 'sick-lid') fish. Endemic means that they are found nowhere else than in Lake Victoria, and presumably evolved there. Depending on whether your ichthyologist is a lumper or a splitter, the number of species of cichlid in Lake Victoria is somewhere between 200 and 500, and a recent authoritative estimate puts it at 450. Of these endemic species, the great majority belong to one tribe, the haplochromines. It looks as though they all evolved, as a single 'species flock', during the last hundred thousand years or so.

As we saw in the Narrowmouth's Tale, the evolutionary splitting of one species into two is called speciation. What surprises us about the young age of Lake Victoria is that it suggests an astonishingly high rate of speciation. There is also evidence that the lake dried up completely about 15,000 years ago, and some people even drew the conclusion that the 450 endemic species must have evolved from a single founder in this astonishingly short time. As we shall see, this is probably an exaggeration. But in any case a little calculation helps to get these short times into perspective. What sort of speciation rate would it take to generate 450 species in 100,000 years? The most prolific pattern of speciation in theory would be a succession of doublings. In this idealised pattern, one ancestral species gives rise to two daughter species, each of those splits into two, then each of those splits into two, and so on. Following this most productive ('exponential') pattern of speciation, an ancestral species could easily generate 450 species in 100,000 years, with what seems like the rather long interval of 10,000 years between speciations within any one lineage. Starting with any one modern cichlid pilgrim and going backwards, there would be only ten rendezvous points in 100,000 years.

Of course, it is highly unlikely that real-life speciation would actually follow the ideal pattern of successive doubling. The opposite extreme would be a pattern in which the founder species successively threw one daughter species after another, with none of the daughter species subsequently speciating. Following this least 'efficient' pattern of speciation, in order to generate 450 species in 100,000 years, the interval between speciation events would need to be a couple of centuries. Even that doesn't sound ridiculously short. The truth surely lies between the two extreme patterns: say one or a few millennia as the average interval between speciation events in any one lineage. When you put it like that, the speciation rate doesn't seem so spectacularly high after all, especially in the light of the sorts of evolutionary rates that we saw in the Galapagos Finch's Tale. Nevertheless, as a sustained feat of speciation, it is very fast and prolific by the standards evolutionists have come to expect, and the cichlid fish of Lake Victoria have become legendary among biologists for this reason.*

Lakes Tanganyika and Malawi are only slightly smaller than

---

* Lake Victoria has been the victim of a man-made catastrophe. In 1954, the British colonial administration, hoping to improve fisheries, introduced the Nile perch (*Lates niloticus*) to the lake. This decision was opposed by biologists, who predicted that the perch would disrupt the lake's unique ecosystems. Their prediction came disastrously true. The cichlids had never evolved to cope with a big predator like the Nile perch. Probably 50 cichlid species have gone for good, and another 130 are critically endangered. In a mere half-century, completely avoidable ignorance has devastated local economies around the lake and irreversibly wiped out a priceless scientific resource.

Victoria – smaller in area, that is. But where Victoria is a wide, shallow basin, Tanganyika and Malawi are Rift Valley lakes: long, narrow and very deep. They are not so young as Victoria. Lake Malawi, which I have already nostalgically mentioned as the site of my first 'seaside' holidays, is between 1 and 2 million years old. Lake Tanganyika is the oldest, at 12–14 million. Despite these differences, all three lakes share the remarkable feature that inspires this tale. All are teeming with hundreds of endemic cichlid fish, unique to the particular lake. Victoria cichlids are a completely different set of species from Tanganyika cichlids, and Malawi cichlids are a completely different set from either. Yet, each of the three flocks of hundreds of species has produced, by convergent evolution in its own lake, an extremely similar range of types. It looks as though a single founding haplochromine species (or very few) entered each infant lake, perhaps through a river. From such small beginnings, successive evolutionary subdivisions – 'speciation events' – generated hundreds of species of cichlids, whose range of types closely paralleled those in each of the other great lakes. This sort of rapid diversification into many different types is called 'adaptive radiation'. Darwin's finches are another famous example of an adaptive radiation, but African cichlids are particularly special because it has happened in triplicate.*

Much of the variation within each lake is concerned with diet. Each of the three lakes has its specialists in plankton feeding, its specialists in grazing algae off rocks, its predators on other fish, its scavengers, its food robbers, its fish-egg eaters. There are even parallels to the cleaner fish habit, which is better known from tropical coral reef fish (see the Polypifer's Tale). Cichlid fish have a complicated system of double jaws. In addition to the 'ordinary' outer jaws that we can see, there is a second set of 'pharyngeal jaws' buried deep in the throat. It is likely that this innovation primed the cichlids for their dietary versatility and hence their ability to diversify repeatedly in the great African lakes.

Despite their greater age, Lakes Tanganyika and Malawi don't have a noticeably larger number of species than Victoria. It is as though each lake achieves a sort of closure, at an equilibrium number of species, that doesn't go on getting larger as time goes by. Indeed, it

---

* The whole topic is treated in detail in Dolph Schluter's recent book, *The Ecology of Adaptive Radiation*.

may even get smaller. Lake Tanganyika, the oldest of the three lakes, has the fewest species. Lake Malawi, of intermediate age, has the most. It seems likely that all three lakes followed the Victoria pattern of extremely rapid speciation from very small beginnings, generating several hundred new endemic species within the first few hundred thousand years.

The Narrowmouth's Tale touched upon the favoured theory of how speciation happens, the geographical isolation theory. It is not the only theory, and more than one may be right in different cases. 'Sympatric speciation', the separation of populations into separate species in the same geographical area, can happen under some conditions, especially in insects where it may even be the norm. There is some evidence for sympatric speciation of cichlid fish in small African crater lakes. But the geographical isolation model of speciation is still the dominant one, and it will prevail through the rest of this tale.

According to the geographical isolation theory, speciation begins with the accidental geographical division of a single ancestral species into separate populations. No longer able to interbreed, the two populations drift apart, or are pushed by natural selection in different evolutionary directions. Then, if they subsequently meet after this divergence, they either can't interbreed or don't want to. They often recognise their own species by some particular feature, and studiously avoid similar species who lack it. Natural selection penalises mating with the wrong species, especially where the species are close enough for it to be a temptation, and close enough for hybrid offspring to survive, to consume costly parental resources, and then turn out to be sterile, like mules. Many zoologists have interpreted courtship displays as aimed mainly against miscegenation. This may be an exaggeration, and there are other important selection pressures bearing upon courtship. But it is still probably correct to interpret some courtship displays, and some bright colours and other conspicuous advertisements, as 'reproductive isolation mechanisms' evolved through selection against hybridisation.

As it happens, a particularly neat experiment was done on cichlid fish by Ole Seehausen, now at the University of Hull, and his colleague Jacques van Alphen at the University of Leiden. They took two related species of Lake Victoria cichlids, *Pundamilia pundamilia* and *P. nyererei* (named after one of Africa's great leaders, Julius Nyerere of Tanzania). The two species are very similar, except that

P. nyererei has a reddish colour, whereas *P. pundamilia* is bluish. Under normal conditions, females in choice tests prefer to mate with males of their own species. But now, Seehausen and van Alphen did their critical test. They gave females the same choice, but in artificial monochromatic light. This does dramatic things to perceived colour, as I remember vividly from schooldays in Salisbury, a city whose streets happened to be lit by sodium lights. Our bright red caps, and the bright red buses, all looked dirty brown. This is what happened to both the red and the blue *Pundamilia* males in Seehausen and van Alphen's experiment. Red or blue in white light, they all went dirty brown. And the result? The females no longer distinguished between them, and mated indiscriminately. Offspring of these matings were fully fertile, indicating that female choice is the only thing that stands between these species and hybridisation. The Grasshopper's Tale gives a similar example. If the two species were a bit more different, their offspring would probably be infertile, like mules. Later still in the process of divergence, isolated populations reach the point where they couldn't hybridise even if they wanted to.

Whatever the basis of the separation, failure to hybridise defines a pair of populations as belonging to different species. Each of the two species is now free to evolve separately, free from contamination by the genes of the other, even though the original geographical barrier to such contamination is no more. Without the initial intervention of geographical barrriers (or some equivalent), species could never become specialised to particular diets, habitats or behaviour patterns. Notice that 'intervention' does not necessarily mean it is geography itself that made the active change – as when a valley floods or a volcano erupts. The same effect is achieved if geographical barriers existed all along, wide enough to impede gene flow, but not so formidable that they are never crossed by occasional founder populations. In the Dodo's Tale we met the idea of sporadic individuals having the luck to cross to a remote island, where they then breed in isolation from their parent population.

Islands like Mauritius or the Galapagos are the classic providers of geographical separation, but islands don't have to mean land surrounded by water. When we are talking about speciation, 'island' comes to mean any kind of isolated breeding area, defined from the animal's point of view. Not for nothing is Jonathan Kingdon's beautiful book on African ecology called *Island Africa*. To a fish, a lake is an island. How, then, could hundreds of new fish species diverge from a single ancestor, if they all live in the same lake?

One answer is that, from the fish's point of view, there are lots of little 'islands' within a large lake. All three of the great East African lakes have isolated reefs. 'Reef' here doesn't mean coral reef, of course, but 'a narrow ridge or chain of rocks, shingle, or sand, lying at or near the surface of the water' (*Oxford English Dictionary*). These lake reefs are covered with algae, and many kinds of cichlids crop them. To such a cichlid, a reef might well constitute an 'island', separated by deep water from the next reef, at a distance large enough to constitute a barrier to gene flow. Even though they are capable of swimming from one to the other, they don't want to. There is genetic evidence to support this, from a study in Lake Malawi that sampled one species of cichlid, *Labeotropheus fuelleborni*. Individuals from opposite ends of a large reef shared the same distribution of genes: there was abundant gene flow along the length of the reef. But when the investigators sampled the same species from other reefs, separated by deep water, they found significant differences in visible coloration and in genes. A gap of two kilometres was enough to cause a measurable genetic separation; and the larger the physical gap, the more the genetic gap. Further evidence comes from a 'natural experiment' in Lake Tanganyika. A violent storm in the early 1970s created a new reef, 14 kilometres from its nearest neighbour. This should have been prime habitat for reef-dwelling cichlids, but when the reef was examined several years later none had arrived. Evidently, from the fish's point of view, there are indeed 'islands' within these large lakes.

In order for speciation to happen, there must be populations that are sufficiently isolated for gene flow between them to be rare; but not so isolated that no founding individuals arrive there at all. The recipe for speciation is 'Genes flow but not much'. That is a section heading from George Barlow's *The Cichlid Fishes*, the book that has been my main inspiration while writing this tale. The section describes yet another genetic study in Lake Malawi of four species of cichlid, inhabiting four neighbouring reefs, roughly one to two kilometres apart. All four species, known as *mbuna* in the local dialect, were present on all four reefs. Within each of the four species, there were genetic differences between the four reefs. A sophisticated analysis of the distributions of genes showed that there was indeed a trickle of gene flow between reefs, but a very slight one – a perfect recipe for speciation.

Here's another way in which speciation might have happened, and one that seems especially plausible for Lake Victoria. Radiocarbon dating of mud suggests that Lake Victoria dried up about 15,000 years ago. *Homo sapiens,* not long predating the dawn farmers of Mesopotamia, could walk dry-footed from Kisumu in Kenya straight across to Bukoba in Tanzania – a journey that today is a 300-kilometre voyage by the MV *Victoria,* a decent-sized ship popularly known as the 'Queen of Africa'. That was an extremely recent drying up, but who knows how many times the Victoria basin has been drained and flooded, flooded and drained in the millennia before that? On the timescale of thousands of years, the lake level may rise and fall like a yo-yo.

Now, hold that thought in mind, together with the theory of speciation by geographical isolation. When the Victoria basin dries up from time to time, what will be left? It could be a desert if the drying up were complete. But a partial drying up would leave a scattering of little lakes and pools, representing the deeper depressions in the basin. Any fish trapped in these little lakes would have the perfect opportunity to evolve away from their colleagues in other little lakes and become separate species. Then, when the basin flooded again and the large lake reconstituted, the newly distinct species would all swim out and join the larger Victoria fauna. When the yo-yo went down the next time, it would be a different set of species that accidentally found itself separated in each of the smaller refugia. Again, what a wonderful recipe for speciation.

Evidence from mitochondrial DNA supports this theory of rising and falling lake levels for the older Lake Tanganyika. Although a deep rift lake, not a shallow basin like Victoria, there is evidence that Lake Tanganyika's level used to be much lower, and it was at that time separated into three medium-sized lakes. The genetic evidence suggests an early segregation of cichlids into three groupings, presumably one for each of the old lakes, followed by further speciations after the formation of the present large lake.

In the case of Lake Victoria, Erik Verheyen, Walter Salzburger, Jos Snoeks and Axel Meyer have done a very thorough genetic study of the mitochondria of haplochromine cichlid fish, not only in the main lake but in the neighbouring rivers, and the satellite lakes Kivu, Edward, George, Albert and others. They showed that Victoria and its smaller neighbours share a monophyletic 'species flock' that began to diverge about 100,000 years ago. This sophisticated piece of research used the methods of parsimony, maximum likelihood and Bayesian analysis that we met in the Gibbon's Tale. Verheyen and colleagues looked at the distribution in all the lakes and neighbouring rivers of 122 'haplotypes' from the mitochondrial DNA of these fish. A haplotype, as we saw in Eve's Tale, is a length of DNA that lasts long enough to be recognised repeatedly in lots of individuals, who might well belong to lots of different species. For simplicity I shall use the word 'gene' as an approximate synonym for haplotype (although purist geneticists would not). The scientists were, temporarily, ignoring the question of species. They were, in effect, imagining genes swimming around in lakes and rivers, and counting the frequency with which they did so.

It is easy to misunderstand the beautiful diagram (see plate 26), with which Verheyen and his colleagues summarised their work. It is tempting to think that the circles represent species clustered around parent species, as in a family tree. Or that they represent small lakes clustered around larger lakes, as in a stylised route map of an (amphibious!) airline's network of destinations. Neither of these is even close to what the diagram represents. The circles are neither species nor geographical hubs. Each one is a haplotype: a 'gene', a particular length of DNA that an individual fish might or might not possess.

Each gene, then, is represented by one circle. The area of the circle conveys the number of individuals, *regardless of species*, added up over all lakes and rivers surveyed, who possessed that particular gene. The small circles indicate a gene that was found in only a single individual. Gene 25, to judge from the area of its circle (the largest one), was found in 34 individuals. The number of circles, or blobs, on the line joining two circles represents the minimal number of mutational changes you need to go from one to the other. You will recognise from the Gibbon's Tale that this is a form of parsimony analysis, but slightly easier than parsimony analysis of distantly related genes, because the intermediates are still around. The small black blobs represent intermediate genes that have not been found in real fish, but can be inferred as probably existing in the course of evolution. It is an unrooted tree that doesn't commit itself to the direction of evolution.

Geography enters into the diagram only in the colour coding. Each circle is a pie chart showing the number of times the gene concerned was found in each of the lakes or rivers surveyed (see the colour key at bottom right of the diagram). Of the numerous genes, those labelled 12, 47, 7 and 56 were found only in Lake Kivu (all red circles). Genes 77 and 92 were found only in Lake Victoria (all blue). Gene 25, the most abundant of all, turned up mostly in Lake Kivu but also in significant numbers in the 'Uganda lakes' (a cluster of small lakes close to each other and to the west of Lake Victoria). The pie chart shows that gene 25 was also found in the Victoria Nile river, in Lake Victoria itself, and in Lake Edward/George (these two small and neighbouring lakes are united for purposes of the count). Once again, bear in mind that the diagram contains no information at all about species. The blue slice of pie in gene 25's circle indicates that two individuals from Lake Victoria contained this gene. We are given no indication at all on whether those two individuals were of the same species as each other, or the same species as any of the Lake Kivu individuals bearing that gene. That is not what this diagram is about. It is a diagram to delight any enthusiast for the selfish gene.

The results were powerfully revealing. Little Lake Kivu emerges as the fountainhead of the entire species flock. Genetic signals show that Lake Victoria was 'seeded' with haplochromine cichlids on two separate occasions from Lake Kivu. The great drying of 15,000 years ago by no means extinguished the species flock, and very probably enhanced it in the way we were just imagining, through the Victoria basin becoming a 'Finland' of lakelets. As for the origin of the older population of cichlids in Lake Kivu itself (it now has 26 species, including 15 endemic haplochromines), the genetic oracle says they came from Tanzanian rivers.

This work is only just beginning. The imagination at first quails, and then is uplifted, by the contemplation of what will be achieved when such methods are routinely applied not just to cichlid fish in African lakes, but to any animals, in any 'archipelago' of habitats.

THE BLIND CAVE FISH'S TALE

Animals of various kinds have found their way into dark caves, where living conditions are obviously very different from outside. Repeatedly, and in many different animal groups including flatworms, insects, crayfish, salamanders and fish, cave dwellers have independently evolved many of the same changes. Some can be thought of as constructive changes – for instance, delayed reproduction, fewer but larger eggs, and increased longevity. Apparently in compensation for their useless eyes, cave animals typically have enhanced senses of taste and smell, long feelers and, in the case of fish, improvements to the lateral line system (a pressure-related sense organ beyond our empathy but deeply meaningful to fish). Other changes are referred to as regressive. Cave dwellers tend to lose their eyes and their skin pigment, becoming blind and white.

The Mexican tetra *Astyanax mexicanus* (also known as *A. fasciatus*) is particularly remarkable because different populations *within* the one species of fish have independently followed streams into caves and very rapidly evolved a common pattern of cave-related regressive changes, which can be directly contrasted with fellow species members still living outside. These 'Mexican blind cave fish' are found only in Mexican caves – mostly limestone caves in a single valley. Once understandably thought to belong to their own separate species, they are now classified as a race of the same species, *Astyanax mexicanus*, which is common in surface waters from Mexico to Texas. The blind race has been found in 29 separate caves and, to repeat, it looks strongly as though at least some of these cave populations evolved their regressive eyes and white coloration independently of each other: surface-dwelling tetras have on many occasions taken up residence in caves, and independently lost their eyes and their colour on each occasion.

Intriguingly, it appears that some populations have been in their caves longer than others, and this shows itself as a gradient in the extent to which they have pushed in the typical cave-specific direction. The extreme is found in the Pachon cave, believed to hold the oldest cave population. At the 'young' end of the gradient is the Micos cave, whose population is relatively unchanged from the normal surface-dwelling form of the species. None of the populations can have been in their caves very long because this is a South American species which could not have crossed into Mexico before the formation of the Isthmus of Panama 3 million years ago – the Great American Interchange. My guess is that the cave populations of tetras are far younger than that.

It is easy to see why dwellers in darkness might never have evolved eyes in the first place; less easy to see why, given that their recent ancestors certainly had normal, functioning eyes, the cave fish should 'bother' to get rid of them. If there is a possibility, however slight, of a cave fish finding itself washed out of its cave into the light of day, wouldn't there be some benefit in keeping the eyes 'just in case'? That isn't how evolution works, but it can be rephrased in respectable terms. Building eyes – indeed, building anything – is not free of cost. Individual fish that divert resources into some other part of the animal's economy would have an advantage over rival fish that retain full-sized eyes.* If a cave-dweller has insufficient probability of needing eyes to offset the economic costs of making them, eyes will disappear. Where natural selection is concerned, even very slight advantages are significant. Other biologists leave economics out of their reckoning. For them, it is sufficient to invoke an accumulation of random changes in eye development, which are not penalised by natural selection because they make no difference. There are many more ways of being blind than of being sighted, so random changes, for purely statistical reasons, tend towards blindness.

And this leads us to the main point of the Blind Cave Fish's Tale. It is a tale of Dollo's Law, which states that evolution is not reversed. Is Dollo's Law disproved by the cave fish's apparent reversal of an evolutionary trend, shrinking again the eyes that grew, so painstakingly, over past evolutionary time? Is there, in any case, some general theoretical reason to expect evolution to be irreversible? The answer to both questions is no. But Dollo's Law has to be correctly understood, and that is the purpose of this tale.

Except in the very short term, evolution cannot be precisely and exactly reversed, but the emphasis is on 'precisely and exactly'. It is very improbable that any particular evolutionary pathway, specified in advance, will be followed. There are too many possible pathways. An exact reversal of evolution is just a special case of a particular evolutionary pathway, specified in advance. With such a large number of possible paths that evolution might follow, the odds are heavily against any one particular path, and that includes an exact reversal of the forward one just travelled. But there is no law against evolutionary reversal as such.

Dolphins are descended from land-dwelling mammals. They returned to the sea and resemble, in many superficial respects, large, fast-swimming fish. But evolution has not reversed itself. Dolphins resemble fish in certain respects, but most of their internal features clearly label them as mammals. If evolution had truly reversed itself, they would simply be fish. Maybe some 'fish' really are dolphins – the

---

* Eyes can be an even more costly extravagance if they become infected or irritated, which is probably why burrowing moles have reduced them as much as possible.

reversion to fish being so perfect and far-reaching that we haven't noticed? Want a bet? That is the sense in which you can bet heavily on Dollo's Law. Especially if you look at evolutionary change at the molecular level.

This interpretation of Dollo's Law could be called the thermo-dynamic interpretation. It is reminiscent of the Second Law of Thermodynamics, which states that entropy (or disorder or 'mixed-upness') increases in a closed system. A popular analogy (or it may be more than an analogy) for the Second Law is a library. Without a librarian energetically reshelving books in their correct places, a library tends to become disordered. The books become mixed up. People leave them on the table, or put them on the wrong shelf. As time goes by, the library's equivalent of entropy inevitably increases. That's why all libraries need a librarian, constantly working to restore the books to order.

The great misunderstanding of the Second Law is to assume that there is a driving urge towards some particular goal state of disorder. It isn't like that at all. It is just that there are far more ways of being disordered than of being ordered. If the books are shuffled at random by sloppy borrowers, the library will automatically move away from the state (or the small minority of states) that anybody would recognise as ordered. There is no drive towards a state of high entropy. Rather, the library meanders in some random direction away from the initial state of high order and, no matter where it wanders in the space of all possible libraries, the vast majority of possible pathways will constitute an increase in disorder. Similarly, of all the evolutionary pathways that a lineage could follow, only one out of a vast number of possible pathways will be an exact reversal of the path by which it has come into being. Dollo's Law turns out to be no more profound than the 'law' that if you toss a coin 50 times, you won't get all heads – nor all tails, nor strict alternation, *nor any other particular, prespecified sequence.* The same 'thermodynamic' law would also state that any *particular* evolutionary pathway in a 'forward' direction (whatever that might mean!) will not be precisely followed twice.

In this thermodynamic sense, Dollo's Law is true but unremark-able. It doesn't deserve the title of law at all, any more than there is a 'law' against tossing a coin 100 times and getting heads every time. One could imagine a 'real law' interpretation of Dollo's Law which stated that evolution could not return to anything that was vaguely like an ancestral state, as a dolphin is vaguely like a fish. This interpretation would indeed be remarkable and interesting but it is (ask any dolphin) false. And I cannot imagine any sensible theoretical rationale that would expect it to be true.

THE FLOUNDER'S TALE

An endearing quality of Chaucer is the naive perfectionism of his General Prologue, where he introduces his pilgrims. It wasn't enough to have a Doctour of Physik on the pilgrimage – he had to be the finest doctor in the land:

> *In all this world ne was ther noon hym lik,*
> *To speke of physik and of surgerye.*

The 'verray, parfit gentil knyght' was, it seemed, unmatched in Christendom for bravery, loyalty and even temper. As for his squire and son, he was 'A lovere and a lusty bacheler . . . wonderly delyvere, and of greet strengthe'. To top it all, he was 'as fressh as is the month of May'. Even the knight's yeoman knew all there was to know of woodcraft. The reader comes to take it for granted that, if a profession is mentioned, its practitioner will automatically turn out to be unrivalled in all England.

Perfectionism is a vice of evolutionists. We are so used to the wonders of Darwinian adaptation, it is tempting to believe there could be nothing better. Actually, it is a temptation that I can almost recommend. A surprisingly strong case can be built for evolutionary perfection, but it must be done with circumspection and sophisticated attention.* Here I shall give just one example of a historical constraint, the so-called 'jet engine effect': imagine how imperfect a jet engine would be if, instead of being designed on a clean drawing board, it had to be changed one step at a time, screw by screw and rivet by rivet, from a propeller engine.

A skate is a flat fish that might have been designed on a drawing board to be flat, resting on the belly, with wide 'wings' reaching symmetrically out to both sides. Teleost flatfish do it in a different way. They rest on one side, either the left (e.g. plaice) or the right (e.g. turbot and flounder). Whichever the side, the shape of the

---

* I have set out the pitfalls in a chapter of *The Extended Phenotype* called 'Constraints on Perfection'.

whole skull is distorted so that the eye on the lower side moves over to the upper side, where it can see. Picasso would have loved them (see plate 27). But, by the standards of any drawing board, they are revealingly imperfect. They have precisely the kind of imperfection you would expect from being evolved rather than designed.

SHARKS AND THEIR KIN

'Out of the murderous innocence of the sea . . .' The context of Yeats's poem was completely other but – I can't help it – the phrase always makes me think of a shark. Murderous, but innocent of deliberate cruelty, just making a living as perhaps the world's most effective killing machine. I know people for whom the great white shark is their worst nightmare. If you are one of them, you may not wish to know that the Miocene shark *Carcharocles megalodon* was three times the size of a great white, with jaws and teeth to scale.

My own recurrent nightmare, having grown up as an exact contemporary of the atomic bomb, is not a shark but a huge, black, futuristic, delta-winged aircraft bristling with high-tech missile launchers, filling the sky with its shade and my heart with foreboding. Almost exactly the shape of a manta ray in fact. The dark shape that roars over the treetops of my dreams, with its twin gun turrets so enigmatically menacing, is a sort of technological cousin to *Manta birostris*. I always found it hard to accept that these seven-metre monsters are harmless filter-feeders, straining plankton through their gills. They are also extremely beautiful.

What of the sawfish, what on earth is that all about? And the hammerhead shark? Hammerheads occasionally attack people, but that is not why they might invade your dreams. It is the bizarre T-shaped head, the eyes set wider than you expect outside science fiction, as though this shark were designed by an artist with a drugged imagination (see plate 28). And the thresher shark, *Alopias*, isn't that another work of art, another candidate for a dream? The upper lobe of the tail is nearly as long as the rest of the body. Threshers use their prodigious tailblades first to herd prey, then to thresh them to death. A thresher, harassed by fishermen in a boat, has been known to decapitate a man with a single swipe of that magnificent tail.

The sharks, rays and other cartilaginous fish or chondrichthyans join us at Rendezvous 21, 460 million years ago, in seas off the icy-cold and barren lands of the Middle Ordovician. The most noticeable difference between the new pilgrims and all the others so far is that

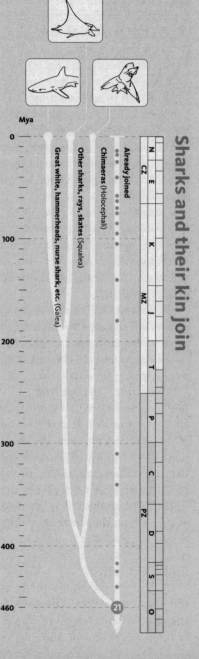

Sharks and their kin join

	N											
	CZ	E		K	MZ	J	T	P	C	PZ	D	S	O

Mya

0 — Already joined

Chimaeras (Holocephali)

Other sharks, rays, skates (Squalea)

Great white, hammerheads, nurse shark, etc. (Galea)

100 —

200 —

300 —

400 —

460 — 21

Sharks and their kin join. The cartilaginous fish, who join us here, include the sharks and rays. Fossils leave no doubt about the early split of the jawed vertebrates into bony fish and these cartilaginous fish. Recent, robust data strongly support this scheme of relationships within the 850 or so species of cartilaginous fish.

Images, left to right: grey reef shark (*Carcharhinus ambly-rhynchos*); manta ray (*Manta birostris*); elephant fish (*Callorhynchus milii*).

sharks have no bone. Their skeleton is made of cartilage. We too use cartilage for special purposes like lining our joints, and all of our skeleton starts out as flexible cartilage in the embryo. Most of it later becomes ossified when mineral crystals, mostly calcium phosphate, incorporate themselves. Except for the teeth, the shark skeleton never undergoes this transformation. Nevertheless, their skeleton is quite rigid enough to sever your leg in a single bite.

Sharks lack the swim bladder that contributes to the success of the bony fish, and many of them have to swim continuously to maintain their desired level in the water. They assist their buoyancy by retaining the waste product urea in the blood and by having a large, oil-rich liver. Incidentally, some bony fish use oil instead of gas in their swim bladder.

If you should be so incautiously affectionate as to stroke a shark, you would find that its whole skin feels like sandpaper, at least if you stroke it 'against the grain'. It is covered with dermal denticles – sharp, tooth-like scales. Not only are they tooth-like, but the formidable teeth of a shark are themselves evolutionary modifications of dermal denticles.

Sharks and rays almost all live in the sea, although a few genera venture up estuaries and rivers. Freshwater shark attacks on humans used to be common in Fiji, but that was when humans were cannibals. All but the choicest cuts were discarded into rivers, and it would seem that sharks were attracted upstream by the smell of leftovers from cannibal feasts. When Europeans arrived they put a stop to cannibalism, but at the same time inadvertently brought new diseases against which the Fijians had not evolved immunity. Corpses of diseased victims were also disposed of in rivers, so sharks continued to be attracted. Nowadays bodies are no longer tossed into rivers, and shark attacks have decreased accordingly. Unlike the bony fish, no sharks have ever shown any inclination to come on land.

The cartilaginous fish are divided into two main groups: the rather weird-looking chimaeras or ratfish, which are not numerous enough to be a significant part of the fauna; and the sharks, skates and rays, which are. Skates and rays are flattened sharks. Dogfish are small sharks, but they are still not very small: no whitebait-sized sharks exist. The spined pygmy shark *Squaliolus laticaudus* grows up to about 20 centimetres. The shark body plan seems to lend itself to large size, and the biggest of all, the whale shark *Rhincodon typus* can be up to 12 metres long and weigh 12 tonnes. Like the second largest, the basking shark *Cetorhinus maximus*, and like the largest whales, the whale shark is a plankton feeder. *Carcharocles megalodon*, already mentioned as the stuff of nightmares, was not – to use a calculated understatement – a filter-feeder. That Miocene monster had teeth, each one as big as your face. It was a voracious predator, like the majority of sharks today, and they have topped the food chains of the sea for hundreds of millions of years with relatively little change.

If manta rays feature in nightmares as bombers, the smaller role of jump-jet fighter might be played by the chimaeras (see plate 29), also known as ratfish or ghost sharks. These strange deep-sea fish occupy the class Holocephali (whole head), where all the rest of the cartilaginous fish, the sharks and rays combined, belong in the Elasmobranchii. They can be recognised by their unusual gill covers, which completely encase the separate gills, providing a single opening for all of them. Unlike sharks and rays, their skin is not covered with dermal denticles but is 'naked'. This may be what gives them their 'ghostly' appearance. Their resemblance to a nightmare plane comes from the fact that their tails are not prominent and they swim by 'flying' with their large pectoral fins. There are only about 35 species of living chimaeras.

Successful as sharks certainly are – and over a spectacularly long time too – teleost fish outnumber them thirtyfold when it comes to species numbers. There have been two major radiations of sharks. The first flourished mightily in the Palaeozoic seas, especially during the Carboniferous Period. This ancient domination of sharks had come to an end by the beginning of the Mesozoic Era (the age of dinosaurs on land). After a lull of about 100 million years, the sharks enjoyed another major resurgence in the Cretaceous, which has continued to this day.

A word association test that mentioned 'shark' would very probably elicit the response 'jaws', so it is appropriate that Concestor 21, perhaps our 200-million-greats-grandparent, is the grand ancestor of all the vertebrates that have true jaws, the gnathostomes. *Gnathos* in Greek means 'lower jaw', and that is specifically what sharks and all the rest of us share. It was one of the triumphs of classical comparative anatomy to demonstrate that jaws evolved from modified parts of the gill skeleton. The next pilgrims to join us, at Rendezvous 22, are the jawless vertebrates, the Agnatha, well endowed with gills but with no lower jaw. Once numerous, diverse and heavily armoured, the Agnatha are now reduced to the eel-shaped lampreys and hagfish.

LAMPREYS AND HAGFISH

Rendezvous 22, where we meet the lampreys and hagfish, occurs somewhere in the warm seas of the early Cambrian, say 530 million years ago, and I would very roughly guess that Concestor 22 was our 240-million-greats-grandparent. The lampreys and hagfish survive as pivotal messengers from the dawn of vertebrates. Although it is convenient to treat them together, as the jawless and limbless fish, I have to admit that many morphologists think that lampreys are closer cousins to us than they are to hagfish. According to this school, we should greet the lamprey pilgrims at Rendezvous 22, and the hagfish at 23. On the other hand, molecular biologists are equally insistent that both join us at one rendezvous, and this is the opinion I am provisionally adopting here. In any case, it is fair to say that neither lampreys nor hagfish do justice to the jawless fish as a whole, most of whom are extinct.

Lampreys and hagfish have a superficially eel-like appearance, with soft bodies – but when the jawless fish dominated the seas, in the Devonian 'Age of Fish', many of them, known as ostracoderms, had hard, bony armour plating, and some had paired fins, unlike lampreys and hagfish. They give the lie to any suggestion that bone is an 'advanced' feature of vertebrates that 'took over' from cartilage. Sturgeons and some other 'bony' fish resemble sharks and lampreys in possessing a skeleton almost entirely made of cartilage, but they are descended from far more bony ancestors – indeed from fish with heavy armour plating – and it is not unlikely that sharks and lampreys are too.

Even more heavily armoured were the placoderms, a wholly extinct group of jaw-bearing and limb-bearing fish of uncertain affinities, who also lived in the Devonian Period, contemporary with some of the jawless ostracoderms and presumably descended from earlier jawless fish. Some of the placoderms were so heavily armoured that even their limbs had a tubular, jointed exoskeleton, superficially similar to a crab's leg. If you encountered one in a poor light and an imaginative frame of mind, you could be forgiven for thinking you had stumbled on a strange kind of lobster or crab. As a

Jawless fish join

Mya

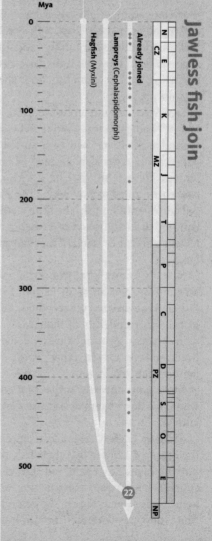

Hagfish (Myxini)

Lampreys (Cephalaspidomorphi)

Already joined

Jawless fish join. There is still a great deal of argument about evolutionary relationships at the base of the vertebrate lineage, particularly with respect to the living jawless fish: the 41 species of lamprey and the 43 species of hagfish. Fossils argue that the earliest divergence occurred between hagfish and the other vertebrates, followed by the divergence of the lamprey lineage. However, the molecular data strongly argue that lampreys and hagfish group together, as shown here.

Images, left to right: New Zealand hagfish (*Eptatretus cirrhatus*); sea lamprey (*Petromyzon marinus*).

rather young undergraduate, I used to dream about discovering a living placoderm – it was my equivalent of the scoring-a-century-for-England fantasy.

Why did both the jawed placoderms and the jawless ostracoderms develop such heavily fortified bodies? What was it about those Palaeozoic seas that demanded such formidable protection? The presumed answer is equally formidable predators, and the obvious candidates, apart from other placoderms, are the eurypterids or sea scorpions, some more than two metres in length – the largest arthropods that ever lived. Whether or not any of the eurypterids had venomous stings like modern scorpions (recent evidence suggests not), they still must have been fearsome predators, capable of driving the Devonian fish, both jawless and jawed, to evolve costly armour plating.

Lampreys are not armoured, and they are easy to eat, as King Henry I had good reason to regret (school history books never fail to remind us that he died of a surfeit of them). Most lampreys are parasitic on other fish. Instead of jaws they have a circular sucker around the mouth, looking a bit like an octopus sucker but with concentric rings of tiny teeth. The lamprey fastens its sucker to the outside of another fish, the little teeth rasp through the skin, and the lamprey sucks the blood of its victim, like a leech. Lampreys have had serious effects on fisheries, for example in the North American Great Lakes.

Nobody knows what Concestor 22 was like but, living as it probably did in the Cambrian Period, long before the Devonian Age of Fish and the dreaded sea scorpions, it probably wasn't armour-plated like the ostracoderms of the jawless fish's heyday. Nevertheless, the ostracoderms seem to be closer cousins to us jawed vertebrates than the lampreys are. In other words, 'before' our pilgrims join the lampreys at Rendezvous 22, we have already incorporated the ostracoderms into our pilgrimage. Our concestor with the ostracoderms, whom we don't number because they are all extinct, was presumably jawless.

Modern hagfish resemble lampreys in their long, eel-like shape, their lack of a lower jaw, their lack of paired limbs, their row of gill port-holes on either side, and their notochord retained into the adult (this stiffening rod, running the length of the back, is in most vertebrates present only in the embryo). But hagfish are not parasites. They rummage with their mouth hole around the bottom of the sea for small invertebrates, or they scavenge on dead fish or whales, often wriggling inside to eat from the inside out. They are exceedingly slimy, and they use their surprising talent for tying themselves in knots in order to get a purchase when burrowing into carcasses.

Vertebrates were once thought to have arisen long after the Cambrian Period. Maybe it was an aspect of our snobbish desire to arrange the animal kingdom on a ladder of progress. Somehow it seemed right and fitting that there was an age where animal life was limited to invertebrates, setting the scene for the eventual arrival of the mighty vertebrates. Zoologists of my generation were taught that the earliest known vertebrate was a jawless fish called *Jamoytius* (named, somewhat freely, after J. A. Moy-Thomas) that lived in the middle of the Silurian Period, 100 million years after the Cambrian when most of the invertebrate phyla arose. Obviously vertebrates must have had ancestors living in the Cambrian, but they were assumed to be invertebrate forerunners of the true vertebrates – protochordates. *Pikaia* has been heavily promoted as the oldest fossil protochordate.* It was a delicious surprise, therefore, when apparently true vertebrate fossils started turning up in the Cambrian strata of China, and the Lower Cambrian at that. This has robbed *Pikaia* of some of its mystique. There were true vertebrates, jawless fish, living before *Pikaia*. The vertebrates hark back to the deep Cambrian.

Not surprisingly given their immense age, these fossils, called *Myllokunmingia* and *Haikouichthys* (although they may belong to the same species) are not in mint condition, and much is still unknown about these primeval fish. They seem to have had most of the features you'd expect from a relative of the lampreys and hagfish, including gills, segmented muscle blocks and a notochord. *Myllokunmingia*, whom we shall meet again in the Velvet Worm's Tale, is perhaps not too far from being a plausible model for Concestor 22.

Rendezvous 22 is a major milestone. From now on, for the first time, all the vertebrates are united in a single pilgrim band. It is a big event because, traditionally, animals were divided into two major groups, the vertebrates and the invertebrates. As a division of convenience, the distinction has always been useful in practice. From a

---

* This Cambrian fossil, originally classified as an annelid worm, was later recognised as a protochordate, in which role it starred in S. J. Gould's *Wonderful Life*.

strictly cladistic point of view, however, the vertebrate/invertebrate distinction is an odd one, nearly as unnatural as the ancient Jewish classification of humanity into themselves and 'gentiles' (literally everybody else). Important though we vertebrates think ourselves, we don't constitute even a whole phylum. We are a subphylum of the phylum Chordata, and the phylum Chordata should be thought of as on a par with, say, the phylum Mollusca (snails, limpets, squids, etc.) or the phylum Echinodermata (starfish, sea urchins, etc.). The phylum Chordata includes other vertebrate-like creatures that nevertheless lack a backbone – for example, the amphioxus, whom we are about to meet at Rendezvous 23.

Strict cladism notwithstanding, there really is something rather special about vertebrates. Professor Peter Holland has made to me the powerful point that there is a massive difference in genome complexity between (all) vertebrates and (all) invertebrates. 'It is, at the genetic level, perhaps the biggest change in our metazoan* ancestry.' Holland thinks the traditional divide between vertebrates and invertebrates needs to be revived, and I see what he means.

The chordates get their name from the already mentioned notochord, the cartilage rod that runs along the back of the animal, in the embryo if not in the adult.[†] Other characteristics of chordates (including vertebrates), which in ourselves are seen only in the embryo, include gill openings near the front end on both sides, and a tail that reaches back beyond the anus. All chordates have a dorsal nerve cord (runs along the back), unlike many invertebrates where the nerve cord is ventral (runs along the belly).

Vertebrate embryos all have a notochord but it is replaced in the adult, to a greater or lesser extent, by the segmented, articulated backbone. In most vertebrates the notochord itself survives into the adult only in fragments, such as the intervertebral discs whose tendency to slip can cause us so much grief. The lampreys and hagfish are unusual among vertebrates in retaining the notochord more or less intact into the adult. In this respect they are, I suppose, borderline vertebrates, but everyone calls them vertebrates anyway.

---

* Metazoa means many-celled animals, and we shall be meeting the term further on in the pilgrimage.
† The word is calculated to confuse, because chord with an h, in modern English, means only something musical, as in *The Lost Chord*, one of my favourite songs. The notochord is a cord, without an h, meaning rope. However, chord is a recognised archaic spelling of cord (rope), and the connection with music may be that *chorda* is the Latin for the string of an instrument.

THE LAMPREY'S TALE

The reason it falls to the lamprey to tell this tale will be revealed at the end. It is a reprise on a theme we have met before: there is a separate gene's-eye view of ancestry and pedigree that is surprisingly independent of the view we get when we think about family trees in more traditional ways.

Haemoglobin is well known as the vitally important molecule that carries oxygen to our tissues and gives our blood its spectacular colour. Human adult haemoglobin is actually a composite of four protein chains called globins, knotted around each other. Their DNA sequences show that the four globin chains are closely related to each other, but they are not identical. Two of them are called alpha globins (each a chain of 141 amino acids), and two are beta globins (each a chain of 146 amino acids). The genes coding for the alpha globins are on our chromosome 11; those coding for the beta globins are on chromosome 16. On each of these chromosomes there is a cluster of globin genes in a row, interspersed with some junk DNA that is never transcribed. The alpha cluster, on chromosome 11, contains seven globin genes. Four of these are pseudogenes – disabled versions of alpha with faults in their sequence, never translated into protein. Two are true alpha globins, used in the adult. The final one is called zeta, and it is used only in embryos. The beta cluster, on chromosome 16, has six genes, some of which are disabled, and one of which is used only in the embryo. Adult haemoglobin, as we've seen, contains two alpha and two beta chains, wrapped around each other to form a beautifully functioning parcel.

Never mind all this complexity. Here's the fascinating point. Careful letter-by-letter analysis shows that the different kinds of globin genes are literally cousins of each other – members of a family. But these distant cousins still co-exist inside you and me. They still sit side by side with their cousins inside every cell of every warthog and every wombat, every owl and every lizard.

On the scale of whole organisms, of course, all vertebrates are cousins of each other too. The tree of vertebrate evolution is the family tree we are all familiar with, its branch-points representing speciation events – the splitting of species into daughter species. In reverse, they are the rendezvous points that punctuate this pilgrimage. But there is another family tree occupying the same timescale, whose branches represent not speciation events but gene duplication events within genomes. And the branching pattern of the globin tree looks very different from the branching pattern of the family tree, if we trace it in the usual, orthodox way, with species branching to form daughter species. There is not just one evolutionary tree in which species divide and give rise to daughter species. Every gene has its own tree, its own chronicle of splits, its own catalogue of close and distant cousins.

The dozen or so different globins inside you and me have come down to us through the entire lineage of our vertebrate ancestors. About half a billion years ago, in a jawless fish perhaps like a lamprey, an ancestral globin gene accidentally split in two, both copies remaining in different parts of that fish's genome. There were then two copies of it, in different parts of the genome of all descendant animals. One copy was destined to give rise to the alpha cluster, on what would eventually become chromosome 11 in our genome, the other to the beta cluster, now on our chromosome 16. There is no point in trying to guess which chromosome either of them sat on in the intermediate ancestors. The locations of recognisable DNA sequences, indeed the number of chromosomes into which the genome is divided, are shuffled and changed with surprisingly gay abandon. Chromosome numbering systems, therefore, do not generalise across animal groups.

As the ages passed, there were further duplications, and doubtless some deletions as well. Around 400 million years ago the ancestral alpha gene duplicated again, but this time the two copies remained near neighbours of each other, in a cluster on the same chromosome. One of them was destined to become the zeta of our embryos, the other became the alpha globin genes of adult humans (further branchings gave rise to the non-functional pseudogenes I mentioned). It was a similar story along the beta branch of the family, but with duplications at other moments in geological history.

Now here's a fascinating point. Given that the split between the alpha cluster and the beta cluster took place half a billion years ago, it will of course not be just our human genomes that show the split, and possess both alpha genes and beta genes in different parts of our genomes. We should see the same within-individual split if we look at the genomes of any other mammals, at birds, reptiles, amphibians or bony fish – for our common ancestor with all of them lived less than 500 million years ago. Wherever it has been investigated, this expectation has proved correct. Our greatest hope of finding a vertebrate that does not share with us the ancient alpha/beta split would be a jawless fish like a lamprey or a hagfish, for they are our most remote cousins among surviving vertebrates. They are the only surviving vertebrates whose common ancestor with the rest is sufficiently ancient that it could have predated the alpha/beta split. Sure enough, these jawless fish are the only known vertebrates that lack the alpha/beta divide. Rendezvous 22 is so ancient, in other words, that it predated the split between alpha and beta globin.

Something like the Lamprey's Tale could be told for each one of our genes, for they all, if you go back far enough, owe their origin to the splitting of some ancient gene. And something like this entire book could be written for each gene. We arbitrarily decided that this should be a human pilgrimage, and we defined our milestones as meeting points with other lineages, which means, in the forward direction, speciation events at which our human ancestors split away from the others. I've already made the point that we could equally have begun our pilgrimage with a modern dugong, or a modern blackbird, and counted a different set of concestors back to Canterbury. But I am now making a more radical point. We could also write a backward pilgrimage for any *gene*.

We could choose to follow the pilgrimage of alpha haemoglobin, or cytochrome-c, or any other named gene. Rendezvous 1 would have been the milestone at which our chosen gene most recently duplicated to make a copy of itself elsewhere in the genome. Rendezvous 2 would have been the previous duplication event, and so on. Each of the rendezvous milestones would have taken place inside some particular animal or plant, just as the Lamprey's Tale has identified a Cambrian jawless fish as the likely receptacle for the split between alpha and beta haemoglobin.

The gene's eye view of evolution keeps forcing itself upon our attention.

LANCELETS

And now here's a tidy little pilgrim, wriggling up all on its own to join the pilgrimage. It is the amphioxus or lancelet. *Amphioxus* used to be its Latin name, but the rules of nomenclature imposed *Branchiostoma* on it. Nevertheless, it had become so well known as *Amphioxus* that the name lives on. The lancelet or amphioxus is a protochordate, not a vertebrate, but it is clearly related to the vertebrates, and placed with them in the phylum Chordata. There are a few other related genera, but they are very similar to *Branchiostoma*, and I shall not distinguish them but call them all, informally, amphioxus.

I call amphioxus tidy because it elegantly lays out the features that proclaim it to be a chordate. It is a living, swimming (well, mostly buried in sand, actually) textbook diagram. There is the notochord running the length of the body, but not a trace of a vertebral column. There is the nerve tube on the dorsal side of the notochord, but no brain unless you count the small swelling at the front end of the nerve tube (where there is also an eye spot), and no skeletal brain case. There are the gill slits at the sides, which are used for filter-feeding, and the segmental muscle blocks along the length of the body, but no trace of limbs. There is the tail, stretching back behind the anus, unlike a typical worm, which has the anus at the posterior tip of the body. Amphioxus is also unlike a worm, but like many fish, in being shaped like a vertical blade, rather than cylindrical. It swims like a fish, with side-to-side undulations of the body, using the fish-like muscle blocks. The gill slits are part of the feeding apparatus, not primarily for breathing at all. Water is drawn in through the mouth and passed out through the gill slits, which act as filters to catch food particles. This is very likely how Concestor 23 used its gill slits, which would mean that gills for breathing came later, as an afterthought. If so, it is a pleasing reversal that, when the lower jaw eventually evolved, it was modified from a part of the gill apparatus.

We are now approaching the point where dating becomes so difficult and controversial that my courage fails me. If forced to put a date on Rendezvous 23, I would guess about 560 million years ago,

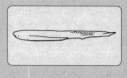

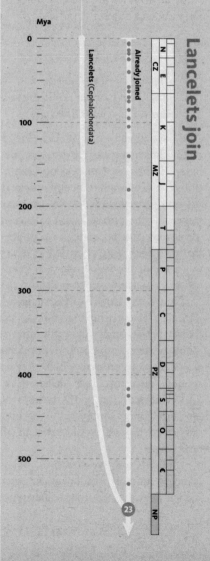

Mya

Lancelets (Cephalochordata)

Already joined

| | | | | | | | | | | | | | | |
|N|CZ|E| |K|MZ|J|T|P|C|D|PZ|S|O|Є|NP|

Lancelets join. The closest living relatives of the vertebrates are the 25 known species of fish-like animals commonly known as lancelets. There is little dispute over this. However, the *dates* of rendezvous points from now on backwards are often disputed (see the Epilogue to the Velvet Worm's Tale).

Image: *Branchiostoma* sp. (formerly *Amphioxus*).

the vintage of our 270-million-greats-grandparent. But I could easily be wrong, and for this reason I shall from now on abandon my attempts to describe the state of the world at the time of the concestor. As for what it looked like, I don't think we shall ever know for certain, but it is not implausible that Concestor 23 really may have been quite like a lancelet (see plate 30). If that is so, it is equivalent to saying that the lancelet is primitive. But that demands an immediate cautionary tale – the Lancelet's Tale.

THE LANCELET'S TALE

> *If just one touch of sunlight more should make his gonads grow*
> *The lancelet's claim to ancestry would get a nasty blow.*
> WALTER GARSTANG (1868–1949)

We have already met Walter Garstang, the distinguished zoologist who idiosyncratically expressed his theories in verse. I quote the couplet above not to develop Garstang's own theme, which, though interesting enough to be the subject of the Axolotl's Tale, is irrelevant to my purpose here.* I am concerned only with the last line, and especially the phrase 'claim to ancestry'. The lancelet, *Branchiostoma* or amphioxus has enough features in common with true vertebrates to have been long regarded as a surviving relative of some remote ancestor of the vertebrates. Or even – which is the real butt of my criticism – as the ancestor itself.

I am being unfair to Garstang, who knew perfectly well that the lancelet, as a surviving animal, could not be literally ancestral. Nevertheless, such talk really does sometimes mislead. Students of zoology delude themselves into imagining that when they look at some modern animal, which they call 'primitive', they are seeing a remote ancestor. This delusion is betrayed by phrases such as 'lower animal', or 'at the bottom of the evolutionary scale', which are not only snobbish but evolutionarily incoherent. Darwin's advice to himself would serve us all: 'Never use the words higher and lower.'

Lancelets are live creatures, our exact contemporaries. They are modern animals who have had exactly the same time as we have in which to evolve. Another telltale phrase is 'a side branch, off the main

---

* In the Garstang poem, 'his' gonads doesn't refer to the lancelet's but to the 'ammocoete' larva of a lamprey.

line of evolution'. All living animals are side branches. No line of evolution is more 'main' than any other, except with the conceit of hindsight.

Modern animals like lancelets, then, should never be revered as ancestors, nor patronised as 'lower', nor, for that matter, flattered as 'higher'. Slightly more surprisingly – and here we come to the second main point of the Lancelet's Tale – it is probably in general safest to say the same of fossils. It is theoretically conceivable that a particular fossil really is the direct ancestor of some modern animal. But it is statistically unlikely, because the tree of evolution is not a Christmas tree or a Lombardy poplar, but a densely branched thicket or bush. The fossil you are looking at probably isn't your ancestor, but it may help you to understand the *kind* of intermediate stage your real ancestors went through, at least in respect of some particular bit of the body, such as the ear, or the pelvis. A fossil, therefore, has something like the same status as a modern animal. Both can be used to illuminate our guesses about some ancestral stage. Under normal circumstances, neither should be treated as though it really is ancestral. Fossils as well as living creatures are usually best treated as cousins, not ancestors.

Members of the cladistic school of taxonomists can become positively evangelical about this, proclaiming the non-specialness of fossils with the zeal of a puritan or a Spanish inquisitor. Some go right over the top. They take the sensible statement, 'It is unlikely that any particular fossil is an ancestor of any surviving species', and interpret it to mean 'There never were any ancestors!' Obviously this book stops short of such an absurdity. At every single moment in history there must have been at least one human ancestor (contemporary with, or identical to, at least one elephant ancestor, swift ancestor, octopus ancestor, etc.), even if any particular fossil almost certainly isn't it.

The upshot is that, on our backward journey towards the past, the concestors we have been meeting have not, in general, been particular fossils. The best we can normally hope for is to put together a list of attributes that the ancestor probably had. We have no fossil of the common ancestor we share with the chimpanzees, even though that was less than 10 million years ago. But we were able to guess, with misgivings, that the ancestor was most likely to have been, in Darwin's famous words, a hairy quadruped, because we are the only ape that walks on its hind legs and has bare skin. Fossils can help us with our inferences, but mostly in the same kind of indirect way that living animals help us.

The moral of the Lancelet's Tale is that it is vastly harder to find an ancestor than a cousin. If you want to know what your ancestors looked like 100 million years ago, or 500 million years ago, it is no use reaching down to the appropriate depth in the rocks and hoping to come up with a fossil labelled 'Ancestor', as if from some Mesozoic or Palaeozoic bran tub. The most we can normally hope for is a series of fossils that, some with respect to one part, others with respect to another part, represent the *kind* of thing the ancestors probably looked like. Perhaps this fossil tells us something about our ancestors' teeth, while that fossil a few million years later gives us an inkling about our ancestors' arms. Any particular fossil is almost certainly not our ancestor but, with luck, some parts of it may resemble the corresponding parts of the ancestor just as, today, the shoulder-blade of a leopard is a reasonable approximation to the shoulder-blade of a puma.

SEA SQUIRTS

The sea squirts seem, at first, unlikely recruits to our human-centred pilgrimage. Previous arrivals have not been too dramatically different from those already on the march. Even the lancelet can plausibly be regarded as a stripped-down fish: lacking major features, to be sure, but you can easily sketch a pathway along which something like a lancelet could evolve into a fish. A sea squirt is something else. It doesn't swim like a fish. It doesn't swim like anything. It doesn't swim. It is far from clear why it deserves the illustrious name of chordate at all. A typical sea squirt is a bag filled with sea water, plus a gut and reproductive organs, anchored to a rock. The bag is topped by two siphons – one for drawing water in, the other for exhaling it. Day and night, water streams in through one siphon and out again through the other. On the way, it passes through the pharyngeal basket, a filtering net that strains out particles of food. Some sea squirts are packed together in colonies, but each member does essentially the same thing. No sea squirt is even faintly reminiscent of a fish, or of any vertebrate, or of the lancelet (see plate 31).

No adult sea squirt, that is. However unchordate-like an adult sea squirt might be, it has a larva that looks like . . . a tadpole. Or like the larva of a lamprey, the ammocoete of Garstang's rhyme on page 374. Like many larvae of sedentary, bottom-dwelling, filter-feeding animals, the tadpole larva of the sea squirt swims in the plankton. It propels itself like a fish by a post-anal tail that undulates from side to side. It has a notochord and a dorsal nerve tube. The larva, though not the adult sea squirt, has the appearance of at least a rudimentary chordate. When it is ready to metamorphose into an adult, the larva fastens itself onto a rock (or whatever is to be its adult resting place) head first, loses its tail, its notochord and most of its nervous system, and settles down for life.

It is even called a 'tadpole larva', and the significance of this was known to Darwin. He gave the sea squirts the following unpromising introduction, under their scientific name of ascidians:

Sea squirts join

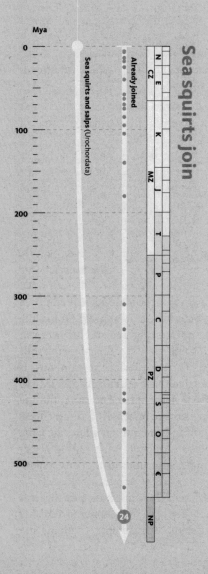

Mya

Sea squirts and salps (Urochordata)

Already joined

N															
	E														
	CZ														
		K													
		MZ													
		J													
		T													
		P													
		C													
		D													
	PZ	S													
		O													
		Є													
		NP													

24

Sea squirts join. Animals with a stiff cartilaginous 'notochord' are classified together as chordates (in humans, remnants of this rod linger as the discs between our vertebrae). It has long been accepted that, among the chordates, the sea squirts and their allies (of which there are about 2,000 described species) are the most distantly related to all the others. This has also been substantiated by recent molecular data.

Image: blue sea squirt (*Rhopalaea crassa*).

They hardly appear like animals, and consist of a simple, tough, leathery sack, with two small projecting orifices. They belong to the Molluscoidea of Huxley – a lower division of the great kingdom of the Mollusca; but they have recently been placed by some naturalists amongst the Vermes or worms. Their larvae somewhat resemble tadpoles in shape, and have the power of swimming freely about.

I should say that neither Molluscoidea nor Vermes are any longer recognised, and sea squirts are no longer placed close to molluscs or worms. Darwin goes on to mention his own satisfaction in discovering such a larva in the Falkland Islands in 1833, and he continues as follows:

M. Kovalevsky has lately observed that the larvae of Ascidians are related to the Vertebrata, in their manner of development, in the relative position of the nervous system, and in possessing a structure closely like the *chorda dorsalis* of vertebrate animals . . . We should then be justified in believing that at an extremely remote period a group of animals existed, resembling in many respects the larvae of our present Ascidians, which diverged into two great branches – the one retrograding in development and producing the present class of Ascidians, the other rising to the crown and summit of the animal kingdom by giving birth to the Vertebrata.

But now we have a division of opinion among experts. There are two theories of what happened: the one Darwin voiced, and a later one, which the Axolotl's Tale has already attributed to Walter Garstang. You remember the message of the axolotl, the message of neoteny. Sometimes the juvenile stage in a life-cycle can develop sex organs and reproduce: it becomes sexually mature, while remaining immature in other aspects of its being. We have previously applied the axolotl's message to Pekinese dogs, to ostriches and to ourselves: we humans appear to some scientists to be juvenile apes who have accelerated their reproductive development and chopped off the adult phase of the life cycle.

Garstang applied the same theory to sea squirts at this much older juncture in our history. The adult phase of our remote ancestor, he suggested, was a sedentary sea squirt, which evolved the tadpole larva as an adaptation to disperse, in the same way as a dandelion seed has a little parachute to carry the next generation far away from the site of its parent. We vertebrates, Garstang suggested, are descended from sea squirt *larvae* – larvae that never grew up: or rather larvae whose reproductive organs grew up but who never turned into sea squirt adults.

A second Aldous Huxley might project fictional human longevity to the point where some super-Methuselah finally settles down on his head and metamorphoses into a giant sea squirt, fastened permanently to the sofa in front of a television. The plot would gain added satirical punch from the popular myth that a sea squirt larva, when it abandons pelagic activity for sedentary adulthood, 'eats its own brain'. Somebody must once have colourfully expressed the more mundane fact that, like a caterpillar in its chrysalis, the metamorphosing sea squirt larva breaks down its larval tissues and recycles them into the adult body. This includes breaking down the head ganglion, which was useful when it was an active swimmer in the plankton. Mundane or not, a literary metaphor as promising as that was never going to pass unnoticed – a meme as fecund would not go unspread. More than once I have seen a reference to the larval sea squirt which, when the time comes, settles down to a sedentary life and 'eats its brain, like an associate professor getting tenure'.

There is a group of modern animals within the sea squirt subphylum called the Larvaceae, which are reproductively adult but resemble sea squirt larvae. Garstang pounced on them, seeing them as a more recent rerun of his ancient evolutionary script. In his view, the larvaceans had ancestors that were bottom-dwelling, sedentary sea squirts, with a planktonic larval phase. They evolved the capacity to reproduce in the larval stage, and then chopped the old adult stage off the end of their life cycle. This could all have transpired rather recently, giving us a fascinating glimpse of what perhaps happened to our ancestors half a billion years ago.

Garstang's theory is certainly an attractive one, and it was much in favour for many years, especially in Oxford under the influence of Garstang's persuasive son-in-law, Alister Hardy. Unfortunately, recent DNA evidence has swung the pendulum in favour of Darwin's original theory. If the larvaceans constitute a recent re-enactment of an ancient Garstang scenario, they should find closer kinship with some modern sea squirts than with others. Alas, this is not so. The oldest split in the entire phylum is that between the larvaceans on the one hand, and all the rest of the phylum on the other. This doesn't conclusively prove that Garstang was wrong but, as the current holder of Alister Hardy's Chair, Peter Holland, has pointed out to me, it weakens his case – and in a way that neither Garstang nor Hardy could possibly have foreseen.

The estimate I have adopted for the date of Concestor 24 is 565 million years ago, which would put it around our 275-million-greats-grandparent, but such estimates are now getting increasingly strained. It may well have looked something like a sea squirt larva. But, *contra* Garstang, it now seems probable that the adult sea squirt evolved later, as Darwin suggested. Darwin tacitly assumed that the adult of that remote species looked like a tadpole. One branch of its descendants stayed tadpole-shaped and evolved into fish. The other branch got tenure, settled down on the sea bottom and became a sedentary filter-feeder, retaining its former adult form only in the larval stage.

AMBULACRARIANS

Our pilgrimage is now a milling horde, having amassed all the vertebrates, together with their primitive chordate cousins, amphioxus and the sea squirts. It comes as quite a surprise that the next pilgrims to join us, our closest relatives among the invertebrates, include those strange creatures – I shall soon refer to them as 'Martians' – the starfish, sea urchins, brittle stars and sea cucumbers. These, together with a largely extinct group called the crinoids or sea lilies, comprise the phylum Echinodermata, the spiny-skinned ones. 'Before' the echinoderms join us, they link arms with a few miscellaneous worm-like groups which, in the absence of molecular evidence, had been placed elsewhere in the animal kingdom. The acorn worms and their kind (Enteropneusta and Pterobranchia) had previously been classified with the sea squirts as protochordates. Molecular evidence now links them, not so very far away, with the echinoderms in a super-phylum called Ambulacraria.

Also now placed in the ambulacrarians is a curious little worm called *Xenoturbella*. Nobody knew where to put little *Xenoturbella* – it seems to lack most of the things that a respectable worm ought to have, like a proper excretory system and a through-flow gut. Zoologists shuffled this obscure little worm from phylum to phylum, and had pretty well given up on it when, in 1997, somebody announced that, despite all appearances, it was a highly degenerate bivalve mollusc, with affinities to cockles. This confident statement came from molecular evidence. *Xenoturbella*'s DNA closely resembled that of a cockle and, as if to clinch it, *Xenoturbella* specimens were found to contain mollusc-type eggs. Terrible warning! In what looks like the

Opposite: **Starfish and their kin join.** We chordates belong to the major branch of animals known as the deuterostomes. Recent molecular studies suggest that all the other 8,100 or so deuterostome species group together. This new group, given the name of Ambulacraria, is quite strongly supported, although there is uncertainty in the position of the distressingly amorphous pair of species in the Xenoturbellida.

Images, left to right: sea apple (*Pseudo-colochirus violaceus*); edible sea urchin (*Echinus esculentus*); common starfish (*Asterias rubens*); brittle star (*Ophiothrix* sp.); feather star (*Cenometra bella*); acorn worm (Enteropneusta).

Starfish and their kin join

Mya

Sea cucumbers (Holothuroidea)

Sea urchins and sand dollars (Echinoidea)

Starfish (Asteroidea)

Brittlestars (Ophiuroidea)

Sea lilies (Crinoidea)

Acorn worms (Enteropneusta)

Pterobranchs (Pterobranchia)

Xenoturbellids (Xenoturbellida)

Already joined

N	E		K		MZ	J		T		P	C		D	PZ	S	O	Є		NP
CZ																			

25

classic nightmare of the modern forensic detective – contamination of the suspect's DNA by that of the murder victim – it has now turned out that the reason *Xenoturbella* contained mollusc DNA and mollusc eggs is that it eats molluscs! The residue of genuine *Xenoturbella* that is left when the mollusc DNA is removed reveals an even more surprising affinity: *Xenoturbella* is a member of the Ambulacraria, possibly the last member to join them 'before' we greet them at Rendezvous 25. Other molecular evidence places this rendezvous somewhere in the late Precambrian, maybe about 570 million years ago. I am guessing that Concestor 25 was approximately our 280-million-greats-grandparent. We have no idea what it looked like, but it surely was more worm-like than starfish-like. There is every indication that the echinoderms evolved their radial symmetry secondarily from left-right symmetrical ancestors – 'Bilateria'.

Echinoderms are a large phylum, with about 6,000 living species and a very respectable fossil record going back to early Cambrian times. Those ancient fossils include some weirdly asymmetrical creatures. Indeed, weird is perhaps the adjective that first occurs to one contemplating the echinoderms. A colleague once described the cephalopod molluscs (octopuses, squids and cuttlefish) as 'Martians'. He made a good point, but I think my candidate for the role might be a starfish. A 'Martian', in this sense, is a creature whose very strangeness helps us to see ourselves more clearly by showing us what we are not.

Earth's animals are mainly bilaterally symmetrical: they have a front end and a rear end, a left side and a right side. Starfish are radially symmetrical, with the mouth right in the middle of the lower surface, and the anus right in the middle of the top surface. Most echinoderms are similar, but heart urchins and sand dollars have rediscovered a modest degree of bilateral symmetry with a front and a rear for purposes of burrowing through the sand. If 'Martian' starfish have sides at all, they have five sides (or, in a few cases, some larger number), not two like most of the rest of us on Earth. Earth's animals mostly have blood. Starfish have piped sea water instead. Earth's animals mostly move about by means of muscles, pulling on bones or other skeletal elements. Starfish move about by means of a unique hydraulic system, using pumped sea water. Their actual propulsive organs are hundreds of small 'tube feet' on their under surface, arrayed in avenues along the five axes of symmetry. Each tube foot looks like a thin tentacle with a little round sucker on the end. On its own it is too small to move the animal, but the whole array pulling together can do it, slowly but powerfully. A tube foot is extended by hydraulic pressure, exerted by a little squeezed bulb at its near end. Each individual tube foot has a cycle of activity rather like a tiny leg. Having exerted its pull, it releases its sucker, picks itself up and swings forward to take a new grip with the sucker, and pull again.

Sea urchins get around by the same method. Sea cucumbers, which are shaped like warty sausages, can move this way too, but burrowing ones move the whole body as earthworms do, by alternately squeezing the body so it elongates forward, then pulling the rear up behind. Brittle stars, which have (usually) five slender, waving arms radiating out from a nearly circular central disc, move by rowing with whole arms, rather than dragging themselves along by tube feet. Starfish too have muscles that swing whole arms about. They use them, for example, to engulf prey and pull mussel shells apart.

'Forward' is arbitrary for these 'Martians', and that includes brittle stars and most urchins as well as starfish. Unlike most Earth life forms, who have a definite front end with a head, a starfish can 'lead' with any one of its five arms. The hundreds of tube feet somehow manage to 'agree' to follow the lead arm at any one time, but the lead role can change from arm to arm. The co-ordination is achieved by a nervous system, but it is a different pattern of nervous system from any others we are accustomed to on this planet. Most nervous systems are based upon a long trunk cable running from front to rear, either along the dorsal side (like our spinal cord) or along the ventral side, in which case it is often double, with a ladder of connections between the left and right sides (as in worms and all arthropods). In a typical Earth creature, the main longitudinal trunk cable has side nerves, often paired in segments repeated serially from front to rear. And it usually has ganglia, local swellings which, when sufficiently large, are dignified with the name of brain. The starfish nervous system is utterly different. As we have come to expect by now, it is radially arranged. There is a complete ring going right round the mouth, from which five (or however many arms there are) cables radiate out, one along each arm. As you would expect, the tube feet along each arm are controlled by the trunk nerve running along it.

In addition to the tube feet, some species also have hundreds of so-called pedicellariae (singular pedicellaria), scattered over the lower surface of the five arms. These have tiny pincers, and are used for catching food, or in defence against small parasites.

Alien 'Martians' though they may appear, starfish and their kind are still our relatively close cousins. Less than four per cent of all animal species are closer cousins to us than starfish are. By far the greater part of the animal kingdom is yet to join our pilgrimage. And they mostly arrive all together, at Rendezvous 26, in one gigantic influx of pilgrims. The protostomes are about to overwhelm even the multitude of pilgrims who are already on the march.

PROTOSTOMES

In the deeps of geological time, and increasingly deprived of the hard support of fossils, we are now entirely reliant on the technique that I referred to in the General Prologue as molecular rangefinding. The upside is that the technique is getting ever more sophisticated. Molecular rangefinding confirms a belief long held by comparative anatomists, or more strictly comparative embryologists, that the greater part of the animal kingdom is deeply divided into two great subkingdoms, the Deuterostomia and the Protostomia.

Here's how embryology comes in. Animals typically pass through a watershed event in their early life called gastrulation. The distinguished embryologist and scientific iconoclast Lewis Wolpert said:

> It is not birth, marriage or death, but gastrulation, which is truly the most important time in your life.

Gastrulation is something that all animals do early in their life. Typically, before gastrulation, an animal embryo consists of a hollow ball of cells, the blastula, whose wall is one cell thick. During gastrulation the ball indents to form a cup with two layers. The opening of the cup closes in to form a small hole called the blastopore. Almost all animal embryos go through this stage, which presumably means it is a very ancient feature indeed. You might expect that so fundamental an opening would become one of the two deep holes in the body, and you'd be right. But now comes the big divide in the animal kingdom, between the Deuterostomia (every pilgrim who arrived before Rendezvous 26, including us) and the Protostomia (the huge throng who are now joining at Rendezvous 26).

In deuterostome embryology, the eventual fate of the blastopore is to become the anus (or at least the anus develops close to the blastopore). The mouth appears later as a separate perforation at the other end of the gut. The protostomes do it differently: in some, the blastopore becomes the mouth, and the anus appears later; in others, the blastopore is a slit that subsequently zippers up in the middle,

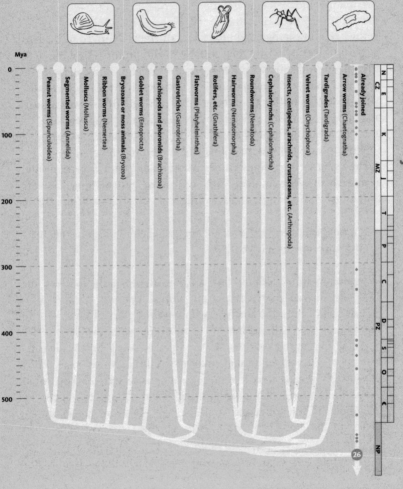

Mya

0

100

200

300

400

500

Peanut worms (Sipunculoidea)

Segmented worms (Annelida)

Molluscs (Mollusca)

Ribbon worms (Nemertea)

Bryozoans or moss animals (Bryozoa)

Goblet worms (Entoprocta)

Brachiopods and phoronids (Brachiozoa)

Gastrotricha (Gastrotricha)

Flatworms (Platyhelminthes)

Rotifers, etc. (Gnathifera)

Hairworms (Nematomorpha)

Roundworms (Nematoda)

Cephalorhynchs (Cephalorhyncha)

Insects, centipedes, arachnids, crustaceans, etc. (Arthropoda)

Velvet worms (Onychophora)

Tardigrades (Tardigrada)

Arrow worms (Chaetognatha)

Already joined

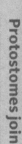

N	E	K		MZ		T	P	C		D	PZ	S	O		C			NP

26

with the mouth at one end and the anus at the other. Protostome means 'mouth first'. Deuterostome means 'mouth second'.

This traditional embryological classification of the animal kingdom has been upheld by modern molecular data. There are indeed two main kinds of animal, the deuterostomes (our lot) and the protostomes (them over there). However, some phyla that used to be included in the deuterostomes have now been moved by molecular revisionists, whom I shall follow, to the protostomes. These are the three so-called lophophorate phyla – the phoronids, brachiopods and bryozoans – now grouped together with the molluscs and annelid worms in the 'Lophotrochozoa' division of the protostomes. For goodness' sake don't bother to remember the 'lophophorates' – I need to mention them here only because zoologists of a certain age might be surprised not to find them among the deuterostomes. There are also some animals that don't belong to either the protostomes or the deuterostomes, but we'll come to them later.

Rendezvous 26 is the biggest of all, more of a gigantic rally of pilgrims than a rendezvous. When does it happen? Such ancient dates are hard to estimate. My attempt of 590 million years is plus or minus a large margin of error. The same goes for the estimate that Concestor 26 is our 300-million-greats-grandparent. The protostomes constitute the great bulk of the pilgrimage of animals. Because our own species is of the deuterostome persuasion, I have given them special attention in this book, and I am portraying the protostomes as joining the pilgrimage all together, at one major rendezvous. Not only the protostomes themselves would see it the other way around – a dispassionate observer would too.

The protostomes have a much greater number of animal phyla than the deuterostomes, including the largest phyla of all. They include the molluscs, with twice as many species as the vertebrates.

Opposite: **Protostomes join.** At this rendezvous the 60,000 or so known deuterostome species join well over a million described protostomes. This protostome phylogeny represents another recent and radical rearrangement brought about by genetics. The two major groupings are now generally accepted, but the order of branching within them is extremely uncertain. The order within the seven lineages on the left ('Lophotrochozoa') is particularly unsure.

Images, left to right: lugworm (*Arenicola* sp.); garden snail (*Helix aspersa*); unknown bryozoan, gastrotrich (*Chaetonotus simrothi*); zebra polyclad flatworm (*Pseudoceros dimidiatus*); Antarctic bdelloid rotifer (*Philodina gregaria*); unknown nematode, leaf-cutter ant (*Atta* sp.); velvet worm (*Peripatopsis moseleyi*); unknown tardigrade.

They include the three great worm phyla: flatworms, roundworms and annelid worms, whose species together outnumber the mammal species perhaps thirtyfold. Above all, the protostome pilgrims include the arthropods: insects, crustaceans, spiders, scorpions, centipedes, millipedes and several other smaller groups. The insects alone constitute at least three-quarters of all animal species, and probably more. As Robert May, the current President of the Royal Society has said, to a first approximation all species are insects.

Before the days of molecular taxonomy, we grouped and divided animals by looking at their anatomy and embryology. Of all the classificatory levels – species, genus, order, class, etc. – phylum had a special, almost mystical status. Animals within one phylum were clearly related to one another. Animals in different phyla were too distinct for any relationships to be taken seriously. The phyla were separated by an all but unbridgeable gulf. Molecular comparison now suggests that the phyla are much more connected than we ever thought they were. In a sense that was always obvious – nobody believed the animal phyla arose separately from primordial slime. They had to be connected to each other, in the same sort of hierarchical patterns as their constituent parts. It was just that the connections were hard to see, lost in deep time.

There were exceptions. The protostome–deuterostome grouping above the phylum level was admitted, based on embryology. And within the protostomes it was widely accepted that the annelid worms (segmented earthworms, leeches and bristle worms) were related to arthropods, both having a segmented body plan. That particular connection now seems to be wrong, as we shall see: nowadays the annelids are partnered with the molluscs. Actually, it was always a bit worrying that marine annelids had a kind of larva that was so similar to the larvae of many marine molluscs that they were given the same name, the 'trochophore' larva. If the annelid–mollusc grouping is right, it means that the segmented body plan was invented twice (by annelids and arthropods), rather than the trochophore larva being invented twice (by annelids and molluscs). The association of annelids with molluscs, and their separation from arthropods, is one of the bigger surprises that molecular genetics has dealt those zoologists brought up on morphologically based taxonomy.

Molecular evidence divides the protostome phyla into two, or perhaps three, main groups: super-phyla, I suppose we could call them. Some authorities have yet to accept this classification, but I shall go along with it while recognising that it could still be wrong. The two super-phyla are called the Ecdysozoa and the Lophotrochozoa. The third super-phylum, which is less widely acknowledged, but which I shall accept rather than lumping them in with the Lophotrochozoa as some prefer, is the Platyzoa.

The Ecdysozoa are named after their characteristic habit of moulting, or ecdysis (from a Greek word meaning roughly to get your kit off). That gives an immediate hint that the insects, crustaceans, spiders, millipedes, centipedes, trilobites and other arthropods are ecdysozoans, and this means that the ecdysozoan faction of the protostome pilgrimage is very large indeed, far more than three-quarters of the animal kingdom.

The arthropods dominate both the land (especially insects and spiders) and the sea (crustaceans and, in earlier times, trilobites). With the exception of the eurypterids, those Palaeozoic sea scorpions* which, we conjectured, terrorised the Palaeozoic fishes, arthropods have not achieved the enormous body size of some extreme vertebrates. This is often attributed to limits set by their method of encasing themselves in an armour-plated exoskeleton, with their limbs in hard jointed tubes. This means they can grow only by ecdysis: casting their outer casing aside at regular intervals and hardening a new, larger one. How the eurypterids managed to exempt themselves from this alleged size limitation is not entirely clear to me.

There is lingering dispute about how the sub-contingents of arthropods are arranged. Some zoologists uphold the earlier view that the insects belong with the myriapods (centipedes, millipedes and their kind), separated off from the crustaceans. The majority now bracket the insects with the crustaceans, pushing the myriapods and spiders off as outgroups. Everyone agrees that spiders and scorpions, together with the terrifying eurypterids, belong together in the group called chelicerates. *Limulus,* the living fossil known, unfortunately, as the horseshoe crab, is also placed in the chelicerates, despite its

---

* There were also giant land scorpions in the Palaeozoic, estimated at a metre long, a fact that I do not greet with equanimity (one of my earliest memories, before I fainted, is of being stung by a modern African scorpion). The largest known trilobite, *Isotelus rex,* reached 72 centimetres in length. Dragonflies with wingspans up to 70 centimetres flourished in the Carboniferous. Today's largest arthropod, the Japanese spider crab *Macrocheira kaempferi* has a 30-centimetre body, and the span between its enormously elongated claw-bearing limbs can be up to four metres.

superficial resemblance to the extinct trilobites, which are separated off in their own group.

Allied to the arthropods within the Ecdysozoa, and sometimes called panarthropods, are two small contingents of pilgrims, the onychophorans and the tardigrades. Onychophorans or velvet worms, such as *Peripatus*, are now classified in the phylum Lobopodia, which has an important fossil contingent, as we shall see in the Velvet Worm's Tale. *Peripatus* itself looks a bit like a caterpillar of rather endearing mien, although in this respect it is outdone by the tardigrades. Whenever I see a tardigrade I want to keep it as a pet. Tardigrades are sometimes called water bears, and they have the cuddly appearance of a baby bear. A very baby bear indeed: you can only just see them without a microscope, waving their eight stubby legs with a charming air of infantile ineptitude.

The other major phylum in the superphylum Ecdysozoa is that of the nematode worms. They too are extremely numerous, a fact made memorable long ago by the American zoologist Ralph Buchsbaum:

> If all the matter in the universe except the nematodes were swept away, our world would still be dimly recognisable . . . we should find its mountains, hills, vales, rivers, lakes, and oceans represented by a film of nematodes . . . Trees would still stand in ghostly rows representing our streets and highways. The location of the various plants and animals would still be decipherable, and, had we sufficient knowledge, in many cases even their species could be determined by an examination of their erstwhile nematode parasites.

I was delighted by this image when I first read Buchsbaum's book, but I must confess, returning to reread it now, I find myself sceptical. Let's just say that nematode worms are extremely numerous and ubiquitous.

Smaller phyla in the Ecdysozoa include various other kinds of worms, including the priapulid or penis worms. These are quite aptly named, although the champion in this vein is the fungus whose Latin name is *Phallus* (wait for Rendezvous 34). It is superficially surprising that the priapulids are now classified so far from the annelid worms.

The lophotrochozoan pilgrims may be outnumbered by the Ecdysozoa, but even they decisively outnumber our own deuterostome pilgrims. The two big lophotrochozoan phyla are the molluscs and the annelids. The annelid worms are not long confusable with the nematode worms, for the annelids are segmented – like the

arthropods, as we have seen. This means that their body is arranged as a series of segments fore and aft, like the trucks of a train. Many body parts, for example nerve ganglia and blood vessels running around the gut, are repeated in every segment along the length of the body. The same is true of arthropods, most obviously millipedes and centipedes because their segments are all pretty much the same as each other. In a lobster or, even more, a crab, many of the segments are different from each other, but you can still clearly see that the body is segmented in the fore-and-aft direction. Their ancestors surely had more uniform segments like a woodlouse or a millipede.*
Annelid worms are like millipedes or woodlice in this respect, although the worms are more closely related to the non-segmented molluscs. The most familiar annelid worms are common or garden (for once the phrase is strictly apt) earthworms. I am privileged to have seen giant earthworms (*Megascolides australis*), in Australia, said to be capable of growing to four metres long.

The Lophotrochozoa include other worm-like phyla, for instance the nemertine worms, not to be confused with the nematodes. The similarity of name is unfortunate and unhelpful, compounded by further confusion with two other worm phyla, the Nematomorpha and the Nemertodermatida. Nema (*nematos*) in Greek means 'thread', while Nemertes was the name of a sea nymph. Unfortunate coincidence, that. On a school marine biology field trip to the Scottish coast with our inspiring zoology teacher Mr I. F. Thomas, we found a bootlace worm, *Lineus longissimus*, a species of nemertine legendarily capable of growing to 50 metres. Our specimen was at least 10 metres long, but I don't remember the exact measurement, and Mr Thomas has sadly lost his photograph of this unforgettable occasion, so it will have to remain as a nemertean version of a fisherman's tall tale.

There are various other more-or-less worm-like phyla, but the biggest and most important phylum of the Lophotrochozoa is the Mollusca: the snails, oysters, ammonites, octopuses and their kind. The mollusc contingent of the pilgrimage mostly creeps at snail's pace, but squids are among the fastest swimmers in the sea, using a form of jet propulsion. They, and their cousins the octopuses, are the most spectacularly proficient colour-changers in the animal

---

* There are some wonderful millipedes – pill millipedes – that look and behave just like woodlice. This is one of my favourite examples of convergent evolution.

kingdom, streets better than the proverbial chameleons, not least because they change in quick time. The ammonites were relatives of the squids who lived in coiled shells that served them as flotation organs, as with the still surviving *Nautilus*. Ammonites once thronged the seas but went finally extinct at the same time as the dinosaurs. I hope they changed colour too.

Another major group of molluscs is the bivalves: oysters, mussels, clams and scallops, with two shells or valves. Bivalves have a single extremely powerful muscle, the adductor, whose function is to close the valves and lock in the closed position against predators. Don't put your foot in a giant clam (*Tridacna*) – you'll never get it back. The bivalves include *Teredo*, the shipworm, which uses its valves as cutting tools to bore through driftwood, wooden ships and the pilings of piers and quays. You will probably have seen their holes, of cleanly circular cross-section. Piddocks do something similar through rock.

Superficially like bivalve molluscs are the brachiopods, the lamp shells. They are also part of the great lophotrochozoan contingent of the protostome pilgrimage, but are not closely related to the bivalve molluscs. We have already met one of them, *Lingula*, in the Lungfish's Tale, as a famous 'living fossil'. There are now only about 350 species of brachiopod, but in the Palaeozoic Era they rivalled the bivalve molluscs.* The resemblance between them is superficial: the bivalve molluscs' two shells are left and right, where the two brachiopod shells are top and bottom. The status of the brachiopod pilgrims, and two allied 'lophophorate' groups called phoronids and bryozoans, is still disputed. As already mentioned, I am following the dominant contemporary school of thought in placing them in the Lophotrochozoa (to which name, indeed, they have contributed). Some zoologists leave them where they used to be, outside the protostomes altogether and in the deuterostomes, but I suspect theirs is a losing battle.

The third major branch of the protostome super-phylum, the Platyzoa, would be joined by some authorities to the Lophotrochozoa. 'Platy' means 'flat', and the name Platyzoa comes from one of the component phyla, the flatworms or Platyhelminthes. 'Helminth' means 'intestinal worm', and while some flatworms are parasitic (tapeworms and flukes), there is also a large group of freeliving

---

* Stephen Gould compared them in a nice essay called 'Ships That Pass in the Night'.

Ceci n'est pas une coquille. Fossil brachiopod (*Doleorthis*) from the Silurian.

flatworms, the turbellarians, which are often extremely beautiful. Recently, some of the animals traditionally classified as flatworms, for example the acoels, have been removed by molecular taxonomists out of the protostomes altogether. We shall meet them presently.

Other phyla are provisionally placed in the Platyzoa, but for the moment it is for want of anywhere more certain to put them, and they are mostly not flat. Belonging to the so-called 'minor phyla', they are fascinating in their own right and each deserves a whole chapter in a textbook of invertebrate zoology. Unfortunately, however, we have a pilgrimage to complete and must press on. Of these minor phyla I shall just mention the rotifers, because they have a tale to tell.

Rotifers are so small that they were originally grouped with single-celled protozoan 'animalcules'. They are actually multicellular and quite complicated in miniature. One group of them, the bdelloid rotifers, are remarkable because no male has ever been seen. This is what their tale is about, and we shall come to it soon.

So this vast flood of protostome pilgrims, composite of tributaries far and wide, truly the dominant stream of animal pilgrims, converges on its rendezvous with the deuterostomes, the junior (by comparison) contingent, whose progress we have hitherto followed for the sufficient reason that it is our own. The grand ancestor of both, Concestor 26 from our human point of view, is extremely hard to reconstruct at such a remote distance of time.

It seems very likely that Concestor 26 was some kind of worm. But that is only to say a long thing, bilaterally symmetrical, with a left and a right side, a dorsal and a ventral side, and a head and a tail end. Indeed, some scientists have given the name Bilateria to all the animals descended from Concestor 26, and I shall use this word. Why is this pattern, the worm form, so common? The most primitive members of all three protostome subgroups, and the most primitive deuterostomes, are all of the form that we should generally call worm-shaped. So let's have a tale about what it means to be a worm.

I wanted to put the worm's tale into the grey and muddy mouth of the lugworm.* Unfortunately, the lugworm spends most of its time in a U-shaped burrow, which is just what we don't need for the tale, as will soon become apparent. We need a more typical worm, which actively crawls or swims in a forward direction: for whom front and rear, left and right, and up and down have a clear meaning. So the lugworm's close cousin, *Nereis* the ragworm, shall take over the role. An 1884 magazine article for anglers said, 'The bait used is that damp kind of centipede called a ragworm.' It is not a centipede, of course, but a polychaete worm. It lives in the sea, where it normally crawls over the bottom but is capable of swimming if necessary.

THE RAGWORM'S TALE

Any animal that moves, in the sense of covering the ground from A to B rather than just sitting in one place and waving its arms or pumping water through itself, is likely to need a specialised front end. It might as well have a name, so let's call it the head. The head hits novelty first. It makes sense to take in food at the end that encounters it first, and to concentrate the sense organs there too – eyes perhaps, some kind of feelers, organs of taste and smell. Then the main concentration of nervous tissue – the brain – had best be near the sense organs, and near the action at the front end, where the food-catching apparatus is. So we can define the head end as the leading end, the one with the mouth, the main sense organs and the brain, if there is one. Another good idea is to void wastes somewhere near the back end, far from the mouth, to avoid re-imbibing what has just been passed out. By the way, although all this makes sense if we think worm, I should remind you that the argument evidently does not apply to radially symmetrical animals such as starfish. I am genuinely puzzled why starfish and their kind opt out of this argument, which is one reason why I referred to them as 'Martians'.

---

* A lugworm, with its grey and muddy mouth
  Sang that somewhere to north or west or south
  There dwelt a gay, exulting, gentle race . . .
                                        W. B. YEATS
                                        (1865–1939)

To return to our primeval worm, having dealt with its fore and aft asymmetry, how about up-down asymmetry? Why is there a dorsal side and a ventral side? The argument is similar, and this one applies to starfish just as much as to worms. Gravity being what it is, there are lots of inevitable differences between up and down. Down is where the sea bottom is, down is where the friction is, up is where the sunlight comes from, up is the direction from which things fall on you. It is unlikely that dangers will threaten equally from below and above, and in any case those dangers are likely to be qualitatively different. So our primitive worm should have a specialised upper or 'dorsal' side and a specialised 'ventral' or lower side, rather than simply not caring which side faces the sea bottom and which side faces the sky.

Put our front-rear asymmetry together with our dorsal-ventral asymmetry, and we have automatically defined a left side and a right side. But unlike the other two axes, we find no general reason to distinguish the left side from the right side: no reason why they should be anything other than mirror images. Danger is not more likely to threaten from the left than the right, or vice versa. Food is not more likely to be found on the left or the right side, though it may well be more likely above or below. Whatever is the best way for a left side to be, there is no general reason to expect any difference for the right. Limbs or muscles that were not mirrored left-right would have the unfortunate effect of driving the animal round in circles, instead of in direct pursuit of some goal.

Perhaps revealingly, the best exception I can think of is fictional. According to a Scottish legend (probably invented for the amusement of tourists, and said to be believed by many of them), the haggis is a wild animal living in the Highlands. It has short legs on one side and long legs on the other, in accordance with its habit of running only one way round the sides of steep Highand hillsides. The prettiest real-life example I can think of is the wonky-eyed jewel squid of Australian waters, whose left eye is much larger than its right. It swims at a 45-degree angle, with the larger, telescopic left eye looking upwards for food, while the smaller right eye looks below for predators. The wrybill is a New Zealand sandpiper whose bill curves markedly to the right. The bird uses it to flick pebbles to the side and expose prey. Striking 'handedness' is to be seen in fiddler crabs, who have one hugely enlarged claw for fighting or, more to the point, displaying their ability to fight. But perhaps the most intriguing story of asymmetry in the animal kingdom was told me by Sam Turvey. Trilobite fossils often display bite marks, indicating narrow escapes from predators. The fascinating thing is that about 70 per cent of these bite marks are on the right-hand side. Either trilobites had an asymmetrical awareness of predators, like the wonky-eyed jewel squid, or their predators had handedness in their attack strategy.

But those are all exceptions, mentioned for their curiosity value and to make a revealing contrast with the symmetrical world of our primitive worm and its descendants. Our crawling archetype has a left and a right side which are mirror images of each other. Organs tend to arise in pairs, and where there are exceptions, such as the wonky-eyed jewel squid, we notice it and comment.

How about eyes? Would the first bilaterian have had eyes? It isn't enough to say that all modern descendants of Concestor 26 have eyes. It isn't enough, because the various kinds of eyes are very diverse: so much so that it has been estimated that 'the eye' has evolved independently more than 40 times in various parts of the animal kingdom.* How do we reconcile this with the statement that Concestor 26 had eyes?

To give intuition a steer, let me say first that what is claimed to have evolved 40 times independently is not light-sensitivity per se, but image-forming optics. The vertebrate camera eye and the crustacean compound eye evolved their optics (working on radically different principles) independently of one another. But both these eyes are descended from one organ in the common ancestor (Concestor 26), which was probably an eye of some kind.

The evidence is genetic, and it is persuasive. In the fruit fly *Drosophila* there is a gene called *eyeless.* Geneticists have the perverse habit of naming genes by what goes wrong when they mutate. The *eyeless* gene normally negates its name by making eyes. When it mutates and fails to have its normal effect on development, the fly has no eyes, hence the name. It is a ludicrously confusing convention. To avoid it, I shall not refer to the *eyeless* gene, but will use the comprehensible abbreviation *ey.* The *ey* gene normally makes eyes, and we know this because when it goes wrong the flies are eyeless. Now the story starts to get interesting. There is a very similar gene in mammals, called *Pax6,* also known as *small eye* in mice and *aniridia*

---

* I have discussed this at length in *Climbing Mount Improbable,* in a chapter called 'The Fortyfold Path to Enlightenment', and I return to it at the end of this book.

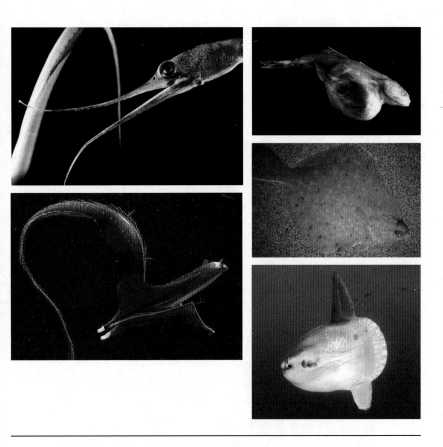

25. Animal shapes are malleable like plasticine
The variety of teleost forms. CLOCKWISE FROM TOP LEFT: Snipe eel
(*Avocettina infans*); black swallower (*Chiasmodon niger*) after meal; plaice
(*Pleuronectes platessa*); sunfish (*Mola mola*); gulper eel (*Eurypharynx
pelecanoides*) (see pages 340–42).

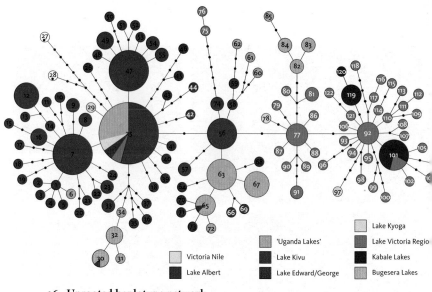

26. Unrooted haplotype network
From Verheyen *et al.* [295] (see page 353).

27. Picasso would have loved them
The skate (*Raja batis*, top) rests on its belly; the flounder (*Bothus lunatus*) on its right side. The eye on that side has migrated over time to the left (upper) side. Drawing by Lalla Ward (see page 359).

Legend (within figure 26):

Lake Kyoga
Lake Victoria Regio
Kabale Lakes
Bugesera Lakes

'Uganda Lakes'
Lake Kivu
Lake Edward/George

Victoria Nile
Lake Albert

28. Designed by an artist with a drugged imagination?
RIGHT: great hammerhead shark (*Sphyrna mokarran*); BELOW: freshwater sawfish (*Pristis microdon*) (see page 360).

29. Ghostly appearance
Elephant fish (*Callorhynchus milii*), showing the characteristic large head and flapping pectoral fins of chimaeras (see page 363).

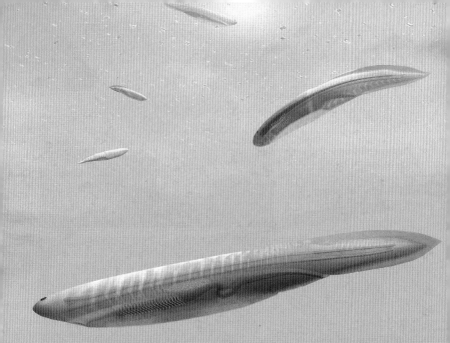

30. Concestor 23
This concestor is thought to have had a notochord
(a stiff cartilaginous rod) extending the length of its
body from below its rudimentary brain. Like the
modern lancelet, it would have had thick myomeres
(V-shaped muscle blocks), and would have filtered
food through its gills. Artistic reconstruction by
Malcolm Godwin (see page 374).

31. Like an associate professor getting tenure?
Adult blue sea squirt (*Rhopalaea crassa*) (see page 377).

32. Carried home by workers in broad, rustling rivers of green
Leaf cutter ants (*Atta* sp.) carrying leaf fragments back to the nest.
Note the small (minim) worker riding on the leaf (see page 406).

33. Wouldn't a Martian split them three against one? (see page 411).

34. Cells 'think' they are in the wrong segment Homeotic mutant fruit fly (see page 430).

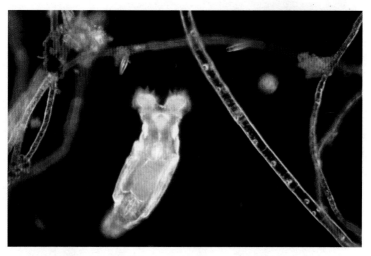

35. An evolutionary scandal
Light micrograph of a bdelloid rotifer (*Philodina gregaria*) found in Antarctica (see page 437).

36. Velvet worm
The modern onychophoran, *Peripatopsis moseleyi* (see page 451).

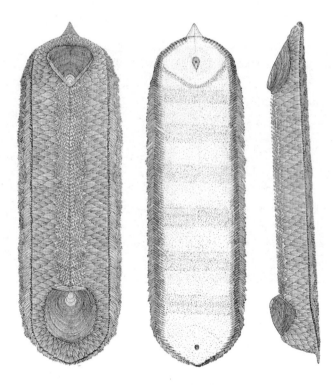

37. Breaks down the mystical reverence for the great phlya
Halkieria evangelista, from Sirius Passet, Greenland, dating from the
Lower Cambrian. Drawing by Simon Conway Morris (see page 453).

(no iris) in humans (again named for the negative effect of its mutant form).

The DNA sequence of the human *aniridia* gene is more similar to the fruit fly's *ey* gene than it is to other human genes. They must be inherited from the shared ancestor which was, of course, Concestor 26. Again, I shall call it *ey*. Walter Gehring and his colleagues in Switzerland did an utterly fascinating experiment. They introduced the mouse equivalent of the *ey* gene into fruit fly embryos, with astounding results. When introduced into the part of a fruit fly embryo that was destined to make a leg, it caused the eventual adult fly to grow an extra 'ectopic' eye on its leg. It was a fly eye, by the way: a compound eye, not a mouse eye. I don't think there is any evidence that the fly could see through it, but it had the unmistakable properties of a respectable compound eye. The instruction given by the *ey* gene seems to be 'grow an eye here, of the kind that you would normally grow'. The fact that the gene is not only similar in mice and flies, but induces the development of eyes in both, is very strong evidence that it was present in Concestor 26; and moderately strong evidence that Concestor 26 could see, even if only the presence versus the absence of light. Perhaps, when more genes have been investigated, the same argument can be generalised from eyes to other bits. In fact, in one sense, this has already been done – we'll deal with it in the Fruit Fly's Tale.

The brain, sitting at the front end for the reasons we have argued, needs to make nervous contact with the rest of the body. In a worm-shaped animal, it is sensible that it should do so via a main cable, a principal nerve trunk, running along the length of the body, probably with side branches at intervals along the body to exercise local control and take in local information. In a bilaterally symmetrical animal like a ragworm or a fish, the trunk nerve must run either dorsal or ventral of the digestive tract, and here we strike one of the main differences between us deuterostomes on the one hand and the protostomes, who have joined us in such strength, on the other. In us, the spinal nerve cord runs along the back. In a typical protostome like a ragworm or a centipede, it is on the ventral side of the gut.

If Concestor 26 was indeed some kind of worm, it presumably followed either the dorsal nerve pattern or the ventral nerve pattern. I can't call them deuterostome and protostome patterns because the two separations don't quite coincide. The acorn worms (those rather obscure deuterostomes who arrived with the echinoderms at

Rendezvous 25) are hard to interpret, but at least on some views they have a ventral nerve cord like a protostome, although for other reasons they are classified as deuterostomes. Let me instead divide the animal kingdom into the dorsocords and the ventricords. The dorsocords are all deuterostomes. The ventricords are mostly proto-stomes, plus some early deuterostomes perhaps including the acorn worms. The echinoderms, with their remarkable reversion to radial symmetry, don't fit into this classification at all. Probably the deuterostomes, as I say, were still ventricords until some time later than Concestor 26.

The difference between dorsocords and ventricords extends to other things than just the position of the main nerve running along the body. Dorsocords have a ventral heart, whereas ventricords have a dorsal heart, pumping blood forward along a main dorsal artery. These and other details suggested in 1820 to the great French zoologist Geoffroy St Hilaire that a vertebrate could be thought of as an arthropod, or an earthworm, turned upside down. After Darwin and the acceptance of evolution, zoologists from time to time suggested that the vertebrate body plan had actually evolved through a worm-like ancestor literally turning upside down.

That is the theory that I want to support here, on balance and with some caution. The alternative, which is that a worm-like ancestor gradually rearranged its internal anatomy while staying the same way up, seems to me less plausible because it would have involved a greater amount of internal upheaval. I believe a change in behaviour came first – suddenly by evolutionary standards – and it was followed by a whole lot of consequential evolutionary changes. As so often, there are modern equivalents to make the idea vivid for us today. The brine shrimp is one example, and we hear its tale next.

THE BRINE SHRIMP'S TALE

Brine shrimps, *Artemia,* and the closely related fairy shrimps are crustaceans that swim on their backs, and therefore have their nerve cord (the 'true' zoological ventral side) on the side that now faces the sky. The upside-down catfish, *Synodontis nigriventris,* is a deutero-stome that does the same thing the other way round. It is a fish that swims on its back, and therefore has its main trunk nerve on the side facing the river bottom, which is the 'true' zoological dorsal side. I don't know why brine shrimps do it, but the catfish swim upside down because they take food from the water surface, or from the undersides of floating leaves. Presumably, individual fish discovered that this was a good source of food and learned to turn over. My conjecture* is that, as the generations went by, natural selection favoured those individuals who learned to perform the trick best, their genes 'caught up' with the learning, and now they never swim any other way.

The brine shrimp's inversion is a recent re-enactment of something that happened, in my view, more than half a billion years ago. An ancient, long-lost animal, some kind of worm with a ventral nerve cord and a dorsal heart like any protostome, turned over and swam, or crawled, upside down like a brine shrimp. A zoologist who happened to be present at the time would have died rather than relabel the main nerve trunk dorsal just because it now ran along the side of the body facing the sky. 'Obviously', all his zoological training would have told him, it was still a ventral nerve cord, corresponding to all the other organs and features that we expect to see on the ventral surface of a protostome. Equally 'obvious' to this Precambrian zoologist, the heart of our inverted worm was, in the deepest sense, a 'dorsal' heart, even though it now beat under the skin nearest the sea bottom.

Given enough time, however – given enough millions of years of swimming or crawling 'upside down' – natural selection would come to reshape all the organs and structures of the body to fit in with the upside-down habit. Eventually, unlike our modern brine shrimp, which has only recently turned over, the traces of the original dorsal/ventral homologies would become obliterated. Later generations of palaeo-zoologists who encountered the descendants of this early maverick, after some tens of millions of years of upside-down habit, would start to redefine their concepts of dorsal and ventral. This is because so many anatomical details would have changed over evolutionary time.

Other animals that swim on their backs are sea otters (especially when engaging in their remarkable habit of smashing shellfish with stones on the belly), and the aptly named backswimmers (all the

---

* Following the theoretical idea known as the Baldwin Effect. Superficially, it sounds like Lamarckian evolution and the inheritance of acquired characteristics. Not so. Learning doesn't imprint itself into the genes. Instead, natural selection favours genetic propensities to learn certain things. After generations of such selection, evolved descendants learn so fast that the behaviour has become 'instinctive'.

time). Backswimmers are a kind of bug,* sometimes known as greater water boatmen, which row themselves underwater with their legs. The related lesser water boatmen do the same kind of thing, but they swim the right way up.

Imagine that the descendants of our modern water boatmen or brine shrimps on the one hand, and the descendants of our modern upside-down catfish on the other, were to maintain their habits of swimming upside down for 100 million years into the future. Isn't it entirely likely that they might each give rise to a whole new sub-kingdom, each body plan so radically reshaped by the upside-down habit that zoologists who didn't know the history would define the brine shrimps' descendants as having a 'dorsal' nerve cord, and the descendants of the catfish as having a 'ventral' nerve cord.

As we saw in the Ragworm's Tale, the world presents important practical differences between up and down, and these would start to imprint themselves, by natural selection, on the sky-pointing side and the floor-pointing side respectively. What had once been the zoologically ventral side would start to look more and more like a zoologically dorsal side, and vice versa. I believe this is exactly what happened somewhere along the line leading to vertebrates, and that is why we now have a dorsal nerve cord and a ventral heart. Modern molecular embryology offers some supporting evidence from the ways in which genes that define the dorso-ventral axis are expressed – genes a bit like the Hox genes that we shall meet in the Fruit Fly's Tale – but the details are beyond our scope here.

The upside-down catfish, recent though its inverted habit undoubtedly is, has already taken one revealing little step in this evolutionary direction.† Its Latin name is *Synodontis nigriventris*. *Nigriventris* means 'dark belly', and it introduces a fascinating vignette at the end of the Brine Shrimp's Tale. One of the main differences between up and down in the world is the predominant direction of light. While not necessarily directly overhead, the sun's rays generally come from above rather than below. Hold your fist up and you'll find, even under an overcast sky, that its upper surface is better lit than its lower surface. This fact opens a key way in which we

---

* Bug means something precise, not just anything small. A bug is an insect of the order Hemiptera.
† So has the nudibranch mollusc (sea slug) *Glaucus atlanticus*. This beautiful creature floats upside down, feeding on Portuguese men-of-war, and it is 'reverse countershaded', just like the catfish.

Turn the fish upside down. Upside-down catfish (*Synodontis nigriventris*) in characteristic pose.

and many other animals can recognise solid, three-dimensional objects. A uniformly coloured curved object, such as a worm or a fish, looks lighter on top, darker below. I am not talking about the hard shadow cast by the body – it is a more subtle effect than that. A gradient of shading, from lighter above to darker below, smoothly betrays the curvature of the body.

It works in reverse. The photograph of moon craters is printed upside down. If your eye (well, to be more precise, your brain) works in the same way as mine, you will see the craters as hills. Turn the

Turn the book upside down. Craters on the far side of the moon.

book upside down, so that the light appears to come from another direction, and the hills will turn into the craters that they truly are.

One of my very first experiments as a graduate student demonstrated that newly hatched baby chicks seem to see the same illusion, straight out of the egg. They peck at photographs of simulated grains, and strongly prefer them if lit as if from above. Turn the photograph over and they shun it. This seems to show that baby chicks 'know' that light in their world normally comes from above. But since they have only just hatched out of the egg, how do they know? Have they learned it during their three days of life? It is perfectly possible, but I tested it experimentally and found it not to be so. I raised chicks and tested them in a special cage in which the only light they ever saw came from below. Experience of pecking grain in this upside-down world would, if anything, teach them to prefer upside-down photographs of solid grains. Instead, they behaved exactly like normal chicks raised in the real world with light coming from above. Apparently because of genetic programming, all the chicks prefer to peck at photographs of solid objects lit from above. The solidity illusion (and hence, if I am right, the 'knowledge' of the predominant direction of light in the real world) seems to be genetically programmed in chicks – what we used to call 'innate' – rather than learned as (I'm guessing) it probably is in us.

Whether learned or unlearned, there is no doubt that the surface shading illusion of solidity is a powerful one. It has provoked a subtle form of camouflage called countershading. Look at any typical fish, out of water on a slab, and you'll notice that the belly is much lighter in colour than the back. The back may be dark brown or grey, while the belly is light grey, verging on white in some cases. And what is all this about? There seems no doubt that it is a form of disguise, based on counteracting the shading gradient that normally gives the game away for curved, solid things like fish. In the best of all possible worlds, a countershaded fish, when viewed in a normal light coming from above, will look perfectly flat. The expected gradient from light above to dark below will be exactly countered by the gradient in the fish's colour from light below to dark above.

Taxonomists often name species from dead specimens in museums.* Presumably that explains *nigriventris* as opposed to *invertus*

---

* Although I have to admit that the habits of the upside-down catfish have long been known. It is portrayed in its customary position in Ancient Egyptian wall paintings and engravings.

or whatever is the Latin for 'upside down'. If you examine the upside-down catfish on a slab, you find that it is *reverse* countershaded. Its belly, which faces the sky, is darker than its back, which faces the bottom. Reverse countershading is one of those wonderfully elegant exceptions that prove the rule. The first catfish to swim upside down would have been horribly conspicuous. Its skin coloration would have conspired with the natural shading from overhead light to make it look preternaturally solid. No wonder the change of habit was followed, in evolutionary time, by a reversal of the usual skin-colour gradient.

Fish are not the only animals to employ countershading as a trick of disguise. My old maestro Niko Tinbergen, before he left the Netherlands for Oxford, had a student called Leen De Ruiter to whom he proposed research on countershading in caterpillars. Caterpillars of many species play exactly the same trick on their (in this case bird) predators as fish do on theirs. These caterpillars are beautifully countershaded, with the result that they look flat when seen in a normal light. De Ruiter took the twigs on which caterpillars were sitting and turned them upside down. Immediately they became far more conspicuous because they suddenly looked much more solid. And birds took them in greater numbers.

If a De Ruiter were to come along and force the catfish to turn over and swim, like any normal fish, with its zoologically dorsal side uppermost, it would suddenly become far more conspicuously solid.* Reverse countershading in upside-down catfish is a single example of a consequential change that has followed, in evolutionary time, their change of habit. In another hundred million years, just think how comprehensively their whole bodies might change. There is nothing sacrosanct about 'dorsal' and 'ventral'. They can reverse, and I think they did reverse in the early ancestry of today's dorsocords. My bet is that Concestor 26 had its main trunk nerve running along the ventral side of the body like any protostome. We are modified worms swimming on our backs, descended from an early equivalent of a brine shrimp which, for some long-forgotten reason, turned over.

---

* You might ask how an experimenter could force a catfish to turn over against its natural preference, and I do not know. But, to add just one tiny vignette, I do know how to make a brine shrimp swim, like any normal crustacean, with its zoologically dorsal side uppermost. Just shine an artificial light on them from below, and they will instantly turn over. Evidently brine shrimps use light as their cue to decide which way is up. I don't know if the catfish use the same cue. They could equally well use gravity.

The more general moral of the Brine Shrimp's Tale is this. Major transitions in evolution may have begun as changes in behavioural habit, perhaps even non-genetic learned changes of habit, which only later were followed by genetic evolution. I fancy that a comparable tale could be told for the first bird ancestor to fly, the first fish to come out on the land, and the first whale ancestor to return to the water (as Darwin speculated with his fly-catching bear). A change of habit by an adventurous individual is later followed by a long evolutionary catch-up and clean-up. That is the most far-reaching lesson from the Brine Shrimp's Tale.

THE LEAF CUTTER'S TALE

Just as humanity did at the time of our Agricultural Revolution, ants independently invented the town. A single nest of leaf cutter ants, *Atta*, can exceed the population of Greater London. It is a complicated underground chamber, up to 6 metres deep and 20 metres in circumference, surmounted by a somewhat smaller dome above ground. This huge ant city, divided into hundreds or even thousands of separate chambers connected by networks of tunnels, is sustained ultimately by leaves cut into manageable pieces and carried home by workers in broad, rustling rivers of green (see plate 32). But the leaves are not eaten directly, either by the ants themselves (though they do suck some of the sap) or by the larvae. Instead they are painstakingly mulched as compost for underground fungus gardens. It is the small round knobs or 'gongylidia' of the fungi that the ants eat and, more particularly, that they feed to the larvae. Cropping by the ants normally stops the fungi forming spore-bearing bodies (the equivalent of the mushrooms that we eat). This deprives fungus experts of the cues they normally use to identify species, and it means that the fungi themselves are dependent on the ants for their propagation. They have apparently evolved to flourish only in the domesticated environment of an ants' nest, which makes it a true example of domestication by an agricultural species other than our own. When a young queen ant flies off to found a new colony, she takes a precious cargo with her: a small culture of the fungus with which to sow the first crop in her new nest. This reminds me of a story of penicillin, perhaps the most important fungus of all. When Florey, Chain and their Oxford colleagues were developing penicillin, at the height of the Second World War, having (typically) failed to interest British companies in manufacturing it, they went to America where (typically again) they met with success. Like a queen ant, they took with them a culture of the precious fungus. On an earlier occasion, a German invasion of Britain was expected. Florey and his junior colleague Heatley deliberately infected their own clothes with the mould, as the best way of secretly preserving the culture.

The energy for running the fungus ant colony is ultimately gathered from the sun by the leaves used to make the compost, a total leaf area that is measured in acres in the case of a large *Atta* colony. Fascinatingly the termites, that other hugely successful group of town-making insects, have also independently discovered fungus agriculture. In their case the compost is made of chewed-up wood. As with ants and their fungus, the termite fungus species is found only in termite nests and it seems to have been 'domesticated'. On those occasions where a termite fungus (*Termitomyces*) is permitted to produce a fruiting body, it sprouts from the side of the mound. It is said to be delicious and is on sale as a delicacy in the markets of Bangkok. A West African species, *Termitomyces titanicus,* is in *The Guinness Book of Records* as the largest mushroom in the world, with a cap diameter of up to one metre.

Several groups of ants have independently evolved the habit of keeping domestic 'dairy' animals in the form of aphids. Unlike other symbiotic insects that live inside ants' nests and don't benefit the ants, the aphids are pastured out in the open, sucking sap from plants as they normally do. As with mammalian cattle, aphids have a high throughput of food, taking only a small amount of nutriment from each morsel. The residue that emerges from the rear end of an aphid is sugar-water – 'honeydew' – only slightly less nutritious than the plant sap that goes in at the front. Any honeydew not eaten by ants rains down from trees infested with aphids, and is plausibly thought to be the origin of 'manna' in the Book of Exodus. It should not be surprising that ants gather it up, for the same reason as the followers of Moses did. But some ants have gone further and corralled aphids, giving them protection in exchange for being allowed to 'milk' the aphids, tickling their rear ends to make them secrete honeydew which the ant eats directly from the aphid's anus.

At least some aphid species have evolved in response to their domestic existence. They have lost some of the normal aphid defensive responses and, according to one intriguing suggestion, some have modified their rear end to resemble the face of an ant. Ants are in the habit of passing liquid food to one another, mouth to mouth, and the suggestion is that individual aphids that evolved this rear-end face-mimicry facilitated being 'milked' and therefore gained protection by ants from predators.

The Leaf Cutter's Tale is a tale of delayed gratification as the basis of agriculture. Hunter-gatherers eat what they gather and eat what they hunt. Farmers don't eat their seed corn; they bury it in the ground and wait months for a return. They don't eat the compost with which they fertilise the soil and don't drink the water with which they irrigate it. Again, it is all done for a delayed reward. And the leaf cutter ant got there first. Consider her ways and be wise.

THE GRASSHOPPER'S TALE

The Grasshopper's Tale treats the vexed and sensitive topic of race.

There is a pair of European grasshopper species, *Chorthippus brunneus* and *C. biguttulus*, which are so similar that even expert entomologists can't tell them apart, yet they never cross-breed in the wild although they sometimes meet. This defines them to be 'good species'. But experiments have shown that you need only allow a female to hear the mating call of a male of her own species caged nearby and she will happily mate with a male of the wrong species, 'thinking', one is tempted to say, that he is the singer. When this happens, healthy and fertile hybrids are produced. It doesn't normally happen in the wild because a female doesn't normally find herself near, but unable to reach, a singing male of her own species at the same time as a male of the wrong species is courting her.

Comparable experiments have been done on crickets, using temperature as an experimental variable. Different species of cricket chirp at different frequencies, but the chirp frequency is also temperature-dependent. If you know your crickets, you can use them as a reasonably accurate thermometer. Fortunately, not only the male's chirping frequency but also the female's perception of it is temperature-dependent: the two vary in lockstep, which normally precludes miscegenation. A female in an experiment, offered a choice of males singing at two different temperatures, chooses the one at her own temperature. The male singing at a different temperature is treated as if he belongs to the wrong species. If you heat up a female, her preference shifts to a 'hotter' song, even if that causes her to prefer a cool male of the wrong species. Once again, this normally doesn't happen in nature. If a female can hear a male at all, he can't be far away, and so is likely to be at approximately the same temperature as she is.

Grasshopper song is temperature-dependent in the same kind of way. Using grasshoppers of the same genus, *Chorthippus*, with which we began (though different species of the genus), German scientists did some technically ingenious experiments. They managed to attach tiny thermometers (thermocouples) and tiny electric heaters to the insects. So miniaturised were these, the experimenters could heat the head of a grasshopper without heating its thorax, or heat the thorax without heating the head. Then they tested females' preferences for songs produced by males stridulating at various temperatures.* They found that what matters for the female song preference is the temperature of the head. But it is the temperature of the thorax that determines the stridulation rate. Fortunately of course, in nature, where there are no experimenters with tiny electric heaters, the head and thorax will normally be at the same temperature, as will the male and the female. So the system works, and hybridisation doesn't happen.

It is quite common to find pairs of related species that never interbreed under natural conditions but that can do so if humans interfere. The case of *Chorthippus brunneus* and *C. biguttulus* is just one example. The Cichlid's Tale told of a comparable case in fish, where monochromatic light abolished the discrimination between a reddish and a bluish species. And it happens in zoos. Biologists normally classify animals that mate under artificial conditions but refuse to mate in the wild as separate species, as has happened with the grasshoppers. But unlike, say, lions and tigers, which can hybridise in zoos to make (sterile) 'ligers' and 'tigrons', those grasshoppers look identical. Apparently the only difference is in their songs. And it is this, and only this, that stops them cross-breeding and therefore leads us to recognise them as separate species. Human beings are the other way round. It requires an almost superhuman

---

* Stridulation is how grasshoppers, and crickets, make sound. Grasshoppers scrape their legs against their wing covers. Crickets scrape the two wing covers against each other. They sound similar, but grasshoppers are generally more buzzy, crickets more musical. Of one nocturnal tree cricket it has been said that if moonlight could be heard, that is how it would sound. Cicadas are quite different. As if buckling a tin lid, they buckle part of the thorax wall, repeatedly and fast, so it sounds like a continuous buzz, usually extremely loud and sometimes patterned in very complex ways, characteristic of the species.

feat of political zeal to overlook the conspicuous differences between our own local populations or races. Yet we happily interbreed across races and are unequivocally and uncontroversially defined as members of the same species. The Grasshopper's Tale is about races and species, about the difficulties of defining both, and what all this has to say about human races.

'Race' is not a clearly defined word. 'Species', as we have seen, is different. There really is an agreed way to decide whether two animals belong in the same species: can they interbreed? Obviously they can't if they are of the same sex, or are too young or too old, or one of them happens to be sterile. But those are pedantries, easy to get around. In the case of fossils, too, which obviously can't breed, we apply the interbreeding criterion in our imaginations. Do we think it *likely* that, if these two animals were not fossils but alive, fertile and of opposite sex, they would be able to interbreed?

The interbreeding criterion gives the species a unique status in the hierarchy of taxonomic levels. Above the species level, a genus is just a group of species that are pretty similar to each other. No objective criterion exists to decide *how* similar they have to be, and the same is true of all the higher levels: family, order, class, phylum and the various 'sub-' or 'super-' names that intervene between them. Below the species level, 'race' and 'subspecies' are used more or less interchangeably and, again, no objective criterion exists that would enable us to decide whether two people should be considered part of the same race or not, nor to decide how many races there are. And of course there is the added complication, absent above the species level, that races interbreed, so there are lots of people of mixed race.

Presumably species, on their way to becoming sufficiently separate to be incapable of interbreeding, usually pass through an intermediate stage of being separate races. Separate races might be regarded as species in the making, except that there is no necessary expectation that the making will continue to its end – to speciation.

The interbreeding criterion works pretty well, and it delivers an unequivocal verdict on humans and their supposed races. All living human races interbreed with one another. We are all members of the same species, and no reputable biologist would say any different. But let me call your attention to an interesting, perhaps even slightly disturbing fact. While we happily interbreed with each other, producing a continuous spectrum of inter-races, we are strangely reluctant to give up our divisive racial language. Wouldn't you expect that if all intermediates are on constant display, the urge to classify people as one or the other of two extremes would wither away, smothered by the absurdity of the attempt, which is continually manifested everywhere we look? Unfortunately, this is not what happens, and perhaps that very fact is revealing.

People who are universally agreed by all Americans to be 'black' may draw less than one-eighth of their ancestry from Africa, and often have a light skin colour well within the normal range for people universally agreed to be 'white'. In the picture of four American politicians (see plate 33), two are described in all newspapers as black, the other two as white. Wouldn't a Martian, unschooled in our conventions but able to see skin shades, be more likely to split them three against one? Surely yes. But in our culture, almost everybody will immediately 'see' Mr Powell as 'black', even in this particular photograph which happens to show him with possibly lighter skin than either Bush or Rumsfeld.

It is an interesting exercise to take a photograph, such as the one in plate 33, of Colin Powell standing next to some representative 'white' men (they must be next to each other so the lighting conditions are the same). From each face, cut a small uniform rectangle, say from the forehead, and place the patches side by side. You will find that there is very little difference between Powell and the 'white' men with whom he is standing. He may be lighter or darker, depending upon the particular cases. But now 'zoom out' and look again at the original photograph. Immediately, Powell will look 'black'. What cues are we picking up on?

Next to a genuinely black man. Colin Powell with Daniel arap Moi.

To ram home the point, do the same 'forehead patch' exercise with Powell standing next to a genuinely black man such as Daniel arap Moi, the recent President of Kenya (see page 411). This time, the forehead patches will look dramatically different. But then, when we 'zoom out' and look at the whole faces, we again 'see' Mr Powell as 'black'. The news story that accompanied this picture of Powell, visiting Moi in May 2001, implied that the same conventions are understood in Africa:

> As the first African-American secretary of state, Powell has been accorded near-messianic treatment in Africa. And perhaps because he is black, there was a profound resonance to Powell's harsh criticism . . .

Why do people so readily swallow the apparent contradiction – and there are numerous similar examples – between the verbal statement, 'he is black', and the picture it accompanies. What is going on here? Various things. First, we are curiously eager to embrace racial classification, even when talking about individuals whose mixed parentage seems to make a nonsense of it, and even where (as here) it is irrelevant to anything that matters.

Second, we tend not to describe people as of mixed race. Instead, we plump for one race or the other. Some American citizens are of pure African descent and some are of pure European descent (leaving aside the fact that, in the longer term, we are all of African descent). Maybe it is convenient for some purposes to call them black and white respectively, and I am not proposing any principled objection to these names. But many people – probably more than most of us realise – have both black and white ancestors. If we are going to use colour terminology, many of us are presumably somewhere in between. Yet society insists on calling us one or the other. It is an example of the 'tyranny of the discontinuous mind', which was the subject of the Salamander's Tale. Americans are regularly asked to fill in forms in which they have to tick one of five boxes: Caucasian (whatever that might mean – it certainly doesn't mean from the Caucasus), African-American, Hispanic (whatever that might mean – it certainly does not mean, as the word seems to suggest, Spanish), Native American or Other. There are no boxes labelled half and half. But the very idea of ticking boxes is incompatible with the truth, which is that many, if not most, people are a complicated mixture of the offered categories and others. My inclination is irritably to refuse to tick any boxes, or to add my own box labelled 'human'. Especially when the rubric uses the mealy-mouthed euphemism 'Ethnicity'.

Third, in the particular case of 'African-Americans', there is something culturally equivalent to genetic dominance in our use of language. When Mendel crossed wrinkled peas with smooth peas, all the first generation progeny were smooth. Smooth is 'dominant', wrinkled is 'recessive'. The first generation progeny all had one smooth allele and one wrinkled, yet the peas themselves were indistinguishable from peas with no wrinkled genes. When an Englishman marries an African, the progeny are intermediate in colour and in most other characteristics. This is unlike the situation in peas. But we all know how society will call such children: 'black' every time. Blackness is not a true genetic dominant like smoothness in peas. But social *perception* of blackness behaves like a dominant. It is a cultural or memetic dominant. That insightful anthropologist Lionel Tiger has attributed this to a racist 'contamination metaphor' within white culture. And no doubt there is also a strong and understandable will on the part of descendants of slaves to identify with their African roots. I have already remarked on this in Eve's Tale – regarding the television documentary where Jamaican immigrants to Britain were emotionally reunited with alleged 'family' in West Africa.

Fourth, there is high inter-observer agreement in our racial categorisations. A man such as Colin Powell, of mixed race and intermediate physical characteristics, is not described as white by some observers and black by others. A small minority will describe him as mixed. All others will without fail describe Mr Powell as black – and the same goes for anybody who shows the *slightest trace* of African ancestry, even if their percentage of European ancestors is overwhelming. Nobody describes Colin Powell as white, unless they are trying to make a political point by the very fact that the word jars against the audience's expectations.

There is a useful technique called 'inter-observer correlation'. It is a measure that is often used in science to establish that there really is a reliable basis for a judgement, even if nobody can pin down what that basis is. The rationale, in the present case, is this. We may not know how people decide whether somebody is 'black' or 'white' (and I hope I have just demonstrated that it isn't because they are black or white!) but there must be *some* sort of reliable criterion lurking there because any two randomly chosen judges will come to the same decision.

The fact that the inter-observer correlation remains high, even over a huge spectrum of inter-races, is impressive testimony to something fairly deep-seated in human psychology. If it holds up cross-culturally, it will be reminiscent of the anthropologists' finding about perception of hue. Physicists tell us that the rainbow, from red through orange, yellow, green and blue to violet is a simple continuum of wavelength. It is biology and/or psychology, not physics, that singles out particular landmark wavelengths along the physical spectrum for special treatment and naming. Blue has a name. Green has a name. Blue-green does not. The interesting finding of anthropologists' experiments (as opposed to some influential anthropological theories, by the way) is that there is substantial agreement over such namings across different cultures. We seem to have the same kind of agreement over judgements of race. It may prove to be even stronger and clearer than for the rainbow.

As I said, zoologists define a species as a group whose members breed with each other under natural conditions – in the wild. It doesn't count if they breed only in zoos, or if we have to use artificial insemination, or if we fool female grasshoppers with caged singing males, even if the offspring produced are fertile. We might dispute whether this is the only sensible definition of a species, but it is the definition that most biologists use.

If we wished to apply this definition to humans, however, there is a peculiar difficulty: how do we distinguish natural from artificial conditions for interbreeding? It is not an easy question to answer. Today, all surviving humans are firmly placed in the same species, and they do indeed happily interbreed. But the criterion, remember, is whether they choose to do so under natural conditions. What are natural conditions for humans? Do they even exist any more? If, in ancestral times, as sometimes today, two neighbouring tribes had different religions, different languages, different dietary customs, different cultural traditions and were continually at war with one another; if the members of each tribe were brought up to believe that the other tribe were subhuman 'animals' (as happens even today); if their religions taught that would-be sexual partners from the other tribe were taboo, 'shiksas', or unclean, there could well be no interbreeding between them. Yet anatomically, and genetically, they could be completely the same as each other. And it would take only a change of religious or other customs to break down the barriers to interbreeding. How, then, might somebody try to apply the interbreeding criterion to humans? If *Chorthippus brunneus* and *C.*

biguttulus are separated as two distinct species of grasshoppers because they prefer not to interbreed although they physically could, might humans, at least in ancient times of tribal exclusivity, once have been separable in the same kind of way? *Chorthippus brunneus* and *C. biguttulus,* remember, in all detectable respects except their song, are identical, and when they are (easily) persuaded to hybridise their offspring are fully fertile.

Whatever we may think as observers of superficial appearances, the human species today is, to a geneticist, especially uniform. Taking such genetic variation as the human population does possess, we can measure the fraction that is associated with the regional groupings that we call races. And it turns out to be a small percentage of the total: between 6 and 15 per cent depending on how you measure it – much smaller than in many other species where races have been distinguished. Geneticists conclude, therefore, that race is not a very important aspect of a person. There are other ways to say this. If all humans were wiped out except for one local race, the great majority of the genetic variation in the human species would be preserved. This is not intuitively obvious and may be quite surprising to some people. If racial statements were as informative as most Victorians, for example, used to think, you would surely need to preserve a good spread of all the different races in order to preserve most of the variation in the human species. Yet this is not the case.

It certainly would have surprised Victorian biologists who, with few exceptions, saw humanity through race-tinted spectacles. Their attitudes persisted into the twentieth century. Hitler was unusual in gaining the power to turn racialist ideas into government policy. Plenty of others, not just in Germany, had the same thoughts but lacked the power. I have previously quoted H. G. Wells's vision of his New Republic (*Anticipations,* 1902), and I do so again because it is such a salutary reminder of how a leading British intellectual, regarded in his time as progressive and left-leaning, could say such horrifying things, only a century ago, and scarcely be noticed doing so.

And how will the New Republic treat the inferior races? How will it deal with the black? . . . the yellow man? . . . the Jew? . . . those swarms of black, and brown, and dirty-white, and yellow people, who do not come into the new needs of efficiency? Well, the world is a world, and not a charitable institution, and I take it they will have to go . . . And the ethical system of these men of the New Republic, the ethical system which will dominate the world state, will be shaped primarily to favour the procreation of what is fine and efficient and beautiful in humanity – beautiful and strong bodies, clear and powerful minds . . . And the method that nature has followed hitherto in the shaping of the world, whereby weakness was prevented from propagating weakness . . . is death . . . The men of the New Republic . . . will have an ideal that will make the killing worth the while.

I suppose we should take comfort from the change that has come over our attitudes during the intervening century. Perhaps, in a negative sense, Hitler can take some credit for this, since nobody wants to be caught saying anything that he said. But what, I wonder, will our successors of the twenty-second century be quoting, in horror, from us? Something to do with our treatment of other species, perhaps?

But that was an aside. We were dealing with the unusually high level of genetic uniformity in the human species, despite superficial appearances. If you take blood and compare protein molecules, or if you sequence genes themselves, you will find that there is less difference between any two humans living anywhere in the world than there is between two African chimpanzees. We can explain this human uniformity by guessing that our ancestors, but not the chimpanzees', passed through a genetic bottleneck not very long ago. The population was reduced to a small number, came close to going extinct, but just pulled through. There is evidence of a fierce bottleneck – perhaps down to a population of 15,000, some 70,000 years ago, caused by a six-year 'volcanic winter' followed by a thousand-year ice age. Like the children of Noah in the myth, we are all descended from this small population, and that is why we are so genetically uniform. Similar evidence, of even greater genetic uniformity, suggests that cheetahs passed through an even narrower bottleneck more recently, around the end of the last Ice Age.

Some people may find the evidence of biochemical genetics unsatisfying because it seems not to square with their everyday experience. Unlike cheetahs, we don't 'look' uniform.* Norwegians, Japanese and Zulus really do look rather dramatically different from one another. With the best will in the world, it is intuitively hard to

---

* As an aside, leopards don't either. But black 'panthers', once thought to be a separate species, differ from spotted leopards at a single genetic locus.

believe what is in fact the truth: that they are 'really' more alike than three chimpanzees who look, to our eyes, much more similar.

This is, of course, a politically sensitive matter, a point I heard being amusingly lampooned by a West African medical researcher at a gathering of about 20 scientists. At the beginning of the conference, the chairman asked each of us around the table to introduce ourselves. The African, who was the only black person there – and he really was black, unlike many 'African-Americans' – happened to be wearing a red tie. He finished his self-introduction by laughingly saying, 'You can easily remember me. I am the one with the red tie.' He was genially mocking the way people bend over backwards to pretend not to notice racial differences. I think there was a *Monty Python* sketch along the same lines. Nevertheless, we can't write off the genetic evidence which suggests that, all appearances to the contrary, we really are an unusually uniform species. What is the resolution to the apparent conflict between appearance and measured reality?

It is genuinely true that, if you measure the total variation in the human species and then partition it into a between-race component and a within-race component, the between-race component is a very small fraction of the total. Most of the variation among humans can be found within races as well as between them. Only a small admixture of extra variation distinguishes races from each other. That is all correct. What is not correct is the inference that race is therefore a meaningless concept. This point has been clearly made by the distinguished Cambridge geneticist A. W. F. Edwards in a recent paper called 'Human genetic diversity: Lewontin's fallacy'. R. C. Lewontin is an equally distinguished Cambridge (Mass.) geneticist, known for the strength of his political convictions and his weakness for dragging them into science at every possible opportunity. Lewontin's view of race has become near-universal orthodoxy in scientific circles. He wrote, in a famous paper of 1972:

> It is clear that our perception of relatively large differences between human races and subgroups, as compared to the variation within these groups, is indeed a biased perception and that, based on randomly chosen genetic differences, human races and populations are remarkably similar to each other, with the largest part by far of human variation being accounted for by the differences between individuals.

This is, of course, exactly the point I accepted above, not surprisingly since what I wrote was largely based on Lewontin. But see how Lewontin goes on:

> Human racial classification is of no social value and is positively destructive of social and human relations. Since such racial classification is now seen to be of virtually no genetic or taxonomic significance either, no justification can be offered for its continuance.

We can all happily agree that human racial classification is of no social value and is positively destructive of social and human relations. That is one reason why I object to ticking boxes in forms and why I object to positive discrimination in job selection. But that doesn't mean that race is of 'virtually no genetic or taxonomic significance'. This is Edwards's point, and he reasons as follows. However small the racial partition of the total variation may be, if such racial characteristics as there are are highly correlated with other racial characteristics, they are by definition informative, and therefore of taxonomic significance.

Informative means something quite precise. An informative statement is one that tells you something you didn't know before. The information content of a statement is measured as reduction in prior uncertainty. Reduction in prior uncertainty, in turn, is measured as a change in probabilities. This provides a way to make the information content of a message mathematically precise, but we don't need to bother with that.* If I tell you Evelyn is male, you immediately know a whole lot of things about him. Your prior uncertainty about the shape of his genitals is reduced (though not obliterated). You now know facts you didn't know before about his chromosomes, his hormones and other aspects of his biochemistry, and there is a quantitative reduction in your prior uncertainty about the depth of his voice, and the distribution of his facial hair and of his body fat and musculature. Contrary to Victorian prejudices, your prior uncertainty about Evelyn's general intelligence or ability to learn, remains unchanged by the news about his sex. Your prior uncertainty about his ability to lift weights or excel at most sports is quantitatively reduced, but only quantitatively. Plenty of females can beat plenty of males at any sport, although the best males can normally beat the best females. Your ability to bet on Evelyn's

---

* As it happens, Lewontin himself was one of the first biologists to use information theory, and indeed he did so in his paper on race, but for a different purpose. He used it as a convenient statistic for measuring diversity.

running speed, say, or the power of his tennis serve, has been slightly raised by my telling you his sex, but it has not reached certainty.

Now to the question of race. What if I tell you Suzy is Chinese, how much is your prior uncertainty reduced? You now are pretty certain that her hair is straight and black (or was black), that her eyes have an epicanthic fold, and one or two other things about her. If I tell you Colin is 'black' this does not, as we have seen, tell you he is black. Nevertheless, it is clearly not uninformative. The high inter-observer correlation suggests that there is a constellation of characteristics that most people recognise, such that the statement 'Colin is black' really does reduce prior uncertainty about Colin. It works the other way around to some extent. If I tell you Carl is an Olympic sprinting champion, your prior uncertainty about his 'race' is, as a matter of statistical fact, reduced. Indeed, you can have a fairly confident bet that he is 'black'.*

We got into this discussion through wondering whether the concept of race was, or had ever been, an information-rich way to classify people. How might we apply the criterion of inter-observer correlation to judging the question? Well, suppose we took standard full-face photographs of 20 randomly chosen natives of each of the following countries: Japan, Uganda, Iceland, Sri Lanka, Papua New Guinea and Egypt. If we presented all 120 people with all 120 photographs, my guess is that every single one of them would achieve 100 per cent success rates in sorting them into six different categories. What is more, if we told them the names of the six countries involved, all 120 subjects, if they were reasonably well educated, would correctly assign all 120 photographs to the correct countries. I haven't done the experiment but I am confident that you will agree with me on what the result would be. It may seem unscientific of me not to bother to do the experiment. But my confidence that you, being human, will agree without doing the experiment, is the very point I am trying to make.

If the experiment were to be done, I do not think Lewontin would expect any other result than the one I have predicted. Yet an opposite prediction would seem to follow from his statement that racial classification has virtually no taxonomic or genetic significance. If there is no taxonomic or genetic significance, the only other way to

---

* Sir Roger Bannister got into terrible hot water, for no good reason that I could discern except people's hair-trigger sensitivities on matters of race, when he said something similar a few years ago.

get a high inter-observer correlation would be a worldwide similarity in cultural bias, and I do not think Lewontin would want to predict that either. In short, I think Edwards is right and Lewontin, not for the first time, wrong. Lewontin did his sums right, of course: he is a brilliant mathematical geneticist. The proportion of the total variation in the human species that falls into the racial partition of variation is, indeed, low. But because the between-race variation, however low a percentage of the total variation, is *correlated*, it is informative in ways that could surely be demonstrated by measuring the inter-observer concordance of judgement.

I must at this point reiterate my strong objection to being asked to fill in forms in which I have to tick a box labelling my 'race' or 'ethnicity', and voice my strong support for Lewontin's statement that racial classification can be actively destructive of social and human relations – especially when people use racial classification as a way of treating people differently, whether through negative or positive discrimination. To tie a racial label to somebody is informative in the sense that it tells you more than one thing about them. It might reduce your uncertainty about the colour of their hair, the colour of their skin, the straightness of their hair, the shape of their eye, the shape of their nose and how tall they are. But there is no reason to suppose that it tells you anything about how well-qualified they are for a job. And even in the unlikely event that it did reduce your statistical uncertainty about their likely suitability for some particular job, it would *still* be wicked to use racial labels as a basis for discrimination when hiring somebody. Choose on the basis of ability, and if, having done so, you end up with an all-black sprinting team, so be it. You have not practised racial discrimination in arriving at this conclusion.

A great conductor, when auditioning instrumentalists for his orchestra, always had them perform behind a screen. They were told not to speak, and they even had to remove their shoes for fear the sound of high heels would betray the sex of the performer. Even if it were statistically the case that women tend to make better harpists, say, than men, this does *not* mean that you should actively discriminate against men when you choose a harpist. Discriminating against individuals purely on the basis of a group to which they belong is, I am inclined to think, always evil. There is near-universal agreement today that the apartheid laws of South Africa were evil. Positive discrimination in favour of 'minority' students on American campuses can fairly, in my opinion, be attacked on the same grounds as apartheid. Both treat people as representative of groups rather than as individuals in their own right. Positive discrimination is sometimes justified as redressing centuries of injustice. But how can it be just to pay back a single individual today for the wrongs done by long-dead members of a plural group to which he belongs?

Interestingly, this kind of singular/plural confusion shows up in a form of words which is tellingly diagnostic of bigots: 'The Jew . . .' instead of 'Jews . . .'

> Your Fuzzy Wuzzy is an excellent fighter, but he can't tell his left from his right. Now, your Pathan . . .

People are individuals, they are individually different, far more different from other members of their group than their groups are from each other. In this, Lewontin is undoubtedly right.

Inter-observer agreement suggests that racial classification is not totally uninformative, but what does it inform *about*? About no more than the characteristics used by the observers when they agree: things like eye shape and hair curliness – nothing more unless we are given further reasons to believe it. For some reason it seems to be the superficial, external, trivial characteristics that are correlated with race – perhaps especially facial characteristics. But why are human races so different in just these superficially conspicuous characteristics? Or is it just that we, as observers, are predisposed to notice them? Why do other species look comparatively uniform whereas humans show differences that, if we encountered them elsewhere in the animal kingdom, might make us suspect we were dealing with a number of separate species?

The most politically acceptable explanation is that the members of any species have a heightened sensitivity to differences among their own kind. On this view, it is just that we *notice* human differences more readily than differences within other species. Chimpanzees whom we find almost identical look just as different, in chimpanzee eyes, as a Kikuyu is different from a Dutchman in our eyes. Expecting to confirm this kind of theory at the within-race level, the eminent American psychologist H. L. Teuber, an expert on the brain mechanisms of facial recognition, asked a Chinese graduate student to study the question, 'Why do Westerners think Chinese people look more alike than Westerners?' After three years' intensive research, the Chinese student reported his conclusion. 'Chinese people really do look more alike than Westerners!' Teuber told the story with much twinkling and wiggling of eyebrows, a sure sign with him that a joke was on the way, so I don't know what the truth of the matter is. But I have no difficulty in believing it, and I certainly don't think it should upset anyone.

Our (relatively) recent worldwide diaspora out of Africa has taken us to an extraordinarily wide variety of habitats, climates and ways of life. It is plausible that the different conditions have exerted strong selection pressures, particularly on externally visible parts, such as the skin, which bear the brunt of the sun and the cold. It is hard to think of any other species that thrives so well from the tropics to the Arctic, from sea level to the high Andes, from parched deserts to dripping jungles, and through everything in between. Such different conditions would be bound to exert different natural selection pressures, and it would be positively surprising if local populations did not diverge as a result. Hunters in the deep forests of Africa, South America and South-East Asia have all independently become small, almost certainly because height is a handicap in dense vegetation. Peoples of high latitude, who, it has been surmised, need all the sun they can get to make vitamin D, tend to have lighter skins than those who face the opposite problem – the carcinogenic rays of the tropical sun. It is plausible that such regional selection would especially affect superficial characteristics like skin colour, while leaving most of the genome intact and uniform.

In theory, that could be the full explanation for our superficial and visible variety, covering deep similarity. But it doesn't seem enough to me. At the very least, I think it might be helped along by an additional suggestion, which I offer tentatively. It takes off from our earlier discussion about cultural barriers to interbreeding. We are indeed a very uniform species if you count the totality of genes, or if you take a truly random sample of genes; but perhaps there are special reasons for a disproportionate amount of variation in those very genes that make it easy for us to *notice* variation, and to distinguish our own kind from others. This would include the genes responsible for externally visible 'labels' like skin colour. Yet again, I want to suggest that this heightened discriminability has evolved by sexual selection, specifically in humans because we are such a culture-bound species. Because our mating decisions are so heavily influenced by cultural tradition, and because our cultures, and sometimes our religions, encourage us to discriminate against outsiders, especially in choosing mates, those superficial differences that helped our ancestors to prefer insiders over outsiders have been enhanced out of all proportion to the real genetic differences between us. No less a thinker than Jared Diamond has supported a similar idea in *The Rise and Fall of the Third Chimpanzee*. And Darwin himself more generally invoked sexual selection in explanation of racial differences.

I want to consider two versions of this theory: a strong and a weak one. The truth could be any combination of the two. The strong theory suggests that skin colour, and other conspicuous genetic badges, evolved actively as discriminators in choosing mates. The weak theory, which can be thought of as leading into the strong version, places cultural differences, such as language and religion, in the same role as geographical separation in the incipient stages of speciation. Once cultural differences have achieved this initial separation, with the consequence that there is no gene flow to hold them together, the groups would subsequently evolve apart genetically, as if geographically separated.

Recall from the Cichlid's Tale that an ancestral population can split into two genetically distinct populations only if given a head start by an initial accidental separation, usually assumed to be geographical. A barrier such as a mountain range reduces gene flow between two populated valleys. So the gene pools in the two valleys are free to drift apart. The separation will normally be abetted by different selection pressures; one valley may be wetter than its neighbour on the other side of the mountains, for instance. But the initial accidental separation, which I have so far assumed to be geographical, is necessary.

Nobody is suggesting that there is anything deliberate about the geographical separation. That isn't what 'necessary' means at all. Necessary just means that, if there didn't happen to be an initial geographical (or equivalent) separation, the various parts of the population would be genetically bound together by sexual mixing between them. Speciation couldn't happen without an initial barrier. Once the two putative species, initially races, have begun to pull apart, genetically speaking, they may then pull even farther apart – even if the geographical barrier subsequently disappears.

There is controversy here. Some people think the initial separation has to be geographical, while others, especially entomologists, emphasise so-called sympatric speciation. Many herbivorous insects eat only one species of plant. They meet their mates and lay their eggs on the preferred plants. Their larvae then apparently 'imprint' on the plant that they grow up eating, and they choose, when adult, the same species of plant to lay their own eggs.* So if an adult female made a mistake and laid her eggs on the wrong species of plant, her daughter would imprint on that wrong plant and would, when the time came, lay her eggs on plants of the same wrong species. Her larvae then would imprint on the same wrong plant, hang around the wrong plant when adult, mate with others hanging around the wrong plant and eventually lay their eggs on the wrong plant.

In the case of these insects, you can see that, in a single generation, gene flow with the parental type could be abruptly cut off. A new species is theoretically free to come into being without the need for geographical isolation. Or – another way of putting it – the difference between two kinds of food plant is, for these insects, equivalent to a mountain range or a river for other animals. It has been argued that this kind of sympatric speciation is commoner among insects than 'true' geographical speciation, in which case, since the majority of species are insects, it could even be that most speciation events are sympatric. Be that as it may, I am suggesting that human culture provides a special way in which gene flow can find itself blocked, which is somewhat analogous to the insect scenario I have just outlined.

In the insect case, plant preferences are handed down from parent to offspring by the twin circumstances of larvae fixating on their food plant, and adults mating and laying eggs on the same food plants. In effect, lineages establish 'traditions' that travel longitudinally down generations. Human traditions are similar, if more elaborate. Examples are languages, religions and social manners or conventions. Children usually adopt the language and the religion of their parents although, just as with the insects and the food plants, there are enough 'mistakes' to make life interesting. Again, as with the insects mating in the vicinity of their preferred food plants, people tend to mate with others speaking the same language and praying to the

---

* Imprinting is the process, often said to have been discovered by Konrad Lorenz, whereby young animals, for instance goslings, take a kind of mental photograph of an object they see during a critical period early in life, and which they follow while young. Usually it will be a parent, but it could be Konrad Lorenz's boots. Later in life, the 'mental photograph' influences choice of mate; this usually means a member of their own species, but they might try to mate with Lorenz's boots. The gosling story isn't as simple as that, but the analogy to the insect case should be clear.

same gods. So different languages and religions can play the role of food plants, or of mountain ranges in traditional geographical speciation. Different languages, religions and social customs can serve as barriers to gene flow. From here, according to the weak form of our theory, random genetic differences simply accumulate on opposite sides of a language or religion barrier, just as they might on opposite sides of a mountain range. Subsequently, according to the strong version of the theory, the genetic differences that build up are reinforced as people use conspicuous differences in appearance as additional labels of discrimination in mate choice, supplementing the cultural barriers that provided the original separation.*

I am certainly not suggesting that humans should be thought of as more than one species. Very much the contrary. What I am suggesting is that human culture – the fact that we depart so strongly from random mating in directions determined by language, religion and other cultural discriminators, has done very odd things to our genetics in the past. Even though, if you take the totality of genes into account, we are a very uniform species, we are astonishingly variable in superficial features which are trivial but conspicuous: discrimination fodder. The discrimination might apply not just to mate choice but to choice of enemies and victims of xenophobic or religious prejudice.

THE FRUIT FLY'S TALE

In 1894 the pioneering geneticist William Bateson published a book called *Materials for the Study of Variation, Treated with Especial Regard to Discontinuity in the Origin of Species.* He compiled a fascinating, almost macabre list of genetic abnormalities, and considered how they might illuminate evolution. He had horses with cloven hooves, antelopes with a single horn in the middle of the head, people with an extra hand, and a beetle with five legs on one side. In his book, Bateson coined the term 'homeosis' for a remarkable type of genetic variation. *Homoio* means 'same' in Greek, and a homeotic mutation (as we would now call it, although 'mutation' had not been coined when Bateson wrote) is one that causes a part of the body to appear in some different part.

* A potential problem, which would need sorting out if the idea were to be pursued, is that the theory of mathematical genetics suggests, for geographical separation and by implication this cultural hypothesis too, that the separation has to be pretty complete for genetic differentiation to be maintained.

Bateson's own examples included a sawfly with a leg growing in the place where an antenna should be. As soon as you hear of this remarkable abnormality, you might suspect, with Bateson, that here must be an important clue to how animals develop normally. You and Bateson would be right, and that is the subject of this tale. That particular homeosis – leg in place of antenna – was later discovered in the fruit fly *Drosophila* and named antennapedia. *Drosophila* ('dew lover') has long been the geneticists' favourite animal. Embryology should never be confused with genetics, but recently *Drosophila* has assumed a starring role in embryology as well as genetics, and this is a tale of embryology.

Embryonic development is controlled by genes, but there are two very different ways in which this might theoretically happen. The Mouse's Tale introduced them as blueprint and recipe. A builder makes a house by placing bricks in positions specified by a blueprint. A cook makes a cake not by placing crumbs and currants in specified positions but by putting ingredients through specified procedures, such as sieving, stirring, beating and heating.* Textbooks of biology are wrong when they describe DNA as a blueprint. Embryos do nothing remotely like following a blueprint. DNA is not a description, in any language, of what the finished body should look like. Maybe on some other planet living things develop by blueprint embryology, but I find it hard to imagine how it would work. It would have to be a very different kind of life. On this planet, embryos follow recipes. Or, to change to another equally un-blueprint-like analogy, which is in some ways more apt than the recipe: embryos construct themselves by following a sequence of origami folding instructions.

The origami analogy fits early embryology better than late. The main organisation of the body is initially laid down by a series of foldings and invaginations of layers of cells. Once the main body plan is safely in place, later stages in development consist largely of growth, as if the embryo were being inflated, in all its parts, like a balloon. It is a very special kind of balloon, however, because different parts of the body inflate at different rates, the rates being carefully controlled. This is the important phenomenon known as allometry. The Fruit Fly's Tale is concerned mostly with the earlier, origami phase of development, not the later, inflationary one.

---

* This favourite analogy was first used by my friend Sir Patrick Bateson, a relative of Sir William, as it happens.

Cells are not laid like bricks to a blueprint, but it is the behaviour of cells that determines embryonic development. Cells attract, or repel, other cells. They change shape in various ways. They secrete chemicals, which may diffuse outwards and influence other cells, even some distance away. Sometimes they die selectively, carving out shapes by subtraction, as if a sculptor were at work. Like termites cooperating to build a mound, cells 'know' what to do by reference to the neighbouring cells with whom they find themselves in contact, and in response to chemicals in gradients of concentration. All cells in the embryo contain the same genes, so it can't be their genes that distinguish one cell's behaviour from another's. What does distinguish a cell is which of the genes are turned on, which usually is reflected in the gene products – proteins – that it contains.

In the very early embryo, a cell needs to 'know' where it lies along two main dimensions: fore and aft (anterior/posterior) and up-down (dorsal/ventral). What does 'know' mean? It initially means that a cell's behaviour is determined by its position along chemical gradients in each of the two axes. Such gradients necessarily start in the egg itself, and are therefore under the control of the mother's genes, not the egg's own nuclear genes. For example, there is a gene called *bicoid* in the *Drosophila* mother's genotype, which expresses itself in the 'nurse' cells that make her eggs. The protein made by the *bicoid* gene is shipped into the egg, where it is concentrated at one end, whence it fades towards the other end. The resulting concentration gradient (and others like it) labels the anterior/posterior axis. Comparable mechanisms at right-angles label the dorsal/ventral axis.

These labelling concentrations persist in the substance of the cells that are produced as the egg subsequently divides. The first few divisions occur without any addition of new material, and the divisions are incomplete: lots of separate nuclei are made, but they are not completely separated by cellular partitions. This multinucleate 'cell' is called a syncitium. Later, partitions form, and the embryo becomes properly cellular. Through all this, as I say, the original chemical gradients persist. It follows that cell nuclei in different parts of the embryo will be bathed in different concentrations of key substances, corresponding to the original two-dimensional gradients, and this will cause different genes to be turned on in different cells (we are now, of course, talking about the embryo's own genes, no longer the mother's). This is how differentiation of cells begins, and projections of the principle lead to further differentiation at later stages of development. The original gradients set up by maternal genes give way to new and more complex gradients set up by the embryo's own genes. Consequent forkings in the lineages of embryonic cells recursively generate further differentiations.

In arthropods there is a larger-scale partitioning of the body, not into cells but into segments. The segments are arrayed in line, from front of head to tip of abdomen. Insects have six head segments, of which the antennae are on segment 2, followed on other segments by the mandibles and then other mouthparts. The segments of the adult head are compressed into a small space, so their fore-and-aft alignment is not too clear, but it can be seen in the embryo. The three thoracic segments (T1, T2 and T3) are more obviously in a line, each bearing a pair of legs. T2 and T3 normally bear wings, but in *Drosophila* and other flies only T2 has wings.* The second pair of 'wings' is modified into halteres, small club-shaped organs on T3, which vibrate and serve as miniature gyroscopes to guide the fly. Some early fossil insects had three pairs of wings, one pair on each of the three thoracic segments. Behind the thoracic segments are a larger number of abdominal segments (11 in some insects, eight in *Drosophila*, depending on how you reckon the genitals at the rear end). Cells 'know' (in the sense already excused) which segment they are in, and they behave accordingly. Each cell is told which segment it is in through the mediation of special control genes called Hox genes, which turn themselves on inside the cell. The Fruit Fly's Tale is mostly a tale of Hox genes.

It would make things neat and easy to explain if I could now tell you that there is one Hox gene for each segment, with all the cells of a given segment having only its own numbered Hox gene turned on. It would be even tidier if the Hox genes were arrayed along the length of a chromosome, in the same order as the segments they influence. Well, it isn't *quite* as tidy as that, but it very nearly is. The Hox genes are indeed arranged in the right order along one chromosome, and that is wonderful – gratuitously so, given what we know of how genes work. But there aren't enough Hox genes for the segments – only eight. And there's a more messy complication that I must get out of the way. The segments of the adult don't exactly correspond to the so-called *para*segments of the larva. Don't ask me why (perhaps the

---

* Some other insects, such as cockroaches and beetles, fly with only T3 wings, having modified the T2 wings into hardened protective wing cases called elytra. Crickets and grasshoppers, as we have heard, further modified the elytra as sound-producing organs.

Designer was having an off day), but each adult segment is made up of the back half of one larval parasegment plus the front half of the next. Unless otherwise stated, I'll use the word segment to mean larval (para) segment. As for the question of how eight Hox genes in a row take charge of some 17 segments in a row, it is partly done by resorting to the chemical gradient trick again. Each Hox gene is mainly expressed in one segment, but it is also expressed, in decreasing concentration as you go backwards, in more posterior segments. A cell knows which segment it is in by comparing the chemical outputs of more than one upstream Hox gene. It is a bit more complicated than that, but there is no need to go into such detail here.

The eight Hox genes are arrayed in two gene complexes, physically separated along the same chromosome. The two are called the Antennapedia Complex and the Bithorax Complex. These names are doubly unfortunate. A complex of genes is named after a single member of that complex, which is no more important than the others. And, worse, the genes themselves are as usual named after what happens when they go wrong, rather than after their normal function. It would be better to call them something like the Front Hox Complex and the Rear Hox Complex. However, we are stuck with the existing names.

The Bithorax Complex consists of the last three Hox genes, which are named, for historical reasons that I shan't go into, *Ultrabithorax*, *Abdominal-A* and *Abdominal-B*. They affect the back end of the animal, as follows. *Ultrabithorax* itself is expressed from segment 8 all the way to the posterior end. *Abdominal-A* is expressed from segment 10 to the end, and *Abdominal-B* is expressed from segment 13 to the end. The products of these genes are made in decreasing concentration gradient as we move towards the back end of the animal, from their various starting points. So, by comparing the concentrations of the products of these three Hox genes, a cell in the posterior part of a larva can tell which segment it is in and act accordingly. It is a similar story for the front end of the larva, where the five Hox genes of the Antennapedia Complex are in charge.

A Hox gene, then, is a gene whose mission in life is to know whereabouts in the body it is, and so inform other genes in the same cell. We are now armed to understand homeotic mutations. When things go wrong with a Hox gene, the cells in a segment are misinformed about which segment they are in, and they make the segment they 'think' they are in. So, for instance, we see a leg growing in the segment that would normally grow an antenna. This makes perfect sense. The cells in any segment are perfectly capable of assembling the anatomy of any other segment. Why should they not? The instructions for making any segment lurk in the cells of every segment. It is the Hox genes, under normal conditions, that call forth the 'correct' instructions for making the anatomy appropriate to each segment. As William Bateson rightly suspected, homeotic abnormality opens a revealing window on how the system normally works.

Recall that flies, unusually among insects, normally have only one pair of wings, plus a pair of gyroscopic halteres. The homeotic mutation *Ultrabithorax* misleads cells in the third thoracic segment into 'thinking' they are in the second thoracic segment. They therefore collaborate to make an extra pair of wings, instead of a pair of halteres (see plate 34). There is a mutant flour beetle (*Tribolium*) in which all 15 segments develop antennae, presumably because the cells all 'think' they are in segment 2.

This brings us to the most wonderful part of the Fruit Fly's Tale. After they had been discovered in *Drosophila*, Hox genes started turning up all over the place: not only in other insects such as beetles, but in almost all other animals that have been looked at, including ourselves. And – this really is almost too good to be true – they very often turn out to be doing the same kind of thing, even down to informing cells which segment they are in and (better still) being arrayed in the same order along chromosomes. Let's now turn to the mammal story, which has been most thoroughly worked out in the laboratory mouse – that *Drosophila* of the mammal world.

Mammals, like insects, have a segmented body plan, or at least a modular, repeated plan that affects the backbone and associated structures. Each vertebra can be thought of as corresponding to one segment, but it isn't just bones that are repeated rhythmically as we move from neck to tail. Blood vessels, nerves, muscle blocks, cartilage discs and ribs, where present, all follow the repetitive, modular plan. As with *Drosophila* the modules, though following the same general plan as each other, are different in detail. And like the insect division into head, thorax and abdomen, vertebrae are grouped into cervical (neck), thoracic (upper back vertebrae with ribs), lumbar (lower back vertebrae without ribs) and caudal (tail). As in *Drosophila*, the cells, whether they are bone cells, muscle cells, cartilage cells or anything else, need to know which segment they are in. And as in *Drosophila*, they know because of Hox genes – Hox genes that recognisably correspond to particular *Drosophila* Hox genes – although, unsurprisingly, given the immensity of time since Concestor 26, they are far from identical. Again as in *Drosophila*, the Hox genes are arranged in the right order on the chromosome. Vertebrate modularity is very different from that of insects, and there is no reason to think that their common ancestor, at Rendezvous 26, was a segmented animal. Nevertheless, the evidence of Hox genes suggests, at the very least, that there is some sort of deep similarity between insect and vertebrate body plans, which was also present in Concestor 26. And, indeed, in other body plans, including those that are not segmented.

In the mouse there is not just one array of Hox genes on one chromosome; there are four separate arrays. The *a* series is on chromosome 6, the *b* series on chromosome 11, *c* on chromosome 15 and *d* on chromosome 2. Resemblances between them show that they have arisen during evolution by duplication: *a4* matches *b4* matches *c4* matches *d4*. There have also been some deletions, for certain slots are missing from each of the four arrays: *a7* and *b7* match each other, but neither the *c* series nor the *d* series has a representative in 'slot' 7. When two, three or four versions of a Hox gene impinge upon one segment, their effects are combined. And, as with *Drosophila*, all mouse Hox genes exert their strongest effect in the first (most anterior) segment in their domain of influence, with a gradient of decreasing expression downstream in more posterior segments.

It gets better. With minor exceptions, each gene from the *Drosophila* array of eight Hox genes resembles its opposite number in the mouse series more than it resembles the other seven genes in the *Drosophila* series. And they are in the same order along their respective chromosomes. Every one of the eight *Drosophila* genes has at least one representative in the mouse series of 13. The detailed gene-for-gene coincidence between *Drosophila* and mouse can only indicate shared inheritance – from Concestor 26, the grand progenitor of all the protostomes and all the deuterostomes. That means the vast majority of animals are descended from an ancestor that had Hox genes arranged in the same linear order as we see in modern *Drosophila* and modern vertebrates. Think of it! Concestor 26 had Hox genes, and in the same order as ours.

As I've already said, it doesn't follow that Concestor 26's body was partitioned into discrete segments. Indeed, it probably wasn't. But there was surely some kind of fore-and-aft gradient running from head to tail that was mediated by the homologous series of Hox genes arranged in the right order along a chromosome. Concestors being dead and beyond the reach of molecular biologists, it now becomes of great interest to look for Hox genes in their modern descendants. Concestor 23 is the ancestor we share with amphioxus. Given that the more distantly related *Drosophila* has the same fore-and-aft series as mammals, it would be positively worrying if amphioxus didn't have it too. My colleague Peter Holland, with his research group, has looked into the matter, and their results are gratifying. Yes, amphioxus's modular body plan is mediated by (14) Hox genes, and yes, they are arranged in the right order along the chromosome. Unlike the mouse, but like *Drosophila,* there is only one series, not four parallel series. Presumably the entire cluster has been duplicated four times somewhere along the line leading from Concestor 23 to modern mammals, followed by some sporadic losses of particular genes.

How about other animals, strategically chosen for what they can tell us about other particular concestors? Hox genes have now been found in every animal that has been looked at except ctenophores and sponges (see Rendezvous 29 and 31 respectively), including sea urchins, *Limulus,* shrimps, molluscs, annelid worms, acorn worms, sea squirts, nematode worms and flatworms. This much we could have guessed, knowing that all these animals are descended from Concestor 26, and we already have good reason to think Concestor 26 had Hox genes, like its descendants *Drosophila* and mouse.

Cnidarians, such as *Hydra* (they aren't due to join us until Rendezvous 28), are radially symmetrical – they don't have an anterior/posterior axis, nor a dorsal/ventral one. They have an oral/aboral (mouth versus opposite-to-mouth) axis. It isn't obvious what, if anything, corresponds to their long axis, so what might we expect their Hox genes to do? It would be tidy if they used them to define the oral/aboral axis, but so far it isn't clear that this is so. Most cnidarians only have two Hox genes anyway, to pit against *Drosophila*'s eight and amphioxus's fourteen. It is agreeable that one of these two genes resembles the anterior complex of *Drosophila,* while the other resembles the posterior complex. Concestor 28, the one we share with them, presumably had the same. Then one of the two duplicated several times during evolution to produce the

Antennapedia Complex, while the other one duplicated within the same animal lineage to produce the Bithorax Complex. That is exactly the kind of way genes increase in the genome (see the Lamprey's Tale). But more research is needed before we shall know what, if anything, the two genes are doing in the planning of the cnidarian body.

Echinoderms are radially symmetrical, like cnidarians, but secondarily so. Concestor 25, which they share with us vertebrates, was bilaterally symmetrical, like a worm. Echinoderms have a variable number of Hox genes – ten in the case of sea urchins. What are these genes doing? Does a relic of the ancestral anterior/posterior axis lurk within the body of a starfish? Or do the Hox genes exert their influence successively along the length of each of the five arms? This might seem to make sense. We know that Hox genes express themselves in the arms, and the legs, of mammals. I don't mean that the arrays of Hox genes from 1 to 13 express themselves in order, from shoulder to fingertips. It is more complicated than that – as well it might be – because the vertebrate limb is not arranged in modules that succeed each other along its length. Instead, there is first one bone (the humerus in the arm, femur in the leg), then two bones (radius and ulna in the arm, tibia and fibula in the leg), then lots of little bones culminating in the fingers and toes. This fan arrangement, inherited from the more obvious fan of our fishy ancestors' fins, doesn't lend itself to straightforward Hox linearity. Even so, Hox genes are involved in the limb development of vertebrates.

By analogy, it would not be surprising if Hox genes were also expressed in starfish or brittle star arms (and even sea urchins can be thought of as starfish who have curled their arms up in a five-pronged arch, meeting at the tips and zipped together down the sides). Moreover, starfish arms, unlike our arms or legs, really are serially modular along their length. The tube feet, with all their associated hydraulic plumbing, are units that repeat in two parallel rows along the length of each arm: just the thing for Hox gene expression! Brittle star arms even look and behave like five worms.

T. H. Huxley referred to 'The great tragedy of science – the slaying of a beautiful hypothesis by an ugly fact.' The true facts about echinoderm Hox genes may not be ugly, but they don't follow the pretty pattern I have just suggested. Something else happens, which has its own rather surprising beauty. Echinoderm larvae are tiny, bilaterally symmetrical swimmers in the plankton. The five-way radially symmetrical bottom-dwelling adult doesn't develop as a transformation of the larva. Instead, it starts as a tiny miniature adult *inside* the larva's body, which grows until eventually the rest of the larva is discarded. Hox genes are expressed in the correct linear order, but not along each arm. Instead the order of expression follows a roughly circular route *around* the baby adult. If we think of the Hox axis as a 'worm', there are not five 'worms', one for each arm. There is a single 'worm' curled around inside the larva. The front end of the 'worm' sprouts arm number 1, the back end of the 'worm' sprouts arm number 5. Homeotic mutations in starfish, then, might be expected to grow too many arms. And, sure enough, mutant starfish with six arms are known, and were recorded in Bateson's book. There are also some species of starfish that have much larger numbers of arms, and they have presumably evolved from homeotic mutant ancestors.

Hox genes have not been found in plants, nor in fungi, nor in the single-celled organisms we used to call protozoa. But now we come to a complication of terminology that must be dealt with before we go any further. 'Hox' was coined as a contraction of 'homeobox', but Hox genes are not synonymous with homeobox genes: they are a subset. Plants and fungi do have homeobox genes but they don't have Hox genes. They must have systems of control genes and chemical gradients, in order to grow into the correct shape. 'MADS box' genes determine the embryology of flowers, and can produce homeotic mutations in flowers in the same way as Hox genes for animals.

'Homeo' comes from Bateson's 'homoeosis', and 'box' refers to a 'box' of 180 code letters that all genes known as homeobox genes have somewhere in their length. The homeobox itself is this diagnostic sequence of 180 code letters, and a 'homeobox gene' is a gene that contains the homeobox sequence somewhere in its length. The name Hox is used not for all homeobox genes but only for the linear arrays of genes that determine position along the length of an animal's body and which have turned out to be homologous in nearly all animals.

The Hox family of homeobox genes was the first to be discovered, but now lots of related families are known. For example, there's a family of genes called ParaHox that was first clearly defined in amphioxus, but which again occurs in all animals except (so far) ctenophores and sponges. It seems that the ParaHox genes are 'cousins' of the Hox genes, in the sense that they correspond to, and are arranged in the same order as, Hox genes. They certainly arose by duplication from the same ancestral set of genes as the Hox genes. Other homeobox genes are more distantly related to Hox and Para-Hox, but form families of their own. The *Pax* family is found in all animals. A particularly notable member of this family is *Pax6*, which corresponds to the gene known as *ey* in *Drosophila*. I've already mentioned that *Pax6* is responsible for telling cells to make eyes. The same gene makes eyes in animals as different as *Drosophila* and mouse, even though the eyes produced are radically different in the two animals. In a similar way to Hox genes, *Pax6* doesn't tell cells *how* to make an eye. It only tells them that here is the *place* to make an eye.

A rather parallel example is the small family of genes called *tinman*. Again *tinman* genes are present in both *Drosophila* and mice. In *Drosophila, tinman* genes are responsible for telling cells to make a heart, and they normally express themselves in just the right place to make a *Drosophila* heart. As we have by now come to expect, *tinman* genes are also involved in telling mouse cells to make a heart in the right place for a mouse's heart.

The whole set of homeobox genes constitutes a very large number, divided into families and subfamilies just as animals themselves are divided into families and subfamilies. It is like the case of haemoglobin, which we examined in the Lamprey's Tale. There we learned that human alpha globin is truly a closer cousin to, say, lizard alpha globin than it is to human beta globin – which is in turn a closer cousin to lizard beta globin. Similarly, human *tinman* is a closer cousin to fruit fly *tinman* than it is to human *Pax6*. It is possible to construct a very full family tree of homeobox genes that exists side by side with the family tree of the animals that contain them. Both family trees are equally valid. Both are true trees of ancestry, formed by splitting events that happened at particular moments in geological history. In the case of the animal family trees, the splitting events are speciations. In the case of the homeobox gene family trees (or the globin genes), the splitting events are gene duplications within genomes.

The tree of animal homeobox genes splits into two great classes, the AntP and the PRD classes. I shall not spell out what these abbreviations stand for, because both are perversely confusing. The PRD class includes the *Pax* genes, and various other subclasses. The AntP class includes the Hox and ParaHox, and again various other subclasses. In addition to these two great classes of animal homeobox genes, there are various more distantly related homeobox genes that are (misleadingly) called 'divergent'. These are found not only in animals but in plants, fungi and 'protozoa' as well.

Only animals have true Hox genes, and they are always used in the same kind of way – to specify information about position in the body, whether or not the body is neatly divided into discrete segments. Although Hox genes have not yet been found in sponges or ctenophores, this doesn't mean they won't be. It would not be surprising to find that all animals have them. This would encourage my colleagues Jonathan Slack, Peter Holland and Christopher Graham, then all at Oxford, who proposed a new definition of the very word 'animal'. Hitherto, animals were defined as opposed to plants, in a rather unsatisfactorily negative way. Slack, Holland and Graham suggested a positive, specific criterion that has the effect of uniting all animals and excluding all non-animals, such as plants and protozoa. The Hox story shows that animals are not a highly varied, unconnected miscellany of phyla, each with its own fundamental body plan acquired and maintained in lonely isolation. If you forget morphology and look only at the genes, it emerges that all animals are minor variations on a very particular theme. What delight to be a zoologist at such a time.

THE ROTIFER'S TALE

The brilliant theoretical physicist Richard Feynman is rumoured to have said, 'If you think you understand quantum theory, you don't understand quantum theory.' I am tempted by an evolutionist's equivalent: 'If you think you understand sex, you don't understand sex.' The three modern Darwinians from whom I believe we have the most to learn – John Maynard Smith, W. D. Hamilton and George C. Williams – all devoted substantial parts of their long careers to wrestling with sex. Williams began his 1975 book *Sex and Evolution* with a challenge to himself: 'This book is written from a conviction that the prevalence of sexual reproduction in higher plants and animals is inconsistent with current evolutionary theory . . . there is a kind of crisis at hand in evolutionary biology . . .' Maynard Smith and Hamilton said similar things. It is to resolve this crisis that all three Darwinian heroes, along with others of the rising generation, laboured. I shall not attempt an account of their efforts, and certainly

I have no rival solution to offer myself. Instead, the Rotifer's Tale displays an under-explored *consequence* of sexual reproduction for our view of evolution.

Bdelloidea is a large class of the phylum Rotifera (see page 395). The existence of the bdelloid rotifers is an evolutionary scandal (see plate 35). Not my own *bon mot* – it rings with the unmistakable tones of John Maynard Smith. Many rotifers reproduce without sex. In this respect they resemble aphids, stick insects, various beetles and a few lizards, and are not particularly scandalous. What stuck in Maynard Smith's craw is that the bdelloids *as a whole* reproduce only asexually – every last one of them, evidently descended from a bdelloid common ancestor that must have lived long enough ago to beget 18 genera and 360 species. Remains in amber suggest that this male-spurning matriarch lived at least 40 million years ago, very probably more. The bdelloids are a highly successful group of animals, astonishingly numerous and a dominant part of the freshwater faunas of the world. Not a single male has ever been found.*

What is so scandalous about that? Well, suppose we take a family tree of the whole animal kingdom. The tips of the twigs, all round the surface of the tree, represent species. Major boughs represent classes or phyla. There are millions of species, which means that the evolutionary tree is far more intricately branched than any woody tree you will ever see. There are only a few tens of phyla, and not that many more classes. The phylum Rotifera is one branch of the tree and it splits into four sub-branches, of which the class Bdelloidea is one. This class branch subdivides, and subdivides further, eventually to yield 360 twigs each representing one species. The same kind of thing is going on for all the other phyla, each with its own classes and so on. The outer twigs of the tree represent the present; slightly inward from the outermost shell of twigs represents a short distance into the past; and so on to the main trunk which represents, say, a billion years ago.

---

* To be scrupulously correct, in nearly 300 years of scientific investigation, there has been a single report of a male bdelloid rotifer made by the Danish zoologist C. Wesenberg-Lund (1866–1955). 'With great hesitation I venture to remark, that twice I saw among the thousands of Philodinidae (*Rotifer vulgaris*) a little creature, unquestionably a male rotifer . . . but both times I failed to get it isolated. It moves round and between the numerous females with extreme rapidity' (understandably, no doubt). Even before the strong evidence of Mark Welch and Meselson (see page 440) zoologists were not inclined to take Wesenberg-Lund's never-repeated observation as sufficient proof of the existence of male bdelloids.

That understood, we now take paint to our grey, wintry tree and colour the tips of certain twigs to label particular features. We might paint red all twigs that represent flying animals – powered flying as opposed to passive gliding, which is much more common. If we now step back and contemplate the whole tree, we shall notice large areas of red separated by even larger areas of grey representing whole major groups of animals that don't fly. Most of the insect twigs, the bird twigs and the bat twigs are red, and they are neighbours of other red twigs. None of the others are. With a few exceptions like fleas and ostriches, three entire classes are made up of flying animals. Redness is distributed in broad patches of uniform red against uniform grey.

Think what this means for evolution. The three patches of red must have begun long ago with three ancestral animals – an early insect, an early bird and an early bat – discovering how to fly. Flying obviously turned out to be a really good idea once it had been discovered, because it persisted and spread through all the descendant branches as the three species eventually gave rise to three large collections of descendant species, almost all of them persisting with their ancestor's ability to fly: the class Insecta, the class Aves and the order Chiroptera.

But now we do the same thing not for flying but for reproducing asexually, without males. (Never without females, by the way. Unlike eggs, sperms are too small to go it alone. Asexual reproduction in animals means dispensing with males.) On our tree of life, we paint the twigs of all asexually reproducing species blue. Now we notice a completely different pattern. Where flying showed up as great swathes of red, asexual reproduction shows up as tiny sporadic dots of blue. An asexual beetle species shows up as a blue twig completely surrounded by grey. Maybe a group of three species in one genus are blue, but neighbouring genera are grey. You see what this means? Asexual reproduction arises from time to time, but it rapidly goes extinct before it has had time to grow into a stout branch with many blue twigs. Unlike flying, the habit of reproducing asexually doesn't persist long enough to give rise to a whole family, order or class of asexual creatures.

With one scandalous exception! Unlike all the other little blips of blue, the bdelloid rotifers are a hefty patch of uninterrupted blue, enough to make the proverbial sailor's trousers. In evolution, what this seems to mean is that an ancestral bdelloid discovered asexual reproduction, just like the odd beetle we talked about. But whereas asexual beetles and hundreds of other asexual species dotted around the tree go extinct long before they can evolve into a larger grouping such as a family or order, let alone a class, the bdelloids seem to have stuck with asexuality and flourished with it through enough evolutionary time to generate an entire asexual class, now numbering 360 species. For the bdelloid rotifers, but not convincingly for any other kind of animal, asexual reproduction is like flying. It seems to be a good and successful innovation for the bdelloids, while in all other parts of the tree it constitutes a fast track to extinction.

The statement that there are 360 species raises a problem. The biological definition of a species is a group of individuals that interbreed with each other and not with others. The bdelloids, being asexual, don't interbreed with anybody; every one is an isolated female, each of whose descendants goes her own sweet way, in genetic isolation from every other individual. So when we say there are 360 species, we can only mean there are 360 types, which we humans recognise as looking sufficiently different from each other that we would expect, if they bred sexually at all, to see them shunning the other types as sexual partners.

Not everyone accepted that the bdelloid rotifers really were asexual. There is a large gap in logic between the negative statement that males have never been seen, and a positive conclusion that there aren't any. As Olivia Judson recounts in her winningly sophisticated zoological comedy *Dr Tatiana's Sex Guide to All Creation,* naturalists have been caught like that before. Apparently asexual species often turn out to have concealed males. The males of certain angler fish are tiny dwarfs who ride about as parasites on the bodies of females. Had they been even smaller, we might have overlooked them altogether, as nearly happened in the case of certain scale insects where the males are, in the words of my colleague Laurence Hurst, 'wee things that stick to the females' legs'. Hurst went on to quote his mentor Bill Hamilton's astute remark:

> How often do you see humans having sex? If you were a Martian looking around, you'd be pretty sure we were asexual.

So it would be nice to come up with some more positive evidence that the bdelloid rotifers really are anciently asexual. Geneticists are becoming increasingly ingenious in the art of reading the patterns of gene distribution in modern animals and making inferences about their evolutionary history. In Eve's Tale, we met Alan Templeton's method of reconstructing early human migrations by picking up 'signals' in the genes of living people. The logic is not deductive. We don't deduce from modern genes that the course of history must have been so-and-so. Instead we say, if the course of history was so-and-so, we should expect to see such-and-such a pattern of gene distribution today. That is what Templeton did for human migrations, and something similar has been done for bdelloid rotifers by David Mark Welch and Matthew Meselson of Harvard University. Mark Welch and Meselson used genetic signals to make inferences not about migration but about asexual reproduction. Again, the form of their logic is not deductive. Instead they reasoned that, if the bdelloids had been purely asexual for many millions of years, we should expect the genes of living bdelloids to show a certain pattern.

What pattern? Mark Welch and Meselson's reasoning was ingenious. First you must understand that the bdelloid rotifers, though asexual, are diploid. That is to say, they are like all sexually reproducing animals in having two copies of each chromosome. The difference is that the rest of us reproduce by making eggs or sperms that have only one copy of each chromosome. The bdelloids produce eggs that have both copies of each chromosome. So a bdelloid's egg cell is like any of her other cells, and a daughter is an identical twin of her mother, give or take the odd mutation. It is these odd mutations that, over the millions of years, have gradually built up in diverging lines, presumably under natural selection, to produce the 360 species that we see today.

An ancestral female, whom I shall call the gynarch, mutated in such a way as to dispense with males and with meiosis, and substitute mitosis as the method of producing eggs.* From then on, throughout the cloned population of females, the fact that the chromosomes were originally paired became irrelevant. Instead of five pairs of chromosomes (or whatever the number was – the equivalent of 23 in us), there were now ten chromosomes – each linked to its erstwhile partner only by a kind of receding memory. Chromosome partners used to meet up and exchange genes every time a rotifer made eggs or sperm. But during the millions of years since the gynarch drove out the males and founded the bdelloid gynodynasty, every chromosome has been drifting apart from its erstwhile partner

---

* Meiosis is the special form of cell division that halves the number of chromosomes in order to make sex cells. Mitosis is the ordinary form of cell division used for making body cells, which duplicates all the chromosomes of a cell.

genetically, as their genes have mutated independently of one another. And this has happened even though they have shared cells in shared bodies all that time. In the good old days of males and sex, it didn't happen. In every generation each chromosome paired off with its opposite number and exchanged genes before making eggs or sperm. This held the pairs of chromosomes in a kind of intermittent embrace, preventing them from drifting apart in their gene content.

You and I have 23 pairs of chromosomes. We have two chromosome 1s, two chromosome 5s, two chromosome 17s and so on. With the exception of the X and Y sex chromosomes, there is no consistent difference between members of a pair. Since they exchange genes every generation, the two chromosome 17s are just chromosome 17s and there is no point in calling them, say, left chromosome 17 and right chromosome 17. But from the moment the rotifer gynarch froze her genome, all that changed. Her left chromosome 5 was passed intact to all her daughters, as was her right chromosome 5, and never the twain met for more than 40 million years. Her 100-greats-grand-daughters still had a left chromosome 5 and a right chromosome 5. Although by then they would have picked up some mutations, all the left chromosomes would be identifiable by their resemblance to each other, inherited from the gynarch's left chrosomome 5.

There are now 360 species of bdelloids, all descended from the gynarch and separated from her by exactly the same length of time as each other. All individuals of all species still have a left and a right copy of each chromosome, inherited with plenty of mutational change down the line, but no gene swapping across from left to right. Each left and right pair within each individual will now be far more different from one another than you'd expect if there had been any sexual activity at all, at any time in their ancestry since the days of the gynarch. They may even be approaching the time when you can no longer recognise that they were originally paired at all.

But now suppose we compare two modern species of bdelloid rotifers, say *Philodina roseola* and *Macrotrachela quadricornifera.* Both belong to the same subgroup of bdelloids, the Philodinidae, and they certainly have a common ancestor who lived much more recently than the gynarch. Given no sex, there has been exactly the same time for the 'left' and the 'right' chromosomes within every individual of either species to drift apart – the time since the gynarch. Left will be very different from right within every individual. But if you compare, say, left chromosome 5 of *Philodina roseola* with left chromosome 5 of *Macrotrachela quadricornifera,* you should find them pretty similar because they haven't had very long to pick up independent mutations. And right compared with right will also yield few differences. We arrive at the remarkable prediction that a chromosomal comparison across once-paired chromosomes *within* individuals should yield a greater difference than a cross-species comparison – of 'left' with 'left', or of 'right' with 'right'. The longer the time since the gynarch, the greater the difference. Given sex, the prediction would be precisely the opposite, essentially because there is no such thing as 'left' or 'right' identity across species, and plenty of gene-swapping between paired chromosomes within species.

Mark Welch and Meselson used these opposite predictions to test the theory that the bdelloids really have been sexless and male-less for a very long time – with stunning success. They looked at modern bdelloids to see whether it was indeed true that paired chromosomes (or chromosomes that had once been paired) were much more unlike each other than they 'should' be, gene for gene, if sexual recombination had been holding them together. They used other rotifers, non-bdelloids who do have sex, as a control for comparison. And the answer is yes. Bdelloid chromosomes are far more different from their pairs than they 'should' be, by an amount compatible with the theory that they gave up sex not just 40 million years ago, which is the age of the oldest amber in which bdelloids have been found, but about 80 million years ago. Mark Welch and Meselson scrupulously bend over backwards to discuss possible alternative interpretations of their results, but these are far-fetched, and I think they are right to conclude that the bdelloid rotifers really are anciently, continuously, universally and successfully asexual. They really are an evolutionary scandal. For perhaps 80 million years they have flourished by doing something that no other group of animals can get away with, except for very short periods before going extinct.

Why would we normally expect asexual reproduction to lead to extinction? Well, that is a big question because it amounts to the question of what is good about sex itself – the question that better scientists than I have spent book after book failing to answer. I shall just point out that the bdelloid rotifers are a paradox within a paradox. In one way, they are like the soldier in the marching platoon whose mother cried out, 'There goes my boy – he's the only one in step.' Maynard Smith called them an evolutionary scandal, but he was the one mainly responsible for pointing out that sex itself, on the face of it, is the evolutionary scandal. At least a naive view of Darwinian theory would predict that sex should be heavily disfavoured by natural selection, outcompeted twofold by asexual reproduction. In that sense the Bdelloidea, far from being a scandal, appear to be the only soldier in step. Here's why.

The problem is the one Maynard Smith dubbed the twofold cost of sex. Darwinism, in its modern form, expects that individuals will strive to pass on as many of their genes as possible. So isn't it just daft to throw half your genes away with every egg or sperm you make, in order to mix the other half with the genes of somebody else? Wouldn't a mutant female who behaves like a bdelloid rotifer, and passes on 100 per cent of her genes to every offspring instead of 50 per cent, do twice as well?

Maynard Smith added that the reasoning breaks down if the male partner works hard, or contributes economic goods, in such a way that a couple can rear twice as many offspring as an asexual loner. In that case, the twofold cost of sex is cancelled out by a doubling in the number of offspring. In a species such as an emperor penguin, with male and female parent contributing approximately equally towards the labour and other costs of childrearing, the twofold cost of sex is abolished, or at least mitigated. In species where economic and labour contributions are unequal, it is nearly always the father who shirks, devoting his energies instead to duffing up other males. This magnifies the cost of sex, up to the full twofold penalty of the original reasoning. This is why Maynard Smith's alternative name, the twofold cost of males, is preferable. In this light – which Maynard Smith himself was largely responsible for shining – it isn't the bdelloid rotifers who are the evolutionary scandal but everything else. More pertinently, the male sex is an evolutionary scandal. Except that it does exist and, indeed, is almost universal throughout the animal kingdom. What is going on? As Maynard Smith wrote, 'One is left with the feeling that some essential feature of the situation is being overlooked.'

The twofold cost is the starting point for masses of theorising by Maynard Smith, Williams, Hamilton and many younger colleagues. The widespread existence of males who don't earn their keep as fathers must mean that there really are very substantial Darwinian benefits to sexual recombination itself. It is not too difficult to think of what they might be in qualitative terms, and lots of possible benefits, some obvious, some esoteric, have been proposed. The problem is to think of a benefit of sufficient quantitative *magnitude* to counteract the massive twofold cost.

To do justice to all the theories would take a book – it has already taken several, including the seminal works I have previously mentioned by Williams and Maynard Smith, and Graham Bell's beautifully written tour de force *The Masterpiece of Nature.* Yet no definitive verdict has emerged. A nice book aimed at a non-specialist audience is Matt Ridley's *The Red Queen.* Though primarily favouring one of the theories on offer, W. D. Hamilton's theory that sex serves an unceasing arms race against parasites, Ridley does not neglect to explain the problem itself and the other answers to it. As for me, I shall swiftly recommend Ridley's book and the others before going straight to the main purpose of this tale, which is to draw attention to an under-appreciated *consequence* of the evolutionary invention of sex. Sex brought into existence the gene pool, made meaningful the species, and changed the whole ball game of evolution itself.

Think what evolution must look like to a bdelloid rotifer. Think how different the evolutionary history of those 360 species must have been from the normal pattern of evolution. We portray sex as raising diversity and so, in a sense, it does: that is the basis of most theories of how sex overcomes its twofold cost. But, paradoxically, it also has a seemingly opposite effect. Sex normally acts as a kind of barrier to evolutionary divergence. Indeed, a special case of this was the basis of Mark Welch and Meselson's research. In a population of mice, say, any tendency to strike out in some enterprising new evolutionary direction is held in check by the swamping effect of sexual mixing. The genes of the would-be enterprising diverger are swamped into conformity by the inertial mass of the rest of the gene pool. That is why geographical isolation is so important to speciation. It takes a mountain range or a difficult sea crossing to allow a newly striking-out lineage to evolve its own way without being dragged back to the inertial norm.

Think how different evolution must be for the bdelloid rotifers. Far from being swamped into normalcy by the gene pool, they don't even *have* a gene pool. The very idea of a gene pool has no meaning if there is no sex.* 'Gene pool' is a persuasive metaphor because the genes of a sexual population are being continually mixed and

---

* People sometimes confusingly say gene pool when they mean genome. The genome is the set of genes within one individual. The gene pool is the set of all genes in all the genomes of a sexually breeding population.

diffused, as if in a liquid. Bring in the time dimension, and the pool becomes a river, flowing through geological time – an image that I developed in *River out of Eden*. It is the binding effect of sex that provides the river with its limiting banks, channelling the species into some kind of evolutionary direction. Without sex, there would be no coherently channelled flow, but a shapeless outward diffusion: less like a river than like a smell, wafting out in all directions from some point of origin.

Natural selection presumably takes place among the bdelloids, but it must be a very different kind of natural selection from the one the rest of the animal kingdom is accustomed to. Where there is sexual mixing of genes, the entity that is carved into shape by natural selection is the gene pool. Good genes tend statistically to help the individual bodies in which they find themselves to survive. Bad genes tend to make them die. In sexually reproducing animals, it is the deaths and reproductions of individual animals that constitute the immediate selective events, but the long-term consequence is a change in the statistical profile of genes in the gene pool. So, it is the gene pool, as I say, that is the object of the Darwinian sculptor's attention.

Moreover, genes are favoured for their capacity to co-operate with other genes in building bodies. That is why bodies are such harmonious engines of survival. The right way to look at this, given sex, is that genes are continually being tried out against different genetic backgrounds. In every generation, a gene is shuffled into a new team of companions, meaning the other genes with which it shares a body on any particular occasion. Genes that are habitually good companions, fitting in well with others and co-operating well with them, tend to be in winning teams – meaning successful individual bodies that pass them on to offspring. Genes that are not good co-operators tend to make the teams in which they find themselves become losing teams – meaning unsuccessful bodies that die before reproducing.

The proximal set of genes with which a gene has to co-operate are the ones with which it shares a body – this body. But in the long term, the set of genes with which it has to co-operate are all the genes of the gene pool, for they are the ones that it repeatedly encounters as it hops from body to body down the generations. This is why I say it is the gene pool of a species that is the entity sculpted into shape by the chisels of natural selection. Proximally, natural selection is the differential survival and reproduction of whole individuals – the individuals that the gene pool throws up as samples of what it can do. Once again, none of this could be said of the bdelloid rotifers. Nothing like the sculpting of the gene pool goes on, for there is no gene pool to sculpt. A bdelloid rotifer has just one big gene.

What I have just called attention to is a consequence of sex, not a theory for the benefit of sex, nor a theory of why sex arose in the first place. But if I ever were to attempt a theory of the benefit of sex; if I were ever to essay a serious assault on the 'essential feature of the situation that is being overlooked', it is hereabouts that I would start. And I would listen again and again to the Rotifer's Tale. These tiny, obscure denizens of puddles and mossy moisture may hold the key to the outstanding paradox of evolution. What's wrong with asexual reproduction, if the bdelloid rotifers have run with it for so long? Or, if it's right for them, why don't the rest of us do it and save the massive twofold cost of sex?

THE BARNACLE'S TALE

When I was at boarding school, it was occasionally necessary to apologise to the housemaster for being late for dinner: 'Sorry I'm late, sir: orchestra practice,' or whatever the excuse might be. On those occasions when there really was no good excuse and we had something to hide, we formed the habit of murmuring, 'Sorry I'm late sir: barnacles.' He always nodded kindly, and I don't know whether he ever physical wondered what this mysterious out-of-school activity might be. It is possible that we were inspired by the example of Darwin, who devoted years of his life to barnacles so single-mindedly that his children were moved to ask, in innocent puzzlement after being shown round the house of some friends, 'Then where does [your father] do his barnacles?' I'm not sure that we knew the Darwin story then, and I suspect that we invented the excuse because there is something about barnacles that seems too implausible to be a bluff. Barnacles are not what they seem. That applies to other animals too. And it is the theme of the Barnacle's Tale.*

---

* The great scientist J. B. S. Haldane offered a completely different Barnacle's Tale, a parable in which philosophical barnacles contemplate their world. Reality, they conclude, is everything they can reach with their filtering arms. They are dimly aware of 'visions', but doubt their physical reality because barnacles on different parts of the rock disagree as to their distance and shape. This clever allegory on the limitations of human thought and the growth of religious superstition is Haldane's tale, not mine, and I shall merely recommend it and pass on. It is in the eponymous essay of *Possible Worlds*.

Contrary to all appearances, barnacles are crustaceans. The ordinary acorn barnacles, which encrust the rocks like miniature limpets, helping your shoes not to slip if you have them and hurting your feet if you don't, are completely unlike limpets internally. Inside the shell, they are distorted shrimps lying on their backs, kicking their legs in the air. Their feet bear feathery combs or baskets with which they filter particles of food out of the water. Goose barnacles do the same thing, but instead of sheltering under a conical shell like an acorn barnacle, they sit on the end of a stout stalk. They get their name from yet another misunderstanding of the true nature of barnacles. Their wet filtering 'feathers' give them the appearance of a baby bird in its egg. In the days when people believed in spontaneous generation, a folk belief grew that goose barnacles hatched into geese, specifically *Branta leucopsis,* the barnacle goose.

Most deceptive of all – indeed probably holding the record for animals not looking remotely like the thing that zoologists know them to be – are the parasitic barnacles, such as *Sacculina. Sacculina* is not what it seems with a vengeance. Zoologists would never have realised that it is in fact a barnacle, but for its larva. The adult is a soft sac that clings to the underside of a crab and sends long, branching, plant-like roots inside to absorb nourishment from the crab's tissues. The parasite not only doesn't look like a barnacle, it doesn't look like a crustacean of any kind. It has completely lost all trace of the armour plating, and all trace of the bodily segmentation that nearly all other arthropods have. It might as well be a parasitic plant or fungus. Yet, in terms of its evolutionary relationships, it is a crustacean, and not just a crustacean but specifically a barnacle. Barnacles are indeed not what they seem.

Fascinatingly, the embryological development of *Sacculina*'s extraordinarily uncrustacean-like body is starting to be understood in terms of the kind of Hox genes that were the subject of the Fruit Fly's Tale. The gene called *Abdominal-A,* which normally supervises the development of a typical crustacean abdomen, is not expressed in *Sacculina.* It looks as though you can turn a swimming, kicking, leggy animal into a shapeless fungoid just by suppressing Hox genes.

By the way, *Sacculina*'s branching root system is not indiscriminate in its invasion of the crab's tissues. It heads first for the crab's reproductive organs, which has the effect of castrating the crab. Is this just an accidental by-product? Probably not. Castration not only sterilises the crab. Like a fat bullock, the castrated crab, instead of

Weird wonder? A whole new Bauplan?
Female *Thaumatoxena andreinii*. Drawing by Henry Disney.

concentrating on becoming a lean, mean, reproducing machine,
diverts resources towards getting larger: more food for the parasite.[*]

To lead into the final tale of this cluster, here's a little fable set in
the future. Half a billion years after vertebrate and arthropod life
completely perished in the mother of all comet collisions, intelligent
life has eventually re-evolved in remote descendants of octopuses.
Octopoid palaeontologists come upon a rich fossil bed dating from
the twenty-first century A D. Not a fair cross-section of contemporary
life, this bounteous shale nevertheless impresses the palaeontologists
with its variety and diversity. Carefully weighing the fossils up with
eight-arm balanced judgement, and expertly sucking the details,
one octopodan scholar goes so far as to suggest that life, during this
pre-catastrophe dawn age, was more extravagantly profligate in its
diversity than it ever would be again, throwing up weird and
wonderful new body plans in gleeful experimentation. You can see
what he means by thinking of your own animal contemporaries and
imagining that a small sampling of them fossilises. Think of the
herculean task facing our future palaeontologist, and empathise with
his difficulties in trying to discern their affinities from imperfect and
sporadic fossil traces.

Just to take one example, how on earth would you classify the
animal above? Evidently a new 'weird wonder', probably deserving to
have a previously unnamed phylum coined in its honour? A whole
new Bauplan, hitherto unknown to zoology?

Well, no. To return from futuristic fantasy to the present, this
weird wonder is actually a fly, *Thaumatoxena andreinii*. Not only
that, it is a fly that belongs to the perfectly respectable family

---

[*] I went to town on such cases of parasites subtly manipulating the intimate physiology of
their hosts in the parasite chapter of *The Extended Phenotype*.

How a fly ought to be? Phorid fly, *Megaselia scalaris* (Loew). Drawing by Arthur Smith.

Phoridae. A more typical member of the Phoridae is pictured above, *Megaselia scalaris.*

What happened to *Thaumatoxena*, the 'weird wonder', is that it took up residence in a termite nest. The demands of life in that claustral world are so different that – probably in rather a short time – it lost all resemblance to a fly. The boomerang-shaped front end is what is left of the head. Then comes the thorax, and you can see the remains of the wings tucked in between the thorax and the abdomen, which is the hairy bit at the back.

The moral is that of the barnacle again. But the parable of the palaeontologist of the future, and his seduction by the rhetoric of weird wonders gleefully carousing in morphospace, was not idly spun. It was intended as a softening-up for the next tale, which is all about the 'Cambrian Explosion'.

THE VELVET WORM'S TALE

If modern zoology admits of anything approaching a full-blown origin myth, it is the Cambrian Explosion. The Cambrian is the first period of the Phanerozoic Eon, the last 545 million years, during which animal and plant life as we know it suddenly became manifest in fossils. Before the Cambrian, fossils were either tiny traces or enigmatic mysteries. From the Cambrian onwards, there has been a clamorous menagerie of multicellular life, more or less plausibly presaging our own. It is the suddenness with which multicellular fossils appear at the base of the Cambrian that prompts the metaphor of explosion.

Creationists love the Cambrian Explosion because it seems, to their carefully impoverished imaginations, to conjure a sort of palaeontological orphanage inhabited by parentless phyla: animals without antecedents, as if they had suddenly materialised overnight from nothing, complete with holes in their socks.* At the other extreme, romantically overheated zoologists love the Cambrian Explosion for its aura of Arcadian Dreamtime, a zoological age of innocence in which life danced to a frenzied and radically different evolutionary tempo: a prelapsarian bacchanalia of leaping improvisation before a fall into the earnest utilitarianism that has prevailed since. In *Unweaving the Rainbow* I quoted the following words of a distinguished biologist who may, by now, have thought better of it:

> Soon after multicelled forms were invented, a grand burst of evolutionary novelty thrust itself outward. One almost gets the sense of multicellular life gleefully trying out all its possible ramifications, in a kind of wild dance of heedless exploration.

If there is one animal, more than any other, that stands for this feverish vision of the Cambrian, it is *Hallucigenia*. Stands? Hallucinations apart, you might suspect that such an unlikely creature never stood in its life. And you would be right. It seems that *Hallucigenia* – and Simon Conway Morris chose its name advisedly – was originally reconstructed upside down. That is why it stands on improbably spiky toothpick stilts. The single row of 'tentacles' along the back were legs, according to the more recent, inverted interpretation. A single row of legs – did it balance as if on a tightrope? No, new fossils discovered in China suggest a second row, and modern reconstructions look as though they might just have been at home in the real world (see opposite page). *Hallucigenia* is no longer classified as a 'weird wonder' of uncertain and probably long-vanished affinities. Instead, together with many other Cambrian fossils, it is now tentatively placed in the phylum Lobopodia, which has modern representatives in the form of *Peripatus* and the other 'onychophorans' or 'velvet worms' whom we met at Rendezvous 26.

In the days when annelid worms were thought to be close relatives of arthropods, the Onychophora were often touted as 'intermediate' – 'bridging the gap' between them, although that is not an entirely helpful concept if you think carefully about how evolution works.

---

* Bertrand Russell, of course.

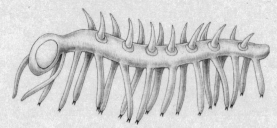

Hallucigenia — modern reconstruction.

The annelids are now placed in the Lophotrochozoa, while the Onychophora are ecdysozoans with the arthropods. *Peripatus,* with its ancient affinities, is well placed among modern pilgrims to tell the tale of the Cambrian Explosion.

The modern Onychophora (see plate 36) are widely distributed in the tropics and especially in the southern hemisphere. *Peripatus, Peripatopsis,* and all the modern onychophorans live on land, in leaf litter and humid places, where they hunt snails, worms, insects and other small prey. In the Cambrian, of course, *Hallucigenia* and the remote forebears of *Peripatus* and *Peripatopsis* lived – along with everybody else – in the sea.

Hallucigenia's connection with the modern Onychophora is still controversial, and we must remember what a lot of imagination necessarily intervenes between a blurred and squashed fossil in a rock, and the reconstruction that is eventually drawn, often in daring colour, on the page. It has even been suggested that *Hallucigenia* might not be a whole animal at all, but a part of some unknown animal. It would not be the first time such a mistake had been made. Some early artists' reconstructions of Cambrian scenes included a swimming jellyfish-like creature, seemingly inspired by tinned pineapple rings, which turned out to be part of the jaw apparatus of the mysterious predatory animal *Anomalocaris* (see page 452). Other Cambrian fossils, for example *Aysheaia,* certainly seem quite like marine versions of *Peripatus,* and this reinforces *Peripatus*'s entitlement to tell this Cambrian tale.

Most fossils, in any era, are the remains of hard parts of animals: vertebrate bones, the carapaces of arthropods or the shells of molluscs or brachiopods. But there are three Cambrian fossil beds – one in Canada, one in Greenland and one in China – where freak conditions, with almost miraculous good fortune for us, preserved

Anomalocaris saron.

soft parts as well. These are the Burgess Shale of British Columbia, Sirius Passet of northern Greenland, and the Chengjiang site of southern China.* The Burgess Shale was first discovered in 1909 and was made famous 80 years later by Stephen Gould in *Wonderful Life*. The Sirius Passet site in northern Greenland was discovered in 1984 but is so far less studied than the other two. In the same year, the Chengjiang fossils were discovered by Hou Xian-guang. Dr Hou is one of those who have collaborated on a beautifully illustrated monograph, *The Cambrian Fossils of Chengjiang, China*, published in 2004 – fortunately for me just before this book went to press.

The Chengjiang fossils are now dated at 525 million years old. That's roughly contemporary with Sirius Passet, and some 10 or 15 million years older than the Burgess Shale, but these outstanding fossil sites have a similar fauna. There are lots of lobopods, many looking more or less like marine versions of *Peripatus*. There are algae, sponges, worms of various kinds, brachiopods looking pretty much like modern ones, and enigmatic animals of uncertain kinship. There are large numbers of arthropods, including crustaceans, trilobites and lots of others that loosely resemble crustaceans or trilobites but may have belonged in their own rather separate groups. The large (over a metre in some cases), apparently predatory *Anomalocaris* and its kind are found in Chengjiang as well as the Burgess Shale. Nobody is quite sure what they were – probably distant relations of the arthropods – but they must have been spectacular. Not all the 'weird wonders' of the Burgess Shale have been found at Chengjiang, for example *Opabinia*, with its famous five eyes.

The Sirius Passet fauna from Greenland includes a beautiful creature called *Halkieria*. It has been thought to be an early mollusc

---

* A fourth site, Orsten ('stink stone') in Sweden, preserves soft bodies in a different way.

but Simon Conway Morris, who has described many of the strange creatures of the Cambrian, believes it has affinities with three major phyla: molluscs, brachiopods and annelid worms. This gladdens my heart because it helps to break down the almost mystical reverence with which zoologists regard the great phyla (see plate 37). If we take our evolution seriously, it has to be the case that, as we go back in time and approach their rendezvous points, they will become more and more like each other, more and more closely related. Whether or not *Halkieria* fits the bill, it would be worrying if there were *not* an ancient animal that united annelids, brachiopods and molluscs. Note the shells, one at each end, in the illustrations in plate 37.

As we saw at Rendezvous 22, Chengjiang has fossils that appear to be true vertebrates, pre-dating the amphioxus-like *Pikaia* of the Burgess Shale and other Cambrian chordates. Traditional zoological wisdom never had vertebrates arising so early. Yet *Myllokunmingia*, of which more than 500 specimens have now been discovered at Chengjiang, looks pretty much like a good jawless fish, such as had previously been thought not to arise until 50 million years later in the middle of the Ordovician. At first, two new genera were described – *Myllokunmingia*, which was described as relatively close to the lampreys, and *Haikouichthys* (alas, not named after the Japanese verse form), which was believed to have hagfish affinities. Some revisionist taxonomists now place the two in one species, *Myllokunmingia fengjiaoa*. This controversial updating of the status of *Haikouichthys* is eloquent of how difficult it is to discern the details of very old fossils. On the following page is a photograph of an individual *Myllokunmingia* fossil, together with a drawing of it made with a camera lucida. I find myself filled with admiration for the patience that goes into reconstructing ancient animals like these.

The pushing of the vertebrates back into the middle of the Cambrian only strengthens the idea of sudden explosion that is the basis of the myth. It really does appear that most of today's major animal phyla first appear as fossils in a narrow span within the Cambrian. This doesn't mean that there were no representatives of those phyla before the Cambrian. But they have mostly not fossilised. How should we interpret this? We can distinguish various combinations of three main hypotheses, rather like the three hypotheses for the explosion of the mammals after the extinction of the dinosaurs.

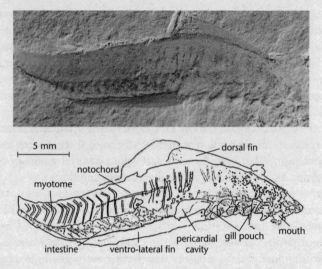

Vertebrates were not supposed to be this old. Fossil *Myllokunmingia fengjiaoa*, Chengjiang. From D-G Shu *et al.* [264].

1. NO REAL EXPLOSION. On this view there was only an explosion of fossilisability, not of actual evolution. The phyla actually go back a long way before the Cambrian, with concestors spread out through hundreds of millions of years in the Precambrian. This view is supported by some molecular biologists who have used molecular clock techniques to date key concestors. For example, G. A. Wray, J. S. Levinton and L. H. Shapiro, in a famous paper of 1996, estimated that the concestor uniting vertebrates and echinoderms lived about a billion years ago, and the concestor uniting vertebrates and molluscs was 200 million years earlier still, more than twice the age of the so-called Cambrian Explosion. Molecular clock estimates have in general tended to push these deep branchings way back into the Precambrian, far further than most palaeontologists are happy with. On this view, fossils were, for unknown reasons, not readily formed before the Cambrian. Perhaps they lacked readily fossilisable hard parts, such as shells, carapaces and bones. After all, the Burgess Shale and the Chengjiang beds are extremely unusual, among all geological layers, in recording soft parts as fossils. Perhaps Precambrian animals, although long existing in a wide range of complex body plans, were simply too small to fossilise. In favour of this idea, there are some small animal phyla that have left no fossils at all after

the Cambrian, until they appear today as live 'orphans'. Why then should we feel entitled to expect fossils before the Cambrian? In any case, some of the Precambrian fossils that have been found, including the Ediacaran fauna (see page 460) and trace fossils of tracks and burrows, indicate the presence of real Precambrian metazoans.

2. MEDIUM-FUSE EXPLOSION. The concestors uniting the various phyla really did live reasonably close to each other in time, but still spread out over several tens of millions of years before the observed explosion of fossils. From the great distance of the present, Chengjiang at 525 million years seems at first sight rather close to a putative concestor at, say, 590 million. But a full 65 million years separates them, which is the same time as has elapsed today since the death of the dinosaurs – the entire time during which modern mammals have radiated and radiated again to produce the spectacularly diverse ranges that we see today. Even 10 million years is a long time in the light of the extremely rapid evolutionary bursts of the Galapagos Finch's Tale and the Cichlid's Tale. It is all too easy, with hindsight, to think that because we recognise two ancient fossils as belonging to different modern phyla, those two fossils must have been as different from each other as modern representatives of the two phyla are. It is too easy to forget that the modern representatives have had half a billion years in which to diverge. There is no good reason to believe that a Cambrian taxonomist, blessedly free of 500 million years' worth of zoological hindsight, would have placed the two fossils in separate phyla. He might have placed them only in separate orders, notwithstanding the then-unknowable fact that their descendants were destined eventually to diverge so far as to warrant separate phylum status.

3. OVERNIGHT EXPLOSION. This third school of thought is, in my opinion, bonkers. Or, to use more parliamentary language, wildly and irresponsibly unrealistic. But I must spend some time on it because it has recently become unaccountably popular, following the rhetoric I attributed to romantically overheated zoologists.

The third school believes that new phyla sprang into existence overnight, in a single macromutational leap. Here are some quotations I have used before, in *Unweaving the Rainbow,* from otherwise reputable scientists.

It was as if the facility for making evolutionary leaps that produced major functional novelties – the basis of new phyla – had somehow been lost when the Cambrian period came to an end. It was as if the mainspring of evolution had lost some of its power . . . Hence, evolution in Cambrian organisms could take bigger leaps, including phylum-level leaps, while later on it would be more constrained, making only modest jumps, up to the class level.

Or this, from the same distinguished scientist from whom we heard at the beginning of the tale.

Early on in the branching process, we find a variety of long-jump mutations that differ from the stem and from one another quite dramatically. These species have sufficient morphological differences to be categorized as founders of distinct phyla. These founders also branch, but do so via slightly closer long-jump variants, yielding branches from each founder of a phylum to dissimilar daughter species, the founders of classes. As the process continues, fitter variants are found in progressively more nearby neighborhoods, so founders of orders, families, and genera emerge in succession.

Those preposterous quotations moved me to retort that it is as though a gardener looked at an old oak tree and remarked, wonderingly:

Isn't it strange that no major new boughs have appeared on this tree for many years? These days, all the new growth appears to be at the twig level!

Here's another quotation, which this time I will attribute because it was published after *Unweaving the Rainbow* and I have therefore not used it before. Andrew Parker's *In the Blink of an Eye* is mainly concerned with advocating his interesting and original theory that the Cambrian Explosion was triggered by animals' sudden discovery of eyes. But before coming to his theory itself, Parker begins by falling hook, line and sinker for the 'wild and irresponsible' version of the Cambrian Explosion myth. He first expresses the myth itself in the most frankly 'explosive' version I have read:

544 million years ago there were indeed three animal phyla with their variety of external forms, but at 538 million years ago there were thirty-eight, the same number that exists today.

He goes on to make it clear that he is not talking about extremely rapid gradualistic evolution compressed into a period of 6 million years, which would be an extreme version of our Hypothesis Two, and just barely acceptable. Nor is he saying, as I would, that near the initial divergence of a pair of (what are destined to become) phyla, they would not have been very different – would, indeed, have passed through successive stages of being a pair of species, then genera, and so on until eventually their separation warranted recognition at the phylum level. No, Parker gives every appearance of regarding his 38 phyla, at 538 million years, as fully fledged phyla that sprang into existence overnight, at the drop of a macromutational hat:

> Thirty-eight animal phyla have evolved on Earth. So only thirty-eight monumental genetic events have taken place, resulting in thirty-eight different internal organisations.

Monumental genetic events are not utterly out of the question. Control genes of the various Hox families that we met in the Fruit Fly's Tale can certainly mutate in dramatic ways. But there's monumental and monumental. A fruit fly with a pair of legs where the antennae should be is about as monumental as it gets, and even then there is a big question mark over survival. There is a powerful general reason for this, which I shall briefly explain.

A mutant animal has a certain probability of being better off as a consequence of its new mutation. 'Better off' means better when compared to the premutated parental type. The parent must have been at least good enough to survive and reproduce, otherwise it wouldn't be a parent. It is easy to see that the smaller the mutation, the more likely it is to be an improvement. 'It is easy to see' was a favourite phrase of the great statistician and biologist R. A. Fisher, and he sometimes used it when it was anything but easy for ordinary mortals to see. In this particular case, however, I think it is genuinely easy to follow Fisher's argument for the case of a simple metric feature – something such as thigh length, which varies in one dimension: some number of millimetres that could grow larger or could grow smaller.

Imagine a set of mutations of increasing magnitude. At one extreme, a mutation of zero magnitude is by definition exactly as good as the parent's copy of the gene which, as we've seen, must have been at least good enough to survive childhood and reproduce. Now imagine a random mutation of small magnitude: the leg, say, gets one millimetre longer or one millimetre shorter. Assuming that the parental gene is not perfect, a mutation that is infinitesimally different from the parental version has a 50 per cent chance of being better and a 50 per cent chance of being worse: it'll be better if it is a step in the right direction, worse if it is a step in the opposite direction, relative to the parental condition. But a very large mutation will probably be worse than the parental version, *even if it is a step in the right direction,* because it will overshoot. To push to the extreme, imagine an otherwise normal man with thighs two metres long.

Fisher's argument was more general than this. When we are talking about macromutational leaps into new phylum territory, we are no longer dealing with simple metric characters like leg length, and we need another version of the argument. The essential point, as I have put it before, is that there are many more ways of being dead than of being alive. Imagine a mathematical landscape of all possible animals. I have to call it mathematical, because it is a landscape in hundreds of dimensions and it includes an almost infinitely large range of conceivable monstrosities, as well as the (relatively) small number of animals that have actually ever lived. What Parker calls a 'monumental genetic event' would be equivalent to a macromutation of huge effect, not just in one dimension as with our thigh example, but in hundreds of dimensions simultaneously. That is the scale of change we are talking about if we imagine, as Parker seems to, an abrupt and immediate change from one phylum to another.

In the multidimensional landscape of all possible animals, living creatures are islands of viability separated from other islands by gigantic oceans of grotesque deformity. Starting from any one island, you can evolve away from it one step at a time, here inching out a leg, there shaving the tip of a horn, or darkening a feather. Evolution is a trajectory through multidimensional space, in which every step of the way has to represent a body capable of surviving and reproducing about as well as the parental type reached by the preceding step of the trajectory. Given enough time, a sufficiently long trajectory leads from a viable starting point to a viable destination so remote that we recognise it as a different phylum, say, molluscs. And a different step-by-step trajectory from the same starting point can lead, through continuously viable intermediates, to another viable destination, which we recognise as yet another phylum, say, annelids. Something like this must have happened for each of the forks leading to each pair of animal phyla from their respective concestors.

The point we are leading up to is this. A random change of sufficient magnitude to initiate a new phylum at one fell swoop will be so large, in hundreds of dimensions simultaneously, that it would have to be preposterously lucky to land on another island of viability. Almost inevitably, a megamutation of that magnitude will land in the middle of the ocean of inviability: probably unrecognisable as an animal at all.

Creationists foolishly liken Darwinian natural selection to a hurricane blowing through a junkyard and having the luck to assemble a Boeing 747. They are wrong, of course, for they completely miss the gradual, cumulative nature of natural selection. But the junkyard metaphor is entirely apt to the hypothetical overnight invention of a new phylum. An evolutionary step of the same magnitude as, say, the overnight transition from earthworm to snail, really would have to be as lucky as the hurricane in the junkyard.

We can, then, with complete confidence, reject the third of our three hypotheses, the bonkers one. That leaves the other two, or some compromise between them, and here I find myself agnostic and eager for more data. As we shall see in the epilogue to this tale, it seems to be increasingly accepted that the early molecular clock estimates were exaggerating when they pushed the major branch points hundreds of millions of years back into the Precambrian. On the other hand, the mere fact that there are few, if any, fossils of most animal phyla before the Cambrian should not stampede us into assuming that those phyla must have evolved extremely rapidly. The hurricane in a junkyard argument tells us that all those Cambrian fossils must have had continuously evolving antecedents. Those antecedents had to be there, but they have not been discovered. Whatever the reason, and whatever the timescale, they failed to fossilise, but they must have been there. On the face of it, it is harder to believe that a whole lot of animals could be invisible for 100 million years than that they could be invisible for only 10 million years. This leads some people to prefer the short-fuse Cambrian Explosion theory. On the other hand, the shorter you make the fuse, the harder it is to believe all that diversification could be crammed into the time available. So this argument cuts both ways and doesn't decisively choose between our two surviving hypotheses.

The fossil record is not completely void of metazoan life before Chengjiang and Sirius Passet. Around 20 million years earlier, almost plumb on the Cambrian/Precambrian boundary, start to appear a variety of microscopic fossils that look rather like tiny shells – together they are known as the 'small shelly fauna'. It came as a surprise to most palaeontologists when some of these were identified as armour plating from lobopods – relatives of the velvet worm. That means that the divergences between different groups of protostomes *must* have occurred in the Precambrian, before the visible 'explosion'.

And there are hints of older animal diversity. Twenty million years before the start of the Cambrian, in the Ediacaran Period of the late Precambrian, there was a worldwide flourishing of a mysterious group of animals called the Ediacaran fauna, named after the Ediacara Hills in South Australia where they were first found. It is hard to know quite what most of them were, but they were among the first large animals to be fossilised. Some of them are probably sponges. Some are a bit like jelly-fish. Others somewhat resemble sea anemones, or sea pens (feather-like relatives of sea anemones). Some look a bit worm-like or slug-like, and could conceivably represent true Bilateria. Others are just plain mysterious. What are we to make of this creature *Dickinsonia*? (See plate 38.) Is it a coral? Or a worm? Or a fungus? Or something completely different from anything that survives today? There is even one tadpole-like fossil from Australia, still not formally described, that is suspected of being a chordate (that's the phylum, remember, to which the vertebrates belong). If this turns out to be right, it would be very exciting, but we must wait and see. In spite of such tantalising straws in the wind, the consensus among zoologists is that the Ediacaran fauna, though intriguing, doesn't help us much one way or the other in tracing the ancestry of most modern animals.

There are also fossil imprints that appear to be the trails or burrows of Precambrian animals. These traces tell us of the early existence of crawling animals large enough to make them. Unfortunately, they don't tell us much about what those animals looked like. There are also some even older, mostly microscopic fossils found at Doushantuo in China which appear to be embryos, though it is not clear what kind of animal they might have grown into. Older still are small, disc-shaped impressions from northwest Canada, dated between about 600 and 610 million years ago, but these animals are, if anything, even more enigmatic than the Ediacaran forms.

This book is hung upon a series of 39 rendezvous points and it seemed desirable to make some sort of guess as to the date of each one. Most of the rendezvous points can now be dated with some confidence, using a combination of datable fossils and molecular clocks calibrated by datable fossils. Not surprisingly, the fossils start to let us down when we reach the older rendezvous points. This means that the molecular methods can no longer be reliably calibrated, and we enter a wilderness of undatability. For completeness I have forced myself to put some sort of date on these wilderness concestors, roughly Concestors 23 to 39. The most recently available evidence seems to me to favour, even if only slightly, a view closer to a medium-fuse explosion. This goes against my earlier bias in favour of no real explosion. When more evidence comes in, as I hope it will, I shall not be in the least surprised if we find ourselves pushed the other way again into the deep Precambrian in our quest for the concestors of modern animal phyla. Or we might be pulled back to an impressively short explosion, in which the concestors of most of the great animal phyla are compressed into a period of 20 or even 10 million years around the beginning of the Cambrian. In this case, my strong expectation would be that even if we correctly place two Cambrian animals in different phyla on the basis of their resemblance to modern animals, back in the Cambrian they would have been much closer to each other than the modern descendants of one are to the modern descendants of the other. Cambrian zoologists would not have placed them in separate phyla but only in, say, separate subclasses.

I wouldn't be surprised to see either of the first two hypotheses vindicated. I'm not sticking my neck out. But I'll eat my hat if any evidence is ever found in favour of Hypothesis Three. There is every reason to suppose that evolution in the Cambrian was essentially the same kind of process as evolution today. All that over-excited rhetoric about the mainspring of evolution running down after the Cambrian; all that euphoric shouting about wild, heedless dances of extravagant invention, with new phyla leaping into existence in a blissful dawn of zoological irresponsibility – now here's something I am prepared to stick my neck out for: all that stuff is just plain dotty.

I hasten to say I have nothing against prose-poetry on the Cambrian. But give me Richard Fortey's version, on page 120 of his beautiful book *Life: An Unauthorised Biography*:

> I can imagine standing upon a Cambrian shore in the evening, much as
> I stood on the shore at Spitsbergen and wondered about the biography
> of life for the first time. The sea lapping at my feet would look and feel

much the same. Where the sea meets the land there is a patch of slightly sticky, rounded stromatolite pillows, survivors from the vast groves of the Precambrian. The wind is whistling across the red plains behind me, where nothing visible lives, and I can feel the sharp sting of wind-blown sand on the back of my legs. But in the muddy sand at my feet I can see worm casts, little curled wiggles that look familiar. I can see trails of dimpled impressions left by the scuttling of crustacean-like animals . . . Apart from the whistle of the breeze and the crash and suck of the breakers, it is completely silent, and nothing cries in the wind . . .

EPILOGUE TO THE VELVET WORM'S TALE
Written with Yan Wong

For much of this book I have tossed rendezvous dates around with insouciance, and even been rash enough, when introducing many of the concestors, to stick a specific number of 'greats' before 'grand-parent'. My dates have mostly been based upon fossils which, as we shall see in the Redwood's Tale, can be dated to a precision com-mensurate with the vast timescales involved. But fossils never helped us much with tracing the ancestry of soft-bodied animals such as flatworms. Coelacanths went missing from the record for the past 70 million years, which was why the discovery of a live one in 1938 was such an exhilarating surprise. The fossil record, even at the best of times, can be a fickle witness. And now, having reached the Cam-brian Period, we are sadly running out of fossils. Whatever inter-pretation we place on 'explosion', everyone agrees that almost all the predecessors of the great Cambrian fauna have, for uncertain reasons, failed to fossilise. As we seek concestors that predate the Cambrian, we find no more help in the rocks. Fortunately, fossils are not our only recourse. In the Elephant Bird's Tale, the Lungfish's Tale and other places, we have made use of the ingenious technique known as the molecular clock. The time has come to explain the molecular clock properly.

Wouldn't it be wonderful if measurable, or countable, evolution-ary changes happened at a fixed rate? We could then use evolution itself as its own clock. And this needn't involve circular reasoning because we could calibrate the evolutionary clock on parts of evolu-tion where the fossil record is good, then extrapolate to parts where it isn't. But how do we measure rates of evolution? And, even if we could measure them, why on earth should we expect that any aspect of evolutionary change should go at a fixed rate like a clock?

There is not the slightest hope that leg length, or brain size, or number of whiskers will evolve at a fixed rate. Such features are important for survival, and their rates of evolution will surely be hideously inconstant. As clocks they are doomed by the very principles of their own evolution. In any case, it is hard to imagine an agreed standard for measuring rates of visible evolution. Do you measure evolution of leg length in millimetres per million years, as percentage change per million years, or what? J. B. S. Haldane proposed a unit of evolutionary rate, the darwin, which is based upon proportional change per generation. Wherever it has been used on real fossils, results vary from millidarwins to kilo-darwins and megadarwins, and nobody is surprised.

Molecular change looks like a much more promising clock. First, because it is obvious what you must measure. Since DNA is textual information written in a four-letter alphabet, there is an entirely natural way to measure its rate of evolution. You just count letter differences. Or, if you prefer, you can go to the protein products of the DNA code and count substitutions of amino acids.* There are reasons to hope that the majority of evolutionary change at the molecular level is neutral rather than being steered by natural selection. Neutral is not the same as useless or functionless – it only means that different versions of the gene are equally good, therefore change from one to the other is not noticed by natural selection. This is good for the clock.

Contrary to my rather ludicrous reputation as an 'ultra-Darwinist' (a slander I would protest more vigorously if the name sounded less of a compliment than it does), I do not think that the majority of evolutionary change at the molecular level is favoured by natural selection. On the contrary, I have always had a lot of time for the so-called neutral theory associated with the great Japanese geneticist Motoo Kimura, or its extension, the 'nearly neutral' theory of his collaborator Tomoko Ohta. The real world has no interest in human tastes, of course, but as it happens I positively *want* such theories to be true. This is because they give us a separate, independent chronicle of evolution, unlinked to the visible features of the creatures around

---

* When the molecular clock was first proposed, by Emile Zuckerkandl and the great Linus Pauling, this was the only method available.

us, and they hold out the hope that some kind of molecular clock might really work.

Just in case the point is misunderstood, I must emphasise that the neutral theory does not in any way denigrate the importance of selection in nature. Natural selection is all-powerful with respect to those visible changes that affect survival and reproduction. Natural selection is the only explanation we know for the functional beauty and apparently 'designed' complexity of living things. But if there are any changes that have no visible effect – changes that pass right under natural selection's radar – they can accumulate in the gene pool with impunity and may supply just what we need for an evolutionary clock.

As ever, Charles Darwin was way ahead of his time with respect to neutral changes. In the first edition of *The Origin of Species*, near the beginning of Chapter 4, he wrote:

> This preservation of favourable variations and the rejection of injurious variations, I call natural selection. Variations neither useful nor injurious would not be affected by natural selection, and would be left a fluctuating element, as perhaps we see in the species called polymorphic.

By the sixth and last edition, the second sentence had an even more modern-sounding addendum:

> . . . as perhaps we see in certain polymorphic species, or would ultimately become fixed . . .

'Fixed' is a genetic technical term and Darwin surely cannot have meant it in quite the modern sense, but it gives me a lovely lead-in to the next point. A new mutation, whose frequency in the population begins near zero by definition, is said to become 'fixed' when it has reached 100 per cent in the population. The rate of evolution that we seek to measure, for purposes of a molecular clock, is the rate at which a succession of mutations of the same genetic locus become fixed in the population. The obvious way for fixation to happen is if natural selection favours the new mutation over the previous 'wild type' allele, and therefore drives it to fixation – it becomes the norm, 'the one to beat'. But a new mutation can also go to fixation even if it is exactly as good as its predecessor – true neutrality. This is nothing to do with selection: it happens by sheer chance. You can simulate the process by tossing pennies, and can calculate the rate at which it will happen. Once a neutral mutation has drifted to 100 per cent, it will become the norm, the so-called 'wild type' at that locus, until another mutation has the luck to drift to fixation.

If there is a strong component of neutrality, we could potentially have a marvellous clock. Kimura himself wasn't particularly concerned with the molecular clock idea. But he believed – it now seems rightly – that the majority of mutations in DNA are indeed neutral – 'neither useful nor injurious'. And, in a remarkably neat and simple piece of algebra, which I shall not spell out here, he calculated that, if this is true, the rate at which genuinely neutral genes should 'ultimately become fixed' is exactly equal to the rate at which the variations are generated in the first place: the mutation rate.

You see how perfect this is for anybody who wants to date bifurcation ('rendezvous') points using a molecular clock. As long as the mutation rate at a neutral genetic locus remains constant over time, the fixation rate will also be constant. You can now compare the same gene in two different animals, say a pangolin and a starfish, whose most recent common ancestor was Concestor 25. Count the number of letters by which the starfish gene differs from the pangolin gene. Assume that half the differences accumulated in the line leading from concestor to starfish, and the other half in the line leading from concestor to pangolin. That gives you the number of ticks of the clock since Rendezvous 25.

But it isn't as simple as that, and the complications are interesting. First, if you listened to the ticking of the molecular clock, it would not be regular like a pendulum clock or a hairspring watch; it would sound like a Geiger counter near a radioactive source. Completely random! Each tick is the fixation of yet another mutation. Under the neutral theory, the interval between successive ticks could be long or it could be short, by chance – 'genetic drift'. In a Geiger counter, the timing of the next tick is unpredictable. But – and this is really important – the *average* interval over a large number of ticks is highly predictable. The hope is that the molecular clock is predictable in the same way as a Geiger counter, and in general this is true.

Second, the tick rate varies from gene to gene within a genome. This was noticed early, when geneticists could look only at the protein products of DNA, not DNA itself. Cytochrome-c evolves at its own characteristic rate, which is faster than histones but slower than globins, which in turn are slower than fibrinopeptides. In the same way, when a Geiger counter is exposed to a very slightly radioactive source such as a lump of granite, versus a highly radio-active source such as a lump of radium, the timing of the next tick is always unpredictable but the average rate of ticking is predictably and dramatically different as you move from granite to radium. Histones are like granite, ticking at a very slow rate; fibrinopeptides are like radium, buzzing like a dementedly randomised bee. Other proteins such as cytochrome-c (or rather the genes that make them) are intermediate. There is a spectrum of gene clocks, each running at its own speed, and each useful for different dating purposes, and for cross-checking with each other.

Why do different genes run at different speeds? What distinguishes 'granite' genes from 'radium' genes? Remember that neutral doesn't mean useless, it means equally good. Granite genes and radium genes are both useful. It is just that radium genes can change at many places along their length and still be useful. Because of the way a gene works, portions of its length can change with impunity without affecting its functioning. Other portions of the same gene are highly sensitive to mutation, and its functioning is devastated if these portions are hit by a mutation. Maybe all genes have a granite portion, which mustn't change much if the gene is to go on working, and a radium portion that can freewheel unchecked as long as the granite portion is not affected. Maybe the cytochrome-c gene has a mixture of granite bits and radium bits; fibrinopeptide genes have a higher proportion of radium bits, while histone genes have a higher proportion of granite bits. There are some problems, or at least complications, with this as an explanation for the differences in tick rates between genes. But what matters for us is that tick rates really do vary between genes, while the rate for any given gene is pretty constant even in widely separated species.

Not completely constant, however, and this brings us to our next problem, which is a serious one. Tick rates are not just vague and sloppy. For any given gene they can be systematically greater in some kinds of creatures than in others, and this introduces a real bias. Bacteria have a much less effective DNA-repair system than our DNA's sophisticated 'proofreading', so their genes mutate at a higher rate and their molecular clocks tick faster. Rodents, too, have slightly sloppy repair enzymes, which might explain why molecular evolution is faster in rodents than in other mammals. Major changes in evolution, like the move to 'warm blood', have the potential to change the mutation rate, which could play havoc with our clock estimates of branch dates. Sophisticated methods are now being developed that can allow for changing mutation rates in different lineages, but these are in their infancy.

Even more worrying, the time of reproduction would seem to offer maximum opportunity for mutation. So species with short life cycles such as fruit flies will pick up mutations at a higher rate per million years than, say, elephants with their long intervals between generations. This would suggest that the molecular clock might count in generations rather than in real time. Actually, however, when molecular biologists looked at rates of change in sequences, using lineages that happened to have a good fossil record for calibration, this isn't what they found. There really did seem to be a molecular clock that measured time in years, not generations. This was nice, but how to explain it?

One suggestion was that, even though the reproductive turnover in elephants is slow compared to fruit flies, during all the years between reproductive events elephant genes are subject to the same bombardment of cosmic rays and other events that can cause mutation as fruit fly genes. Admittedly, fruit fly genes are hopping into a new fly once a fortnight, but why should cosmic rays care about that? Well, genes sitting in one elephant for ten years are being hit by the same number of cosmic rays as genes hopping through a succession of 250 fruit flies during the same period. There may be something in this theory, but it probably isn't a sufficient explanation. It really is true that most mutations occur when a new generation is being made, so we seem to need another explanation for the molecular clock's apparent ability to tell the time in years rather than in generations.

Here's where Kimura's colleague Tomoko Ohta made a clever contribution: her *nearly* neutral theory. Kimura, as I said, calculated from his fully neutral theory that the rate of fixation of neutral genes should equal the mutation rate. This remarkably simple conclusion depended on an elegant piece of 'cancelling out' in the algebra. And the quantity that cancelled out was the population size. Population size comes into the equation, but it ends up both above and below the line, so it conveniently vanishes in a puff of mathematical smoke, and fixation rate emerges as equal to mutation rate. But *only* if the genes concerned really are completely neutral. Ohta revisited Kimura's algebra but she allowed her mutations to be nearly neutral instead of completely neutral. And this made all the difference. Population size no longer cancelled out.

This is because – as has long been calculated by mathematical geneticists – in a large population, slightly harmful genes are more likely to be eliminated by natural selection before they have a chance to drift to fixation. In a small population, luck is more likely to carry a slightly harmful gene to fixation before natural selection 'notices'. To push to the extreme, imagine a population almost entirely wiped out by some catastrophe, with only half a dozen individuals remaining. It would not be very surprising if, by chance, all six happened to have the slightly deleterious gene. In that case, we have fixation – 100 per cent of the population. That's an extreme, but the mathematics shows the same effect more generally. Small populations favour the drifting to fixation of genes that would be eliminated in a large population.

So, as Ohta pointed out, population size no longer cancels out of the algebra. On the contrary, it stays in just the right place to do the molecular clock theory a bit of good. Now, back to our elephants and fruit flies. Large animals with long life cycles, such as elephants, also tend to have small populations. Small animals with short life cycles, such as fruit flies, tend to have large populations. This isn't just a vague effect, it is a pretty lawful one, and it holds for reasons that are not hard to imagine. So even if fruit flies have short generation times which would tend to speed the clock up, they also have a large population, which slows it down again. Elephants may have a slow clock as far as mutations are concerned, but their small populations speed the clock up again in the fixation department.

Professor Ohta has evidence that truly neutral mutations, as in junk DNA or in 'synonymous' substitutions* seem to tick in generation time as opposed to real time: creatures with short generation times show accelerated DNA evolution if you measure it in real time. Conversely, mutations that actually change something, and therefore fall foul of natural selection, tick away more or less constantly in real time.

Whatever the theoretical reason, it does seem to be the case in practice that, with known exceptions that we can usually allow for (by carefully choosing our clock genes, and avoiding species such as rodents with exceptional rates of mutation), the molecular clock has proved itself a workable instrument. To use it, we need to draw the evolutionary tree that relates the set of species we are interested in,

---

* The DNA code being 'degenerate', any one amino acid can be specified by more than one 'synonymous' mutation. A mutational change resulting in an exact synonym makes no difference at all to the final outcome.

and estimate the amount of evolutionary change in each lineage. This is not as simple as just counting differences between the genes of two modern species and dividing by two. We need to use the advanced tree-building techniques of maximum likelihood and Bayesian phylogenetics that we met in the Gibbon's Tale. Anchored with some known fossil dates for calibration, we can then make a good guess at the dates of rendezvous points on the tree.

Carefully deployed in this way, the molecular clock has produced some stunning results. Molecular clock datings of the human/chimpanzee common ancestor centre around 6 million years plus or minus a million years or so. When first announced, this date caused near outrage among palaeoanthropologists, who had dated the split at around 20 million years. Nowadays, just about everybody accepts the molecular short date. The clock's best success story is perhaps the dating of the radiation of placental mammals, as described in the Great Cretaceous Catastrophe. After excluding rodents for their abnormal mutation rates, we find that several molecular clock estimates agree in placing the concestor of all mammals far back in the Cretaceous. One clock study of DNA from modern placental mammals, for instance, placed the concestor at more than 100 million years ago, right in the thick of dinosaur hegemony. When such dates were first announced, they were at odds with the fossil evidence, which seemed to show a much later 'explosion' of mammals and a dearth of earlier mammal fossils. But the molecular clock dates have now been vindicated by recently discovered fossil mammals from 125 million years ago, and the early dates are becoming widely accepted. Success stories abound, and they have contributed to the dates used throughout this book.

Complacency alert! Listen to those alarm bells ringing.

Molecular clocks ultimately depend on calibration by fossils. Radiometrically calculated dates for fossils are accepted with the respect that biology rightly bestows upon physics (see the Redwood's Tale). One strategically located fossil that confidently places a lower bound on the dating of an important evolutionary branch point can be used to calibrate a whole lot of molecular clocks dotted around the genomes of a range of animals dotted around the phyla. But when we get back to Precambrian territory where the supply of fossils gives out, we have to depend on relatively young fossils to calibrate great-great-grandfather clocks that are then used to estimate much older dates. And that spells trouble.

Fossils suggest 310 million years for the date of Rendezvous 16, the junction point between mammals and sauropsids (birds, crocodiles, snakes, etc.). This one date provides the master calibration for many molecular clock datings of much older branch points. Now, any date estimate has a certain margin of error, and in their scientific papers scientists try to remember to place 'error bars' on each of their estimates. A date is quoted plus or minus, say, 10 million years. That's all very well when the dates we seek with the molecular clock are in the same ballpark as the fossil dates used to calibrate it. When there is a great mismatch between ballparks, the error bars can grow alarmingly. The implication of a wide error bar is that if you tweak some small assumption, or slightly alter some small number that you feed into the calculation, the impact on the final result could be dramatic. Not plus or minus 10 million years but plus or minus half a billion years, say. Wide error bars mean that the estimated date is not robust against measurement error.

In the Velvet Worm's Tale itself, we saw various molecular clock estimates that placed important branch points in the deep Precambrian, for example 1,200 million years for the split between vertebrates and molluscs. More recent studies, using sophisticated techniques that allow for possible variations in mutation rates, bring the estimates down to dates in the 600-million-year range: a dramatic shortening – accommodated in the error bars of the original estimate, but that is small consolation.

Although I am a firm supporter of the molecular clock idea in general, I think its estimates of very early branch points need to be treated with caution. Extrapolating backwards from a 310-million-year-old calibration fossil to a rendezvous point more than twice as old is fraught with danger. For example, it is possible that the rate of molecular evolution in the vertebrates (which enters into our calibration calculation) is not typical of the rest of life. They are thought to have undergone two rounds of doubling of their entire genome. The sudden creation of large numbers of duplicate genes may affect the selection pressure on nearly neutral mutations. Some scientists (I am not one of them, as I have already made clear) believe that the Cambrian marked a great shift in the whole process of evolution. If they are right, the molecular clock would need a radical recalibration before it should be let loose in the Precambrian.

In general, as we go back further in time and the supply of fossils peters out, we enter a realm of almost complete conjecture.

Nevertheless, I am hopeful of future studies. The dazzling fossils of the Chengjiang and similar formations may greatly extend the range of calibration points into regions of the animal kingdom hitherto off limits.

Meanwhile, recognising that we are wandering in an ancient wilderness of conjecture, Yan Wong and I have adopted the following rough strategy in trying to estimate dates from here on in the pilgrimage. We have provisionally accepted 1,100 million years for Rendezvous 34, the junction of animals and fungi. This is a date commonly used in the scientific literature, and it is compatible with the oldest fossil plant, a red alga from 1,200 million years ago. We then space out Concestors 27 to 34 roughly in the ratios indicated by molecular clock studies. However, if we have got Rendezvous 34 badly wrong, then our dates from here on in the pilgrimage could be overestimates by many tens or even hundreds of millions of years. Please bear this in mind as we enter that wilderness of undatability. I am so unconfident of dates in this vicinity that from now on I shall give up the already rather quaint conceit of estimating the number of 'greats' to put before grandparent. That number will soon be getting into the billions. The order of joinings at the successive rendezvous points is more certain, but even that could be wrong too.

ACOELOMORPH FLATWORMS

When we were talking about the protostomes, descendants of Concestor 26, I grouped the flatworms, Platyhelminthes, firmly within them. But now we have an interesting little complication. Recent evidence quite strongly suggests that the Platyhelminthes are a fiction. I'm not saying flatworms themselves don't exist, of course. But they are a heterogeneous collection of worms who should not be united under one name. Most of them are true protostomes and we met them at Rendezvous 26, but a few of them are quite separate and don't join us until here at Rendezvous 27. This we are dating at 630 million years ago, although out in these remote reaches of geological time these datings become more and more uncertain.

Six hundred and thirty million years is quite a lot older than the 590-million-year date we adopted for Rendezvous 26. Perhaps the long gap can be explained by the 'Snowball Earth' episode which, according to one imaginative theory, preceded the Cambrian. The idea is that, for reasons that are obscure but may have to do with the fashionable and possibly overrated mathematical theory of chaos, the entire Earth was gripped by a global ice age from about 620 million years ago to about 590 million years, rather neatly filling the large gap between Rendezvous 27 and 26. There was plenty of glaciation. But whether or not the glaciations engulfed the entire planet is a contentious question, and one that I shall pass over.

What all flatworms have in common, apart from their eponymous flatness, is that they lack an anus and they lack a coelom. The coelom of a typical animal, such as you or me or an earthworm, is the body cavity. This doesn't mean the gut: the gut, though a cavity, is topologically part of the outside world, the body being a topological doughnut, the hole in the middle of the ring being the mouth, the anus and the gut that connects them. The coelom, by contrast, is the cavity within the body in which the intestines, the lungs, heart, kidney and so on all sit. Platyhelminths don't have a coelom. Instead of a body cavity in which the guts slop about, flatworm guts and other internal organs are embedded in solid tissue called parenchyma. This may seem a trivial distinction, but the coelom is

Acoelomorph flatworms join

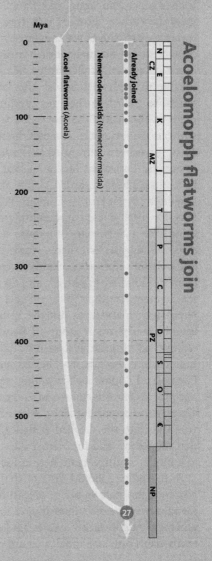

Mya

		N				
	CZ	E				
		K				
	MZ	J				
		T				
		P				
		C				
	PZ	D				
		S				
		O				
		Є				
	NP					

Acoel flatworms (Acoela)

Nemertodermatids (Nemertodermatida)

Already Joined

27

Acoelomorph flatworms join.
The vast majority of bilaterally
symmetrical animals are
protostomes or deuterostomes.
However, recent molecular data
hive off a pair of flatworm groups
as neither protostomes nor deuterostomes but earlier branching
lineages. These are the classes Acoela
(about 320 described species) and
Nemertodermatida (ten described
species), together known as the
acoelomorph flatworms. This is
likely to be quickly accepted by
taxonomists. Current evidence
hints that the Acoela and the
Nemertodermatida are sister
groups, as shown here.

Image: unknown acoel flatworms
on bubble coral.

embryologically defined and lies deep in the collective unconscious of zoologists.

Lacking an anus, how do flatworms expel wastes? Through the mouth if there is nowhere else. The gut may be a simple sac or, in larger flatworms, it branches into a complicated system of blind alleys, like the air tubes in our lungs. Our lungs, too, could theoretically have had an 'anus' – a separate hole for the air to leave by, with its waste carbon dioxide. Fish sort of do the equivalent, for their respiratory stream of water enters by one hole, the mouth, and leaves by others, the gill apertures. But our lungs are tidal, and so is the digestive system of flatworms. Flatworms lack lungs or gills and breathe through their skins. They also lack a system of circulating blood, so their branched gut presumably serves to transport nutriment to all parts of the body. In a few turbellarians, especially those with an exceptionally complex branched gut, an anus (or lots of anuses) has been reinvented after a long absence.

Because flatworms lack a coelom and mostly lack an anus, they have always been regarded as primitive – the most primitive of the bilaterally symmetrical animals. It was always assumed that the ancestor of all deuterostomes and protostomes was probably something like a flatworm. But now, as I began by saying, molecular evidence suggests that there are two unconnected kinds of flatworms, and only one of these two kinds is genuinely primitive. The genuinely primitive kind are the Acoela and the Nemertodermatida. The Acoela are named for their lack of coelom which, for them and the Nemertodermatida but not the Platyhelminthes proper, is a primitive lack. The main group of flatworms proper, the flukes, tapeworms and turbellarians, are now thought to have lost their anus and their coelom secondarily. They passed through a stage of being more like normal Lophotrochozoa, then reverted to being like their earlier ancestors again, *sans* anus and *sans* coelom. They joined our pilgrimage at Rendezvous 26, along with all the rest of the protostomes. I won't go into the detailed evidence, but will accept the conclusion that the Acoela and the Nemertodermatida are different and join us as a tiny incoming stream here at Rendezvous 27.

At this point I should describe these tiny worms that are joining us but, though I hate to say it, at least by comparison with most of the wonders we have seen, there is not a lot to describe. They live in the sea and they not only lack a coelom but lack a proper gut too – a situation that's viable only in animals that are very small, which they are.

Some of them supplement their diet by giving house room to plants, and hence benefiting indirectly from their photosynthesis. Members of the genus *Waminoa* have symbiotic dinoflagellates (unicellular algae) and live off their photosynthesis. Another acoel, *Convoluta,* has a similar relationship with a single-celled green alga, *Tetraselmis convolutae.* Symbiotic algae presumably make it possible for these little worms to be less little. The worms seem to take steps to make life easier for their algae, and hence themselves, crowding at the surface to give them as much light as possible. Professor Peter Holland writes to me that *Convoluta roscoffensis*

> . . . are amazing animals to see in their natural habitat. They appear as a green 'slime' on certain beaches in Brittany, the slime really being thousands of acoels plus their endosymbiotic algae. And as you creep up on the 'slime' it hides! (by disappearing into the sand). Very strange to see.

The Acoela are still with us, and therefore must be treated as modern animals, but their form and simplicity suggest that they might not be greatly changed since the time of Concestor 27. Modern acoel worms might be a reasonable approximation to the ancestor of all bilaterally symmetrical animals.

Our gathered pilgrims now include all the phyla recognised as Bilateria, which means the great bulk of the animal kingdom. The name refers to their bilateral symmetry, and is intended to exclude the two main radially symmetrical phyla, grouped together as the Radiata, who are now about to join the pilgrimage: the Cnidaria (sea anemones, corals, jellyfish, etc.) and the Ctenophora (comb jellies). Unfortunately for this simple terminology, starfish and their kin, which zoologists are sure are descended from Bilateria, are also radially symmetrical, at least in the adult phase. Echinoderms are assumed to have become secondarily radial when they took to a bottom-living existence. They have bilaterally symmetrical larvae, and are not closely related to the 'truly' radiate animals such as jellyfish. Reflexively, not all the cnidarians (sea anemones and their kind) are (quite) radially symmetrical, and some zoologists think they too had bilaterally symmetrical ancestors.

All in all, Bilateria is an unfortunate name by which to unite the descendants of Concestor 27 and separate them from those pilgrims still to join. Another possible criterion is 'triploblasty' (three layers of cells) versus 'diploblasty' (two). At a crucial stage in their embryology, cnidarians and ctenophores build their bodies out of two main layers of cells ('ectoderm' and 'endoderm'), the Bilateria out of three (they add 'mesoderm' in the middle). Even this is open to dispute, however. Some zoologists believe 'Radiata' also have mesodermal cells. I think the sensible thing is not to worry about whether Bilateria and Radiata are really good words to use, nor diploblastic and triploblastic, but just concentrate on who are the next pilgrims to join.

Even this is subject to dispute. Nobody doubts that the cnidarians are a unitary group of pilgrims who all join up with each other 'before' they join anyone else. And nobody doubts the same of the ctenophores. The question is, in what order do they join each other and join us? All three logical possibilities have been supported. To make matters worse, there is a tiny phylum, the Placozoa, containing only a single genus, *Trichoplax*, and nobody knows for sure where to put *Trichoplax*. I shall follow the school of thought that says the cnidarians are the first to join us at Rendezvous 28, then the ctenophores at Rendezvous 29, then *Trichoplax* at Rendezvous 30. All this will become resolved definitively when more molecular data become available. This will be soon but, I fear, not soon enough for this book. Be warned that Rendezvous 28 and 29, as well as 30 and 31, could turn out to be in the wrong order.

CNIDARIANS

Our pilgrim band of worms and their descendants has now swelled to very large numbers, and we all pass on back to Rendezvous 28 where we are joined by the cnidarians (the c is silent). They include the freshwater hydras and the more familiar marine sea anemones, corals and jellyfish, all very different from worms. Unlike the Bilateria, they are radially symmetrical about a central mouth. They have no obvious head, no front or rear, no left or right, only an up or down.

What is the date of the rendezvous? Well, who knows? In order to draw rendezvous points in proportional positions in the diagrams that accompany them, it is necessary to set a date. But out here in deep time, there is so much uncertainty that we can do little but space our dates out to the nearest 50 or even 100 million years. Anything smaller would convey a false sense of precision. Some authorities would disagree by hundreds of millions of years.

Because they are among our most distant animal cousins (some were once even confused with plants), the cnidarians are often regarded as very primitive. Of course this doesn't follow – they have had the same time to evolve since Concestor 28 as we have. But it is true that they lack many of the features that we regard as advanced in an animal. They have no long-distance sense organs, their nervous system is a diffuse network, not urbanised into brain, ganglia or major nerve trunks, and their digestive organ is a single, usually uncomplicated cavity with only one opening, the mouth, which also does duty as anus.

On the other hand, there aren't many animals who could claim to have redrawn the map of the world. Cnidarians make islands: islands you can live on; islands big enough to need, and accommodate, an airport. The Great Barrier Reef is more than 2,000 kilometres long. It was Charles Darwin himself who worked out how such coral reefs are formed, as we shall see in the Polypifer's Tale. Cnidarians also include the most dangerously venomous animals in the world, the extreme example being the box jellyfish, which oblige prudent Australian bathers to wear nylon bodystockings. The weapon cnidarians use is remarkable for various reasons, in addition to its formidable power. Unlike a snake's fangs, or the sting of a scorpion or a hornet, the jellyfish sting emerges from inside a cell as a miniature harpoon. Well, thousands of cells, called cnidocytes (or sometimes nematocysts, but this is strictly just one variety of cnidocyte), each with its own cell-sized harpoon called a cnida. *Knide* is Greek for nettle, and it gives the Cnidaria their name. Not all of them are as dangerous to us as box jellyfish, and many are not even painful. When you touch the tentacles of a sea anemone, the 'sticky' feeling on your finger is the clutch of hundreds of tiny harpoons, each on the end of its own little thread, which attaches it to the anemone.

The cnidarian harpoon is probably the most complicated piece of apparatus inside any cell anywhere in the animal or plant kingdoms. In the resting state, waiting to be launched, the harpoon is a coiled tube inside the cell, under pressure (osmotic pressure, if you want the details) waiting to be released. The hair trigger is indeed a tiny hair, the cnidocil, projecting outwards from the cell. When triggered, the cell bursts open, and the pressure turns the entire coiled mechanism inside out with great force, shooting into the body of the victim and injecting poison. Once triggered in this way, the harpoon cell is spent. It cannot be charged up again for re-use. But, as with most kinds of cell, new ones are being made all the time.

Probably the most complicated piece of apparatus inside any cell. Cross-section of a cnidarian harpoon.

All cnidarians have cnidae, and only cnidarians have them. That is the next remarkable thing about them: they provide one of very few examples of an utterly unambiguous, single diagnostic characteristic of any major animal group. If you see an animal without any cnidae, it is not a cnidarian. If you see an animal with a cnida, it is a cnidarian. Actually, there is one exception, and it is as neat a case as you could want of an exception proving a rule. Sea slugs of the molluscan group called nudibranchs (they joined us along with almost everybody else at Rendezvous 26) often have beautifully

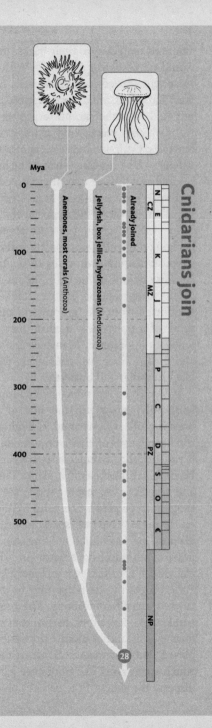

Cnidarians join. The order of branching of the cnidarians (jelly-fishes, corals, sea anemones and the like) and the ctenophores (comb jellies) is effectively unresolved. Most authors position either one or the other (or sometimes both) as the closest living relative of the bilaterally symmetrical animals. Certain molecular data hint that cnidarians may occupy this position. Unfortunately, the form and branching of sub-groups within the 9,000 or so cnidarian species is also disputed, but the fundamental division between lineages with or without an evolved medusa stage in their life cycle (see text) is widely accepted.

Images, left to right: white-spotted rose anemone (*Urticina lofotensis*); leptomedusan jellyfish (*Aequorea* sp.).

Mya

Cnidarians join

0

Anemones, most corals (Anthozoa)

Jellyfish, box jellies, hydrozoans (Medusozoa)

Already joined

N
E
CZ

100

K
MZ
J

200

T

P

300

C

D
PZ

400

S

O

500

€

NP

28

coloured tentacles on their backs, the kind of coloration that makes would-be predators back off. With good reason. In some species, these tentacles contain cnidocytes, identical to those of true cnidarians. But only Cnidaria are supposed to have cnidae, so what is going on? As I said, the exception proves the rule. The slug eats jellyfish, from which it passes cnidocytes, intact and still working, to its own tentacles. Commandeered weapons, they are still capable of firing, in defence of the sea slug – hence the bright warning coloration.

Cnidarians have two alternative body plans: the polyp and the medusa. A sea anemone or a *Hydra* is a typical polyp: sedentary, mouth uppermost, the opposite end fixed to the ground like a plant. They feed by waving tentacles about, harpooning small prey, then bringing the tentacle, complete with prey, to the mouth. A jellyfish is a typical medusa, swimming through the open sea by pulsing muscular contractions of the bell. The mouth of a jellyfish is in the centre, on the lower side. So you can think of a medusa as a polyp that has freed itself from the bottom and turned over to swim. Or you can think of a polyp as a medusa that has settled on its back with its tentacles uppermost. Many species of cnidarian have both polypoid and medusoid forms, alternating them through the life cycle, a bit like caterpillar and butterfly.

Polyps often reproduce by budding vegetatively, like plants. A new baby polyp grows on the side of a freshwater *Hydra*, eventually breaking off as a separate individual: a clone of the parent. Many marine relatives of *Hydra* do something similar, but the clone doesn't break off and assume a separate existence. It stays attached and becomes a branch, as in a plant. These 'colonial hydrozoans' branch and branch again, making it easy for us to understand why they were thought to be plants. Sometimes more than one kind of polyp grows on the same polyp tree, specialised for different roles, such as feeding, defence, or reproduction. You can think of them as a colony of polyps, but there is a sense in which they are all parts of one individual, for the tree is a clone: all the polyps have the same genes. Food caught by one polyp may be used by others, since their gastric cavities are all continuous. The branches of the tree and its main trunk are hollow tubes that you can think of as a shared stomach – or maybe as a kind of circulatory system playing the role that in us is played by blood vessels. Some of the polyps bud off tiny medusae, which swim away like miniature jellyfish to reproduce sexually and disperse the genes of the parent polyp tree to distant places.

A group of cnidarians called the siphonophores have taken the colonial habit to an extreme. We can think of them as polyp trees which, instead of being fixed to a rock or a piece of seaweed, hang down either from one or a cluster of swimming medusae (which are, of course, members of the clone) or to a float at the surface. The Portuguese man-of-war *Physalia* has a large gas-filled float with a vertical sail on top. A complicated colony of polyps and tentacles dangles beneath. It doesn't swim but gets about through being blown by the wind. The smaller *Velella* is a flat, oval raft with a diagonally placed vertical sail. It too uses the wind to disperse, and its English names are Jack-sails-by-the-wind or by-the-wind-sailor. You often find the dried-up little rafts with their sails on the beach, where they usually lose their blue colour and seem to be made of whitish plastic. *Velella* resembles the true Portuguese man-of-war in that both sail by the wind. *Velella* and its relative *Porpita* are not siphonophore colonies, however, but single, highly modified polyps, hanging down from a float rather than sticking up from a rock (see plate 39).

Many siphonophores can adjust their depth in the water, rather as bony fish do with their swim bladders, by secreting gas into the float, or releasing it. Some have a combination of floats and swimming medusae, and all have polyps and tentacles dangling beneath. The siphonophores are treated by E. O. Wilson, founder of the science of sociobiology, as one of the four pinnacles of social evolution (the others being the social insects, the social mammals and ourselves). This, then, is another superlative that one can attach to the Cnidaria. Except that, since the members of a colony are clones, genetically identical to each other, it is by no means clear that we should call them a colony rather than a single individual.

Hydrozoans see the medusa as a way for their genes to hop occasionally from one stable living place to another. Jellyfish could be said to take the medusoid form seriously, as what living is all about. Corals, by contrast, take sedentary living to the extreme lengths of building a hard, solid house that is destined to stay there for thousands of years. We shall take their tales in order.

THE JELLYFISH'S TALE

Jellyfish ride the ocean currents as Jack-sails-by-the-wind. They don't pursue their prey, as a barracuda or squid might. Instead, they rely on their long, trailing, armed tentacles to trap planktonic creatures that are unlucky enough to bump into them. Jellyfish do swim, with the languorous heartbeat of the bell, but they are not swimming in any particular direction, at least as we would understand direction. Our understanding, however, is limited by our two dimensional trammels: we crawl over the surface of the land, and even when we take off into the third dimension it is only in order to crawl a bit faster in the other two. But in the sea, the third dimension is the most salient. It is the dimension in which travelling has the most effect. In addition to the steep pressure gradient with depth, there is a light gradient, complicated by a gradient of colour balance. But the light disappears anyway as day gives way to night. As we shall see, a planktonic animal's preferred depth changes dramatically with the 24-hour cycle.

During the Second World War, sonar operators looking for submarines were puzzled by what seemed to be a false bottom of the sea that rose towards the surface every evening, and sank back down again the next morning. It turned out to be the bulk of the plankton, millions of tiny crustaceans and other creatures, rising to feed near the surface at night, then sinking at morning. Why should they do this? The best guess seems to be that during the hours of daylight they are vulnerable to visually hunting predators such as fish and squids, so they seek the dark safety of the depths by day. Why, then, come to the surface at night, for it is a long journey that must consume a lot of energy? One student of the plankton has compared it to a human daily walking 25 miles each way, just to get breakfast.

The reason for visiting the surface is that food ultimately comes from the sun, via plants. The surface layers of the sea are unbroken green prairies, with microscopic single-celled algae in the role of waving grass. The surface is where the food ultimately is, and that is where the grazers, and those that feed on the grazers, and those that in turn feed on them, must be. But if it is safe to be there only by night because of visually hunting predators, a diurnal migration is exactly what the grazers and their small predators must undertake. And apparently they do. The 'prairie' itself doesn't migrate. If there were any sense in doing so, it should swim against the animal tide, for its whole *raison d'etre* is to catch sunlight at the surface during the day, and avoid being eaten.

Whatever the reason, most of the animals in the plankton migrate down for the day and up for the night. The jellyfish, or many of them, follow the herds, like lions and hyenas tracking the wildebeest across the Mara and Serengeti plains. Although, unlike lions and hyenas, jellyfish don't target individual prey, even blindly trailing tentacles will benefit by following the herds, and this is one of the reasons jellyfish swim. Some species increase their catch rate by zigzagging about, again not individually targeting prey, but increasing the area swept by those tentacles with their batteries of lethal harpoons. Others just migrate up and down.

A different kind of migration has been described for the massed jellyfish of 'Jellyfish Lake' on Mercherchar, one of the Palau Islands (an American colony in the western Pacific). The lake, which communicates underground with the sea and is therefore salty, is named after its huge population of jellyfish. There are several kinds, but the dominant one is *Mastigias,* an estimated 20 million of them in a lake 2.5 kilometres long and 1.5 kilometres wide. All the jellyfish spend the night near the western end of the lake. When the sun rises in the east, they all swim straight towards it and therefore the eastern end of the lake. They stop before they reach the shore, for an interestingly simple reason. The trees fringing the shore cast a deep shadow, cutting off so much of the sun's light that the jellyfish's sun-seeking automatic pilot starts to drive them towards the now brighter west. As soon as they come out from the trees' shadow, however, they turn east again.

This internal conflict traps them around the line of the shadow, with the consequence (which I dare not think is more than co-incidence) of keeping them a safe distance from the dangerously predatory sea anemones that line the shore itself. In the afternoon, the jellyfish follow the sun back to the western end of the lake, where the whole armada again becomes trapped at the shadow line of the trees (see plate 40). When it becomes dark, they swim vertically up and down at the western end of the lake, until the dawn sun lures their automatic guidance system back towards the east. I don't know what they might gain from this remarkable twice-daily migration. The published explanation satisfies me too little to bear repetition. For now, the lesson of the tale must be that the living world offers much that we don't *yet* understand, and that is exciting in itself.

THE POLYPIFER'S TALE

All evolving creatures track changes in the world: changes in the weather, in temperature, rainfall and – more complicated because they hit back in evolutionary time – changes in other evolving lines such as predators and prey. Some evolving creatures alter, by their very presence, the world in which they live, and to which they must adapt. The oxygen we breathe was not there before green plants put it there. At first a poison, it provided radically changed conditions that most animal lineages were forced first to tolerate, and then to depend upon. On a shorter timescale, the trees in a mature forest inhabit a world that they themselves have created, over hundreds of years – the time it takes to transform bare sand into climax forest. A climax forest is, of course, also a complex and rich environment to which huge numbers of other plant and animal species have become adapted.

Because the word 'coral' is used both for an organism and for the hard material that it builds, I shall indulge a fancy and adopt from Darwin the older word 'polypifer' for the coral organism that tells this tale. Pronounce it 'pol-lip-if-er', with the stress on lip. Coral organisms, or polypifers, transform their world, over a timespan of hundreds of thousands of years, by building on the dead skeletons of their own past generations to construct huge underwater mountains: wave-resisting ramparts. Before they die, corals combine with countless other corals to condition the world in which future corals will live. And not just future corals, but future generations of an enormous and intricate community of animals and plants. The idea of community will be the main message of this tale.

The picture reproduced in plate 41 shows Heron Island, the one island of the Great Barrier Reef that I have visited (twice). The houses dotted around the near end of the little island give an idea of scale. The huge pale area surrounding the island itself is the reef, of which the island is just the highest tip, covered with sand made of crushed coral (much of it having passed through fish guts) in which vegetation of limited variety grows, supporting a similarly limited fauna of land animals. For objects that are entirely made by living creatures, coral reefs are big, and core drillings show some of them to be many hundreds of metres deep. Heron Island is just one of the more than 1,000 islands and nearly 3,000 reefs that constitute the Great Barrier Reef, arcing round the north-east side of Australia for 2,000 kilometres. The Great Barrier Reef is often said – with what veracity I don't know – to be the only evidence of life on our planet that is large enough to be visible from outer space. It is also said to be home to 30 per cent of the world's sea creatures, but again I am not sure quite what that means – what is being counted? Never mind, the Great

Barrier Reef is an utterly remarkable object, and it has been entirely built by the small sea anemone-like animals called corals or polypifers. The living polypifers occupy only the surface layers of a coral reef. Beneath them, to a depth of hundreds of metres in some oceanic atolls, are the skeletons of their predecessors, compacted to limestone.

Nowadays only corals build reefs, but in earlier geological eras they had no such monopoly. Reefs have at various times been built by algae, sponges, molluscs and tube worms too. The great success of coral organisms themselves seems to stem from their association with microscopic algae, which live inside their cells and photosynthesise in the sunlit shallows, to the eventual benefit of the corals. These algae, called zooxanthellae, have a variety of different coloured pigments for trapping light, which accounts for the vividly photogenic appearance of coral reefs. It is not surprising that corals were once thought to be plants. They get much of their food in the same way as plants, and they compete for light as plants do. It is only to be expected that they would take on similar shapes. Moreover, their struggles to overshadow, and not be overshadowed, lead to the whole community of corals taking on something of the appearance of a forest canopy. And, like any forest, a coral reef is also home to a large community of other creatures.

Coral reefs hugely increase the 'ecospace' of an area. As my colleague Richard Southwood puts it in his book *The Story of Life:*

> Where there would otherwise be a surface of rock or sand with a column of water above it, the reef provides a complex three-dimensional structure with a great amount of extra surface, with many cracks and small caves.

Forests do the same kind of thing, inflating the effective surface area available for biological activity and colonisation. Increased ecospace is the kind of thing we expect to find in complex ecological communities. Coral reefs are home to a huge variety of animals of all kinds, nestling in every corner and nook of the prodigious ecospace provided.

Something similar happens in the organs of a body. The human brain increases its effective area – and hence its functional capacity – by elaborate folding. It may be no accident that the 'brain coral' so strikingly resembles it.

Darwin himself was the first to understand how coral reefs are formed. His debut scientific book (after his travel book on the *Voyage of the Beagle*) was the treatise on *Coral Reefs* that he published when only 33. Here is Darwin's problem as we would see it today, although he did not have access to most of the information that is relevant either to posing the problem or solving it. Darwin, indeed, was as astoundingly prescient in his theory of coral reefs as he was to be in his more famous theories of natural selection and sexual selection.

Corals can live only in shallow water. They depend upon the algae in their cells, and the algae of course need light. Shallow water is also good for the planktonic prey with which corals supplement their diet. Corals are denizens of shorelines, and you can indeed find shallow 'fringing reefs' around tropical coasts. But what is puzzling about corals is that you can also find them surrounded by very deep water. Oceanic coral islands are the summits of lofty underwater mountains made by generations of dead corals. Barrier reefs are an intermediate category, following the line of a coast, but farther out than fringing reefs, and with deeper water between them and the shore. Even in the case of remote coral islands completely isolated in the deep ocean, the living corals are always in shallow water, close to the light where they and their algae can thrive. But the water is shallow only by courtesy of the generations of earlier corals on which they sit.

Darwin, as I say, didn't have all the information needed to realise the extent of the problem. It is only because people have drilled down into reefs, and found compacted coral to great depths, that we now know that coral atolls are the summits of towering underwater mountains made of ancient coral. In Darwin's time the prevailing theory was that atolls were superficial encrustations of coral on top of submerged volcanoes that lay only just below the surface. On this theory there was no problem to solve. Corals grew only in shallow water, and it was the volcanoes that gave them the perch they needed to find shallow water. But Darwin didn't believe it, even though he had no way of knowing the dead coral was so deep.

Darwin's second feat of prescience was his theory itself. He suggested that the sea bottom was continually subsiding in the vicinity of the atoll (while rising in other places, as he vividly knew from finding marine fossils high in the Andes). This was, of course, long before the theory of plate tectonics. Darwin was inspired by his mentor, the geologist Charles Lyell, who believed that parts of the

Earth's crust rose and sank relative to one another. Darwin proposed that as the sea bottom subsided, it took the coral mountain down with it. Corals grew on top of the subsiding undersea mountain, just keeping pace with the subsidence in such a way that the summit was always near the surface of the sea, in the zone of light and prosperity. The mountain itself was just layer upon layer of dead corals which had once prospered in the sun. The oldest corals, at the base of the underwater mountain, probably began as a fringing reef of some forgotten piece of land or long-dead volcano. As the land gradually submerged beneath the water, the corals later became a barrier reef, at an increasing distance away from the receding coastline. With further subsidence the original land disappeared altogether, and the barrier reef became the basis for a prolonged extension of the under-water mountain for as long as the subsidence continued. Remote oceanic coral islands got their start perched on the top of volcanoes, the base of which slowly subsided in the same way. Darwin's idea is still substantially supported today, with the addition of plate tectonics to explain the subsidence.

A coral reef is a textbook example of a climax community, and this shall be the climax of the Polypifer's Tale. A community is a collection of species that have evolved to flourish in each other's presence. A rainforest is a community. So is a bog. So is a coral reef. Sometimes the same kind of community springs up in parallel in different parts of the world where the climate favours it. 'Mediterranean' communities have arisen not just around the Mediterranean Sea itself, but on the coasts of California, Chile, south-western Australia and the Cape region of South Africa. The particular species of plants found in these five regions are different, but the plant communities themselves are as characteristically 'Mediterranean' as, say, Tokyo and Los Angeles are recognisably 'urban sprawl'. And an equally characteristic fauna goes with the Mediterranean vegetation.

Tropical reef communities are like that. They vary in detail but are the same in essentials, whether we are talking about the South Pacific, the Indian Ocean, the Red Sea or the Caribbean. There are also temperate-zone reefs, which are somewhat different, but one very particular thing the two have in common is the remarkable phenomenon of cleaner fish – a wonder which epitomises the sort of subtle intimacy that can arise in a climax ecological community.

A number of species of small fish, and some shrimps, ply a

prosperous trade, harvesting nutritious parasites, or mucus, off the surfaces of larger fish, and in some cases even entering their mouths, picking their teeth and exiting through the gills. This argues for an astonishing level of 'trust',* but here my interest is more focused on cleaner fish as an example of a 'role' in a community. Individual cleaners typically have a so-called 'cleaning station', to which larger fish come to be serviced. The advantage of this to both parties is presumably a saving of time that might otherwise be spent searching for a cleaner or searching for a client. Site-tenacity also allows repeated meetings between individual cleaners and clients, which allows the all-important 'trust' to build up. These cleaning stations have been compared to barbers' shops (see plate 42). It has been claimed – though the evidence has more recently been disputed – that if all the cleaners are removed from a reef, the general health of the fish on the reef nosedives.

In different parts of the world, the local cleaners have evolved independently, and are drawn from different groups of fish. On the Caribbean reefs, the cleaner trade is mostly filled by members of the goby family, which typically form small groups of cleaners. In the Pacific, on the other hand, the best-known cleaner is a wrasse, *Labroides*. *L. dimidiatus* runs its 'barber's shop' by day, while *L. bicolor*, so George Barlow, my colleague from Berkeley days tells me, services the nocturnal guild of fishes who take refuge in caves during the day. Such divvying-up of a trade among species is typical of a mature ecological community. Professor Barlow's book, *The Cichlid Fishes*, gives examples of freshwater species in the great lakes of Africa that have taken convergent steps towards the cleaner habit.

On tropical coral reefs, the almost fantastic levels of co-operation achieved between cleaner fish and 'client' is symbolic of the way an ecological community can sometimes simulate the intricate harmony of a single organism. Indeed, the resemblance is seductive – too seductive. Herbivores depend on plants; carnivores depend on herbivores; without predation, population sizes would spiral out of control with disastrous results for all; without scavengers like burying beetles and bacteria, the world would sate with corpses, and manure would never be recycled into the plants. Without particular 'key-stone' species, whose identity is sometimes quite surprising, the

---

* The evolutionary problems of evolving 'trust' are interesting, but I have already dealt with the matter in *The Selfish Gene*, so must refrain from repeating myself here.

whole community would 'collapse'. It is tempting to see each species as an organ in the super-organism that is the community.

To describe the forests of the world as its 'lungs' does no harm, and it might do some good if it encourages people to preserve them. But the rhetoric of holistic harmony can degenerate into a kind of dotty, Prince Charles-style mysticism. Indeed, the idea of a mystical 'balance of nature' often appeals to the same kind of airheads who go to quack doctors to 'balance their energy fields'. But there are profound differences between the way the organs of a body and the species of a community interact with each other in their respective domains to produce the appearance of a harmonious whole.

The parallel must be treated with great caution. Yet it is not completely without foundation. There is an ecology within the individual organism, a community of genes in the gene pool of a species. The forces that produce harmony among the parts of an organism's body are not wholly unlike the forces that produce the illusion of harmony in the species of a coral reef. There is balance in a rainforest, structure in a reef community, an elegant meshing of parts that recalls co-adaptation within an animal body. In neither case is the balanced unit favoured *as a unit* by Darwinian selection. In both cases the balance comes about through selection at a lower level. Selection doesn't favour a harmonious whole. Instead, harmonious parts flourish in the presence of each other, and the illusion of a harmonious whole emerges.

Carnivores flourish in the presence of herbivores, and herbivores flourish in the presence of plants. But what about the other way around? Do plants flourish in the presence of herbivores? Do herbivores flourish in the presence of carnivores? Do animals and plants need enemies to eat them in order to flourish? Not in the straightforward way suggested by the rhetoric of some ecological activists. No creature normally benefits from being eaten. But grasses that can withstand being cropped better than rival plants really do flourish in the presence of grazers – on the principle of 'my enemy's enemy'. And something like the same story might be told of victims of parasites – and predators, although here the story is more complicated. It is still misleading to say that a community 'needs' its parasites and predators like a polar bear needs its liver or its teeth. But the 'enemy's enemy' principle does lead to something like the same result. It can be right to see a community of species, such as a coral reef, as a kind of balanced entity that is potentially threatened by removal of its parts.

This idea of community, as made up of lower-level units that flourish in the presence of each other, pervades life. Even within the single cell, the principle applies. Most animal cells house communities of bacteria so comprehensively integrated into the smooth working of the cell that their bacterial origins have only recently become understood. Mitochondria, once free-living bacteria, are as essential to the workings of our cells as our cells are to them. Their genes have flourished in the presence of ours, as ours have flourished in the presence of theirs. Plant cells by themselves are incapable of photosynthesis. That chemical wizardry is performed by guest workers, originally bacteria and now re-labelled chloroplasts. Plant eaters, such as ruminants and termites, are themselves largely incapable of digesting cellulose. But they are good at finding and chewing plants (see the Mixotrich's Tale). The gap in the market offered by their plant-filled guts is exploited by symbiotic micro-organisms that possess the biochemical expertise necessary to digest plant material efficiently. Creatures with complementary skills flourish in each other's presence.

What I want to add to that familiar point is that the process is mirrored at the level of every species' 'own' genes. The entire genome of a polar bear or a penguin, of a caiman or a guanaco, is an ecological community of genes that flourish in each other's presence. The immediate arena of this flourishing is the interior of an individual's cells. But the long-term arena is the gene pool of the species. Given sexual reproduction, the gene pool is the habitat of every gene as it is recopied and recombined down the generations.

CTENOPHORES

The ctenophores, who join us at Rendezvous 29, are some of the most beautiful of all the animal pilgrims. A superficial resemblance has led them to be wrongly classified as jellyfish. They used to be placed in the same phylum, which was known as the Coelenterata, celebrating their shared characteristic, the fact that the main body cavity is also the digestive chamber. They also have a simple nerve net, like the Cnidaria, and their bodies are likewise built from (disputably) only two layers of tissue. The balance of modern evidence suggests, however, that the cnidarians are closer cousins to us than they are to the ctenophores: another way of saying that the cnidarians join the pilgrimage 'before' the ctenophores do. I don't feel confident enough of this, however, to quote a date for the event.

Ctenophore in Greek means 'comb-bearer'. The 'combs' are prominent rows of hair-like cilia, whose beating propels these delicate creatures in place of the pulsating muscles that do the same for the superficially similar jellyfish. It is not a fast system of propulsion, but it presumably serves adequately, not for actively chasing prey but for the same kind of undirected improvement in capture rate that the jellyfish achieve. Because of their resemblance to jellyfish, and their delicate jelly-like consistency, the ctenophores are known in English as comb jellies. There aren't many species of them – only about 100 – but the total number of individuals is not small, and they beautify, by any standards, all the oceans of the world. Waves of synchronised motion pass up the comb rows in eerie iridescence.

Ctenophores are predatory but, like jellyfish, they rely on prey passively bumping into their tentacles. Although their tentacles look like those of jellyfish, they have no cnidocytes. Instead, they have their own make of 'lasso cells', which discharge a kind of glue instead of sharp, poisonous harpoons. Perhaps we could see a ctenophore as a kind of alternative way of being a jellyfish. Some of them, however, are far from bell-shaped. The ravishingly beautiful *Cestum veneris* is one of those rare animals whose English and Latin names mean exactly the same thing, Venus's girdle, and no wonder: the body is a

Ctenophores join

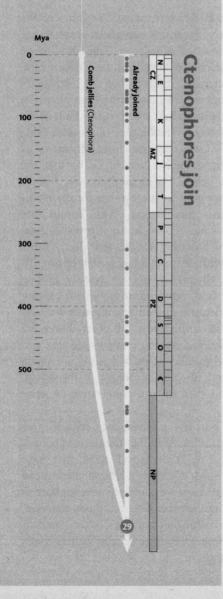

N	E	K	MZ	J	T	P	C	PZ	D	S	O	Є

Mya

Comb jellies (Ctenophora)

Already joined

29

Ctenophores join. The bilaterally symmetrical animals, together with the cnidarians and ctenophores, are sometimes referred to as the 'Eumetazoa'. Following some molecular studies, the 100 known species of ctenophore are here placed as the most distant relatives of the rest, but this position is not definitive.

Image: *Beroe* sp.

long, shimmering, ethereally beautiful ribbon, too good for a goddess (see plate 43). Notice that, although the Venus's girdle is long and thin like a worm, the 'worm' has no head or tail end, but is mirrored about the middle, where the mouth is – the 'buckle' of the girdle. It is still radially (or strictly biradially) symmetrical.

PLACOZOANS

Here is an enigmatic little animal: *Trichoplax adhaerens*, the only known species in its entire phylum, the Placozoa – which, of course, doesn't necessarily mean it is the only one. I should mention that in 1896 a second placozoan from the Gulf of Naples was described and named *Treptoplax reptans*. It has never been found again, however, and most experts think that particular specimen was *Trichoplax* itself. Molecular evidence may well uncover other species soon.

Trichoplax lives in the sea and doesn't look like anything much, not symmetrical in any direction; a bit like an amoeba, except that it is made of lots of cells instead of just one; a bit like a very small flatworm, except that it doesn't have any obvious front or back end, nor left or right side. A tiny, irregularly shaped mat, perhaps three millimetres across, *Trichoplax* creeps over the surface on a little upside-down carpet of beating cilia. It feeds on single-celled creatures, mostly algae, even smaller than itself, which it digests through its lower surface without taking them inside its body.

There is not much in its anatomy to connect *Trichoplax* with any other kind of animal. It has two main cell layers, like a cnidarian or a ctenophore. Sandwiched between the two main layers are a few contractile cells that work as its nearest approach to muscles. The animal shortens these strings to change its shape. Strictly the two main cell layers should probably not be called dorsal and ventral. The upper layer is sometimes called protective and the lower layer digestive. Some authors claim that the digestive layer invaginates to form a temporary cavity for digestive purposes, but not all observers have seen this and it may not be true.

Trichoplax has had a somewhat confused history in the zoological literature, as is recounted by T. Syed and B. Schierwater in a recent paper. When first described in 1883, *Trichoplax* was thought to be very primitive; it has now recovered that honoured status. Unfortunately, it bears a superficial resemblance to the so-called planula larva of some cnidarians. In 1907 a German zoologist called Thilo Krumbach thought he saw *Trichoplax* where he had previously seen planula larvae, and he regarded the little creatures as modified planulae. That

Placozoans join

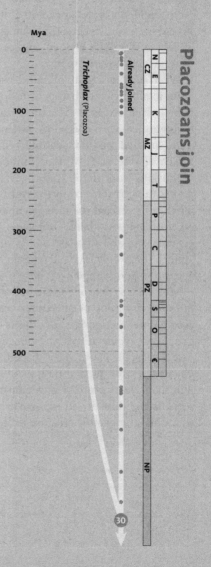

Mya

Trichoplax (Placozoa)

Already joined

| N | E | | K | | MZ | J | | T | | P | | C | | | | D | PZ | S | O | | Є | | NP |

30

Placozoans join. As with Rendez-vous 28 and 29, the order of Rendezvous 30 and 31 is pretty well unresolved. Rendezvous 30 could be with the placozoans (represented by its single species, *Trichoplax*), or it could be with the sponges. At present this ordering is essentially arbitrary. It would be entirely unsurprising if Rendezvous 30 and 31 had to swap.

Image: *Trichoplax adhaerens.*

wouldn't have mattered too much but for the death in 1922 of W. Kükenthal, the editor of the authoritative multi-volume *Handbuch der Zoologie*. Unluckily for *Trichoplax*, Kükenthal's understudy as editor was the very same Thilo Krumbach. *Trichoplax* was duly billed as a cnidarian in Kükenthal & Krumbach, and this was copied by the French equivalent, the *Traité de Zoologie*, edited by P. P. Grassé (who, incidentally, retained anti-Darwinian sympathies long after he should have known better). It was also picked up from the *Handbuch* by Libbie Henrietta Hyman, author of the leading American multi-volume work *Invertebrates*.

With such a weight of multi-volumed authority bearing down upon it, what chance had poor little *Trichoplax*, especially given that nobody had looked at the animal itself for more than half a century? It languished as an alleged cnidarian larva until the molecular revolution opened up the possibility of discovering its real affinities. Whatever else it is, it is definitely not a cnidarian. Preliminary indications from rRNA studies (see Taq's Tale) suggest that *Trichoplax* is more distant from the rest of the animal kingdom than any other group except the sponges, and it may be that even the sponges are closer to us than *Trichoplax* is. Trichoplax has the smallest genome and the simplest bodily organisation of any multicellular animal. It has only four cell types in its body, compared to more than 200 in us. And it appears to have a single Hox gene.

Molecular genetic evidence indicates provisionally that this lonely little pilgrim joins us at Rendezvous 30, perhaps 780 million years ago, 'before' the sponges. But this is really anyone's guess. It could be that Rendezvous 30 and 31 (sponges) should be reversed, in which case *Trichoplax* is our most distant cousin among the true animals. Understandably, there is now some strong lobbying for *Trichoplax* to join that select company of organisms whose genome is completely sequenced. I think it will happen, in which case we should soon know what this strange little creature really is.

SPONGES

Sponges are the last pilgrims to join us who are members of the Metazoa, the truly multicellular animals. Sponges haven't always been dignified as Metazoa, but were written off as 'Parazoa' – a name for a kind of second-class citizen of the animal kingdom. Nowadays the same class distinction is fostered by placing the sponges in the Metazoa, but coining the word Eumetazoa for all the rest *except* sponges (some authors also except *Trichoplax,* the little animal we met at Rendezvous 30).

People are occasionally surprised to learn that sponges are animals rather than plants. Like plants, they don't move. Well, they don't move their whole body. Neither plants nor sponges have muscles. There is movement at the cellular level, but that is true of plants too. Sponges live by passing a ceaseless current of water right through the body, from which they filter food particles. Consequently, they are full of holes, which is what makes them so good at holding water in the bath.

Bath sponges, however, don't give a good idea of the typical body form, which is a hollow pitcher with a big opening at the top and lots of smaller holes all round the sides. As is easy to tell by putting a little dye in the water outside the pitcher of a living sponge, water is drawn in through the small holes around the sides, and expelled into the main hollow interior, from which it flows out through the main entrance of the pitcher. The water is driven by special cells called choanocytes, which line the chambers and canals of the walls of a sponge. Each choanocyte has a waving flagellum (like a cilium, only larger) surrounded by a deep collar. We shall meet the choanocytes again, as they are important for our evolutionary story.

Sponges have no nervous system and a relatively simple internal structure. Although they have several different kinds of cells, those cells don't organise themselves into tissues and organs the way ours do. Sponge cells are 'toti potent', which means that every cell is capable of becoming any of the sponge's repertoire of cell types. This is not true of our cells. A liver cell is not capable of giving rise to a kidney cell or a nerve cell. But sponge cells are so flexible that any

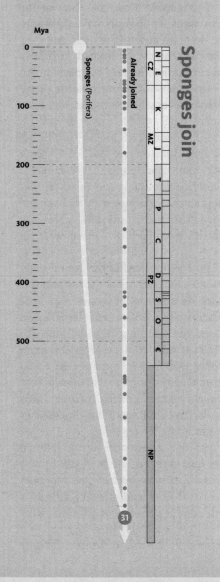

Sponges join

Mya					
0		N		CZ	
		E			
100		K		MZ	
200		J			
		T			
300		P			
		C		PZ	
400		D			
		S			
500		O			
		C			
		NP			

Sponges (Porifera)

Already joined

31

Sponges join. Since the time of Linnaeus, the animals ('metazoans') have been classified as one of the kingdoms of life. The approximately 10,000 described species of sponge are usually seen as a very early diverging branch, a position confirmed by molecular data (though *Trichoplax* may have diverged even earlier). A minority of molecular taxonomists think that there are two lineages of sponges, one more closely related to the rest of the Metazoa than the other – this implies that the earliest metazoans really did look like sponges and would have been classified as such – but this is highly controversial.

Image: yellow tube sponge (*Aplysina fistularis*).

isolated cell is capable of growing a whole new sponge (and there's more to it, as we shall see in the Sponge's Tale).

Unsurprisingly, therefore, sponges make no distinction between 'germ line' and 'soma'. In the Eumetazoa, germ-line cells are those that are capable of giving rise to reproductive cells and whose genes are therefore in principle immortal. The germ line is a small minority of cells residing in ovaries or testes, and insulated from the need to do anything else but reproduce. Soma is that part of the body that is not germ line – somatic cells are destined not to pass their genes on indefinitely. In a eumetazoan such as a mammal, a subset of cells is set aside early in embryology as germ line. The rest of the cells, the cells of the soma, may divide a few times to make liver or kidney, bone or muscle, but then their dividing career comes to an end.

Cancer cells are the sinister exception. They have somehow lost the ability to stop dividing. But as Randolph Nesse and George C. Williams, the authors of *The Science of Darwinian Medicine*, point out, we should not be surprised. On the contrary, the surprising thing about cancer is that it is not more common than it is. Every cell in the body, after all, is descended from an unbroken line of billions of generations of germ-line cells that have not stopped dividing. Suddenly being asked to become a somatic cell like a liver cell, and learn the art of *not* dividing, has never happened before in the entire history of the cell's ancestors! Don't be confused. Of course the bodies that housed the cell's ancestors had livers. But germ-line cells – by definition – are not descended from liver cells.

All sponge cells are germ-line cells – all potentially immortal. They have several different cell types, but they are deployed in development in a different way from most multicellular animals. Eumetazoan embryos form cell layers that fold and invaginate in complicated 'origami' ways to build the body. Sponges don't have that kind of embryology. Instead they self-assemble – each of their toti-potent cells has an affinity for hooking up to other cells, as though they were autonomous protozoa with sociable tendencies. Nevertheless, modern zoologists include the sponges as members of the Metazoa, and I am following this trend. They are probably the most primitive living group of multicellular animals, giving us a better idea of the early Metazoa than any other modern animals.

As with other animals, each species of sponge has its own characteristic shape and colour. The hollow pitcher is only one of many forms. Others are variants on it, systems of hollow cavities connected to one another. Sponges characteristically toughen their structures with collagen fibres (that's what makes bath sponges spongy) and with mineral spicules: crystals of silica or calcium carbonate, the shape of which is often the most reliable diagnostic of the species. Sometimes the spicule skeleton can be intricate and beautiful, as in the glass sponge, *Euplectella* (see plate 44).

The date of Rendezvous 31 is given as 800 million years on the phylogeny diagram, but the usual despairing warnings for such ancient datings apply. The evolution of multicellular sponges from single-celled protozoa is one of the landmark events in evolution – the Origin of the Metazoa – and we shall examine it in the next two tales.

THE SPONGE'S TALE

The 1907 issue of the *Journal of Experimental Zoology* contains a paper on sponges by H. V. Wilson of the University of North Carolina. The research was classic, and the paper describing it recalls a golden age when scientific papers were written in a discursive style that you could understand, and at a length that made it possible to visualise a real person doing real experiments in a real laboratory.

Wilson took a living sponge and separated its cells by forcing them through a fine sieve – a piece of 'bolting cloth'. The disassembled cells were passed into a saucer of sea water, where they formed a red cloud, mostly consisting of single cells. The cloud settled down into a sediment at the bottom of the saucer, where Wilson observed them with his microscope. The cells behaved like individual amoebas, crawling over the bottom of the saucer. When these amoeboid crawlers met others of their kind, they joined up to form growing agglomerations of cells. Eventually, as Wilson and others showed in a series of papers, such agglomerations grow to become whole new sponges. Wilson also tried mashing up sponges of two different species and mixing the two suspensions together. The two species were of different colours, so he could easily see what happened. The cells chose to agglomerate with their own species and not the other. Oddly, Wilson reported this result as a 'failure', since he was hoping – for reasons I don't understand, and which perhaps reflect the different theoretical preconceptions of a zoologist of nearly a century ago – that they would form a composite sponge of two different species.

The 'sociable' behaviour of sponge cells as exhibited by such experiments perhaps sheds light on the normal embryonic development of individual sponges. Does it also give us some sort of hint of how the first multicellular animals (metazoans) *evolved* from single-celled ancestors (protozoans)? The metazoan body is often called a colony of cells. In keeping with this book's pattern of using some tales as modern re-enactments of evolutionary happenings, could the Sponge's Tale be telling us something about the remote evolutionary past? Could the behaviour of the crawling and agglomerating cells in Wilson's experiments represent some sort of re-enactment of how the first sponge arose – as a colony of protozoans?

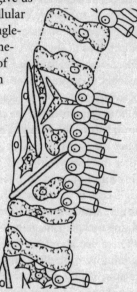

Almost certainly it was not the same in detail. But here is a hint. The most characteristic cells of sponges are the choanocytes, which they use for generating currents of water. The picture shows a portion of the wall of a sponge, with the inside

Sociable cells. Portion of sponge wall showing choanocytes, with their distinctive collars and flagella.

of the cavity to the right. The choanocytes are the cells that line the cavity of the sponge. 'Choano-' comes from the Greek for 'funnel', and you can see the little funnels or collars, made up of many fine hairs known as microvilli. Each choanocyte has a beating flagellum, which draws water through the sponge, while the collar catches nutrient particles in the stream. Take a good look at those choanocytes, for we shall meet something rather like them at the next rendezvous. And then, in the light of that, the following tale will complete our speculation about the origin of multicellularity.

CHOANOFLAGELLATES

The choanoflagellates are the first protozoans to join our pilgrimage, and they do so at Rendezvous 32, which, very tentatively on molecular evidence with worryingly large extrapolations, we date at 900 million years. Look at the picture below. Do the little flagellated cells remind you of anything? Yes, they are very similar to the choanocytes lining the water canals of sponges. It has long been suspected that either they represent a hangover from a sponge ancestor, or they are the evolutionary descendants of sponges that have degenerated to single cells or very few cells. Molecular genetic evidence suggests the former, which is why I am considering them as separate pilgrims, joining our pilgrimage here.

There are about 140 species of choanoflagellates. Some are free-swimming, propelling themselves along with the flagellum. Others are attached by a stalk, sometimes several together in a colony, as in the picture. They use their flagellum to drive water into the funnel, where food particles such as bacteria are trapped and engulfed. In this respect they are different from the choanocytes of sponges. In a sponge, each flagellum is used not to drive food into the individual funnel of the choanocyte, but in co-operation with other choanocytes to draw a current of water in through holes in the walls of the sponge and out through the sponge's main opening. But anatomically each individual choanoflagellate, whether it is in a colony or not, is suspiciously similar to a sponge choanocyte. This fact will bulk large in the Choanoflagellate's Tale,

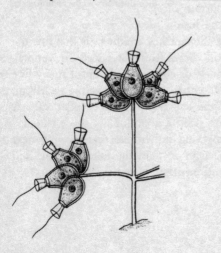

Is this how it was? A colony of choanoflagellates.

Choanoflagellates join

CZ		**MZ**				**PZ**							**NP**
N	E	K	J	T	P	C	D	S	O	Є			

Mya

0

100

200

300

400

500

Choanoflagellates (Choanoflagellata)

Already joined

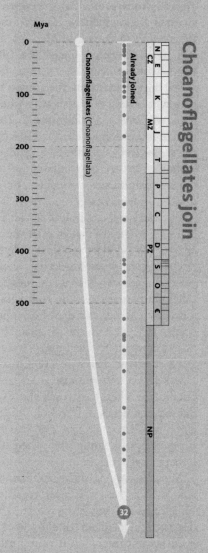

32

Choanoflagellates join. The 120 or so known species of choanoflagellate are conventionally seen as close relatives of the animals, a position strongly supported by both morphological and molecular data.

Image: *Codosiga gracilis.*

which resumes the topic begun by the Sponge's Tale: the origin of multicellular sponges.

THE CHOANOFLAGELLATE'S TALE

Zoologists have long enjoyed speculating about how multi-cellularity evolved from protozoan ancestors. The great nineteenth-century German zoologist Ernst Haeckel was one of the first to propose a theory of the origin of the Metazoa, and some version of his theory is still much favoured today: the first metazoan was a colony of flagellate protozoa.

We met Haeckel in the Hippo's Tale, in connection with his prescient linking of hippos with whales. He was a passionate Darwinian, who made a pilgrimage to Darwin's house (which the great man found irksome). He was also a brilliant artist, a dedicated atheist (he sardonically called God a 'gaseous vertebrate'), and a particular enthusiast for the now unfashionable theory of recapitulation: 'Ontogeny recapitulates phylogeny', or 'The developing embryo climbs up its own family tree.'

You can see the appeal of the idea of recapitulation. The life story of every young animal is a telescoped re-enactment of its (adult) ancestry. We all start as a single cell: that represents a protozoan. The next stage in development is a hollow ball of cells, the blastula. Haeckel suggested that this represents an ancestral stage, which he called the blastaea. Next in embryology, the blastula invaginates, like a ball punched in as a dent from one side, to form a cup lined by a double layer of cells, the gastrula. Haeckel imagined a gastrula-stage ancestor, which he called the gastraea. A cnidarian, such as a hydra or a sea anemone, has two layers of cells, like Haeckel's gastraea. In Haeckel's recapitulationist view, cnidarians stop climbing up their family tree when they reach the gastrula stage, but we soldier on. Subsequent stages in our embryology resemble a fish with gill slits and a tail. Later we lose our tail. And so on. Each embryo stops climbing up its family tree when it reaches its appropriate evolutionary stage.

Appealing as it is, the recapitulation theory has become unfashionable – or rather it is now regarded as a small part of what is sometimes but not always true. The whole matter is thoroughly discussed in Stephen Gould's book *Ontogeny and Phylogeny*. We must leave it there, but it is important for us to see where Haeckel was coming from. From the point of view of the origin of the Metazoa, the interesting stage in Haeckel's theory is the blastaea: the hollow ball of cells that, in his view, was the ancestral stage now reprised in embryology as the blastula. What modern creature can we find that resembles a blastula? Where shall we find an adult creature that is a hollow ball of cells?

Setting aside the fact that they are green and photosynthesise, the group of colonial algae called the Volvocales seemed almost too good to be true. The eponymous member of the group is the largest, Volvox itself, and Haeckel could hardly have wished for a neater model blastaea than Volvox. It is a perfect sphere, hollow like a blastula, with a single layer of cells, each resembling a unicellular flagellate (which happens to be green).

Haeckel's theory did not have the field to itself. In the mid-twentieth century a Hungarian zoologist called Jovan Hadzi proposed that the first metazoan was not round at all, but elongated like a flatworm. His contemporary model for the first metazoan was an acoelomorph worm of the kind we met at Rendezvous 27. He derived it from a ciliate protozoan (we shall meet them at Rendezvous 37) with many nuclei (which some of them have to this day). It crawled along the bottom with its cilia, as some small flatworms do today. Cell walls appeared between the nuclei, turning an elongate protozoan with one cell but many nuclei (a 'syncitium'), into a creeping worm with many cells, each with its own nucleus – the first metazoan. On Hadzi's view the round metazoans such as cnidarians and ctenophores secondarily lost their elongated worm shape and became radially symmetrical, while most of the animal kingdom continued to expand upon the bilateral worm shape in ways that we see all around us.

Hadzi's ordering of the rendezvous points would, therefore, be very different from ours. The rendezvous with the cnidarians and ctenophores would come earlier in the pilgrimage than the rendezvous with the acoelomorph flatworms. Unfortunately, modern molecular evidence goes against Hadzi's ordering. Most zoologists today support some version of the Haeckel 'colonial flagellate' theory against the Hadzi 'syncitial ciliate' theory. But attention has switched away from the Volvocales, elegant as they are, and to the group whose tale this is, the choanoflagellates.

One type of colonial choanoflagellate is so sponge-like it is even called Proterospongia. The individual choanoflagellates (or should we stick our necks out and call them choanocytes?) are embedded in a matrix of jelly. The colony is not a ball, which would not have pleased Haeckel, although he appreciated the beauty of the choano-flagellates, as his wonderful drawings of them show. *Proterospongia* is a colony of cells of a type almost indistinguishable from those that dominate the interior of a sponge. The choanoflagellates marginally get my vote as the most plausible candidates for a recent re-enact-ment of the origin of the sponges, and ultimately of the whole group of Metazoa.

The choanoflagellates would once have been lumped with all the remaining organisms who have not yet joined our pilgrimage, as 'Protozoa'. Protozoa doesn't work any more as the name for a phylum. There are lots of different ways of being a single-celled organism (or, as some would prefer, acellular – having a body not divided into constituent cells). Different members of the group formerly known as Protozoa will now be joining our pilgrimage in DRIPs and drabs, separated by major contingents of multicellular creatures such as fungi and plants. I shall continue to use the word protozoan as an informal name for a single-celled eukaryote.

DRIPs

There is a small group of single-celled parasites known as either Mesomycetozoea or Ichthyosporea, mostly parasites of fish and other freshwater animals. The name Mesomycetozoea suggests an association with both fungi and animals, and it is true that their rendezvous with us animals is our last before we all join the fungi. This fact is now known from molecular genetic studies, which unite what had hitherto been a rather miscellaneous set of single-celled parasites, both with each other and with animals and fungi.[*]

Both 'Mesomycetozoea' and 'Ichthyosporea' are quite hard to remember, and there is disagreement over which of them to prefer. This may be why a practice has grown up of using the nickname DRIPs – an acronym from the initial letters of the only four genera known to the discoverers of the group. The genera that provide the D, the I and the P are *Dermocystidium, Ichthyophonus* and *Psorospermium*. The R was always a bit of a cheat because it is not a Latin name. It stood for 'Rosette agent', a commercially important parasite of salmon, now formally named *Sphaerothecum destruens*. So I suppose the acronym should really have been amended to DIPS, or DIPSs in the plural. But DRIPs with an s for plural seems to have stuck. And now, with what seems like the workings of nomenclatural providence, another organism, whose name happens to begin with R, has recently been discovered to be a DRIP too. This is *Rhinosporidium seeberi,* a parasite of human noses. So we can redesign the name DRIPS, with all five letters being comfortably acronymical, and try to ignore the embarrassing question of whether it is singular or plural.

Rhinosporidium seeberi was first discovered in 1890, and it has long been known as the cause of rhinosporidiosis, an unpleasant disease of the human, indeed mammalian, nose, but its affinities were a mystery. At different times it has been moved from protozoan pillar to

[*] Confusingly (that's putting it mildly), the name Mesomycetozoa, as opposed to Mesomycetozoea (can you spot the difference?), has been used for a more inclusive group. This seems positively designed to confuse, like the Hominoidea, Hominidae, Homininae, Homimini complex of names for our own relatives, and I prefer to boycott them all.

DRIPs join

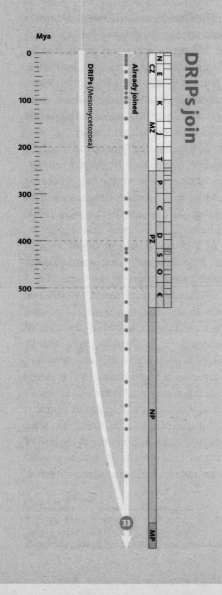

Mya

DRIPs (Mesomycetozoea)

Already joined

N E | CZ
K | MZ
J
T
P
C
D | PZ
S
O
€

NP

MP

33

DRIPs join. The closest single-celled relatives of the animals are the choanoflagellates and the DRIPs. It is currently uncertain whether these two groups are each other's closest relatives (so collapsing Rendezvous 32 and 33 into one), or whether the 30 or so described DRIPs species are the most distantly related to all the others. The most extensive molecular study to date supports the latter scheme, which we therefore follow.

Image: *Ichthyophonus hoferi.*

fungal post, but molecular studies now show it to be the fifth DRIP. Fortunately for pun-haters, *R. seeberi* doesn't seem to cause the nose to drip. On the contrary, it blocks the nostril with polyp-like growths. Rhinosporidiosis is mainly a disease of the tropics, and doctors have long suspected that people catch it by bathing in freshwater rivers or lakes. Since all other known DRIPs are parasites of freshwater fish, crayfish or amphibians, it seems likely that freshwater animals constitute the primary host of *R. seeberi* too. The discovery that it is a DRIP might be helpful to doctors in other ways. For example, attempts to treat it with antifungal agents have failed, and we can now get an inkling as to why: it is not a fungus.

Dermocystidium appears as cysts in the skin or gills of carp, salmonids, eels, frogs and newts. *Ichthyophonus* causes economically important systemic infections of more than 80 species of fish. *Psorospermium*, which, incidentally, was originally discovered by our old friend Ernst Haeckel, infects crayfish (not fish at all, of course, but crustaceans), and again has economically important effects on crayfish stocks. And *Sphaerothecum*, as we have seen, infects salmon.

DRIPs organisms themselves would be dismissed as unremarkable, but for their evolutionarily aristocratic status – their branch point, after all, is the deepest in the animal kingdom, their rendezvous with us the oldest. We don't know what Concestor 33 looked like, except insofar as single-celled organisms all look pretty much alike to our jaded multicellular eyes. It was not a parasite like a DRIP – not of fish, amphibians, crustaceans or humans, that's for sure, for they all still lay unimaginably far in the future.

The one adjective that is always applied to DRIPs is 'enigmatic', and who am I to break with tradition? If a DRIP were to tell its enigmatic tale, I suspect that it would be a tale of how, now that we have reached such ancient rendezvous points, it is nearly arbitrary which of our single-celled cousins happen to have survived. Not accidentally, it is also pretty arbitrary which single-celled organisms scientists have chosen to examine at the level of molecular genetics. People have looked at DRIPs because some of them are commercially important parasites of fish, and others, we now know, bung up our noses. There could be single-celled organisms that are just as pivotal on the family tree of life, but which nobody has bothered to look at because they parasitise, say, komodo dragons rather than salmon or people.

Nobody, however, could overlook the fungi. We are about to greet them.

FUNGI

At Rendezvous 34 we animals are joined by the second of the three great multicellular kingdoms, the fungi. The third consists of the plants. It might at first be surprising that fungi, which seem so plant-like, are more closely related to animals than they are to plants, but molecular comparison leaves little doubt. And perhaps that is not too surprising. Plants import energy from the sun into the biosphere. Animals and fungi, in their different ways, are parasites on the plant world.

The fungi are a very large and important influx of pilgrims, with 69,000 species so far described out of an estimated total of 1.5 million. Mushrooms and toadstools give the wrong impression – these conspicuous plant-like structures are the spore-producing tips of the iceberg. Most of the business part of the organism that made the mushroom is under the ground: a spreading network of threads called hyphae. The collection of hyphae belonging to one individual fungus is called the mycelium. The total length of mycelium of an individual fungus may be measured in kilometres, and may spread through a substantial area of soil.

A single mushroom is like a flower growing on a tree. But the 'tree', instead of being a tall, vertical structure, is spread out like the strings of a giant tennis racket underground, in the surface layers of the soil. Fairy rings are a vivid reminder of this. The circumference of the ring represents the extent of growth of a mycelium, spreading outwards from a central starting point, perhaps originally a single

Opposite: **Fungi join.** Molecular taxonomy reveals fungi to be closer to animals than to plants. The two largest groups, the Ascomycota (about 40,000 described species) and Basidiomycota (about 22,000), are usually considered closest relatives, with recent studies finding the 160 arbuscular mycorrhizal fungi to be their sister group. The groupings and branching order of the rest of the 3,000 or so fungi are not well established, particularly the number of separate branches previously lumped together in the 'Zygomycota', and the position of the microsporidians.

Images, left to right: common morel (*Morchella esculenta*); stinkhorn (*Phallus impudicus*); arbuscular mycorrhizal fungus (*Glomus* sp.) from root of bluebell (*Hyacinthoides nonscripta*); pin mould (*Mucor* sp.); *Rhizoclostamium* sp.; *Enterocytozoon bieneusi*.

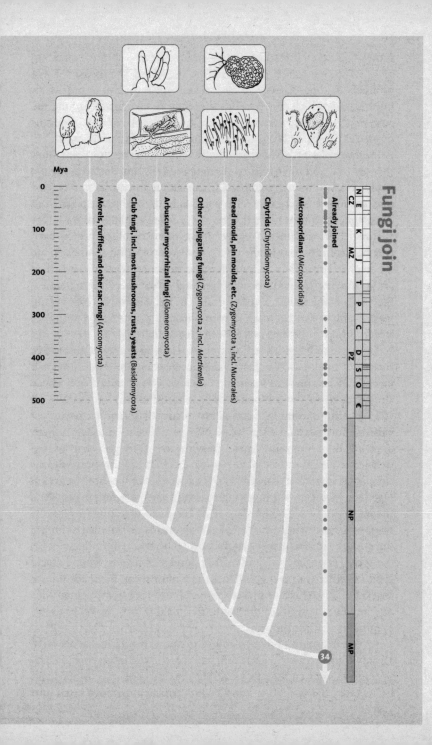

Fungi join

	N	E	K	J	T	P	C	D	S	O	€		
	CZ		MZ				PZ					NP	MP

Mya

0 — 100 — 200 — 300 — 400 — 500

Already joined

Microsporidians (Microsporidia)

Chytrids (Chytridiomycota)

Bread mould, pin moulds, etc. (Zygomycota 1, incl. Mucorales)

Other conjugating fungi (Zygomycota 2, incl. *Mortierella*)

Arbuscular mycorrhizal fungi (Glomeromycota)

Club fungi, incl. most mushrooms, rusts, yeasts (Basidiomycota)

Morels, truffles, and other sac fungi (Ascomycota)

34

spore. The circular leading edge – feeding edge – of the expanding
mycelium, the frame of the racket, is where the broken-down
products of digestion are richest. These are a source of nutriment
for the grass, which consequently grows more luxuriantly around the
ring. Where there are fruiting bodies (mushrooms, or any of dozens
of species of related fungi) they tend to grow up in the ring too.

Hyphae may be divided into cells by cross-walls. But sometimes
they are not, and the nuclei containing the DNA are dotted along the
hypha in a syncitium, meaning a tissue with many nuclei not divided
into separate cells (we met other syncitia in the early development of
Drosophila, and in Hadzi's theory of the origin of the Metazoa). Not
all fungi have a thread-like mycelium. Some, such as yeasts, have
reverted to single cells which divide and grow in a diffuse mass. What
the hyphae (or yeast cells) are doing is digesting whatever it is they
are burrowing through: dead leaves and other decaying matter (in
the case of soil fungi), curdled milk (in the case of cheese-making
fungi), grapes (in the case of wine-making yeasts), or the grape-
treader's toes (if he happens to suffer from athlete's foot).

The key to efficient digestion is to expose a large area of absorptive
surface to the food. We achieve that by chewing the food into small
pieces and passing the fragments through a long coiled gut whose
already large area is compounded by a forest of tiny projections, or
villi, covering its lining. Each villus in turn has a brush border of
hair-like micro-villi, so the total absorptive area of an adult human
intestine is millions of square centimetres. A fungus such as the
well-named *Phallus* (see plate 45) or the field mushroom *Agaricus
campestris* spreads its mycelium over a similar area of soil, secreting
digestive enzymes and digesting the soil material where it lies. The
fungus doesn't walk about devouring food and digesting it inside its
body as a pig or a rat would. Instead it spreads its 'intestines', in the
form of thread-like mycelia, right through the food and digests it on
the spot. From time to time hyphae come together to form a single
solid structure with recognisable form: a mushroom (or toadstool, or
bracket). This structure manufactures spores that float high and far
on the wind, spreading the genes for making new mycelium and,
eventually, new mushrooms.

As you'd expect of a new influx of 100,000 pilgrims, they have
already joined up with each other in large sub-contingents 'before'
they meet us at Rendezvous 34. All the major sub-groups of fungi end
in 'mycete', the Greek for 'mushroom', which sometimes turns into

'mycota'. We have already met 'mycete' in Mesomycetozoea, the name for DRIPs that implies some sort of intermediate status between animals and fungi. The two largest and most important of these sub-contingents of the fungus pilgrims are the ascomycetes (Ascomycota) and the basidiomycetes (Basidiomycota).

The ascomycetes include some famous and important fungi, such as *Penicillium*, the mould from which the first antibiotic was accidentally discovered and largely ignored by Fleming until Florey, Chain and their colleagues rediscovered it 13 years later. Incidentally, it is rather a pity that the name antibiotic has stuck. These agents attack bacteria, not viruses, and if only they had been called antibacterials instead of antibiotics, patients might stop demanding that doctors prescribe them (uselessly and even counterproductively) for viral infections. Another Nobel Prize-winning ascomycete is *Neurospora crassa*, the mould with which Beadle and Tatum developed the 'one gene one enzyme hypothesis'. Then there are the human-friendly yeasts that make bread, wine and beer, and the unfriendly *Candida* from which we get unpleasant diseases, such as vaginitis (thrush). Edible morels and the highly prized truffles are ascomycetes. Truffles are traditionally found with the aid of female pigs, who are strongly attracted by the smell of what appears to be alpha-androstenol, a male sex pheromone secreted by boars. It isn't clear why truffles produced this dead giveaway, but it may – in some interesting way yet to be worked out – account for their gastronomic appeal to us.

Most of the edible and notoriously inedible or hallucinogenic fungi are basidiomycetes: mushrooms, chanterelles, boletuses, shiitakes, ink caps, death caps, stinkhorns, bracket fungi, toadstools and puffballs. Some of their spore-producing bodies can reach impressive sizes. Basidiomycetes are also of economic importance as causes of the plant diseases known as rusts and smuts. Some basidiomycetes and ascomycetes, as well as all members of a specialised group called the glomeromycetes, collaborate with plants to supplement their root hairs with mycorrhizae, a most remarkable story, which I'll briefly relate.

We saw that the villi in our intestines and the mycelium threads of a fungus are thin, to increase the surface area for digestion and absorption. In just the same way, plants have numerous fine root hairs to increase the surface area for absorption of water and nutrients from the soil. But it is an amazing fact that most of what appear to be root hairs are no part of the plant itself. Instead, they are provided by symbiotic fungi, whose mycelium both resembles and works like true root hairs. These are the mycorrhizae, and close examination reveals that there are several independently evolved ways in which the mycorrhizal principle has been implemented. Much of plant life on our planet is utterly dependent on mycorrhizae.

In an even more impressive feat of symbiotic co-operation, basidiomycetes and – independently evolved again – ascomycetes form associations with algae or cyanobacteria to create lichens, those remarkable confederacies which can achieve so much more than either partner on its own, and can produce body forms so dramatically different from the body form of either partner. Lichens (pronounced LIE-kins) are sometimes mistaken for plants, and that isn't so far from the truth – for plants too, as we shall see at the Great Historic Rendezvous, originally made a compact with photosynthetic micro-organisms for their food production. Lichens can loosely be thought of as plants-in-the-making, forged from two organisms. The fungus could almost be said to 'farm' captured crops of photosynthesisers. The metaphor gains from the fact that in some lichens the partnership is largely co-operative, and in others the fungus is more exploitative. Evolutionary theory predicts that the lichens in which the reproduction of the fungus and the photosynthesiser go hand-in-hand generally form co-operative relationships. Lichens in which the fungus just captures available photosynthetic organisms from the environment are predicted to have more exploitative relationships. And this seems to be the case.

What especially fascinates me about lichens is that their phenotypes (see the Beaver's Tale) look nothing like a fungus – nor indeed like an alga. They constitute a very special kind of 'extended phenotype', wrought of a collaboration of two sets of gene products. In my vision of life, explained in other books, such a collaboration is not in principle different from the collaboration of an organism's 'own' genes. We are all symbiotic colonies of genes – genes co-operating to weave phenotypes about them.

AMOEBOZOANS

Joining us at Rendezvous 35 is a little creature that once had the distinction of being, in the popular and even scientific imagination, the most primitive of all, little more than naked 'protoplasm': *Amoeba proteus.* On this view, Rendezvous 35 would be the final encounter of our long pilgrimage. Well, we still have a way to go, and *Amoeba* has, when compared to bacteria, quite an advanced, elaborate structure. It is also surprisingly large, visible to the naked eye. The giant amoeba *Pelomyxa palustris* can be as much as half a centimetre across.

Amoebas famously have no fixed shape – hence the species name *proteus,* after the Greek god who could change his form. They move by streaming their semi-liquid interior, either as a more or less coherent single blob, or by thrusting out pseudopodia. Sometimes they 'walk' on those temporarily extruded 'legs'. They eat by engulfing prey, throwing pseudopodia around it and enclosing it in a spherical bubble of water. Being engulfed by an amoeba would be a nightmarish experience, if you weren't too small to have nightmares. The spherical bubble or vacuole can be thought of as part of the outside world, lined by a portion of the amoeba's 'outer' wall. Once in the vacuole, the food is digested.

Some amoebas live inside animal guts. For example, *Entamoeba coli* is extremely common in the human colon. It is not to be confused with the (much smaller) bacterium *Escherichia coli* on which it probably feeds. It is harmless to us, unlike its near relative *Entamoeba histolytica,* which destroys the cells lining the colon and causes amoebic dysentery, familiarly known (in British English) as Delhi Belly, or (in American English) as Montezuma's Revenge.

Three rather different groups of amoebozoans are called slime moulds because they have independently evolved similar habits (plus another unrelated group of 'slime moulds', the acrasids, which will join us at Rendezvous 37). Of the amoebozoan ones, the best known are the cellular slime moulds or dictyostelids. They have been the life work of the distinguished American biologist J. T. Bonner, and what follows is largely drawn from his scientific memoir *Life Cycles.*

Amoebas join

N	E	K	J	T	P	C	D	S	O	€					
CZ		MZ					PZ				NP			MP	

Mya

0
100
200
300
400
500

Amoebas and most slime moulds (Amoebozoa)

Already joined

35

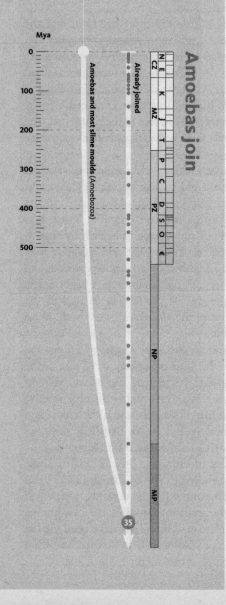

Amoebas join. The word 'amoeba' is a description rather than a strict classification because many unrelated eukaryotes exhibit an amoeboid form. The Amoebozoa include the classic amoebas, such as the *Amoeba proteus* shown here, as well as most of the slime moulds – about 5,000 known species in total.

Image: *Amoeba proteus.*

Cellular slime moulds are social amoebas. They literally blur the distinction between a social group of individuals and a single multi-cellular individual. In part of their life cycle, separate amoebas creep through the soil, feeding on bacteria and reproducing, as amoebas will, by dividing in two, feeding some more, then dividing again. Then, rather abruptly, the amoebas switch into 'social mode'. They converge on aggregation centres, from which chemical attractants radiate outwards. As more and more amoebas stream in on an attraction centre, the more attractive it becomes, because more of the beacon chemical is released. It is a bit like the way planets form from aggregating debris. The more debris accumulates in a given attraction centre, the more its gravitational attraction. So after a while, only a few attraction centres remain, and they become planets. Eventually the amoebas in each major attraction centre unite their bodies to form a single multicellular mass, which then elongates into a multicellular 'slug'. About a millimetre long, it even moves like a slug, with a definite front and back end, and is capable of steering in a coherent direction – for example towards light. The amoebas have suppressed their individuality to forge a whole organism.

After crawling around for a while, the slug initiates the final phase of its life cycle, the erection of a mushroom-like 'fruiting body'. It begins the process by standing on its 'head' (the front end as defined by its crawling direction), which becomes the 'stalk' of the miniature mushroom. The inner core of the stalk becomes a hollow tube made of swollen cellulose carcasses of dead cells. Now cells around the top of the tube pour into the tube like, in Bonner's simile, a fountain flowing in reverse. The result is that the tip of the stalk rises into the air, with the originally posterior end of the stalk at the top. Each of the amoebas in the originally posterior end now becomes a spore encased in a thick protective coat. Like the spores of a mushroom, they are now shed, each one bursting out of its coat a free-living, bacteria-devouring amoeba, and the life cycle begins again.

Bonner gives an eye-opening list of such social microbes – multi-cellular bacteria, multicellular ciliates, multicellular flagellates and multicellular amoebas, including his beloved slime moulds. These creatures might represent instructive re-enactments (or pre-enact-ments) of our kind of metazoan multicellularity. But I suspect that they are all completely different, and the more fascinating because of it.

PLANTS

Rendezvous 36 is where we meet the true lords of life, the plants. Life could get along without animals and without fungi. But abolish the plants, and life would rapidly cease. Plants sit, indispensably, at the base – the very foundation – of nearly every food chain. They are the most noticeable creatures on our planet, the first living things any visiting Martian would remark. By far the largest single organisms that ever lived are plants, and an impressive percentage of the world's biomass is locked up in plants. This doesn't just happen to be so. Some such high proportion follows necessarily from the fact that almost* all biomass comes ultimately from the sun via photosynthesis, most of it in green plants, and the transaction at every link of the food chain is only about 10 per cent efficient. The surface of the land is green because of plants, and the surface of the sea would be green too if its floating carpet of photosynthesisers were macroscopic plants instead of micro-organisms too small to reflect noticeable quantities of green light. It is as though plants are going out of their way to cover every square centimetre with green, leaving none uncovered. And that is pretty much what they are doing, for a very sensible reason.

A finite number of photons reaches the planet's surface from the sun, and every last photon is precious. The total number of photons that can be garnered from its star by a planet is limited by its surface area, with the complication that only one side is facing its star at any one time. From a plant's point of view, a square centimetre of the Earth's surface that is anything but green amounts to a negligently wasted opportunity to sweep up photons. Leaves are solar panels, as flat as possible to maximise photons caught per unit expenditure. There is a premium on placing your leaves in such a position that they are not over-shadowed by other leaves, especially somebody else's leaves. This is why forest trees grow so tall. Tall trees that are not in a forest are out of place, probably because of human interference. It is a complete waste of effort to grow tall if you are the only

---

* The reason for this little hedge will emerge when we get to Canterbury.

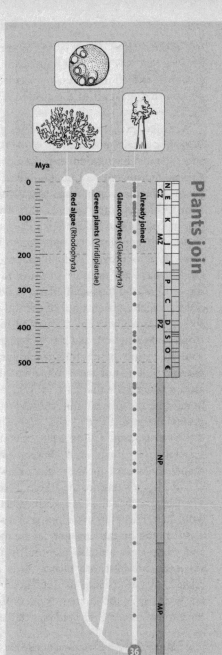

Plants join

	N	E	K									
	CZ			MZ		T	P	C	D	S	O	Є
									PZ			

Mya

0

100

200

300

400

500

Red algae (Rhodophyta)

Green plants (Viridiplantae)

Glaucophytes (Glaucophyta)

Already joined

NP

MP

36

Plants join. The plants comprise about 13 species of glaucophyte (single-celled algae, with chloroplasts whose morphology is very similar to free-living cyanobacteria), 5,000 or so species of red alga, and about 30,000 species of 'green plant'. The green plants include many single-celled and colonial green algae, such as *Volvox,* as well as the more familiar mosses, ferns, conifers, flowering plants and the like. The order of branching of these three groups is reasonably well established, but the position of the plants in the eukaryotic phylogeny in general is disputed (see Rendezvous 37).

Images, left to right: dulse (*Rhodymenia palmata*); volvox (*Volvox aurelia*); giant sequoia (*Sequoiadendron giganteum*).

tree around. It is much better to spread out sideways like grasses because that way you trap more photons per unit of effort put into growing. As for forests, it is no accident that they are so dark. Every photon that makes it to the ground represents failure on the part of the leaves above.

With few exceptions, such as Venus flytraps, plants don't move. With few exceptions, such as sponges, animals do. Why the difference? It must have to do with the fact that plants eat photons while animals (ultimately) eat plants. We need that 'ultimately', of course, because the plants are sometimes eaten at second or third hand, via animals eating other animals. But what is it about eating photons that makes it a good idea to sit still with roots in the ground? What is it about eating plants, as opposed to being a plant, that makes it a good idea to move? Well, I suppose given that plants stay still, animals have got to move in order to eat them. But why do plants stay still? Maybe it has something to do with the need to be rooted in order to suck nutrients out of the soil. Maybe there is too unbridgeable a distance between the best shape to be if you want to move (solid and compact), and the best shape to be if you want to expose yourself to lots of photons (high surface area, hence straggly and unwieldy). I'm not sure. Whatever the reason, of the three great groups of megalife that have evolved on this planet, two of them, the fungi and the plants, stay mostly still as statues, while the third group, the animals, do most of the scurrying about, most of the active go-getting. Plants even make use of animals to do their scurrying for them, and flowers, with their beauteous colours, shapes and scents, are the instruments of this manipulation.[*]

The pilgrims that we meet here at Rendezvous 36 are not all green. The deepest divide[†] among them is between the red algae on the one hand and green plants (including green algae) on the other. Red algae are common on the seashore. So are the various kinds of green algae, and green algae are also plentiful in freshwater. The most familiar seaweeds, however, are brown algae and these are more distantly related: they don't join us until Rendezvous 37. Of those whom we are greeting at the current rendezvous, the most familiar and the most impressive are the land plants. Plants conquered the land earlier

---

[*] I would have included a tale about this if I had not already done it in two chapters of *Climbing Mount Improbable*, 'Pollen Grains and Magic Bullets', and 'A Garden Inclosed'.
[†] Apart from the rather insignificant 13 species of single-celled glaucophytes, which seem to be the outgroup.

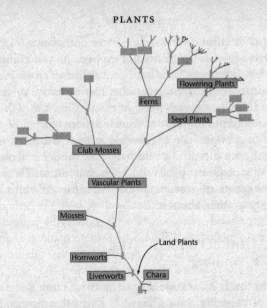

Flowering Plants

Ferns

Seed Plants

Club Mosses

Vascular Plants

Mosses

Land Plants

Hornworts

Liverworts   Chara

If only Darwin and Hooker had had a computer. A tree of green plants, from the Deep Green program, *http://ucjeps.berkeley.edu/map2.html*. The program runs on Mac or virus-compatible PC (enable Java in your browser). The root of the tree is at the bottom of the picture.

than animals did. That is almost obvious, for without plants to eat, what would it profit an animal to be there? Plants probably didn't move directly from the sea onto the land but, like animals, went via freshwater. For an artist's impression of Concestor 36 see plate 46.

As usual, when we greet a large army of pilgrims, we find them already marching in complicated sub-groupings who have joined each other 'before' they rendezvous with us. In the case of the green plants, I strongly recommend a stunningly well-produced computer program called Deep Green, which, at the time of writing, is available on the internet. When you launch Deep Green, you see a rooted phylogenetic tree. Some of the branches have a name at the tip, the name of a plant or a group of plants. Some of them have no name and are pointing 'off the page'. The beauty of this program is that you may seize the tree with your mouse, and drag it, in the most delightfully natural and intuitive way, to see more of the tree. As you drag, you watch twigs sprout before your eyes, and as you swivel the tree round, you see a whole lot of new names appear on the screen, together with many new, unnamed branches. You then explore the tree as far as you like: it seems to go on for ever, which tells you what a huge diversity of green plants has evolved. As you climb through the branches, hand over hand as it seems, like a Darwinian monkey in evolutionary tree heaven, remember that each fork you encounter represents a true rendezvous point in exactly the sense of this book. It would be wonderful to have an animal version, too.

I ended a previous tale by remarking what delight it is to be a zoologist at such a time. I could have said the same about being a botanist. What a pleasure it would be to demonstrate Deep Green to Joseph Hooker – in the company of his close friend Charles Darwin. I almost weep to think about it.

THE CAULIFLOWER'S TALE
Written with Yan Wong

The tales in this book are intended to be about more than the private concerns of the teller. Like Chaucer's, they are supposed to reflect upon life in general – in his case human life, in our case life. What has the cauliflower to tell the huge assemblage of pilgrims at the great get-together after Rendezvous 36 when the plants join the animals? An important principle that applies to every plant and every animal. It could be seen as a continuation of the Handyman's Tale.

The Handyman's Tale was about brain size, and it made great play with the logarithmic way of making scatter plots to compare different species. Larger animals seemed to have proportionally smaller brains than small animals. More specifically, the slope of a log-log graph of body mass against brain mass was pretty much exactly $\frac{3}{4}$. This fell, you will remember, between two intuitively comprehensible slopes: $\frac{1}{1}$ (brain mass simply proportional to body mass) and $\frac{2}{3}$ (brain area proportional to body mass). The observed slope for log brain mass against log body mass turned out to be not just vaguely higher than $\frac{2}{3}$ and lower than $\frac{1}{1}$. It was exactly $\frac{3}{4}$. Such precision of data seems to demand equal precision from theory. Can we think up some rationale for the $\frac{3}{4}$ slope? It isn't easy.

To add to the problem, or perhaps give us a hint, biologists have long noticed that lots of other things besides brain size follow this precise $\frac{3}{4}$ relationship. In particular, the energy use of various organisms – the metabolic rate – follows a $\frac{3}{4}$ rule, and this was raised to the status of a natural law, Kleiber's Law, even though there was no known rationale for it. The graph opposite plots log metabolic rate

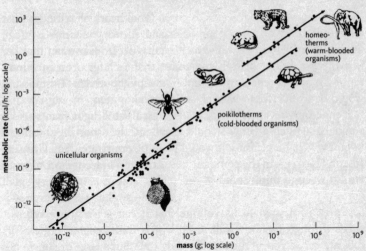

Law holds over twenty orders of magnitude. Kleiber's Law plot, adapted from West, Brown and Enquist [304].

against log body mass (the Handyman's Tale spells out the rationale for log-log plots).

The really astonishing thing about Kleiber's Law is that it holds good from the smallest bacterium to the largest whale. That's about 20 orders of magnitude. You need to multiply by ten 20 times – or add 20 noughts – in order to get from the smallest bacterium to the largest mammal, and Kleiber's Law holds right across the board. It works for plants and single-celled organisms too. The diagram shows that the best fit is obtained with three parallel lines. One line is for micro-organisms, a second for cold-blooded large creatures ('large' here means anything heavier than about a millionth of a gram!) and the third is for warm-blooded large creatures (mammals and birds). All three lines have the same slope ($\frac{3}{4}$) but they are of different height: not surprisingly, warm-blooded creatures have a higher metabolic rate, size for size, than cold-blooded creatures.

For years no one could think of a really convincing reason for Kleiber's Law, until a piece of brilliant collaborative work between a physicist, Geoffrey West, and two biologists, James Brown and Brian Enquist. Their derivation of the precise $\frac{3}{4}$ law is a piece of mathematical magic that is hard to translate into words, but it is so ingenious and important that it is worth the effort.

Tissues have a supply problem.
The complex supply system of a cauliflower.

The theory of West, Enquist and Brown, henceforth WEB, takes off from the fact that the tissues of large organisms have a supply problem. That is what blood systems in animals and vascular tubing in plants are all about: transporting 'stuff' to and from tissues. Small organisms don't face the problem to the same extent. A very small organism has such a large surface area compared to its volume that it can get all the oxygen it needs through its body wall. Even if it is multicellular, none of its cells are very far from the outside body wall. But a large organism has a transport problem because most of its cells are far away from the supplies they need. They need to pipe stuff from place to place. Insects literally pipe air into their tissues in a branching network of tubes called tracheae. We too have richly branched air tubes, but they are confined to special organs, the lungs, which have a correspondingly richly branched blood network to take the oxygen from the lungs to the rest of the body. Fish do a similar thing with gills: area-intensive organs designed to increase the interface between water and blood. The placenta does the same kind of thing for maternal blood and foetal blood. Trees use their richly dividing branches to supply their leaves with water drawn from the ground, and pump sugars back from the leaves to the trunk.

The cauliflower above, freshly bought from a local greengrocer and cut in half, shows what a typical stuff-transporting system looks like. You can see how much effort a cauliflower puts in to provide a supply network for its surface covering of 'flower buds'.*

Now, we could imagine that such supply networks – tubes of air, pipes of blood or sugar solution, or whatever they are – might compensate perfectly for increased body size. If they did, a typical cell of a modest cauliflower would be exactly as well supplied as a typical cell of a giant redwood, and the metabolic rate of the two cells would be the same. Since the number of cells in an organism is

---

*Buds that have been grotesquely modified in this case by artificial selection under domestication, but the principle still stands.

proportional to its mass, the scatter plot of total metabolic rate against body mass, with both axes on a log scale, would fall on a line with a slope of 1. Yet what we actually observe is a slope of $\frac{3}{4}$. Small organisms have a higher metabolic rate than they 'should' have, for their mass, compared with large organisms. What this means is that the metabolic rate of a cauliflower cell is higher than the metabolic rate of an equivalent cell in a redwood, and the metabolic rate of a mouse is higher than the metabolic rate of a whale.

At first sight, this seems strange. A cell is a cell is a cell, and you might think there is an ideal metabolic rate which would be the same for a cauliflower as for a redwood, a mouse as for a whale. Perhaps there is. But what seems to happen is that the difficulty of delivering water, or blood, or air, or whatever 'stuff' it is, seems to place a limit on achieving that ideal. There has to be a compromise. The WEB theory explains the compromise, and why it ends up delivering exactly a slope of $\frac{3}{4}$, and it does so in precise, quantitative detail.

The theory consists of two key points. The first is that the branching tree of pipes, which delivers stuff to a given volume of cells, itself occupies some volume, competing for space with the cells that it is supplying. Towards the tips of the supply network, the pipes occupy a substantial space in their own right. And if you double the number of cells that need to be supplied, the network volume more than doubles because more pipes are needed to plumb the network into the main system, pipes which themselves occupy space. If you want to double the number of supplied cells whilst only doubling the space occupied by the pipes, you need a more sparsely distributed plumbing network. The second key point is that, whether you are a mouse or a whale, the most efficient transport system – the one that wastes the least energy in moving stuff around – is one that takes up a fixed percentage of the volume of your body. That's how the mathematics works out, and it is also an empirically observed fact.[*] For example, mammals, whether mice, humans, or whales, have a volume of blood (i.e. the size of the transport system) which occupies between six and seven per cent of their body.

Taking these two points together, it means that if we wish to double the volume of cells to be supplied, but still keep the most

---

[*] The actual percentage might differ a little depending on, say, whether you are warm-blooded or cold-blooded.

efficient transport system, we need a more sparsely distributed supply network. And a more sparse network means that less stuff is supplied per cell, meaning that the metabolic rate must go down. But by how much, precisely, must it go down?

WEB calculated the answer to this question. Wonderful to say, the mathematics predicts a straight line with a slope of exactly $\frac{3}{4}$ for the graph of log metabolic rate against log body size! More recent work has built upon their initial theory, but the essential aspects still remain. Kleiber's Law – whether in plants, animals, or even at the level of transport within a single cell – has finally found its rationale. It can be derived from the physics and geometry of supply networks.

THE REDWOOD'S TALE

People argue about the one place in the world that you must visit before you die. My candidate is Muir Woods, just north of the Golden Gate Bridge. Or, if you leave it too late, I can't imagine a better place to be buried (except I doubt that it's allowed, nor should it be). It is a cathedral of greens and browns and stillness, the nave lofted by the world's tallest trees, *Sequoia sempervirens*, the Pacific coast redwoods, whose padded bark damps out the echoes that would fill a man-made cathedral. The related species *Sequoiadendron giganteum* (see plate 47), found inland on the foothills of the Sierra Nevada range, is typically slightly shorter but more massive. The largest single living creature in the world, the General Sherman tree, is a *giganteum* over 30 metres in circumference and over 80 metres tall, with an estimated weight of 1,260 tonnes. Its age is not known for certain but the species is known to survive more than 3,000 years. General Sherman's age could be ascertained exactly to the nearest year if we cut it down – a major undertaking, for the bark alone is about a metre thick.* Let us hope this will never happen, in spite of Ronald Reagan's notorious opinion, when he was Governor of California: 'If you've seen one, you've seen them all.'

How is it that we can know the age of a large tree, even one as old as General Sherman, accurately to the nearest year? We count the rings in its stump. Ring counting, in a more sophisticated form, has given rise to the elegant technique of dendrochronology, by which

---

* Actually we don't need to cut it down. A core sample would be good enough.

archaeologists working on a timescale of centuries can precisely date any wooden artefact.

It has been left to this tale to explain how, throughout our pilgrimage, we have been able to date historical specimens on an absolute timescale. Tree rings are very accurate but only for the nearest reaches of the historical record. Fossils are dated by other methods, mostly involving radioactive decay, and we shall come to them, along with other techniques, in the course of the tale.

Annual rings in a tree result from the unsurprising fact that a tree puts on more growth in some seasons than in others. But, by the same token, whether in summer or winter, trees grow more in a good year than in a poor year. Good years are two a penny, and so are bad years, so one tree ring is no good for identifying a particular year. But a sequence of years has a fingerprint pattern of wide and narrow rings, which labels that sequence in different trees over a wide area. Dendrochronologists compile catalogues of these labelled signature patterns. Then a fragment of wood, perhaps from a Viking longship buried in mud, can be dated by matching its ring pattern against previously collected libraries of signatures.

The same principle is used in dictionaries of melodies. Suppose you have a tune in your head and you can't remember its name. How might you look it up? Various principles are used, of which the simplest is the Parsons code. Turn your tune into a series of ups and downs (the first note is ignored because, obviously, it can be neither up nor down). Here, for instance, is the pattern of a favourite tune, the Londonderry Air, or Air from County Derry, which I have just typed into the Melodyhound website:

UUUDUUDDDDDUUUUUDDDUD

Melodyhound correctly sleuthed my tune (calling it 'Danny Boy' – the name by which Americans know it because of some twentieth-century words that were set to it). At first it seems surprising that a tune should be identified by such a short sequence of symbols, telling only the direction of movement, not the distance, and with no indication of durations of notes. But it really works. For the same kind of reason, a fairly short consecutive pattern of tree rings suffices to identify a particular sequence of yearly growth rings.

In a newly felled tree, the outer ring represents the present. The past can be exactly reckoned by counting inwards. So absolute dates can be put upon ring pattern signatures in recent trees whose date of felling is recorded. By looking for overlaps – signature patterns near the core of a young tree that match the pattern in the outside layers of an older tree – we can put absolute dates on ring patterns in older trees too. By daisy-chaining the overlaps backwards, it is in principle possible to put absolute dates on very old wood indeed – in principle even from the Petrified Forest of Arizona, if only there were a continuous series of petrified intermediates – if only! By this technique of overlapping jigsaws, libraries of fingerprint patterns can be built up and consulted to recognise wood that is older than the oldest tree we ever see alive. The changing thickness of tree rings can also, incidentally, be used not just for dating wood but for reconstructing year-to-year climate and ecological patterns dating from long before meteorological records were kept.

Dendrochronology is limited to the relatively recent time domains inhabited by archaeologists. But tree growth is not the only process that spurts and slows on an annual cycle, or on some other regular or even irregular cycle. Any such process can in principle be used for dating, aided by the same ingenious trick of daisy-chaining overlapped patterns. And some of these techniques work over a longer period than dendrochronology itself. Sediments are laid down on the sea bottom at an uneven rate, and in stripes which we can think of as equivalent to tree rings. These stripes can be counted, and signatures recognised, in core samples extracted by deep cylindrical probes.

Another example, which we encountered in the Epilogue to the Elephant Bird's Tale, is palaeomagnetic dating. As we saw there, the Earth's magnetic field reverses from time to time. What had been magnetic north suddenly becomes magnetic south for some thousands of years, then flips again. This has happened 282 times during the last 10 million years. Although I say 'flips' and 'suddenly', it is sudden only by geological standards. Entertaining as it might be if a polar flip today turned every plane and ship around in its tracks, that isn't the way things work. The 'flip' actually takes a few thousand years, and is much more complicated than the flip word suggests. The magnetic North Pole in any case seldom coincides exactly with the true, geographic North Pole (around which the Earth spins). It wanders around the polar region over the years. At present the magnetic North Pole is located near Bathurst Island in northern Canada, about 1,000 miles from the true North Pole. During a 'flip' there is an interregnum of magnetic confusion, with large and complicated variations in field strength and direction, sometimes involving the temporary appearance of more than one magnetic north and more than one magnetic south. Eventually the confusion stabilises again, and when the dust settles it may turn out that the previous magnetic north is now near the true South Pole and vice versa. Stability, with wandering, then resumes for perhaps a million years until the next flip.

A thousand years in geology's sight is but an evening gone. The time spent 'flipping' is negligible compared to the time spent in the rough vicinity of either the true North or the true South Pole. Nature, as we saw earlier, keeps an automatic record of such events. In molten volcanic rock, certain minerals behave like little compass needles. When the molten rock solidifies, these mineral needles constitute a 'frozen' record of the Earth's magnetic field at the moment of solidification (by a rather different process, palaeomagnetism can be observed in sedimentary rock, too). After a 'flip', the miniature compass needles in the rocks point in the opposite direction compared with before the flip. It's like tree rings all over again, except that the stripes are not a year apart but of the order of a million years. Once again, patterns of stripes can be matched up with other patterns, and a continuous chronology of magnetic flips can be daisy-chained together. Absolute dates can't be calculated by counting stripes because, unlike tree rings, the stripes represent unequal durations. Nevertheless, the same signature pattern of stripes can be picked up in different places. This means that if some other method of absolute dating (see pages 531–536) is available for one of the places, magnetic stripe patterns, like the Parsons code for a melody, can be used to recognise the same time zone in other places. As with tree rings and other dating methods, the full picture is built up from fragments gathered in different places.

Tree rings are good for dating recent relics to the nearest year. For older dates, with inevitably less fine pinpointing, we exploit the wellunderstood physics of radioactive decay. To explain this, we begin with a digression.

All matter is made of *atoms*. There are more than 100 types of atoms, corresponding to the same number of *elements*. Examples of elements are iron, oxygen, calcium, chlorine, carbon, sodium and hydrogen. Most matter consists not of pure elements but of *compounds*: two or more atoms of various elements bonded together, as in calcium carbonate, sodium chloride, carbon monoxide. The binding of atoms into compounds is mediated by *electrons*, which are tiny particles orbiting (a metaphor to help us understand their real behaviour, which is much stranger) the central *nucleus* of each atom. A nucleus is huge compared to an electron but tiny compared to an electron's orbit. Your hand, consisting mostly of empty space, meets hard resistance when it strikes a block of iron, also consisting mostly of empty space, because forces associated with the atoms in the two solids interact in such a way as to prevent them passing through each other. Consequently iron and stone seem solid to us because our brains most usefully serve us by constructing an illusion of solidity.

It has long been understood that a compound can be separated into its component parts, and recombined to make the same or a different compound with the emission or consumption of energy. Such easy-come easy-go interactions between atoms constitute chemistry. But, until the twentieth century, the atom itself was thought to be inviolate. It was the smallest possible particle of an element. A gold atom was a tiny speck of gold, qualitatively different from a copper atom, which was a minimal particle of copper. The modern view is more elegant. Gold atoms, copper atoms, hydrogen atoms and so on are just different arrangements of the same fundamental particles, just as horse genes, lettuce genes, human genes and bacterial genes have no essential 'flavour' of horse, lettuce, human or bacteria but are just different combinations of the same four DNA letters. In the same way as chemical compounds have long been understood to be arrangements put together from a finite repertoire of 100 or so atoms, so each atomic nucleus turns out to be an arrangement of two fundamental particles, the *protons* and *neutrons*. A gold nucleus is not 'made of gold'. Like all other nuclei, it is made of protons and neutrons. An iron nucleus differs from a gold nucleus, not because it is made of a qualitatively different kind of stuff called iron, but simply because it contains 26 protons (and 30 neutrons), instead of gold's 79 protons (and 118 neutrons). At the level of a single atom there is no 'stuff' that has the properties of gold or iron. There are just different combinations of protons, neutrons and electrons. Physicists go on to tell us that protons and neutrons are themselves composed of yet more fundamental particles, the quarks, but we shall not follow them to such depths.

Protons and neutrons are almost the same size as each other, and much larger than electrons. Unlike a neutron, which is electrically neutral, each proton has one unit of electric charge (arbitrarily designated positive), which exactly balances the negative charge of one electron 'in orbit' around the nucleus. A proton can be transformed into a neutron if it absorbs an electron, whose negative charge neutralises the proton's positive one. Conversely, a neutron can transform itself into a proton by expelling a unit of negative charge – one electron. Such transformations are examples of nuclear reactions, as opposed to chemical reactions. Chemical reactions leave the nucleus intact. Nuclear reactions change it. They usually involve much larger exchanges of energy than chemical reactions, which is why nuclear weapons are so much more devastating, weight for weight, than conventional (i.e. chemical) explosives. The alchemists' quest to change one metallic element into another failed only because they tried to do it by chemical rather than nuclear means.

Each element has a characteristic number of protons in its atomic nucleus, and the same number of electrons in 'orbit' around the nucleus: one for hydrogen, two for helium, six for carbon, 11 for sodium, 26 for iron, 82 for lead, 92 for uranium. It is this number, the so-called atomic number, which (acting via the electrons) largely determines an element's chemical behaviour. The neutrons have little effect on an element's chemical properties, but they do affect its mass and they do affect its nuclear reactions.

A nucleus typically has roughly the same number of neutrons as protons, or a few more. Unlike the proton count, which is fixed for any given element, the neutron count varies. Normal carbon has six protons and six neutrons, giving a total 'mass number' of 12 (since the mass of electrons is negligible and a neutron weighs approximately the same as a proton). It is therefore called carbon 12. Carbon 13 has one extra neutron, and carbon 14 two extra neutrons, but they all have six protons. Such different 'versions' of an element are called 'isotopes'. The reason all three of these isotopes have the same name, carbon, is that they have the same atomic number, 6, and therefore all have the same chemical properties. If nuclear reactions had been discovered before chemical reactions, perhaps the isotopes would have been given different names. In a few cases, isotopes are different enough to earn different names. Normal hydrogen has no neutrons. Hydrogen 2 (one proton and one neutron) is called deuterium. Hydrogen 3 (one proton and two neutrons) is called tritium. All behave chemically as hydrogen. For example, deuterium combines with oxygen to make a form of water called heavy water, famous for its use in the manufacture of hydrogen bombs.

Isotopes, then, differ only in the number of neutrons they have, along with the fixed number of protons that characterise the element. Among the isotopes of an element, some may have an unstable nucleus, meaning it has an occasional tendency to change at an unpredictable instant, though with predictable probability, into a different kind of nucleus. Other isotopes are stable: their probability of changing is zero. Another word for unstable is radioactive. Lead has four stable isotopes and 25 known unstable ones. All isotopes of the very heavy metal uranium are unstable – all are radioactive. Radioactivity is the key to the absolute dating of rocks and their fossils: hence the need for this digression to explain it.

What actually happens when an unstable, radioactive element changes into a different element? There are various ways in which this can happen, but the two best known are called alpha decay and beta decay. In alpha decay the parent nucleus loses an 'alpha particle', which is a pellet consisting of two protons and two neutrons stuck together. The mass number therefore drops by four units, but the atomic number drops by only two units (corresponding to the two protons lost). So the element changes, chemically speaking, into whichever element has two fewer protons. Uranium 238 (with 92 protons and 146 neutrons) decays into thorium 234 (with 90 protons and 144 neutrons).

Beta decay is different. One neutron in the parent nucleus turns into a proton, and it does so by ejecting a beta particle, which is a single unit of negative charge or one electron. The mass number of the nucleus remains the same because the total number of protons plus neutrons remains the same, and electrons are too small to bother with. But the atomic number increases by one because there is now one more proton than before. Sodium 24 transforms itself, by beta decay, into magnesium 24. The mass number has remained the same, 24. The atomic number has increased from 11, which is uniquely diagnostic of sodium, to 12, which is uniquely diagnostic of magnesium.

A third kind of transformation is neutron-proton replacement. A stray neutron hits a nucleus and knocks one proton out of the nucleus, taking its place. So, as in beta decay, there is no change in the mass number. But this time the atomic number has decreased by one because of the loss of one proton. Remember that the atomic number is simply the number of protons in the nucleus. A fourth way in which one element can turn into another, which has the same effect on atomic number and mass number, is electron capture. This is a kind of reversal of beta decay. Whereas in beta decay a neutron turns into a proton and expels an electron, electron capture transforms a proton into a neutron by neutralising its charge. So the atomic number drops by one, while the mass number remains the same. Potassium 40 (atomic number 19) decays to argon 40 (atomic number 18) by this means. And there are various other ways in which nuclei can be radioactively transformed into other nuclei.

One of the cardinal principles of quantum mechanics is that it is impossible to predict exactly when a particular nucleus of an unstable element will decay. But we can measure the statistical likelihood that it will happen. This measured likelihood turns out to be utterly characteristic of a given isotope. The preferred measure is the half-life. To measure the half-life of a radioactive isotope, take a lump of the stuff and count how long it takes for exactly one half of it to decay into something else. The half-life of strontium 90 is 28 years. If you have 100 grams of strontium 90, after 28 years you'll have only 50 grams left. The rest will have turned into yttrium 90 (as it happens, which in turn changes into zirconium 90). Does this mean that after another 28 years you'll have no strontium left? Emphatically no. You'll have 25 grams left. After another 28 years the amount of strontium will have halved again, to 12.5 grams. Theoretically, it never reaches zero but only approaches it by successively halved steps. This is the reason we have to talk about the half-life rather than the 'life' of a radioactive isotope.

The half-life of carbon 15 is 2.4 seconds. After 2.4 seconds you'll be left with half of your original sample. After another 2.4 seconds you'll have only a quarter of your original sample. After another 2.4 seconds you are down to an eighth, and so on. The half-life of uranium 238 is nearly 4.5 billion years. This is approximately the age of the solar system. So, of all the uranium 238 that was present on Earth when it first formed, about half now remains. It is a wonderful and very useful fact about radioactivity that half-lives of different elements span such a colossal range, from fractions of seconds to billions of years.

We are approaching the point of this whole digression. The fact that each radioactive isotope has a particular half-life offers an opportunity to date rocks. Volcanic rocks often contain radioactive isotopes, such as potassium 40. Potassium 40 decays to argon 40 with a half-life of 1.3 billion years. Here, potentially, is an accurate clock.

But it's no use just measuring the amount of potassium 40 in a rock. You don't know how much there was when it started! What you need is the ratio of potassium 40 to argon 40. Fortunately, when potassium 40 in a rock crystal decays, the argon 40 (a gas) remains trapped in the crystal. If there are equal amounts of potassium 40 and argon 40 in the substance of the crystal, you know that half the original potassium 40 has decayed. It is therefore 1.3 billion years since the crystal was formed. If there's, say, three times as much argon 40 as potassium 40, only one quarter (half of a half) of the original potassium 40 remains, so the age of the crystal is two half lives or 2.6 billion years.

The moment of crystallisation, which in the case of volcanic rocks is the moment when the molten lava solidified, is the moment when the clock was zeroed. Thereafter, the parent isotope steadily decays and the daughter isotope remains trapped in the crystal. All you have to do is measure the ratio of the two amounts, look up the half-life of the parent isotope in a physics book, and it is easy to calculate the age of the crystal. As I said earlier, fossils are usually found in sedimentary rocks, while dateable crystals are usually in volcanic rocks, so fossils themselves have to be dated indirectly by looking at volcanic rocks that sandwich their strata.

A complication is that often the first product of the decay is itself another unstable isotope. Argon 40, the first product of decay of potassium 40, happens to be stable. But when uranium 238 decays it passes through a cascade of no fewer than 14 unstable intermediate stages, including nine alpha decays and seven beta decays, before it finally comes to rest as the stable isotope lead 206. By far the longest half-life of the cascade (4.5 billion years) belongs to the first transition, from uranium 238 to thorium 234. An intermediate step in the cascade, from bismuth 214 to thallium 210, has a half-life of only 20 minutes, and even that is not the fastest (i.e. most probable). The later transitions take negligible time compared to the first, so the observed ratio of uranium 238 to the finally stable lead 206 can be set against a half-life of 4.5 billion years to calculate the age of a particular rock.

The uranium/lead method and the potassium/argon method, with their half-lives measured in billions of years, are useful for dating fossils of great age. But they are too coarse for dating younger rocks. For these, we need isotopes with shorter half-lives. Fortunately a range of clocks is available with a wide selection of isotopic half-lives.

You choose your half-life to give best resolution for the rocks with which you are working. Better yet, the different clocks can be used as checks on each other.

The fastest radioactive clock in common use is the carbon 14 clock, and this brings us full circle to the teller of this tale, for wood is one of the main materials subjected to carbon 14 dating by archaeologists. Carbon 14 decays to nitrogen 14 with a half-life of 5,730 years. The carbon 14 clock is unusual in that it is used to date the actual dead tissues themselves, not volcanic rocks sandwiching them. Carbon 14 dating is so important for relatively recent history – much younger than most fossils, and spanning the range of history normally called archaeology – that it deserves special treatment.

Most of the carbon in the world consists of the stable isotope carbon 12. About one million-millionth part of the world's carbon consists of the unstable isotope carbon 14. With a half-life measured in only thousands of years, all the carbon 14 on Earth would long since have decayed to nitrogen 14 if it were not being renewed. Fortunately, a few atoms of nitrogen 14, the most abundant gas in the atmosphere, are continually being transformed, by bombardment of cosmic rays, into carbon 14. The rate of creation of carbon 14 is approximately constant. Most of the carbon in the atmosphere, whether carbon 14 or the more usual carbon 12, is chemically combined with oxygen in the form of carbon dioxide. This gas is sucked in by plants, and the carbon atoms used to build their tissues. To plants, carbon 14 and carbon 12 look the same (plants are only 'interested' in chemistry, not the nuclear properties of atoms). The two varieties of carbon dioxide are imbibed approximately in proportion to their availability. Plants are eaten by animals, which may be eaten by yet other animals, so carbon 14 is dispersed in a known proportion relative to carbon 12 throughout the food chain during a time which is short compared to the half-life of carbon 14. The two isotopes exist in all living tissues in approximately the same proportion as in the atmosphere, one part in a million million. To be sure, they occasionally decay to nitrogen 14 atoms. But this constant rate is offset by their continuous exchange, via the links of the food chain, with the ever-renewed carbon dioxide of the atmosphere.

All this changes at the moment of death. A dead predator is cut off from the food chain. A dead plant no longer takes in fresh supplies of carbon dioxide from the atmosphere. A dead herbivore no longer eats fresh plants. The carbon 14 in a dead animal or plant continues to decay to nitrogen 14. But it is not replenished by fresh supplies from the atmosphere. So the ratio of carbon 14 to carbon 12 in the dead tissues starts to drop. And it drops with a half-life of 5,730 years. The bottom line is that we can tell when an animal or plant died by measuring the ratio of carbon 14 to carbon 12. This is how it was proved that the Turin Shroud cannot have belonged to Jesus – its date is medieval. Carbon 14 dating is a wonderful tool for dating the relics of relatively recent history. It is of no use for more ancient dating because almost all the carbon 14 has decayed to nitrogen 14, and the residue is too tiny to measure accurately.

There are other methods of absolute dating, and new ones are being invented all the time. The beauty of having so many methods is partly that they collectively span such an enormous range of time-scales. It is also that they can be used as a cross-check on each other. It is extremely hard to argue against datings that are corroborated across different methods.

UNCERTAIN

> *The Microbe is so very small*
> *You cannot make him out at all,*
> *But many sanguine people hope*
> *To see him through a microscope.*
> *His jointed tongue that lies beneath*
> *A hundred curious rows of teeth;*
> *His seven tufted tails with lots*
> *Of lovely pink and purple spots,*
> *On each of which a pattern stands,*
> *Composed of forty separate bands;*
> *His eyebrows of a tender green;*
> *All these have never yet been seen –*
> *But Scientists, who ought to know,*
> *Assure us that they must be so . . .*
> *Oh! let us never, never doubt*
> *What nobody is sure about.*
> HILAIRE BELLOC *(1870–1953)*
> From *More Beasts for Worse Children (1897)*

Hilaire Belloc was a brilliant versifier but a prejudiced man. If there is an element of anti-scientific prejudice above, let us not play up to it. There is much that we are unsure about in science. Where science scores over alternative world views is that we know our uncertainty, we can often measure its magnitude, and we work optimistically to reduce it.

At Rendezvous 37 we enter a world of microbes and also a realm of uncertainty: uncertainty not so much about the microbes themselves as about the order in which we are to greet them. I thought of making a guess and sticking to it, but that would be unfair on the other rendezvous points, about which we can be at least somewhat more certain. If this book's publication were delayed a year or two, the chances of resolution would be good. But for now, let us treat Belloc's verse as a Cautionary Tale for Scientists. We know whom we

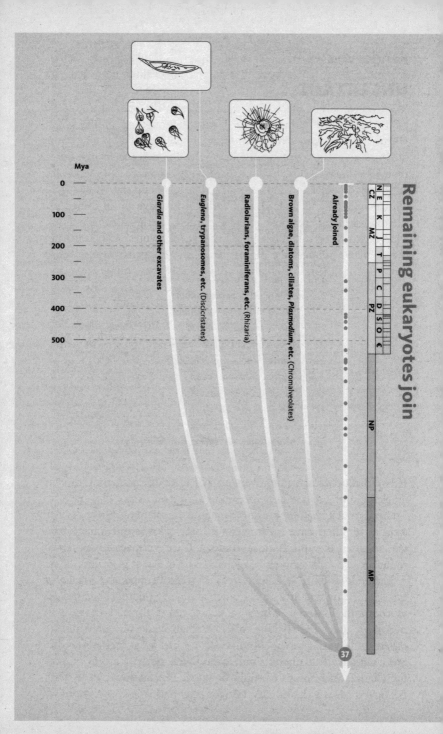

Remaining eukaryotes join

Mya

0

100

200

300

400

500

Giardia and other excavates

Euglena, trypanosomes, etc. (Discicristates)

Radiolarians, foraminiferans, etc. (Rhizaria)

Brown algae, diatoms, ciliates, *Plasmodium*, etc. (Chromalveolates)

Already joined

CZ
N E K J MZ T P C D S O €
PZ
NP
MP

37

are to meet at the next one or two or three rendezvous points, but we don't know in what order, and we don't know how many separate rendezvous points there are.

This uncertainty affects all the 'eukaryotes' who are yet to join the pilgrimage. This important word will be explained in the Great Historic Rendezvous. For the moment, know simply that one of the most momentous events in the history of life was the formation of the eukaryotic cell. Eukaryotic cells are the large and complex cells, with walled nuclei and mitochondria, that make up the bodies of all animals, plants and, indeed, all pilgrims who have so far joined us. That is, all living creatures except the true bacteria and the archaea, which used to be called bacteria. These 'prokaryotes' will constitute the final two rendezvous points, and, as it happens, we are more certain about them. I shall arbitrarily number these final two 38 and 39. This means that the remainder of the eukaryotes all join us together at rendezvous 37, which is one of the possible theories at present. But please bear in mind that this is a toss-up: our final rendezvous, with the true bacteria, could be anything from 39 to 42.

Part of the problem is rooting. We met this in the Gibbon's Tale. A star diagram such as the one on the following page is compatible with many different evolutionary trees, and that means many different ways of organizing our rendezvous.

Before we move on to the main point, notice with proper humility, the tiny line labelled 'animals'. If you can't find them, look at the branch labelled 'opisthokonts' on the bottom left, where you'll find us as the sister group to the choanoflagellates. That is where you and I belong, together with the entire populace of pilgrims who joined us up to and including Rendezvous 31.

Clearly there are many places where we could sling the root. The fact that the two most strongly supported hypotheses (indicated by the dotted arrows) are at two such distantly separated extremes contributed to the sapping of my confidence. But it gets worse. The positioning of the root is only the first of our problems. The second problem is that five of the lines meet at a single point in the middle.

Opposite: **Remaining eukaryotes join.** The high-level phylogeny of the remaining 50,000 or so described species of eukaryote is currently unresolved (see text). Faded lines indicate the current high level of uncertainty. The chromalveolate branch is often subdivided into the chromista (heterokonts) and the alveolates.

Images, left to right: *Giardia lamblia; Euglena acus;* foraminiferan (*Globigerina* sp.); leather kelp (*Ecklonia radiata*).

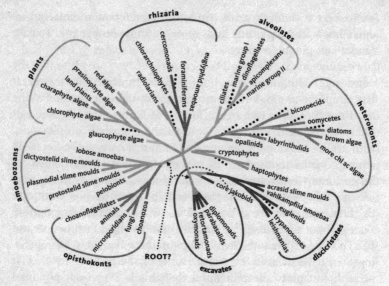

Notice, with proper humility, where you and I belong. Unrooted phylogram or star diagram of all life, based on the consensus of currently available molecular and other studies. Adapted from Baldauf [13].

This doesn't mean that anybody thinks all those five groups burst forth from a single ancestor at the same moment and are all equally close cousins to each other. All it means is yet more uncertainty. We don't know which of the five are closer cousins to each other, so, rather than commit ourselves to what may be an error, justly to be lampooned by a latter-day Belloc, we draw them all as radiating out from a single point. The point where the five lines meet should eventually be resolved into a series of forking lines. Each one of those lines is potentially a place where we could sling our root.

By now it will be clear why I backed away from committing myself to the details of the next few rendezvous points. Actually, if you look at the diagram, you'll notice that I was even a bit rash committing myself to Rendezvous 36 as the place where the plants join us. The line of the plants is one of the five that radiates out from the centre of the star. Since decisions hereabouts are still so arbitrary, I decided to treat the plants as though they had a separate rendezvous with us, but only because they are such a huge and important group that they seemed to deserve a separate pilgrim band of their own. What I did, in effect, was draw out a line from the middle of the star diagram. We could make a similarly arbitrary decision over how to resolve the remaining trichotomy, but my courage finally deserts me. I shall leave it buried within the uncertainty of Rendezvous 37, the blind-date rendezvous.

Instead of committing myself to the order in which they join us, I shall simply go through the remaining groups of eukaryotes, briefly describing them. The Rhizaria include various groups of single-celled eukaryotes, some green and photosynthesising, some not. Among them are the foraminiferans and the radiolarians, notable for their beauty, never better captured than in the drawings of Ernst Haeckel, the eminent German zoologist who seems to keep cropping up through this book. The alveolates include some further beautiful creatures, including the ciliates and the dinoflagellates. Among the ciliates, or so it would seem, is *Mixotricha paradoxa,* whose tale we shall soon hear. The 'so it would seem' and the *'paradoxa'* form the substance of the tale, whose thunder I shall not steal here.

The heterokonts are another mixed group. They include some further beautiful unicellular creatures, such as the diatoms, again memorably illustrated by Haeckel. But this group has also independently discovered multicellularity, in the form of the brown algae. These are the largest and most prominent of all seaweeds, with giant kelps reaching 100 metres in length. The brown algae include the wracks of the genus *Fucus*, the various species of which segregate themselves in strata up the beach, each being best suited to a particular zone of the tide cycle. *Fucus* might well be the genus upon whom the leafy sea dragon (see its tale) is modelled.

The discicristates include photosynthetic flagellates, such as the green *Euglena*, and parasitic ones, such as *Trypanosoma*, which causes sleeping sickness. There are also the acrasid slime moulds, which are not closely related to the dictyostelid slime moulds whom we met at Rendezvous 35. As so often in this long pilgrimage, we marvel at the capacity of life to reinvent similar body forms for similar ways of life. 'Slime moulds' pop up in two or even three different pilgrim bands; so do 'flagellates', so do 'amoebas'. Probably we should think of 'the amoeba' as a way of life, like 'the tree'. 'Trees', meaning very large plants stiffened with wood, pop up in many separate plant families. It looks as though the same is true of 'amoebas' and 'flagellates'. It is certainly true of multicellularity, which has arisen in animals, fungi, plants, brown algae and various other places, such as slime moulds.

The last major group of our unresolvable star consists of the excavates. These are single-celled creatures that would once have been called flagellates and united with *Trypanosoma*, the sleeping sickness organism. Now separated off, the excavates include the nasty gut parasite *Giardia*, the nasty sexually transmitted vaginal parasite *Trichomonas*, and various fascinatingly complicated single-celled creatures found only in the guts of termites. And that is the cue for a tale.

THE MIXOTRICH'S TALE

Mixotricha paradoxa means 'unexpected combination of hairs', and we shall see why in a moment. It is a micro-organism that lives in the gut of an Australian termite, 'Darwin's termite', *Mastotermes darwiniensis*. Pleasingly, though not necessarily for the human inhabitants, one of the main places where it flourishes is the town of Darwin in northern Australia.

Termites bestride the tropics like a distributed colossus. In tropical savannahs and forests, they reach population densities of 10,000 per square metre, and are estimated to consume up to a third of the total annual production of dead wood, leaves and grass. Their biomass per unit area is double that of migrating herds of wildebeest on the Serengeti and Masai Mara, but is spread across the entire tropics.

If you ask the source of the termites' alarming success, it is twofold. First, they can eat wood, which includes cellulose, lignin and other matter that animal guts normally can't digest. I'll return to this. Second, they are highly social and gain great economies from division of labour among specialists. A termite mound has many of the attributes of a single large and voracious organism, with its own anatomy, its own physiology and its own mud-fashioned organs, including an ingenious ventilation and cooling system. The mound itself stays in one place, but it has a myriad mouths and six myriad legs, and these range over a foraging area the size of a football pitch.

Termites' legendary feats of co-operation are possible, in a Darwinian world, only because the majority of individuals are sterile but closely related to a minority who are very fertile indeed. Sterile workers act like parents towards their younger siblings, thereby freeing the queen to become a specialised egg factory, and a grotesquely efficient one at that. Genes for worker behaviour are passed to future generations via the minority of the workers' siblings who

38. What are we to make of this?
Dickinsonia costata, part of the Ediacaran fauna (see page 460).

39. Sail by the wind
The blue button or blue sea star *Porpita* has a central gas-filled float surrounded by tentacles. Like its close relative *Velella*, *Porpita* is now thought to be a highly modified polyp rather than a colony (see page 481).

40. Jelly armada
Mastigias jellyfish amassed at the water surface, Palau, western Pacific (see page 483).

41. There aren't many animals who could claim to have redrawn the map of the world
Heron Island, on the Great Barrier Reef (see page 484).

42. Trust, at the barber's shop of the sea
Cleaner fish (*Labroides dimidiatus*) working on a rosy goatfish (*Parupeneus rubescens*), Red Sea (see page 488).

43. Too good for a goddess
Venus's girdle (*Cestum veneris*) (see page 493).

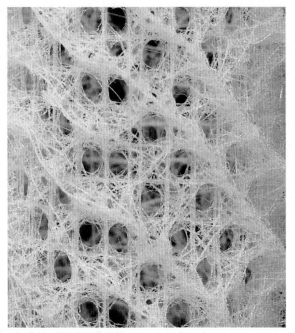

44. Venus's flower basket
Detail of spicule skeleton of the glass sponge, *Euplectella aspergillum* (see page 500).

45. Orgy of mushrooms
Stinkhorn mushroom, *Phallus impudicus*, a basidiomycete (see page 512).

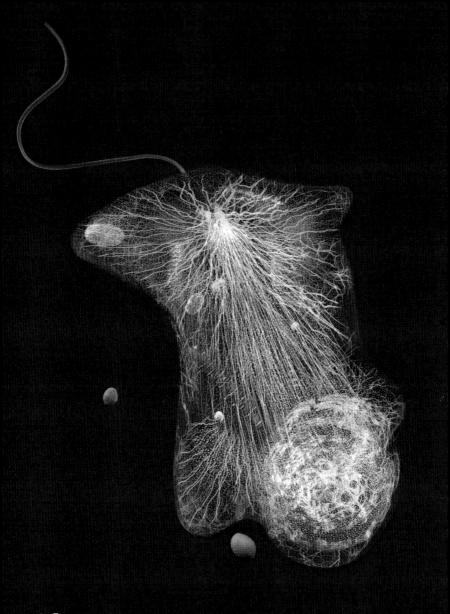

46. Concestor 36
The single-celled eukaryote shows a nucleus at bottom right, surrounded by sheets of endoplasmic reticulum. The cell structure is maintained by a cytoskeleton (the network of white threads). The concestor would probably have moved both by using its whip-like flagellum and by extending parts of its body. Artistic reconstruction by Malcolm Godwin (see page 521).

47. If you've seen one…
Sequoiadendron gigan-teum, Sequoia National Park, California (see page 526).

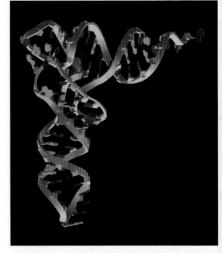

48. Free to tie itself in knots
Computer graphic of transfer RNA, paired with itself to make a miniature double helix (see page 589).

49. Reappearance of Ichthyosauri

This cartoon by Henry de la Beche lampooned a view expressed by Charles Lyell that the Earth's periodic changes in climate and associated wildlife might lead to a future world in which *Iguanodon* would once more roam the woods and ichthyosaurs reappear in the seas (see page 601).

50. Almost indecently eager to evolve eyes

Some examples of eyes. CLOCKWISE FROM TOP LEFT: *Nautilus pompilius* (pinhole eye); fossil trilobite (*Phacops*, compound eye made of calcite lenses – some of these can be seen in position in the top part of the eye); simulian blackfly (*Simulium damnosum*, compound eye); stoplight parrotfish (*Sparisoma viride*, fish eye); great horned owl (*Bubo virginianus*, corneal eye) (see page 603).

are destined to reproduce (helped by the majority of their siblings who are destined to be sterile). You will appreciate that the system works only because it is a strictly non-genetic decision whether a young termite shall become a worker or a reproducer. All young termites have a genetic ticket to enter an environmental lottery which decides whether they become reproductives or workers. If there were genes for being unconditionally sterile, they obviously could not be passed on. Instead, they are *conditionally* switched-on genes. They are passed on when they find themselves in queens or kings because copies of the very same genes cause workers to labour for that end and forgo reproduction themselves.

The analogy of insect colony to human body is often made, and it is not a bad one. The majority of our cells subjugate their individuality, devoting themselves to assisting the reproduction of the minority that are capable of it: 'germ-line' cells in the testes or ovaries, whose genes are destined to travel, via sperms or eggs, into the distant future. But genetic relatedness is not the only basis for subjugation of individuality in fruitful division of labour. Any sort of mutual assistance, where each side corrects a deficiency in the other, can be favoured by natural selection on both. To see an extreme example, we dive inside the gut of an individual termite, that seething and, as I assume, noisome chemostat which is the world of the mixotrich.

Termites, as we have seen, enjoy an additional advantage over bees, wasps and ants: their prodigious feats of digestion. There is almost nothing that termites can't eat, from houses to billiard balls to priceless First Folios. Wood is potentially a rich food source but it is denied to almost all animals because cellulose and lignin are so indigestible. Termites and certain cockroaches are the outstanding exception. Termites are, indeed, related to cockroaches, and Darwin's termite, like other so-called 'lower' termites, is a sort of living fossil. One could imagine it halfway between cockroaches and advanced termites.

In order to digest cellulose, you need enzymes called cellulases. Most animals can't make cellulases, but some micro-organisms can. As Taq's Tale will explain, bacteria and archaea are biochemically more versatile than the rest of the living kingdoms put together. Animals and plants perform a fraction of the biochemical mix of tricks available to bacteria. For digesting cellulose, herbivorous mammals all rely upon microbes in their guts. Over evolutionary time, they have entered into a partnership in which they make use of chemicals such as acetic acid which, to the microbes, are waste products. The microbes themselves gain a safe haven with plenty of raw materials for their own biochemistries, preprocessed and ready-chopped into small, manageable pieces. All herbivorous mammals have bacteria in the lower gut, which the food reaches after the mammal's own digestive juices have had a go at it. Sloths, kangaroos, colobus monkeys and especially cud-chewing ruminants have independently evolved the trick of also keeping bacteria in the upper portion of the gut, which precedes the mammal's own main digestive efforts.

Unlike mammals, termites are capable of manufacturing their own cellulases, at least in the case of the so-called 'advanced' termites. But up to one-third of the net weight of a more primitive (i.e. more cockroach-like) termite, such as Darwin's termite, consists of its rich gut fauna of microbes, including eukaryotic protozoa as well as bacteria. The termites locate and chew the wood into small, manageable chips. The microbes live on the wood chips, digesting them with enzymes unavailable to the termites' own biochemical toolkit. Or you could say the microbes have become tools in the termites' toolkit. As with the cattle, it is the waste products of the microbes that the termites live on. I suppose we could say that Darwin's termite and the other primitive termites farm microorganisms in their guts.* And this brings us, eventually, to the mixotrich, whose tale this is.

Mixotricha paradoxa is not a bacterium. Like many of the microbes in termite guts, it is a large protozoan, half a millimetre long or more, and large enough to contain hundreds of thousands of bacteria inside itself – as we shall see. It lives nowhere except in the gut of Darwin's termite, where it is a member of the mixed community of microbes that thrive on the wood chips milled by the termite's jaws. Micro-organisms populate the termite's gut as richly as the termites themselves populate the mound, and as termite mounds populate the

---

* There are two main processes by which energy is extracted from food fuel: anaerobic (without oxygen) and aerobic (with oxygen). Both are chemical sequences in which fuel, rather than being burned, is coaxed into trickling out its energy in a way that can be efficiently used. The most common anaerobic sequence yields pyruvate as its major product, and this is the starting point of the most common aerobic cascade. Termites go out of their way to deprive their guts of free oxygen, thereby forcing their microbes to use only the anaerobic process, using wood fuel to produce pyruvate that the termite can then use for aerobic energy release.

savannah. If the mound is a town of termites, each termite gut is a town of micro-organisms. We have here a two-level community. But – and now we come to the crux of the tale – there is a third level, and the details are utterly remarkable. *Mixotricha* itself is a town. The full story was revealed by the work of L. R. Cleveland and A. V. Grimstone, but it is especially the American biologist Lynn Margulis who has drawn our attention to *Mixotricha*'s significance for evolution.

When J. L. Sutherland first examined *Mixotricha* in the early 1930s she saw two kinds of 'hairs' waving on its surface. It was almost completely carpeted by thousands of tiny hairs, beating to and fro. She also saw a few very long, thin, whip-like structures at the front end. Both seemed familiar to her, the small ones as 'cilia', the large ones as 'flagella'. Cilia are common in animal cells, for instance in our nasal passages, and they cover the surface of those protozoans called, not surprisingly, ciliates. Another traditionally recognised group of protozoans, the flagellates, have much longer, whip-like 'flagella' (singular 'flagellum' and, unlike cilia, they often are). Cilia and flagella share an identical ultrastructure. Both are like multistranded cables, and the strands have exactly the same signature pattern: nine pairs in a ring surrounding one central pair.

Cilia, then, can be seen as just smaller and more numerous flagella, and Lynn Margulis goes so far as to abandon the separate names and call them all by her own name of 'undulipodia', reserving 'flagella' for the very different appendages of bacteria. Nevertheless, according to the taxonomy of Sutherland's day, protozoans were supposed to have either cilia or flagella but not both.

This is the background to Sutherland's naming of *Mixotricha paradoxa:* 'unexpected combination of hairs'. *Mixotricha,* or so it seemed to Sutherland, has *both* cilia and flagella. It violates protozoological protocol. It has four large flagella at the front end, three pointing forwards and one backwards, in the manner characteristic of a particular, previously known group of flagellates called the Parabasalia. But it also has a dense coat of waving cilia. Or so it seemed.

As it has turned out, *Mixotricha*'s 'cilia' are even more unexpected than Sutherland realised, and they don't violate precedent in the way she feared. It's a pity she didn't get the chance to see *Mixotricha* alive, instead of fixed on a slide. Mixotrichs swim too smoothly to be swimming with their own undulipodia. In the words of Cleveland and Grimstone, flagellates normally 'swim at varying speeds, turning from side to side, changing direction, and sometimes coming to rest'. The same is true of ciliates. Mixotricha glides along smoothly, usually in a straight line, never stopping unless physically blocked. Cleveland and Grimstone concluded that the smooth gliding movement is caused by the waving of the 'cilia' but – a far more exciting conclusion, this – they demonstrated with the electron microscope that the 'cilia' are not cilia at all. They are bacteria. Each one of the hundreds of thousands of tiny hairs is a single spirochaete – a bacterium whose entire body is a long, wiggling hair. Some important diseases, such as syphilis, are caused by spirochaetes. They normally swim freely, but *Mixotricha*'s spirochaetes are stuck to its body wall, exactly as though they were cilia.

They don't move like cilia, however: they move like spirochaetes. Cilia move with an actively propulsive rowing stroke, followed by a recovery stroke in which they bend so as to present less resistance to the water. Spirochaetes undulate in a completely different and very characteristic manner, and that is just what *Mixotricha*'s 'hairs' do. Amazingly, they seem to be co-ordinated with each other, moving in waves that begin at the front end of the body and travel backwards. Cleveland and Grimstone measured the wavelength (the distance between wave-crests) as about a hundredth of a millimetre. This suggests that the spirochaetes are somehow 'in touch' with each other. Probably they are literally in touch: responding directly to the movement of neighbours, with a delay that determines the wavelength. I don't think it is known why the waves pass from front to back.

What is known is that the spirochaetes are not just jammed haphazardly into the mixotrich's skin. On the contrary, the mixotrich has, in a repeat pattern all over its surface, a complicated apparatus for holding spirochaetes and, what's more, pointing them in a posterior direction so that their undulating movements drive the mixotrich forwards. If these spirochaetes are parasites, it is hard to think of a more remarkable example of a host being 'friendly' to its parasites. Each spirochaete has its own little emplacement, called a 'bracket' by Cleveland and Grimstone. Each bracket is tailor-made to hold one spirochaete, or sometimes more than one. No cilium could ask for more. It becomes quite tricky to draw the line between 'own' body and 'alien' body in such cases. And that, to anticipate, is one of the main messages of this tale.

The resemblance to cilia goes further. If you look with a powerful

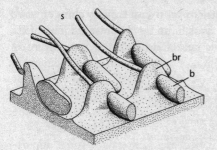

Arrangement of pill bacteria (b), brackets (br) and spirochaetes (s) on the surface of the mixotrich. From Cleveland and Grimstone [49].

microscope into the very fabric of a ciliate protozoan, such as *Paramecium*, you'll find that every cilium has a so-called basal body at its root. Now, amazingly, although the 'cilia' of *Mixotricha* are not cilia at all, they do appear to have basal bodies. Each spirochaete-toting bracket has at its base one basal body, shaped rather like a vitamin pill. Except that . . . well, having learned about *Mixotricha*'s idiosyncratic way of doing things, what would you guess those 'basal bodies' actually are? Yes! They too are bacteria. A completely different kind of bacteria – not spirochaetes but oval, pill-shaped bacteria.

Over large parts of the body wall there is a one-to-one relationship between bracket, spirochaete and basal bacterium. Each bracket has one spirochaete stuck through it, and one pill bacterium at its base. Looking at this, it is easy to understand why Sutherland saw 'cilia'. She naturally expected to see basal bodies wherever there are cilia . . . and when she looked, lo and behold, there were the 'basal bodies'. Little could she know, both 'cilia' and 'basal bodies' were hitch-hiking bacteria. As for the four 'flagella', the only true undulipodia the mixotrich possesses, they seem to be used not for propulsion at all, but as rudders for steering the craft as it is propelled by the thousands of spirochaete 'galley slaves'. Much as I'd like to claim it, by the way, that evocative phrase is not my own. It was coined by S. L. Tamm, who found, after Cleveland and Grimstone's work on *Mixotricha*, that other termite-gut protozoa do the same trick, but instead of spirochaetes, their galley slaves are ordinary bacteria with flagella.

Now for the other bacteria in the mixotrich, the pill-shaped ones that look like basal bodies – what are they doing? Are they contributing to the economy of their host? Are they getting something out of the relationship themselves? Probably yes, but it has not been shown definitely. They may well be making cellulases that digest wood. For of course, the mixotrichs subsist on the tiny chips of wood in the termite's gut, originally broken up by the powerful jaws of the termite. We have here a triple-decker dependency, reminiscent of Jonathan Swift's verse:

> *So, naturalists observe, a flea*
> *Has smaller fleas that on him prey;*
> *And these have smaller still to bite 'em;*
> *And so proceed ad infinitum.*
> *Thus every poet, in his kind,*
> *Is bit by him that comes behind.*

By the way, Swift's scansion in the middle lines is (surprisingly) so ungainly that we can understand why Augustus De Morgan came behind for another bite, giving us the rhyme in the form that most of us know today:

> *Great fleas have little fleas upon their backs to bite 'em,*
> *And little fleas have lesser fleas, and so ad infinitum.*
> *And the great fleas themselves, in turn, have greater fleas to go on;*
> *While these again have greater still, and greater still, and so on.*

And finally we come to the strangest part of the Mixotrich's Tale, the climax towards which the narrative has been directed. This whole story of vicarious biochemistry, the borrowing by greater creatures of the biochemical talents of lesser ones inside them, is charged with evolutionary *déjà vu*. The message of the mixotrich to the rest of the pilgrims is this: *This has all happened before.* We have reached the Great Historic Rendezvous.

THE GREAT HISTORIC RENDEZVOUS

Rendezvous, in this book, has a special meaning, following the central metaphor of a backward pilgrimage. But there is one cataclysmic event, arguably the most decisive event in the history of life, which really was a rendezvous, literally a historic rendezvous that actually took place in the true, forward direction of history. This was the origin of the eukaryotic (nucleated) cell: the high-tech, miniature machine that is the microfoundation of all large-scale and complex life on this planet. To distinguish it from all the other metaphorical backwards rendezvous points, I call it the Great Historic Rendezvous. The word 'historic' has here a double meaning: it means 'of major importance', and it also means 'forward chronology' as opposed to backward.

I have referred to the Great Historic Rendezvous as one event because of what now appears to be its single momentous consequence, the evolution of the eukaryotic cell, with its nucleus to contain the chromosomes, its complicated ultrastructure of membranes, and its self-reproducing miniature organelles, such as mitochondria and (in plants) chloroplasts. But it was actually two or three events, perhaps widely spaced in time. Each one of these historic rendezvous events was a merging with bacterial cells to form a larger cell. The Mixotrich's Tale, as a recent re-enactment, has prepared us to see the kind of thing that happened.

Perhaps 2 billion years ago, an ancient single-celled organism, some kind of proto-protozoan, entered into a strange relationship with a bacterium: a relationship similar to that between *Mixotricha* and its bacteria. As with *Mixotricha*, the same thing happened more than once, with different bacteria, the events possibly separated by hundreds of millions of years. All our cells are like individual mixotrichs, stuffed with bacteria which have become so transformed by generations of co-operation with the host cell that their bacterial origins are almost lost to sight. As with the mixotrich, only more so, the bacteria have become so intimately enmeshed in the life of the eukaryotic cell that it was a major scientific triumph to detect that they were there at all. I am fond of the Cheshire Cat simile, used by

Sir David Smith, one of our leading experts on symbiosis, for the co-operative living together of once-distinct elements in cells.

> In the cell habitat, an invading organism can progressively lose pieces of itself, slowly blending into the general background, its former existence betrayed only by some relic. Indeed, one is reminded of Alice in Wonderland's encounter with the Cheshire Cat. As she watched it, 'it vanished quite slowly, beginning with the tail, and ending with the grin, which remained some time after the rest of it had gone'. There are a number of objects in a cell like the grin of the Cheshire Cat. For those who try to trace their origin, the grin is challenging and truly enigmatic.

What are the biochemical tricks that these once-free bacteria brought into our lives: tricks that they perform to this day, tricks without which life would instantly cease? The two most important ones are photosynthesis, which uses solar power to synthesise organic compounds, and oxygenates the air as a by-product; and oxidative metabolism, which uses oxygen (ultimately from plants) to slow-burn the organic compounds and redeploy the energy that originally came from the sun.* These chemical technologies were developed before the Great Historic Rendezvous by (different) bacteria – and, in a sense, bacteria are still the only game in town. All that has changed is that they now practise their biochemical arts in the purpose-built factories called eukaryotic cells.

Photosynthetic bacteria used to be called blue-green algae, a terrible name since most of them aren't blue-green and none of them are algae. Most are green, and it is better to call them green bacteria, although some are reddish, yellowish, brownish, blackish or, yes, in some cases bluish-green. 'Green' also is sometimes used as a word for photosynthetic, and in that sense, too, green bacteria is a good name. Their scientific name is cyanobacteria. They are true bacteria rather than Archaea, and they seem to be a good monophyletic group. In other words, all of them (and nothing else) are descended from a single ancestor which would, itself, have been classified as a cyanobacterium.

The green colour of algae, and of cabbages, pine trees and grasses, comes from small green bodies called chloroplasts within their cells.

---

* Bacteria (including Archaea) also have a monopoly (apart from lightning strikes and human industrial chemists) on nitrogen fixation.

Chloroplasts are distant descendants of once free-living green bacteria. They still have their own DNA, and they still reproduce by asexual division, building up to a substantial population within each host cell. As far as a chloroplast is concerned, it is a member of a reproducing population of green bacteria. The world in which it lives and reproduces is the interior of a plant cell. From time to time its world suffers a minor upheaval when the plant cell divides into two daughter cells. Roughly half the chloroplasts find themselves in each daughter cell, and they soon resume their normal existence of reproducing to populate their new world with chloroplasts. All the while, the chloroplasts use their green pigment to trap photons from the sun and channel the sun's energy in the useful direction of synthesising organic compounds from carbon dioxide and water supplied by the host plant. The oxygen wastes are partly used by the plant and partly exhaled into the atmosphere through holes in the leaves called stomata (singular 'stoma'). The organic compounds synthesised by the chloroplasts are ultimately made available to the host plant cell.

Interestingly reminiscent of the Mixotrich's Tale, some chloroplasts show evidence of having entered plant cells indirectly, by piggybacking inside other eukaryotic cells, which would presumably have been called algae. The evidence is that some chloroplasts have a double membrane. Presumably the inner one is the wall of the original bacterium, the outer one the wall of the alga. As with *Mixotricha*, we can see recent re-enactments in the many examples of single-celled green algae being incorporated in the cells or tissues of fungi and animals, for example the algae that inhabit corals. Those chloroplasts that have a single membrane presumably entered directly, not on the coat-tails of algae.

All the free oxygen in the atmosphere comes from green bacteria, whether free-living or in the form of chloroplasts. And, as mentioned before, when it first appeared in the atmosphere oxygen was a poison. Indeed, some people colourfully say it still is a poison, which is why doctors advise us to eat 'anti-oxidants'. In evolution, it was a brilliant chemical coup to discover how to use oxygen to extract (originally solar) energy from organic compounds. This discovery, which can be seen as a sort of reverse photosynthesis, was entirely made by bacteria, but a different kind of bacteria. As with photosynthesis itself, bacteria still have a monopoly on the technology except that, again as with photosynthesis, eukaryotic cells like ours give house room to these oxygen-loving bacteria, who now travel under the name of mitochondria. We have become so dependent on oxygen, via the biochemical wizardry of mitochondria, that the statement that it is a poison makes sense only when uttered in a tone of self-conscious paradox. Carbon monoxide, the deadly poison in car exhausts, kills us by competing with oxygen for the favours of our oxygen-carrying haemoglobin molecules. Depriving somebody of oxygen is a swift way to kill them. Yet our own cells, unaided, wouldn't know what to do with oxygen. It is only mitochondria, and their bacterial cousins, that do.

As with chloroplasts, molecular comparison tells us the particular group of bacteria from which mitochondria are drawn. Mitochondria sprang from the socalled alpha-proteo bacteria and they are therefore related to the rickettsias that cause typhus and other nasty diseases. Mitochondria themselves have lost much of their original genome, and have become completely adapted to life inside eukaryotic cells. But, like chloroplasts, they still reproduce autonomously by division, making populations within each eukaryotic cell. Although mitochondria have lost most of their genes, thay haven't lost all of them, and this is fortunate for molecular geneticists, as we have seen throughout this book.

Lynn Margulis, who is largely responsible for promoting the idea – now all but universally accepted – that mitochondria and chloroplasts are symbiotic bacteria, has tried to do the same thing with cilia. Inspired by possible re-enactments such as we saw in the Mixotrich's Tale, she traces cilia back to spirochaete bacteria. Unfortunately, in view of the beauty and persuasiveness of the mixotrich parallel, the evidence that cilia (undulipodia) are symbiotic bacteria is found unpersuasive by almost everybody who was persuaded by Margulis's evidence in the case of mitochondria and chloroplasts.

Because the Great Historic Rendezvous is a true rendezvous in the forward historical direction, our pilgrimage, from now on, should strictly be a split pilgrimage. We should follow the separate backward pilgrimages of the various participants in the eukaryotic compact until they are finally reunited in the deep past, but I think that would make for a gratuitously complicated journey. Both chloroplasts and mitochondria have their affinities with the eubacteria, not the other prokaryotic group, the Archaea. But our nuclear genes are slightly closer to Archaea, and the next rendezvous in our backwards story is with them.

ARCHAEA

After the uncertainty about what happened at Rendezvous 37, and indeed about how many rendezvous are concealed behind that figleaf of a heading, it is a relief to return to a rendezvous about which most people now agree. All the eukaryotic pilgrims – at least their nuclear genes – are next joined by the archaeans, formerly called Archae-bacteria. Whether it is Rendezvous 38, 39, 40 or 41 might be up for grabs (or, rather, up for research in the next couple of years). But it is agreed that the prokaryotes or, as some would still call them, bacteria, are of two very different kinds – the eubacteria and the archaeans. And the prevailing view is that the Archaea are closer cousins to us than they are to the Eubacteria, which is why I have placed the two rendezvous in the order I have. But it has to be remembered that, owing to the odd circumstances of the Great Historic Rendezvous, bits of our cells are closer to the eubacteria, even if our nuclei are closer to the archaeans.

My Oxford colleague Tom Cavalier-Smith, whose view of the early evolution of life is informed by his great knowledge of microbial diversity, has coined the name Neomura to embrace both archaeans and eukaryotes but exclude the Eubacteria. He also uses 'bacteria' to embrace eubacteria and archaeans but *not* eukaryotes. Bacteria is for him, therefore, a 'grade' name whereas Neomura is a clade. The clade to which the eubacteria belong is simply life, since it includes the Archaea and the eukaryotes too.

Cavalier-Smith believes that the Neomura arose only 850 million years ago, which is a more recent dating than I have dared to contemplate. He thinks the archaeans evolved their peculiar biochemical features within the bacteria as an adaptation to thermophily. Thermophily comes from the Greek for 'love of heat', which in practice usually means living in hot springs. He believes that these heat-loving bacteria – 'thermophiles' – then split into two. Some became hyperthermophiles (liking it very hot indeed) and gave rise to the modern Archaea. Others left the hot springs and, under cooler conditions, became the eukaryotes by absorbing other prokaryotes and making use of them, in the manner of the Mixotrich's Tale.

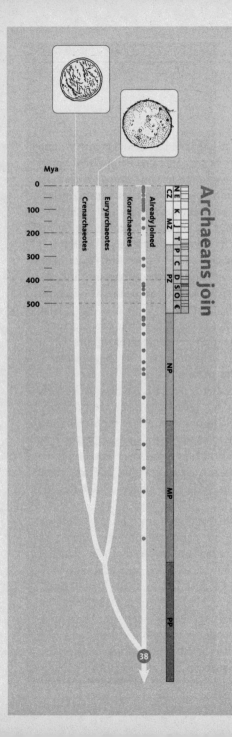

Archaeans join

N	E	K	J	T	P	C	D	S	O	E					
CZ		MZ					PZ				NP		MP		PP

Crenarchaeotes

Euryarchaeotes

Korarchaeotes

Already joined

Mya

0 — 100 — 200 — 300 — 400 — 500 —

38

Archaeans join. Most experts believe the archaeans are the sister group to the eukaryotes, on the basis of nuclear DNA, as well as certain details of biochemistry and cellular morphology. However, if mitochondrial DNA were used, the closest relatives would be the α-proteobacteria because that is what mitochondria themselves once were (see the Great Historic Rendezvous). Archaeans are usually recognised as falling into two groups: the crenarchaeotes and the euryarchaeotes. DNA sequences from hot springs suggest one other, early-diverging branch, the korarchaeotes, but none have actually been seen. No species numbers are given: it isn't clear what 'species' means in asexual organisms.

Images, left to right: *Desulfurococcus mobilis; Methanococcoides burtonii.*

If he is right, we know the conditions in which Rendezvous 38 takes place: in a hot spring, or perhaps in a volcanic upwelling from the bottom of the sea. But of course he may not be right, and it has to be said that his view is far from the consensus.

It was the great American microbiologist Carl Woese of the University of Illinois who discovered and defined the Archaea (then called Archaebacteria) in the late 1970s. The deep separation from other bacteria was controversial at first because it was so different from previous ideas. But it is now very widely accepted, and Woese has been justly honoured with prizes and medals, including the highly prestigious Crafoord Prize and the Leeuwenhoek Medal.

The Archaea include species that thrive in different kinds of extreme conditions, whether it is very high temperatures, or very acid, alkaline or salty water. The archaeans as a group seem to 'push the envelope' of what life can tolerate. Nobody knows whether Concestor 38 was such an extremophile, but it is an intriguing possibility.

EUBACTERIA

When the pilgrimage began, our time machine ground away in bottom gear and we thought in terms of tens of thousands of years. We changed up through the gears, upgrading our imaginations to cope with millions, then hundreds of millions of years as we accelerated back to the Cambrian, picking up animal pilgrims along the way. But the Cambrian is alarmingly recent. For the great majority of its career on this planet life has been nothing but prokaryotic life. We animals are a recent afterthought. For the home stretch to Canterbury, our time machine has to go into hyperdrive to save the book from intolerable *longueur*. With what may seem almost indecent haste, our pilgrims, now including the eukaryotes and the archaeans, speed backwards to what I am assuming is one last rendezvous – Rendezvous 39 with the Eubacteria. But it might be more than one, and we might be closer to some eubacteria than others. Such uncertainty is why the tree on the opposite page is drawn unrooted.

Bacteria, as we have already seen and as Taq's Tale will agree, are supremely versatile chemists. They are also the only non-human creatures known to me who have developed that icon of human civilisation, the wheel. *Rhizobium* tells the tale.

Opposite: **Eubacteria join.** An unrooted tree (see the Gibbon's Tale), with crosses marking two tentative positions for the true root. The tip of each branch represents the present day. Traditionally, the eubacteria have been considered the sister group to the rest of life, equivalent to dangling the root from Concestor 39 (cross A). However, with no outgroups, there is no firm evidence to support this. Another possibility is that the root lies within the eubacteria (e.g. cross B), which would mean more rendezvous points. Within the eubacteria, there is considerable disagreement about phylogenetic relationships. The groups used here are generally accepted; their interrelationships are not. This particularly applies to the cyanobacteria.

Images, clockwise from top: *Escherichia coli* 0111; *Chlamydia* sp.; *Leptospira interrogans*; chloroplast from unknown plant; *Thermus aquaticus*; *Staphylococcus aureus*.

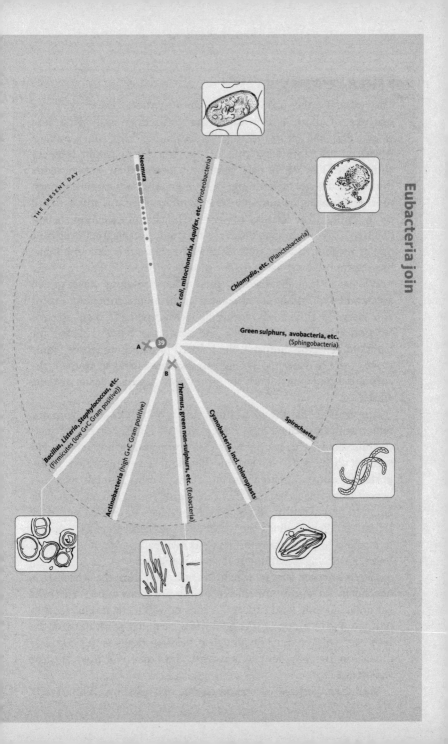

Eubacteria join

THE PRESENT DAY

Neomura

E. coli, mitochondria, *Aquifex*, etc. (Proteobacteria)

Chlamydia, etc. (Planctobacteria)

Green sulphurs, avobacteria, etc. (Sphingobacteria)

Spirochaetes

Cyanobacteria, incl. chloroplasts

Thermus, green non-sulphurs, etc. (Eobacteria)

Actinobacteria (high G+C Gram positive)

Bacillus, *Listeria*, *Staphylococcus*, etc. (Firmicutes (low G+C Gram positive))

A ✕ 39

B ✕

THE *RHIZOBIUM*'S TALE

The wheel is the proverbial human invention. Take apart any machine of more than rudimentary complexity and you'll find wheels. Ship and aeroplane propellors, spinning drills, lathes, potters' wheels – our technology runs on the wheel and would seize up without it. The wheel may have been invented in Mesopotamia during the fourth millennium BC. We know it was elusive enough to *need* inventing because the New World civilisations still lacked it by the time of the Spanish conquest. The alleged exception there – children's toys – seems so bizarre as to prompt suspicion. Could it be one of those myths that spreads purely because it is so memorable, like the Inuit having 50 words for snow?

Whenever humans have a good idea, zoologists have grown accustomed to finding it anticipated in the animal kingdom. Examples pervade this book, including echo-ranging (bats), electro-location (the Duckbill's Tale), the dam (the Beaver's Tale), the parabolic reflector (limpets), the infrared heat-seeking sensor (some snakes), the hypodermic syringe (wasps, snakes and scorpions), the harpoon (cnidarians) and jet propulsion (squids). Why not the wheel?

It is possible that the wheel impresses us only by contrast with our rather undistinguished legs. Before we had engines driven by fuels (fossilised solar energy), we were easily outpaced by animal legs. No wonder Richard III offered his kingdom for four-footed transportation out of his predicament. Perhaps most animals wouldn't benefit from wheels because they can already run so fast on legs. After all, until very recently, all our wheeled vehicles have been pulled by leg power. We developed the wheel not to go faster than a horse, but to enable a horse to transport us at its own pace – or a bit less. To a horse, a wheel is something that slows you down.

Here's another way in which we risk overrating the wheel. It is dependent for maximum efficiency on a prior invention – the road (or other smooth, hard surface). A car's powerful engine enables it to beat a horse or a dog or a cheetah on a hard, flat road. But run the race over wild country or ploughed fields, perhaps with hedges or ditches in the way, and it is a rout: the horse will leave the car wallowing.

Well then, perhaps we should change our question. Why haven't animals developed the road? There is no great technical difficulty. The road should be child's play compared with the beaver dam or the bower-bird's ornamented arena. There are even some digger wasps that tamp soil hard, picking up a stone tool to do so. Presumably the same skills could be used by larger animals to flatten a road.

But it raises an unexpected problem. Even if roadbuilding is technically feasible, it is a dangerously *altruistic* activity. If I as an individual build a good road from A to B, you may benefit from the road just as much as I do. Why should this matter? Because Darwinism is a selfish game. Building a road that might help others will be penalised by natural selection. A rival individual benefits from my road just as much as I do, but he does not pay the cost of building. Freeloaders, who use my road and don't bother to build their own, will be free to concentrate their energy on out-reproducing me, while I slave away on the road. Unless special measures are taken, genetic tendencies towards lazy, selfish exploitation will thrive at the expense of industrious roadbuilding. The upshot will be that no roads get built. With the benefit of foresight, we can see that everybody will be worse off. But natural selection, unlike us humans with our big, recently evolved brains, has no foresight.

What is so special about humans that we have managed to overcome our antisocial instincts and build roads that we all share? Oh, there is so much. No other species comes remotely close to a welfare state, to an organisation that takes care of the old, that looks after the sick and the orphaned, that gives to charity. On the face of it these things present a challenge to Darwinism, but this is not the place to go into that. We have governments, police, taxation, public works to which we all subscribe whether we like it or not. The man who wrote, 'Sir, You are very kind, but I'd prefer not to join your Income Tax Scheme', heard back, we may be sure, from the Inland Revenue. Unfortunately, no other species has invented the tax. They have, however, invented the (virtual) fence. An individual can secure his exclusive use of a resource if he actively defends it against rivals.

Many species of animals are territorial, not just birds and mammals, but fish and insects too. They defend an area against rivals of the same species, often so as to sequester a private feeding ground, or a private courtship bower or nesting area. An animal with a large territory might benefit by building a network of good, flat roads across the territory from which rivals were excluded. This is not impossible, but such animal roads would be too local for long-distance, high-speed travelling. Roads of any quality would be limited to the small area that an individual can defend against genetic rivals. Not an auspicious beginning for the evolution of the wheel.

But now, finally, we come to the teller of this tale. There is one revealing exception to my premise. Some very small creatures *have* evolved the wheel in the fullest sense of the word. The wheel may even have been the first locomotor device ever evolved, given that for most of its first 2 billion years, life consisted of nothing but bacteria. Many bacteria, of which *Rhizobium* is typical, swim using thread-like spiral propellors, each driven by its own continuously rotating propellor shaft. It used to be thought that these 'flagella' were wagged like tails, the appearance of spiral rotation resulting from a wave of motion passing along the length of the flagellum, as in a wriggling snake. The truth is much more remarkable. The bacterial flagellum* is attached to a shaft that rotates freely and indefinitely in a hole that runs through the cell wall. This is a true axle, a freely rotating hub. It is driven by a tiny molecular motor which uses the same biophysical principles as a muscle. But a muscle is a reciprocating engine, which, after contracting, has to lengthen again to prepare for a new power stroke. The bacterial motor just keeps on going in the same direction: a molecular turbine.

The fact that only very small creatures have evolved the wheel suggests what may be the most plausible reason why larger creatures have not. It's a rather mundane, practical reason, but nonetheless important. A large creature would need big wheels which, unlike man-made ones, would have to grow *in situ* rather than being separately fashioned out of dead materials and then mounted. For a large, living organ, growth *in situ* demands blood or something equivalent, and probably something equivalent to nerves too. The problem of supplying a freely rotating organ with blood vessels (not to mention nerves) that don't tie themselves in knots is too vivid to need spelling out. There might be a solution, but we need not be surprised that it has not been found.

Human engineers might suggest running concentric ducts to carry blood through the middle of the axle into the middle of the wheel. But what would the evolutionary intermediates have looked like?

---

* The bacterial flagellum, as we have seen, is completely different in structure from the eukaryotic (or protozoan) flagellum or 'undulipodium' that we met in the Mixotrich's Tale. Unlike the eukaryotic arrangement of 9+2 microtubules, the bacterial flagellum is a hollow tube made of the protein flagellin.

A true axle, a freely rotating hub . . . driven by a tiny molecular motor.

Evolutionary improvement is like climbing a mountain. You can't jump from the bottom of a cliff to the top in a single leap. Sudden, precipitous change is an option for engineers, but in nature the summit of the evolutionary mountain can be reached only via a gradual ramp upwards from the starting point. The wheel may be one of those cases where the engineering solution can be seen in plain view, yet be unattainable in evolution because it lies on the other side of a deep valley: unevolvable by large animals but within the reach of bacteria because of their small size.

In an imaginative piece of lateral thinking, Philip Pullman, in his epic of childhood fiction *His Dark Materials,* solves the problem for big animals in a completely unexpected but very biological way. He invents a species of benevolent, trunked animal, the mulefa, who have evolved symbiotically with a species of gigantic tree that sheds hard, circular, wheel-like seed pods. The feet of the mulefa have a horny, polished spur which fits into a hole in the centre of a seed pod, which then works as a wheel. The trees gain from the arrangement because whenever – as eventually must happen – a wheel wears out and has to be discarded, the mulefa disperse the seeds inside. The trees have evolved to return the favour by making the pods perfectly circular, with a suitable hole for the mulefan axle right in the centre, into which they secrete a high-grade lubricating oil. The mulefa's four legs are placed in a diamond pattern. The fore and aft legs are in the midline, and they are the ones that slot into the wheels. The other two legs, halfway along the body and to the sides, have no wheels and are used to punt the animal along like an old-fashioned boneshaker bicycle without pedals. Pullman cleverly notes that the whole system is made possible only by a geological peculiarity of the world on which these creatures live. Basalt happens to form in long, ribbon-like lines over the savannah, which serve as unmade but hard roads.

Short of Pullman's ingenious symbiosis, we may provisionally accept the wheel as one of those inventions that, even if it were a good idea in the first place, cannot evolve in large animals: either because of the prior need for a road, or because the problem of the twisted blood vessels could never be solved, or because the inter-mediates to a final solution would never be good for anything. Bacteria were able to evolve the wheel because the world of the very small is so very different and presents such different technical prob-lems.

As it happens, the bacterial flagellar motor itself has recently, in the hands of a species of creationists who call themselves 'Intelligent Design Theorists', been elevated to the status of icon of alleged unevolvability. Since it manifestly exists, the conclusion of their argument is different. Whereas I proposed unevolvability as an explanation for why large animals like mammals don't grow wheels, creationists have seized upon the bacterial flagellar wheel as some-thing that cannot exist and yet does – so it must have come about by supernatural means!

This is the ancient 'Argument from Design', also called the 'Argu-ment from Paley's Watchmaker', or the 'Argument from Irreducible Complexity'. I have less kindly called it the 'Argument from Personal Incredulity' because it always has the form: 'I personally cannot imagine a natural sequence of events whereby X could have come about. Therefore it must have come about by supernatural means.' Time and again scientists have retorted that if you make this argu-ment, it says less about nature than about the poverty of your imagination. The 'Argument from Personal Incredulity' would lead us to invoke the supernatural every time we see a good conjuror whose tricks we cannot fathom.

It is perfectly legitimate to propose the argument from irreducible complexity as a possible explanation for the lack of something that doesn't exist, as I did for the absence of wheeled mammals. That is very different from evading the scientist's responsibility to explain something that *does* exist, such as wheeled bacteria. Nevertheless, to be fair, it is possible to imagine validly using some version of the argument from design, or the argument from irreducible complexity.

Future visitors from outer space, who mount archaeological digs of our planet, will surely find ways to distinguish designed machines such as planes and microphones, from evolved machines such as bat wings and ears. It is an interesting exercise to think about how they will make the distinction. They may face some tricky judgements in the messy overlap between natural evolution and human design. If the alien scientists can study living specimens, not just archaeological relics, what will they make of fragile, highly strung racehorses and greyhounds, of snuffling bulldogs who can scarcely breathe and can't be born without Caesarian assistance, of blear-eyed Pekinese baby surrogates, of walking udders such as Friesian cows, walking rashers such as Landrace pigs, or walking woolly jumpers such as Merino sheep? Molecular machines – nanotechnology – crafted for human benefit on the same scale as the bacterial flagellar motor, may pose the alien scientists even harder problems.

Francis Crick, no less, has speculated semi-seriously in *Life Itself* that bacteria might not have originated on this planet but been seeded from elsewhere. In Crick's fantasy, they were sent in the nose-cone of a rocket by alien beings, who wanted to propagate their form of life, but shrank from the technically harder problem of transporting themselves and relied, instead, upon natural evolution to finish the job once the bacterial infection had taken root. Crick, and his colleague Leslie Orgel, who originally suggested the idea with him, supposed that the bacteria had originally evolved by natural processes on the home planet, but they could equally, while in the mood for science fiction, have added a touch of nanotechnological artifice to the mix, perhaps a molecular gearwheel like the flagellar motor which we see in *Rhizobium* and many other bacteria.

Crick himself – whether with regret or relief it is hard to say – finds little good evidence to support his own theory of Directed Panspermia. But the hinterland between science and science fiction constitutes a useful mental gymnasium in which to wrestle with a genuinely important question. Given that the illusion of design conjured by Darwinian natural selection is so breathtakingly powerful, how do we, in practice, distinguish its products from deliberately designed artefacts? Another great molecular biologist, Jacques Monod, began his *Chance and Necessity* in similar terms. Could there be genuinely persuasive examples of irreducible complexity in nature: complex organisation made of many parts, the loss of any one of which would be fatal to the whole? If so, might this suggest genuine design by a superior intelligence, say from an older and more highly evolved civilisation on another planet?

It is possible that an example of such a thing might eventually be discovered. But the bacterial flagellar motor, alas, is not it. Like so many previous allegations of irreducible complexity, from the eye on, the bacterial flagellum turns out to be eminently reducible. Kenneth Miller of Brown University deals with the whole question in a *tour de force* of clear exposition. As Miller shows, the allegation that the component parts of the flagellar motor have no other functions is simply false. As one example, many parasitic bacteria have a mechanism for injecting chemicals into host cells called the TTSS (Type Three Secretory System). The TTSS makes use of a subset of the very same proteins that are used in the flagellar motor. In this case they are used not for providing rotatory motion of a circular hub, but for making a circular hole in a host's cell wall. Miller summarises:

> Stated directly, the TTSS does its dirty work using a handful of proteins from the base of the flagellum. From the evolutionary point of view, this relationship is hardly surprising. In fact, it's to be expected that the opportunism of evolutionary processes would mix and match proteins to produce new and novel functions. According to the doctrine of irreducible complexity, however, this should not be possible. If the flagellum is indeed irreducibly complex, then removing just one part, let alone 10 or 15, should render what remains 'by definition nonfunctional.' Yet the TTSS is indeed fully functional, even though it is missing most of the parts of the flagellum. The TTSS may be bad news for us, but for the bacteria that possess it, it is a truly valuable biochemical machine.
>
> The existence of the TTSS in a wide variety of bacteria demonstrates that a small portion of the 'irreducibly complex' flagellum can indeed carry out an important biological function. Since such a function is clearly favored by natural selection, the contention that the flagellum must be fully assembled before any of its component parts can be useful is obviously incorrect. What this means is that the argument for intelligent design of the flagellum has failed.

Miller's indignation at 'Intelligent Design Theory' receives a boost from an interesting source: his deep religious convictions, which are more fully articulated in *Finding Darwin's God*. Miller's God (if not Darwin's) is the God revealed in – or perhaps synonymous with – the deep lawfulness of nature. The creationist quest to demonstrate God through the negative route of the Argument from Personal Incredulity turns out, as Miller shows, to assume that God capriciously *violates* his own laws. And this, to those – like Miller – of a thoughtfully religious disposition, is a cheap and demeaning sacrilege.

As a non-religious person I can sympathetically buttress Miller's argument with a parallel one of my own. If not sacrilegious, the intelligent design style of argument from personal incredulity is *lazy*. I have satirised it in an imagined conversation between Sir Andrew Huxley and Sir Alan Hodgkin, both sometime presidents of the Royal Society, who shared the Nobel Prize for working out the molecular biophysics of the nerve impulse.

> 'I say, Huxley, this is a terribly difficult problem. I can't see how the nerve impulse works, can you?'
>
> 'No, Hodgkin, I can't, and these differential equations are fiendishly hard to solve. Why don't we just give up and say that the nerve impulse propagates by nervous energy?'
>
> 'Excellent idea, Huxley, let's write the letter to *Nature* now: it'll only take one line, then we can turn to something easier.'

Andrew Huxley's elder brother Julian made a similar point when, long ago, he satirised vitalism, then usually epitomised by Henri Bergson's name of *élan vital*, as tantamount to explaining that a railway engine was propelled by *élan locomotif*.[*] My censure of laziness, and Miller's of sacrilege, do not apply to the hypothesis of directed panspermia. Crick was talking about superhuman, not supernatural, design. The difference really matters. On Crick's world view, superhuman designers of bacteria, or of the means to seed Earth with them, would themselves have originally evolved by some local equivalent of Darwinian selection on their own planet. Crucially, Crick would always seek what Daniel Dennett calls a 'crane': would never resort – as Henri Bergson would – to a 'skyhook'.

The main objection to the irreducible complexity argument amounts to a demonstration that the allegedly irreducible complex entity, the flagellar motor, the blood-clotting cascade, the Krebs cycle, or whatever it might be, is actually reducible. The personal incredulity was simply wrong. To this we add the reminder that, even if we can't *yet* think of a step-by-step pathway by which the

---

[*] It is depressing to reflect that Henri Bergson – a vitalist – represents the nearest approach to a scientist in the entire list of 100 winners of the Nobel Prize for Literature. The nearest competitor is Bertrand Russell, but he won it for his humanitarian writings.

complexity might have evolved, the eager slide to assuming that it is therefore supernatural is either sacrilegious or lazy, according to taste.

But there is another objection that needs to be mentioned: the 'arch and scaffolding' of Graham Cairns-Smith. Cairns-Smith was writing in a different context, but his point works here too. An arch is irreducible in the sense that if you remove part of it, the whole collapses. Yet it is possible to build it gradually by means of scaffolding. The subsequent removal of the scaffolding, so that it no longer appears in the visible picture, does not entitle us to a mystified and obscurantist attribution of supernatural powers to the masons.

The flagellar motor is common among bacteria. *Rhizobium* was chosen to tell the tale because of a second claim to impress us with the versatility of bacteria. Farmers sow plants of the pea family, Leguminosae, as a part of most good crop rotation schemes for one very good reason. Leguminous plants can use raw nitrogen straight out of the air (it is by far the most abundant gas in our atmosphere) rather than having to suck up nitrogen compounds from the soil. But it isn't the plants themselves that fix atmospheric nitrogen and turn it into usable compounds. It is symbiotic bacteria – specifically *Rhizobium* – housed for the purpose in special nodules provided for them, with every indication of inadvertent solicitude, on the roots of the plants.

Such contracting out of ingenious chemical tricks to chemically much more versatile bacteria is an extremely common pattern throughout animals and plants. It is the main message of Taq's Tale.

TAQ'S TALE
Written with Yan Wong

Having reached our most ancient rendezvous, having gathered into our pilgrimage all of life as we know it, we are in a position to survey its diversity. At the deepest level, the diversity of life is chemical. The trades plied by our fellow pilgrims span a range of skills in the arts of chemistry. And, as we have seen, it is the bacteria, including archaeans, who display the fullest spread of chemical skills. Bacteria taken as a group are the master chemists of this planet. Even the chemistry of our own cells is largely borrowed from bacterial guest workers, and it represents a fraction of what bacteria are capable of. Chemically, we are more similar to some bacteria than some bacteria are to other bacteria. At least as a chemist would see it, if you wiped out all life except bacteria, you'd still be left with the greater part of life's range.

The particular bacterium that I choose to tell the tale is *Thermus aquaticus,* known fondly to molecular biologists as Taq. Different bacteria seem alien to us for different particular reasons. *Thermus aquaticus,* as its name suggests, likes to be in hot water. *Very* hot water. As we saw at Rendezvous 38, many of the archaeans are thermophiles and hyperthermophiles, but the archaeans don't have a monopoly on this way of life. Thermophiles and hyper-thermophiles are not taxonomic categories, but something more like trades or guilds, like Chaucer's Clerk, Miller and Physician. They make their living in places where nothing else can: the scald-ing-hot springs of Rotorua and Yellowstone Park, or the volcanic vents on mid-ocean ridges. *Thermus* is a eubacterial hyper-thermophile. It can survive with little problem in near-boiling water – although it prefers a more balmy 70°C, or so. It doesn't quite hold the world temperature record – there are deep-sea archaeans that thrive at up to 115°C, well above the normal boiling point of water.*

Thermus is famous in molecular biology circles for being the source of the DNA duplication enzyme known as Taq polymerase. Of course, all organisms have enzymes to duplicate DNA, but *Thermus* has had to evolve one that can withstand near-boiling temperatures. This is useful for molecular biologists because the easiest way to ready DNA for duplication is to boil it, separating it into its two constituent strands. Repeated boiling and cooling of a solution containing both DNA and Taq polymerase duplicates – or 'amplifies' – even the most minute quantities of original DNA. The method is called the 'polymerase chain reaction', or PCR, and it is brilliantly clever.

Thermus's fame as a wizard of the biochemistry laboratory is justification enough to let it tell this tale. But, as it happens, there may be another reason that *Thermus* is particularly well placed to present the instructively alien perspective of bacteria. *Thermus* be-longs to the small group of bacteria known as the Hadobacteria. In his taxonomic scheme mentioned at Rendezvous 39, Tom Cavalier-Smith suggests that the Hadobacteria, together with their cousins the

---

* Again, if it seems surprising that water can be found so far above its normal boiling point, remember that water boils hotter at high pressure.

green non-sulphur bacteria, may be the earliest branching bacterial group. If so, their group is as distant a cousin of the rest of life as it is possible to be.

According to this view, *Thermus* and its relatives are out on a limb. All the rest of the bacteria share an ancestor with each other and with the rest of life, which *Thermus* doesn't share. If upheld, this means the following. Just as any bacterium might lump 'the rest of life' into one 'cadet branch' of the family of life, so, within the bacteria, *Thermus* can lump 'the rest of the bacteria' into one branch of the bacteria. This, together with its penchant for being boiled, was my reason for singling out *Thermus* to tell a tale of life's diversity. But whereas the evidence for the special status of *Thermus* is not particularly secure, there is no doubt that the great majority of life's diversity at the fundamental level of chemistry is microbial, and a substantial majority of it is bacterial. The tale of life's diversity, insofar as it is mostly chemical diversity, is rightfully told by a bacterium, and it might as well be Taq.

Traditionally, and understandably, the tale was told from the point of view of big animals – us. Life was divided into the animal kingdom and the vegetable kingdom, and the difference seemed pretty clear. The fungi counted as plants because the more familiar of them are rooted to the spot and don't walk away while you try to study them. We didn't even know about bacteria until the nineteenth century, and when they were first seen through powerful microscopes people didn't know where to put them in the scheme of things. Some thought of them as miniature plants, others as miniature animals. Yet others put the light-trapping bacteria in the plants (as 'blue-green algae') and the rest in the animals. Much the same was done with the 'protists' – single-celled eukaryotes that are not bacteria and are much larger than bacteria. The green ones were called Protophyta and the rest Protozoa. A familiar example of a protozoan is *Amoeba*, once thought to be close to the grand ancestor of all life – how wrong we were, for an *Amoeba* is scarcely distinguishable from a human when viewed through the 'eyes' of bacteria.

All that was in the days when living organisms were classified by their visible anatomy, in which bacteria are much less diverse than animals or plants and it was pardonable to put them down as primitive animals and plants. It was another matter entirely when we began to classify creatures using the much richer information provided by their molecules, and when we looked at the range of

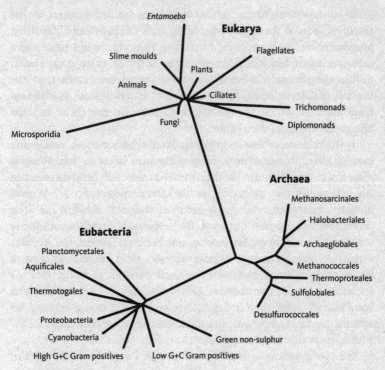

The deepest divisions of life. Tree of life, showing division into three major domains, based on recent molecular work. Adapted from Gribaldo and Philippe [113].

chemical 'trades' perfected by microbes. Here's approximately how things look today.

If animals and plants are treated as a pair of kingdoms, by the same standards there are dozens of microbial 'kingdoms', whose uniqueness entitles them to the same status as animals and plants. The diagram above shows the tip of the iceberg. Not only have some deep-rooted branches been omitted, but I've shown only those that live in accessible places and can be cultured in the laboratory. Indeed, simply trawling new locations for DNA and not bothering to inquire which organisms they come from can find entire new microbial kingdoms. The ever-resourceful Craig Venter and his team claim to have found at least 1,800 new species of microbes by a shotgun analysis of DNA floating around in the Sargasso Sea. Animals, plants and fungi constitute just three small branches of the tree of life. What distinguishes these three familiar kingdoms from the others is that the organisms in them are large, being built of many cells. The other kingdoms are almost entirely microbial. Why do we not unite them into one microbial kingdom, on a par with the three great multi-cellular kingdoms? One reason, and a sound one, is that, at the biochemical level, many of the microbial kingdoms are as different from each other, and from the big three, as the three familiar kingdoms are from each other.

It would be worthless to argue in detail whether there 'really' are, say, 20 kingdoms on this scale of difference, or 25 or 100. What is clear from the diagram is that these dozens fall into three main super-kingdoms – 'domains' in the terminology of Carl Woese, already mentioned as the originator of this new view of life. The three domains are first our own, the eukaryotes, in whose company we have been travelling for most of our journey. Second, the Archaea – the microbes we met at Rendezvous 38 – who, on the old view of life, would be lumped in with the third domain, the true (or Eu-) bacteria. It is the members of this third, eubacterial domain who have joined us for the last leg of our pilgrimage. It is a privilege to share these final steps with the most ubiquitous and efficient DNA-propagators that have ever existed.

The star diagram itself is not, of course, based upon the sorts of features we can see and touch. If you want to compare organisms, you have to choose features that they all approximately share. You can't compare legs if most species don't have legs. Legs, heads, leaves, collar bones, roots, hearts, mitochondria – each is restricted to a subset of creatures. But DNA is universal, and there's a handful of particular genes that all living creatures share with each other, with only minor, countable differences. These are what we must use for large-scale comparison. Perhaps the best example is provided by the codes that go to make ribosomes.

Ribosomes are cellular machines that read RNA messages (them-selves transcribed from DNA genes) and churn out proteins. Ribo-somes are vital to all cells, and are universally present. They are themselves largely made of RNA – called rRNA and completely separate from the RNA message 'tapes' that the ribosomes read and translate into protein. rRNA is itself originally specified by DNA genes. The sequence of rRNA may be read directly, or as the DNA genes that code it: rDNA. Either way, I shall call it rDNA. rDNA is particularly useful for direct comparison between any creature and any other because they all have it. rDNA is used not only because of its ubiquity. Just as important, it shows the right amount of genetic variation – sufficiently similar between all living species that there is something to compare, yet not so extremely similar as to leave no differences to count. Using the methods of the Gibbon's Tale, we can use rDNA to piece together the whole tree of life, and work out the vast evolutionary distances within, and even between, the major domains. We must take care. rDNA is fully vulnerable to 'long branch attraction' and other such pitfalls. But with the assistance of other genes too, and the use of rare genomic changes – insertions and deletions of large chunks of DNA – a tentative tree can be drawn. That is what we have on page 569. Certainly, some branches in this tentative tree are uncertain, particularly within the Eubacteria, and this may reflect their tendency to exchange DNA between themselves – a problem we have not met in any eukaryotes. Nevertheless, researchers have found a core group of bacterial genes that are rarely swapped, so it is conceivable that we may one day agree upon an unimpeachable order of branching within the tree of life. I look forward to it.

Taxonomic distance, as measured by comparing genomes, is one way of looking at diversity. Another is to look at the range of ways of life, the range of 'trades' that our pilgrims ply. At first sight, different bacteria might seem more similar in this regard than, say, a lion is to a buffalo, or a mole to a koala. To big animals like us, burrowing underground for worms seems very different, as a way of life, from chewing leaves up a gum tree. But from the chemical point of view of our bacterial storyteller, all moles, koalas, lions and buffalos are doing much the same thing. All are deriving their energy by breaking down complex molecules ultimately put together by energy from the sun captured by plants. Koalas and buffalos eat the plants directly. Lions and moles get their solar energy at one remove, by eating other animals that (ultimately) eat plants.

The primary source of outside energy is the sun. The sun, through symbiotic green bacteria inside plant cells, is the only begetter of energy for all the life we can see with the unaided eye. Its energy is trapped by green solar panels (leaves) and used to drive uphill the synthesis of organic compounds, such as sugar and starch in plants. In a series of energy-coupled downhill and uphill chemical reactions, the rest of life is then powered by the energy originally trapped from the sun by plants. Energy flows through the economy of life, from the sun to plants to herbivores to carnivores to scavengers. At every step of the way, not only between creatures but within them, every transaction in the energy economy is wasteful. Inevitably, some of it is dissipated as heat and never recovered. Without the massive inflow of energy from the sun, life would, or so the textbooks used to say, grind to a halt.

That is still mostly true. But those textbooks reckoned without the bacteria and archaeans. If you are a sufficiently ingenious chemist, it is possible to dream up alternative schemes of energy flow on this planet, which do not start with the sun. And if a useful piece of chemistry can be dreamed up, the chances are that a bacterium got there first: maybe even before they discovered the solar energy trick, and that was more than 3 billion years ago. There has to be some kind of external source of energy, but it doesn't have to be the sun. There is chemical energy locked up in lots of substances, energy that can be released by the right chemical reactions. Sources economically worth mining by living creatures include hydrogen, hydrogen sulphide, and some iron compounds. We will revisit the mining way of life at Canterbury.

Although our tales are not, in the main, told in the first person, let us make an exception for the last word of all our tales, and give it to *Thermus aquaticus*:

> Look at life from our perspective, and you eukaryotes will soon cease giving yourselves such airs. You bipedal apes, you stump-tailed tree-shrews, you desiccated lobe-fins, you vertebrated worms, you Hoxed-up sponges, you newcomers on the block, you eukaryotes, you barely distinguishable congregations of a monotonously narrow parish, you are little more than fancy froth on the surface of bacterial life. Why, the very cells that build you are themselves colonies of bacteria, replaying the same old tricks we bacteria discovered a billion years ago. We were here before you arrived, and we shall be here after you are gone.

CANTERBURY

As befits the destination of a 4-billion-year pilgrimage, our Canterbury has a patina of mystery. It is the singularity known as the origin of life, but we could better call it the origin of heredity. Life itself is not clearly defined, a fact that contradicts intuition and traditional wisdom. Ezekiel, Chapter 37, in which the prophet was commanded down into the valley of the bones, identifies life with breath. I cannot resist quoting the passage ('bone to his bone' – such wonderful economy of language).

> So I prophesied as I was commanded: and as I prophesied, there was a noise, and behold a shaking, and the bones came together, bone to his bone.
> And when I beheld, lo, the sinews and the flesh came up upon them, and the skin covered them above: but there was no breath in them.
> Then said he unto me, Prophesy unto the wind, prophesy, son of man, and say to the wind, Thus saith the Lord God: Come from the four winds, O breath, and breathe upon these slain, that they may live.

And, of course, the winds did. A great army breathed and stood up. Breath, for Ezekiel, defines the difference between dead and alive. Darwin himself implied the same in one of his more eloquent passages, the concluding words of *The Origin of Species* (emphasis added):

> Thus, from the war of nature, from famine and death, the most exalted object which we are capable of conceiving, namely, the production of the higher animals, directly follows. There is grandeur in this view of life, with its several powers, *having been originally breathed* by the Creator into a few forms or into one; and that, whilst this planet has gone cycling on according to the fixed law of gravity, from so simple a beginning endless forms most beautiful and most wonderful have been, and are being, evolved.

Darwin rightly reversed Ezekiel's order of events. The breath of life came first and created the conditions under which bones and sinews, flesh and skin would eventually evolve. Incidentally, the phrase 'by the Creator' is not present in the first edition of the *Origin*. It was added in the second edition, probably as a sop to the religious lobby. Darwin later regretted this in a letter to his friend Hooker:

> I have long since regretted that I truckled to public opinion and used the Pentateuchal term of creation, by which I really meant 'appeared' by some wholly unknown process. It is mere rubbish thinking at present of the origin of life; one might as well think of the origin of matter.

Darwin probably (and in my view rightly) saw the origin of primitive life as a relatively (and I stress relatively) easy problem compared with the one he solved: how life, once begun, developed its amazing diversity, complexity and powerful illusion of good design. Nevertheless, Darwin did later (in another letter to Hooker) venture a guess about the 'wholly unknown process' that started it all. He was led to it through wondering why we don't see life originating again and again.

> It is often said that all the conditions for the first production of a living organism are now present, which could ever have been present. But if (and oh! what a big if!) we could conceive in some warm little pond, with all sorts of ammonia and phosphoric salts, light, heat, electricity, &c., present, that a proteine compound was chemically formed ready to undergo still more complex changes, at the present day such matter would be instantly absorbed, which would not have been the case before living creatures were found.

The doctrine of spontaneous generation had only lately been experimentally attacked by Pasteur. It had long been believed that rotting meat spontaneously generated maggots, that goose barnacles spontaneously generated goslings and even that dirty laundry placed with wheat generated mice. Perversely, the spontaneous generation theory was supported by the Church (following Aristotle in this as in so much else). I say perversely because, at least with hindsight, spontaneous generation was as direct a challenge to divine creation as evolution would ever be. The idea that flies or mice could spring spontaneously into existence hugely underestimates the stupendous achievement that the creation of flies or mice would be: an insult to the Creator, one might have thought. But the science-free mindset fails to grasp how complex and inherently improbable a fly or a mouse is. Darwin was perhaps the first to appreciate the full magnitude of this mistake.

As late as 1872 in a letter to Wallace, the co-discoverer of natural selection, Darwin could still find it necessary to express his scepticism about 'Rotifers and Tardigrades being spontaneously generated', as had been suggested in a book, *The Beginnings of Life*, which he otherwise admired. His scepticism was on target as usual. Rotifers and tardigrades are complex life forms beautifully fitted to their respective ways of life. For them to be spontaneously generated would imply that they became fit and complex 'by a happy accident, and this I cannot believe'. Happy accidents of such magnitude were anathema to Darwin, as they should have been to the Church for a different reason. The whole rationale of Darwin's theory was, and is, that adaptive complexity comes about by slow and gradual degrees, step by step, no single step making too large a demand on blind chance as explanation. The Darwinian theory, by rationing chance to the small steps needed to supply variation for selection, provides the only realistic *escape* from sheer luck as the explanation of life. If rotifers could spring into existence just like that, Darwin's life-work was unnecessary.

But natural selection itself had to have a beginning. In this sense alone, some kind of spontaneous generation must have happened, if only once. The beauty of Darwin's contribution was that the single spontaneous generation we must postulate did not have to synthesise anything complicated like a maggot or a mouse. It only had to make . . . well, now we approach the heart of the problem. If not breath, what was the vital ingredient that first enabled natural selection to get going and lead eventually, after epics of cumulative evolution, to maggots, mice and men?

The details lie buried, perhaps beyond recovery, at our ancient Canterbury, but we can give the key ingredient a minimalist name to express the *kind* of thing it must have been. That name is heredity. We should be seeking not the origin of life, which is vague and undefined, but the origin of heredity – true heredity, and that means something very precise. I have previously invoked fire to help explain it.

Fire rivals breath as imagery for life. When we die, the fire of life goes out. Our ancestors who first tamed it probably thought fire a living thing, a god even. Staring into flames or embers, especially at night when the campfire warmed and protected them, did they commune in imagination with a glowing, dancing soul? Fire stays alive as long as you feed it. Fire breathes air; you can suffocate it by cutting off its oxygen supply, you can drown it with water. Wild fire devours the forest, driving animal prey before it with the speed and ruthlessness of a pack of wolves in (literally) hot pursuit. As with wolves, our ancestors could capture a fire cub as a useful pet, tame it, feed it regularly and clear away its ashy excreta. Before the art of firemaking was discovered, society would have prized the lesser art of husbanding a captured fire. Perhaps a live scion of the home fire was carried in a pot for barter to a neighbouring group whose own fire had unfortunately died.

Wild fires would have been observed giving birth to daughter fires, spitting sparks and live cinders up on the wind, like dandelion puffs, to land and seed the dry grass at a distance. Did ergastrine philosophers theorise that fire cannot spontaneously generate, but must always be born of a parent fire, either wild fire out on the plains, or domestic fire fenced in by hearthstones? And did the first firemaking sticks therefore rub out a world view?

Our ancestors might even have imagined a population of reproducing wild fires, or a pedigree of descent among domestic fires traced from a glowing ancestor bought from a distant clan and traded on to others. But still there was no true heredity. Why not? How can you have reproduction and a pedigree, yet no heredity? This is the lesson fire has for us here.

True heredity would mean the inheritance not of fire itself but of *variations* among fires. Some fires are yellower than others, some redder. Some roar, some crackle, some hiss, some smoke, some spit. Some have tinges of blue or green amongst the flames. Our ancestors, if they had studied their domesticated wolves, would have noticed a telling difference between dog pedigrees and fire pedigrees. With dogs, like begets like. At least some of what distinguishes one dog from another is handed down by its parents. Of course some comes in sideways too: from food, disease and accident. With fires, all the variation comes from the environment, none descends from the progenitive spark. It comes from the quality and dampness of the fuel, from the lie and strength of the wind, from the drawing qualities of the hearth, from the soil, from traces of copper and potassium that add touches of blue-green and lilac to sodium's yellow flame. Unlike a dog, nothing about the quality of an adult fire arrives via the spark that gave it birth. Blue fires don't beget blue fires. Crackling fires don't inherit their crackle from the parent fire that threw up their initiating spark. Fires exhibit reproduction without heredity.

The origin of life was the origin of true heredity; we might even say the origin of the first gene. By first gene, I hasten to insist, I don't mean first DNA molecule. Nobody knows whether the first gene was made of DNA, and I bet it wasn't. By first gene I mean first replicator. A replicator is an entity, for example a molecule, that forms lineages of copies of itself. There will always be errors in copying, so the population will acquire variety. The key to true heredity is that each replicator resembles the one from which it was copied more than it resembles a random member of the population. The origin of the first such replicator was not a probable event, but it only had to happen once. Thereafter, its consequences were automatically self-sustaining and they eventually gave rise, by Darwinian evolution, to all of life.

A length of DNA or, under certain conditions, the related molecule RNA is a true replicator. So is a computer virus. So is a chain letter. But all these replicators need a complicated apparatus to assist them. DNA needs a cell richly equipped with pre-existing biochemical machinery highly adapted to read and copy the DNA code. A computer virus needs a computer with some sort of data link to other computers, all designed by human engineers to obey coded instructions. A chain letter needs a good supply of idiots, with evolved brains educated at least enough to read. What is unique about the first replicator, the one that sparked life, is that it had no ready supply of anything evolved, designed or educated. The first replicator worked *de novo, ab initio,* without precedent, and without help other than from the ordinary laws of chemistry.

A powerful source of help to a chemical reaction is a catalyst, and catalysis in some form was surely involved in the origin of replication. A catalyst is an agent that speeds up a chemical reaction while not being consumed by it. All biological chemistry consists of catalysed reactions, the catalysts usually being the large protein molecules called enzymes. A typical enzyme offers the shaped cavities of its three-dimensional form as receptacles for the ingredients of one chemical reaction. It lines them up for each other, enters into temporary chemical liaison with them, matchmakes with an aimed precision that they would be unlikely to discover in open diffusion.

Catalysts, by definition, are not consumed in the chemical reaction they boost, but they may be produced. An autocatalytic reaction is a reaction that manufactures its own catalyst. As you can imagine, an autocatalytic reaction is reluctant to start but, once started, it takes off on its own – like wild fire indeed, for fire has some of the properties of an autocatalytic reaction. Fire is not strictly a catalyst but it is self-generating. Chemically, it is an oxidation process that gives off heat, and needs heat to push it over a threshold to start. Once started, it continues and spreads as a chain reaction because it generates the heat needed to restart itself. Another famous chain reaction is an atomic explosion, in this case not a chemical reaction but a nuclear one. Heredity began as a lucky initiation of an auto-catalytic, or otherwise self-regenerating, process. It immediately took off and spread like a fire, eventually leading to natural selection – and all that was to follow.

We too oxidise carbonaceous fuel to generate heat, but we don't burst into flames because we do our oxidation in a controlled way, step by step, trickling the energy into useful channels instead of dissipating it as undisciplined heat. Such controlled chemistry, or metabolism, is as universal a feature of life as heredity. Theories of the origin of life need to account for both heredity and metabolism, but some writers have mistaken the priority. They have sought a theory of metabolism's spontaneous origin, and somehow hoped that heredity would follow, like other useful devices. But heredity, as we shall see, is not to be thought of as a useful device. Heredity has to be first on the scene because, before heredity, usefulness itself had no meaning. Without heredity, and hence natural selection, there would have been nothing to be useful for. The very idea of usefulness cannot begin until the natural selection of hereditary information does.

The earliest theories of the origin of life that are taken seriously today are those of A. I. Oparin in Russia and J. B. S. Haldane in England, writing in the 1920s in ignorance of each other. Both emphasised metabolism rather than heredity. Both tumbled to the important fact that the atmosphere of Earth before life would have to be 'reducing' if life were to arise. This rather unhelpful technical term means that the atmosphere lacked free oxygen. Organic compounds (compounds of carbon), when there is free oxygen around, are vulnerable to being burned or otherwise oxidised to carbon dioxide. It seems odd to us, who die within minutes if deprived of oxygen, but life could not originate on any planet with free oxygen in its atmo-sphere. As I've already explained, oxygen would have been a deadly poison to our earliest ancestors. Everything we know about other planets makes it almost certain that Earth's original atmosphere was a reducing one. Free oxygen came late. It was a polluting waste product of green bacteria, at first swimming free and later incorporated into plant cells. At some point our ancestors evolved the ability to cope with oxygen, and later came to depend upon it.

Incidentally, having said that oxygen is produced by green plants and algae, it is an oversimplification to leave it at that. It is true that plants give off oxygen. But when a plant dies, its decay, in chemical reactions equivalent to burning all its carbonaceous materials, would use up an amount of oxygen equal to all the oxygen released by that plant during its lifetime. There would therefore be no net gain in atmospheric oxygen, but for one thing. Not all dead plants decay. Some of them are laid down as coal (or equivalents), where they are removed from circulation. If all the fossil fuels in the world were burned by humanity, much of the oxygen in the atmosphere would be replaced by carbon dioxide, restoring the ancient status quo. This is not likely to happen in the near future. But we should not forget that the only reason we have oxygen to breathe is that most of the carbon in the world is tied up underground. We burn it all at our peril.

Oxygen atoms were always present in that early atmosphere, but they were not free as oxygen gas. They were tied up in compounds such as carbon dioxide and water. Carbon today is mostly locked up in living bodies or – a far greater proportion – in rocks, such as chalk, limestone and coal, which come from the remains of once-living bodies. In Canterbury times, those same carbon atoms would mostly have been in the atmosphere as compound gases, such as carbon dioxide and methane. Nitrogen, today's main atmospheric gas, would in a reducing atmosphere have been compounded with hydrogen as ammonia.

Oparin and Haldane realised that a reducing atmosphere would have been friendly to the spontaneous synthesis of simple organic compounds. Here are Haldane's own words, which I quote for the sake of his famous concluding phrase:

> Now, when ultraviolet light acts on a mixture of water, carbon dioxide, and ammonia, a vast variety of organic substances are made, including sugars and apparently some of the materials from which proteins are built. This fact has been demonstrated in the laboratory by Baly of

Liverpool and his colleagues. In this present world such substances, if left about, decay – that is to say, they are destroyed by micro-organisms.* But before the origin of life they must have accumulated till the primitive oceans reached the consistency of hot dilute soup.

This was written in 1929, more than 20 years before the much cited experiment of Miller and Urey, which, one would think, from Haldane's account, to have been a kind of repeat of Baly's. However, E. C. C. Baly was not concerned with the origin of life. His interest was photosynthesis, and his achievement was to synthesise sugars by beaming ultraviolet rays into water containing dissolved carbon dioxide in the presence of a catalyst, such as iron or nickel. It was Haldane, rather than Baly himself, who, with characteristic brilliance,† was anticipating something remarkably like the Miller–Urey experiment and reading it backwards into Baly's work.

What Miller, under Urey's direction, did was take two flasks, one above the other, connected by two tubes. The lower flask contained heated water to represent the primaeval ocean. The upper flask housed the mocked-up primordial atmosphere (methane, ammonia, water vapour and hydrogen). Through one of the two tubes, vapour rose from above the heated 'ocean' in the lower flask and was fed into the top of the 'atmosphere' in the upper flask. The other tube returned downwards from 'atmosphere' to 'ocean'. On the way it passed through a spark chamber ('lightning') and a cooling chamber, where vapour condensed to form 'rain' which replenished the 'ocean'.

After only a week in this recycling simulacrum, the ocean had turned yellow-brown and Miller analysed its content. As Haldane would have predicted, it had become a soup of organic compounds, including no fewer than seven amino acids, the essential building blocks of proteins. Among the seven were three – glycine, aspartic acid and alanine – from the list of 20 found in living things. Later experiments along Miller's lines, but substituting carbon dioxide or carbon monoxide for methane, have achieved similar results. We can draw the robust conclusion that biologically important small molecules, including amino acids, sugars and, significantly, the building blocks of DNA and RNA, spontaneously form when various versions

* This was the point Darwin was making in his 'warm little pond' letter.
† Sir Peter Medawar, no slouch himself, described Haldane as the cleverest man he ever knew.

of the Oparin/Haldane primitive Earth are simulated in the laboratory.

Before Oparin and Haldane, speculators about the origin of life had assumed that the first organisms must have been plants of some kind, perhaps green bacteria. People were used to the idea that life depends upon photosynthesis, the sunlight-driven manufacture of organic compounds, accompanied by a release of oxygen. Oparin and Haldane, with their reducing atmosphere, saw plants as arriving on the scene later. Early life arose in a sea of pre-existing organic compounds. There was soup to eat, and no need to photosynthesise – at least until the soup ran out.

For Oparin, the vital step was the origin of the first cell. And, to be sure, cells, like organisms, have the important property that they never arise spontaneously but always from other cells. It was pardonable to see the origin of the first 'cell' (metaboliser) as a synonym for the origin of life, rather than the first 'gene' (replicator), as I would. Among more modern theorists with the same bias, the distinguished theoretical physicist Freeman Dyson is aware of it and defends it. The majority of recent theorists, including Leslie Orgel in California, Manfred Eigen and his colleagues in Germany, and Graham Cairns-Smith in Scotland – more of a lone maverick, but by no means to be written off – give priority to self-replication, both chronologically and in terms of centrality: rightly so, in my opinion.

What would heredity without a cell look like? Don't we have a chicken-and-egg problem? Certainly so if we take heredity to demand DNA, for DNA can't replicate without a large supporting cast of molecules, including proteins that can only be made by DNA-coded information. But just because DNA is the main self-replicating molecule we know, it doesn't follow that it is the only one we could imagine, or the only one that has ever existed in nature. Graham Cairns-Smith has persuasively argued that the original replicators were inorganic mineral crystals, with DNA a late usurper, stepping into the starring role when life had evolved to the point where such a *Genetic Takeover* became possible. I won't expound his case here, partly because I have already given it my best shot in *The Blind Watchmaker,* but also for a larger reason. Cairns-Smith makes the clearest case I have read that replication was primary, and that DNA must have had a forerunner of some kind whose nature is unknown, save that it exhibited true heredity. I think it is a shame if this unassailable part of his case becomes tied in people's minds to his more controversial and speculative case for mineral crystals as the forerunner.

I have nothing against the mineral crystals theory, which is why I expounded it before, but what I really want to emphasise is the primacy of replication, and the strong likelihood that there was a late takeover by DNA from some forerunner. I can make the point most forcefully by deliberately switching in this book to a different *particular* theory of what that forerunner might have been. Whatever its ultimate merits as the original replicator, RNA is certainly a better candidate than DNA, and it has been cast as forerunner by a number of theorists in their so-called 'RNA World'. To introduce the RNA World theory, I need to digress on enzymes. If the replicator is the star of life's show, the enzyme is the co-star, more than just supporting cast.

Life depends utterly on the virtuoso ability of enzymes to catalyse biochemical reactions in a very fussy way. When I first learned about enzymes at school, the conventional (and in my view mistaken) wisdom that science should be taught by homely example meant that we spat into water to demonstrate the power of the salivary enzyme amylase to digest starch and make sugar. From this we gained the impression that an enzyme is like a corrosive acid. Biological washing powders, which use enzymes to digest dirt out of clothes, give the same impression. But these are destructive enzymes, working to dismember large molecules into their smaller components. Constructive enzymes are involved in synthesising large molecules from smaller ingredients, and they do so by behaving as 'robotic matchmakers', as I shall explain.

The interior of a cell contains a solution of thousands of molecules, atoms and ions of many different kinds. Pairs of these could combine with each other in almost infinitely varied ways, but on the whole they do not. So there is a huge repertoire of potential chemistry waiting to happen in a cell, but most of it doesn't happen. Hold that in mind while reflecting on the following. A chemistry lab has hundreds of bottles on its shelves, all securely stoppered so their contents don't meet each other unless a chemist desires it, in which case a sample from one bottle is added to a sample from another. You could say that the shelves in a chemistry lab also house a huge repertoire of potential chemistry waiting to happen. And again most of it doesn't happen.

But imagine taking all the bottles off all the shelves and tipping them into a single vat full of water. A preposterous act of scientific vandalism, yet such a vat is pretty much what a living cell is, although admittedly with a lot of membranes that complicate the picture. The hundreds of ingredients of thousands of potential chemical reactions are not kept in separate bottles until required to react together. Instead, they are all mixed up together in the same shared space, all the time. But still they wait, largely unreactive, until required to react, as though separated in virtual bottles. There are no virtual bottles, but there are enzymes working as robotic matchmakers, or we might even call them robotic lab assistants. Enzymes discriminate, much as a radio tuner does when it puts a particular wireless set in touch with a particular transmitter while ignoring the hundreds of other signals simultaneously bombarding its aerial with a babel of carrier frequencies.

Suppose there is an important chemical reaction in which ingredient A combines with ingredient B to yield product Z. In a chemistry lab we achieve this by taking the bottle labelled A off the shelf, and the bottle labelled B from another shelf, mixing their contents in a clean flask, and providing other necessary conditions, such as heat or stirring. We achieve the specific reaction we want by taking only two bottles off the shelf. In the living cell lots of A molecules and lots of B molecules are among the huge variety of molecules floating around in the water, where they may meet, but seldom combine even if they do. In any case, they are no more likely to meet than thousands of other possible combinations. Now we introduce an enzyme called abzase, which is specifically shaped to catalyse the A+B=Z reaction. There are millions of abzase molecules in the cell, each one acting as a robotic lab assistant. Each abzase lab assistant grabs one A molecule, not off a shelf but floating free in the cell. It then grabs a B molecule as it drifts by. It holds the A firmly in its grip so that it faces in a particular direction. And it holds the B equally firmly so that it abuts the A, in just the right position and orientation to bond with the A and make a Z. The enzyme may do other things too – the equivalent of the human lab assistant wielding a stirrer or lighting a Bunsen burner. It may form a temporary chemical alliance with A or B, exchanging atoms or ions that will eventually be paid back, so the enzyme ends up as it started, thus qualifying as a catalyst. The result of all this is that a new Z molecule forms in the shaped 'grip' of the enzyme molecule. The lab assistant then releases the new Z into the water and waits for another A to come by, whereupon it grabs it and the cycle resumes.

If there were no robotic lab assistant, a drifting A would occasionally bump into a drifting B under the right conditions to bond. But this lucky occurrence would be rare, no more common than the occasional chance-matched encounters that either A or B might make with lots of other potential partners. A might bump into C and make Y. Or B might bump into D and make X. Small amounts of Y and X are being made all the time by lucky drift. But it is the presence of the lab assistant enzyme abzase that makes all the difference. In the presence of abzase, Z is churned out in (from the cell's point of view) industrial quantities: an enzyme typically multiplies the spontaneous rate of reaction by a factor varying between a million and a trillion. If a different enzyme, acyase, were introduced, A would be combined with C instead of B, again at racing conveyor-belt speed, to make a lavish supply of Y. It is the very same A molecules we are talking about, not confined to a bottle but free to combine with either B or C, depending on which enzyme is present to grab them.

The production rates of Z and Y will therefore depend on, among other things, how many of each of the two rival lab assistants, abzase and acyase, are floating about in the cell. And *that* depends on which of two genes in the nucleus of the cell is turned on. It is, however, a little more complicated than that: even if a molecule of abzase is present, it may be inactivated. One way this can happen is that another molecule comes and sits in the active 'cavity' of the enzyme. It is as though the lab assistant's robotic arms were temporarily handcuffed. The handcuffs remind me, by the way, to issue the ritual warning that, as always with metaphors, there is a risk that 'robotic lab assistant' might mislead. An enzyme molecule doesn't actually have arms to reach out and seize ingredients such as A, let alone submit to handcuffs. Instead, it has special zones in its own surface for which A, say, has an affinity, either because of a snug physical fit to a shaped cavity, or due to some more recondite chemical property. And this affinity can be temporarily negated in ways that resemble the calculated throwing of an off-switch.

Most enzyme molecules are special-purpose machines which make only one product: a sugar, say, or a fat; a purine or a pyrimidine (building blocks of DNA and RNA), or an amino acid (twenty of them are building blocks of natural proteins). But some enzymes are more like programmable machine tools that take in a punched paper tape to determine what they do. Outstanding among these is the ribosome, briefly explained in Taq's Tale, a large and complicated machine tool constructed from both protein and RNA, which makes proteins themselves. Amino acids, the building blocks of proteins, have already been made by special-purpose enzymes and are floating around in the cell, available to be picked up by the ribosome. The punched paper tape is RNA, specifically 'messenger RNA' (mRNA). The messenger tape, which itself has copied its message from DNA in the genome, feeds into the ribosome and, as it passes through the 'reading head', the appropriate amino acids are assembled into a protein chain in the order specified by the tape using the genetic code.

How this specification works is known, and it is unspeakably wonderful. There is a set of small transfer RNAs (tRNA), each about 70 building blocks long. Each of the tRNAs attaches itself selectively to one, and only one, of the twenty kinds of natural amino acids. At the other end of the tRNA molecule is an 'anti-codon', a triplet precisely complementing the short mRNA sequence (codon) that specifies the particular amino acid according to the genetic code. As the tape of mRNA moves through the reading head of the ribosome, each codon of the mRNA binds to a tRNA with the right anti-codon. This causes the amino acid dangling off the other end of the tRNA to be brought into line, in the 'matchmaking' position, to attach to the growing end of the newly forming protein. Once the amino acid is attached, the tRNA peels off in search of a new amino acid molecule of its preferred type, while the mRNA tape inches forward another notch. So the process continues and the protein chain is extruded step by step. Amazingly, one physical tape of mRNA can cope with several ribosomes at once. Each of these ribosomes moves its reading head along a different portion of the tape's length, and each extrudes its own copy of the newly minted protein chain.

As each new protein chain is completed, when the mRNA feeding its ribosome has completely gone through that ribosome's reading head, the protein detaches itself. It coils up into a complicated three-dimensional structure whose shape is determined, through the laws of chemistry, by the sequence of amino acids in the protein chain. That sequence was itself determined by the order of code symbols along the length of the mRNA. And that order was, in turn, determined by the complementary sequence of symbols along the DNA, which constitutes the master database for the cell.

The coded sequence of DNA therefore controls what goes on in the cell. It specifies the sequence of amino acids in each protein, which determines the protein's three-dimensional shape, which in turn gives that protein its particular enzymatic properties. Importantly, the control may be indirect in that, as we saw in the Mouse's Tale, genes determine which other genes shall be turned on and when. Most genes in any one cell are not switched on. This is why of all the reactions that could be going on in the 'vat full of mixed ingredients', only one or two actually do go on at any one time: the ones whose specific 'lab assistants' are active in the cell.

After that digression on catalysis and enzymes, we now turn from ordinary catalysis to the special case of autocatalysis, some version of which probably played a key role in the origin of life. Think back to our hypothetical example of molecules A and B combining to make Z under the influence of the enzyme abzase. What if Z itself is its own abzase? I mean, what if the Z molecule happens to have just the right shape and chemical properties to seize one A and one B, bring them together in the correct orientation, and combine them to make a new Z, just like itself? In our previous example we could say that the amount of abzase in the solution would influence the amount of Z produced. But now, if Z actually is one and the same molecule as abzase, we need only a single molecule of Z to seed a chain reaction. The first Z grabs As and Bs and combines them to make more Zs. Then these new Zs grab more As and Bs to make still more Zs and so on. This is autocatalysis. Under the right conditions the population of Z molecules will grow exponentially – explosively. This is the kind of thing that sounds promising as an ingredient for the origin of life.

But it is all hypothetical. Julius Rebek and his colleagues at the Scripps Institute in California made it real. They explored some fascinating examples of autocatalysis in real chemistry. In one of their examples, Z was amino adenosine triacid ester (AATE), A was amino adenosine and B was pentafluorophenyl ester, and the reaction took place not in water but in chloroform. Needless to say, none of these particular chemical details, and certainly not the long names, need to be remembered. What matters is that the product of the chemical reaction is its own catalyst. The first molecule of AATE is reluctant to form but, once formed, an immediate chain reaction is set in train as more and more AATE synthesises itself by serving as its own catalyst. As if that weren't enough, this brilliant series of experiments went on to demonstrate true heredity in the sense defined here. Rebek and his team found a system in which more than one variant of the autocatalysed substance existed. Each variant catalysed the synthesis of itself, using its preferred variant of one of the ingredients. This raised the possibility of true competition in a population of entities showing true heredity, and an instructively rudimentary form of Darwinian selection.

Rebek's chemistry is highly artificial. Nevertheless, his story beautifully illustrates the principle of autocatalysis, according to which the product of a chemical reaction serves as its own catalyst. It is something like autocatalysis that we need for the origin of life. Could RNA, or something like RNA under the conditions of the early Earth, have autocatalysed its own synthesis Rebek-style, and in water instead of chloroform?

The problem is a formidable one, as explained by the German Nobel Prize-winning chemist Manfred Eigen. He pointed out that any self-replication process is subject to degradation by copying error – mutation. Imagine a population of replicating entities in which there is a high probability of error in every copying event. If a coded message is to hold its own against the ravages of mutation, at least one member of the population in any one generation must be identical to its parent. If there are ten code units ('letters') in an RNA chain, for example, the average error rate per letter must be less than one in ten: we can then expect that at least some members of the offspring generation will have the full complement of ten correct code letters. But if the error rate is greater, there will be a relentless degradation as the generations go by, simply because of mutation alone, no matter how strong the selection pressure. This is called an error catastrophe. Error catastrophes in advanced genomes form the main theme of Mark Ridley's provocative book *Mendel's Demon*,[*] but here we are concerned with the error catastrophe that threatened the origin of life itself.

Short chains of RNA and, indeed, DNA can spontaneously self-replicate without an enzyme. But the error rate per letter is far higher

[*] This excellent book has suffered the common fate of being renamed in mid-Atlantic. If you want to find it in the USA, look for *The Cooperative Gene*. Why do publishers *do* this? It causes so much confusion. I hasten to say that I have nothing against *The Cooperative Gene* as a title. Genes certainly are co-operative (see *The Selfish Gene*). *Mendel's Demon*, too, is a fine title, though *Genetic Meltdown* might have suited the book's message even better. Matt Ridley (no relation to Mark, except as established by Y-chromosome analysis) tells me that although the hardback of his *Nature via Nurture* is already so named in the USA, the paperback is to be renamed – wait for it – *The Agile Gene*.

than when an enzyme is present. And this means that long before a sufficient length of gene could be built up to make the protein for a working enzyme, the fledgling gene would have been destroyed by mutation. That is the Catch-22 of the origin of life. A gene big enough to specify an enzyme would be too big to replicate accurately without the aid of an enzyme of the very kind that it is trying to specify. So the system apparently cannot get started.

The solution to the Catch-22 that Eigen offers is the theory of the hypercycle. It uses the old principle of divide and rule. The coded information is subdivided into sub-units small enough to lie below the threshold for an error catastrophe. Each sub-unit is a mini replicator in its own right, and it is small enough for at least one copy to survive in each generation. All the sub-units co-operate in some important larger function, large enough to suffer an error catastrophe if catalysed by a single large chemical rather than being subdivided.

As I have so far described the theory, there is a danger that the whole system would be unstable because some sub-units would self-replicate faster than others. This is where the clever part of the theory kicks in. Each sub-unit flourishes in the presence of the others. More specifically, the production of each is catalysed by the presence of another, such that they form a cycle of dependency: a 'hypercycle'. This automatically prevents any one element from racing ahead. It cannot do so because it depends on its predecessor in the hypercycle.

John Maynard Smith pointed out the similarity of a hypercycle to an ecosystem. Fish numbers depend on the population of *Daphnia* (waterfleas) on which they feed. In turn, fish numbers affect the population of fish-eating birds. The birds provide guano, which assists blooms of algae on which the *Daphnia* flourish. The whole cycle of dependency is a hypercycle. Eigen and his colleague Peter Schuster propose some kind of molecular hypercycle as the solution to the Catch-22 riddle of the origin of life.

I'm going to leave the hypercycle theory at this point and return to the suggestion, which is fully compatible with it, that RNA, in the early days when life was just beginning and proteins did not yet exist, might have served as its own catalyst. This is the RNA World theory. To see how plausible it is, we need to look at why proteins are good at being enzymes but bad at being replicators; at why DNA is good at replicating but bad at being an enzyme; and finally why RNA might just be good enough at both roles to break out of the Catch-22.

Three-dimensional shape is largely what matters for enzyme activity. Proteins are good at being enzymes because they can assume almost any shape you want in three dimensions, as an automatic consequence of their amino acid sequence in one dimension. It is the chemical affinities of amino acids for other amino acids in different parts of the chain that determine the particular knot into which the protein chain ties itself. So the three-dimensional shape of a protein molecule is specified by the one-dimensional sequence of amino acids, and that is itself specified by the one-dimensional sequence of code letters in a gene. In principle (practice is a different matter, and formidably difficult) it should be possible to write down a sequence of amino acids that would spontaneously coil itself up into almost any shape you like: not just shapes that make good enzymes, but any arbitrary shape you choose to specify.* It is this protean talent that qualifies proteins to act as enzymes. There is a protein capable of selecting any one out of the hundreds of potential chemical reactions that could go on in a cell full of jumbled ingredients.

Proteins, then, make wonderful enzymes, capable of tying themselves into knots of any desired shape (see plate 48). But they are lousy replicators. Unlike DNA and RNA, whose component elements have specific pairing rules (the 'Watson–Crick pairing rules' discovered by those two inspired young men), amino acids have no such rules. DNA, by contrast, is a splendid replicator but a lousy candidate for the enzyme role in life. This is because, unlike proteins with their near infinite variety of three-dimensional shapes, DNA has only one shape, the famous double helix itself. The double helix is ideally suited to replication because the two sides of the stairway peel easily away from one another, each being then exposed as a template for new letters to join, following the Watson–Crick pairing rules. It is not much good for anything else.

RNA has some of the virtues of DNA as a replicator and some of the virtues of protein as a versatile shaper of enzymes. The four letters of RNA are sufficiently similar to the four letters of DNA that either set can serve as a template for the other. On the other hand, RNA does not easily form a long double helix, which means that it is somewhat inferior to DNA as a replicator. This is partly because the double helix system lends itself to proof-correction. When the DNA

* Indeed, there are lots of different amino acid sequences that will yield the same shape, which is one reason to doubt naive calculations of the astronomical 'improbability' of a particular protein chain, obtained by raising 20 to the power of its length.

double helix splits and each single helix immediately serves as a template for its complement, errors can instantly be spotted, and corrected. Each daughter chain remains attached to its 'parent', and comparison between the two permits instant error detection. Proofreading based on this principle reduces mutation rates to the order of one in a billion, which is what makes large genomes like ours possible. RNA, lacking this kind of proofreading, has mutation rates that are thousands of times greater than DNA. This means that only simple organisms with small genomes, such as some viruses, can use RNA as their primary replicator.

But the lack of a double helix structure has its upside as well as its downside. Because the RNA chain doesn't spend all its time paired with its complementary chain but breaks away from the complement as soon as it is formed, it is free to tie itself in knots like a protein. Just as the protein does it by virtue of the chemical affinities of amino acids for other amino acids in different parts of the same chain, RNA does it using the ordinary Watson–Crick base-pairing rules, the same ones as are used to make copies of RNA. Putting it another way, lacking a partner chain to pair with in a double helix like DNA, RNA is free to 'pair' with odd bits of itself. RNA finds small stretches of itself with which it can pair, either in a miniature double helix or in some other shape. The pairing rules insist that these stretches have to be going in opposite directions. An RNA chain therefore has a tendency to fall into a series of hairpin bends.

The repertoire of three-dimensional shapes into which an RNA molecule is capable of throwing itself may not be as great as the repertoire of large protein molecules. But it is large enough to encourage the thought that RNA might furnish a versatile armoury of enzymes. And, to be sure, many RNA enzymes, called ribozymes, have been discovered. The conclusion is that RNA has some of the replicator virtues of DNA and some of the enzyme virtues of proteins. Maybe, before the coming of DNA, the arch-replicator, and before the coming of proteins, the arch-catalysts, there was a world in which RNA alone had enough of both virtues to stand in for both experts. Perhaps an RNA fire ignited itself in the original world, and then later started to make proteins that turned around and helped synthesise RNA, and later DNA too, which took over as the dominant replicator. That is the hope of the RNA World theory. It receives indirect support from a lovely series of experiments initiated by Sol Spiegelman of Columbia University, and repeated in various forms by others over the years. Spiegelman's experiments use a protein enzyme, which might be thought to be cheating, but they produce such spectacular results, illuminating such important links in the theory, that you can't help feeling it was worth it anyway.

First, the background. There is a virus called Qβ. It is an RNA virus, which means that, instead of DNA, its genes are entirely made of RNA. It uses an enzyme to replicate its RNA, called Qβ replicase. In the wild state, Qβ is a bacteriophage (phage for short) – a parasite of bacteria, specifically of the gut bacterium *Escherichia coli*. The bacterial cell 'thinks' the Qβ RNA is a piece of its own messenger RNA, and its ribosomes process it exactly as though it were, but the proteins that it manufactures are good for the virus instead of for the host bacterium. There are four such proteins: a coat protein to protect the virus; a glue protein to stick it to the bacterial cell; a so-called replication factor, which I'll mention again in a moment; and a bomb protein to destroy the bacterial cell when the virus has finished replicating, thereby releasing some tens of thousands of viruses, each to travel in its little protein coat until it bumps into another bacterial cell and renews the cycle. I said I would return to the replication factor. You might think that this must be the enzyme Qβ replicase, but actually it is smaller and simpler. All that the little viral gene itself does is make a protein that sews together three other proteins which the bacterium is making anyway for its own (completely different) purposes. When these are stitched together by the virus's own little protein, the composite so formed is the Qβ replicase.

Spiegelman was able to isolate from this system just two components, Qβ replicase and Qβ RNA. He put them together in water with some small-molecule raw materials – the building blocks for making RNA – and watched what happened. The RNA seized small molecules and built copies of itself using Watson–Crick pairing rules. It managed this feat without any bacterial host, and without the protein coat or any other part of the virus. That in itself was a nice result. Notice that protein synthesis, which is part of the normal action of this RNA in the wild, has been completely taken out of the loop. We have a stripped-down RNA replication system making copies of itself without bothering to make protein.

Then Spiegelman did something wondrous. He set a form of evolution in motion in this wholly artificial test-tube world, with no cells involved at all. Imagine his set-up as a long row of test tubes, each containing Qβ replicase and raw building blocks but no RNA.

He seeded the first tube with a small amount of Qβ RNA, and it duly replicated lots more copies of itself. He then drew out a small sample of the liquid, and put a drop of it into the second tube. This seed RNA now set about replicating in the second tube, and when this had been going on for a while Spiegelman drew out a drop from the second tube and seeded the third virgin tube. And so on. This is like the spark from our fire seeding a new fire in the dry grass, and the new fire seeding another, and so on in a chain of seedings. But the result was very different. Whereas fires don't inherit any of their qualities from the seed, Spiegelman's RNA molecules did. And the consequence was . . . evolution by natural selection in its most basic and stripped-down form.

Spiegelman sampled the RNA in his tubes as the 'generations' went by and monitored its properties, including its potency in infecting bacteria. What he found was fascinating. The evolving RNA became physically smaller and smaller and, at the same time, less and less infective when bacteria were offered to samples of it. After 74 generations* the typical RNA molecule in a tube had evolved to a small fraction of the size of its 'wild ancestor'. The wild RNA had been a necklace about 3,600 'beads' long. After 74 generations of natural selection, the average inhabitant of a test tube had reduced itself down to a mere 550: no good at infecting bacteria but brilliant at infecting test tubes. What had happened was clear. Spontaneous mutations in the RNA had occurred all along the line, and the mutants that survived were well fitted to do so in the test-tube world, as opposed to the natural world of bacteria waiting to be parasitised. The main difference was presumably that the RNA in the tube world could dispense with all the coding devoted to making the four proteins needed to make the coat, the bomb and the other requirements for survival of the wild virus as a working parasite of bacteria. What was left was the bare minimum required to replicate in the featherbedded world of test tubes full of Qβ replicase and raw materials.

This bare minimum survivor, less than a tenth the size of its wild ancestor, has become known as Spiegelman's Monster. Being smaller, the streamlined variant reproduces more rapidly than its competitors, and therefore natural selection gradually increases its rep-

---

* That's tube generations, of course: the number of RNA generations would be more, because RNA molecules are replicating many times within each tube generation.

resentation in the population (and population, by the way, is exactly the right word, even though we are talking about free-floating molecules, not viruses or organisms of any kind).

Amazing to relate, almost the same Spiegelman monster repeatedly evolves when the experiment is run over again. Moreover, Spiegelman and Leslie Orgel, one of the leading figures in research on the origin of life, performed further experiments in which they added a nasty substance, such as ethidium bromide, to the solution. Under these conditions, a different monster evolves, one that is resistant to ethidium bromide. Different chemical obstacle courses foster evolution towards different specialist monsters.

Spiegelman's experiments used natural 'wild type' Qβ RNA as a starting point. M. Sumper and R. Luce, working in the laboratory of Manfred Eigen, obtained a truly stunning result. Under some conditions, a test tube containing *no RNA at all*, just the raw materials for making RNA plus the Qβ replicase enzyme, can spontaneously generate self-replicating RNA which, under the right circumstances, will evolve to become similar to Spiegelman's Monster. So much, incidentally, for creationist fears (or hopes, we might rather say) that large molecules are too 'improbable' to have evolved. Such is the simple power of cumulative natural selection (so far is natural selection from being a process of blind chance) Spiegelman's Monster takes only a few days to build itself up from scratch.

These experiments are still not direct tests of the RNA World hypothesis of the origin of life. In particular, we still have the 'cheat' of Qβ replicase being present throughout. The RNA World hypothesis pins its hopes on RNA's own catalytic powers. If RNA can catalyse other reactions, as it is known to do, might it not catalyse its own synthesis? Sumper and Luce's experiment dispensed with RNA but provided the Qβ replicase. What we need is a new experiment that dispenses with the Qβ replicase too. Research continues, and I expect exciting results. But now I want to switch to a newly fashionable line of thought, fully compatible with the RNA World, and with many others among the current theories of the origin of life. What is new is the suggested location in which the crucial events first took place. Not 'warm little pond' but 'hot deep rock' – an exciting theory which amounts to this: our pilgrims, to complete their journey and locate their Canterbury, are now going to have to bore deep underground, into the primordial rock. The main inspirer of the theory is another maverick, Thomas Gold, originally an astronomer but versatile enough to deserve the now-rare accolade 'general scientist', and distinguished enough to have been elected to both the Royal Society of London and the American National Academy of Sciences.

Gold believes that our emphasis on the sun as energetic prime mover of life may be misplaced. Perhaps we have yet again been misled by what happens to be familiar: yet again assigned to ourselves and our kind of life a centrality in the scheme of things that we do not deserve. There was a time when textbooks asserted that all life depended ultimately on sunlight. Then, in 1977, the startling discovery was made that volcanic vents on the floors of deep oceans support a strange community of creatures, living without benefit of sunlight. Heat from red-hot lava raises the water temperature to more than 100°C, still well below boiling point at the colossal pressures of those depths. The surrounding water is very cold, and the temperature gradient drives various kinds of bacterial metabolism. These thermophile bacteria, including sulphur bacteria who make use of hydrogen sulphide streaming from the volcanic vents, constitute the base of elaborate food chains, higher links of which include blood-red tube worms up to three metres long, limpets, mussels, starfish, barnacles, white crabs, prawns, fish and other annelid worms capable of thriving at 80°C. There are bacteria, as we have seen, which can take such Hadean temperatures in their stride, but no other animal is known to do so, and these polychaete worms have accordingly been dubbed Pompeii worms. Some of the sulphur bacteria are given house room by animals, for example by mussels, and by the huge tube worms, who take special biochemical steps, using haemoglobin (hence their blood-red colour) to feed sulphide to their own bacteria. These colonies of life, based on bacterial extraction of energy from hot volcanic vents, astonished everybody, first by their very existence, and then by their abundant richness, which contrasted startlingly with the near-desert conditions of the surrounding sea bottom.

Even after this sensational discovery, most biologists continue to believe that life is centred on the sun. The creatures of the deep-sea smoker communities, fascinating though they might be, are assumed by most of us to be a rare and unrepresentative aberration. Gold believes otherwise. He thinks hot, dark, high-pressure depths are where life fundamentally belongs and where it originated. Not necessarily in the sea, but perhaps in the rocks, deep underground. We who live at the surface, in the light and the cool and the fresh air, we are the anomalous aberrations! He points out that 'hopanoids', organic molecules made in bacterial cell walls, are ubiquitous in rocks, and quotes an authoritative estimate of between 10 trillion and 100 trillion tonnes of hopanoids in the rocks of the world. This comfortably exceeds the trillion tonnes or so of organic carbon in surface-dwelling life.

Gold notes that the rocks are seamed with cracks and fissures, which, though small to our eyes, provide more than a billion trillion cubic centimetres of hot, wet space suitable for life on the bacterial scale of existence. Heat energy, and the chemicals of the rocks themselves, would be enough to sustain bacteria in huge numbers. Gold notes that many bacteria thrive at temperatures up to 110°C, and this would permit them to live down to depths of between 5 and 10 kilometres, a distance that would take them less than a thousand years to travel. It is impossible to verify his estimate, but he thinks the biomass of bacteria in the hot, deep rocks might exceed the biomass of the surface sun-based life with which we are familiar.

Turning to the question of the origin of life, Gold and others have pointed out that thermophily – love of high temperatures – is not a rare oddity among bacteria and archaeans. It is common: so common, and so widely distributed around bacterial family trees, that it might well be the primitive state from which our familiar cool forms of life have evolved. With respect to both chemistry and temperature, the conditions on the surface of the primitive Earth – some scientists call it the Hadean Age – were more like those in Gold's hot deep rocks than they were like today's surface conditions. A persuasive case can indeed be made that when we dig down into the rocks we are digging backwards in time, and rediscovering something like the conditions of life's scalding Canterbury.

The idea has been further championed recently by the Anglo-Australian physicist Paul Davies, whose book *The Fifth Miracle* summarises new evidence discovered since Gold's paper of 1992. Various drilling samples have been found to contain hyper-thermophile bacteria, alive and reproducing, amid scrupulous precautions to preclude contamination from the surface. Some of these bacteria have been successfully cultured . . . in a modified pressure cooker! Davies, like Gold, believes life may have originated deep underground, and that the bacteria which still live there may be relatively unchanged relics of our remote ancestors. This idea is especially appealing for our pilgrimage because it offers us the hope of meeting something like the earliest bacteria, rather than the more familiar bacteria, modified for modern conditions of light, cold and oxygen. Having endured ridicule at first, the hot deep rock theory of the origin of life is now verging on the positively fashionable. Whether it will turn out to be right must await more research, but I confess to hoping that it will.

There are many other theories that I have not gone into. Maybe one day we shall reach some sort of definite consensus on the origin of life. If so, I doubt if it will be supported by direct evidence because I suspect that it has all been obliterated. Rather, it will be accepted because somebody produces a theory so elegant that, as the great American physicist John Archibald Wheeler said in another context:

> . . . we will grasp the central idea of it all as so simple, so beautiful, so compelling that we will say to each other, 'Oh, how could it have been otherwise! How could we all have been so blind for so long!'

If that isn't how we finally realise we know the answer to the riddle of life's origin, I don't think we ever shall know it.

THE HOST'S RETURN

The genial host, having guided Chaucer and the other pilgrims from London to Canterbury and stood impresario to their tales, turned around and led them straight back to London. If I now return to the present, it must be alone, for to presume upon evolution's following the same forward course twice would be to deny the rationale of our backward journey. Evolution was never aimed at any particular endpoint. Our backwards pilgrimage has been a series of swelling mergers, as we were swallowed up in ever more inclusive groupings: the apes, the primates, the mammals, the vertebrates, the deutero-stomes, the animals and so on back to the arch ancestor of all life. If we turn around and move forward now, we cannot retrace our steps. That would imply that evolution, were it to be rerun, would follow the same course, putting those same mergers into reverse gear in the form of splits. The stream of life would branch in all the 'right' places. Photosynthesis and an oxygen-based metabolism would be rediscovered, the eukaryotic cell would reconstitute itself, cells would club together in neometazoan bodies. There would be a new split between plants on the one hand, and animals plus fungi on the other; a new split between protostomes and deuterostomes; the backbone would be rediscovered, and so would eyes, ears, limbs, nervous systems . . . Eventually a swollen-brained biped would emerge, with skilled hands guided by forward-looking eyes, culminating in the proverbial cricket team to beat the Australians.

My disavowal of aimed evolution underlay my original choice to do history backwards. And yet in my opening lines I confessed to an ear for a rhyme that would lead me into cautious flirtation with recurring patterns, with lawfulness and forward directionality in evolution. So although my return as host will not be a retracing of steps, I shall be publicly wondering whether something a little bit like a retracing might not be appropriate.

Rerunning Evolution

The American theoretical biologist Stuart Kauffman put the question well in a 1985 article:

One way to underline our current ignorance is to ask, if evolution were to recur from the Precambrian when early eukaryotic cells had already been formed, what organisms in one or two billion years might be like. And, if the experiment were repeated myriads of times, what properties of organisms would arise repeatedly, what properties would be rare, which properties were easy for evolution to happen upon, which were hard? A central failure of our current thinking about evolution is that it has not led us to pose such questions, although the answers might in fact yield deep insight into the expected character of organisms.

I especially like Kauffman's statistical proviso. He envisages not just one thought experiment but a statistical sample of thought experiments in quest of general laws of life, as opposed to local manifestations of particular lifes. The Kauffman question is akin to the science fiction question of what life on other planets might be like – except that on other planets the starting and prevailing conditions would be different. On a large planet, gravity would impose a whole new set of selection pressures. Animals the size of spiders could not have spidery limbs (they'd break under the weight) but would need the support of stout, vertical columns, like the tree trunks on which our elephants stand. Conversely, on a smaller planet, animals the size of elephants but of gossamer build could skitter and leap over the surface like jumping spiders. Those expectations about body build will apply to the whole statistical sample of high-gravity worlds and the whole statistical sample of low-gravity worlds.

Gravity is a given condition of a planet, which life cannot influence. So is its distance from its central star. So is its speed of rotation, which determines day length. So is the tilt of its axis, which, on a planet like ours with its near-circular orbit, is the main determinant of seasons. On a planet with a far-from-circular orbit like Pluto, the dramatically changing distance from the central star would be a much more significant determiner of seasonality. The presence, distance, mass and orbit of a moon or moons exert a subtle but strong influence on life via the tides. All these factors are givens, uninfluenced by life and therefore to be treated as constant in successive reruns of the Kauffman thought experiment.

Earlier generations of scientists would have treated the weather and the chemical composition of the atmosphere as givens too. Now we know that the atmosphere, especially its high oxygen and low carbon content, is conditioned by life. So our thought experiment must allow for the possibility that in successive reruns of evolution the atmosphere might vary under the influence of whatever life forms evolve. Life could thereby influence the weather, and even major climatic episodes, such as ice ages and droughts. My late colleague W. D. Hamilton, who was right too many times to be laughed away, suggested that clouds and rain are themselves adaptations manufactured by micro-organisms for their own dispersal.

So far as we know, the innermost workings of the Earth remain unaffected by the froth of life on its surface. But thought experiments in the rerunning of evolution should acknowledge possible differences in the course of tectonic events, and hence the histories of continental positions. It is an interesting question whether episodes of volcanism and earthquakes, and bombardments from outer space, should be assumed to be the same on successive Kauffman reruns. It is probably wise to treat tectonics and celestial collisions as important variables that can be averaged out if we imagine a sufficiently large statistical sample of reruns.

How shall we set about answering the Kauffman question? What would life be like if the 'tape' were rerun a statistical number of times? Immediately we can recognise a whole family of Kauffman questions, of steadily increasing difficulty. Kauffman chose to reset the clock at the moment when the eukaryotic cell was assembled from its bacterial components. But we could imagine restarting the process two or three eons earlier, with the origin of life itself. Or, at the other extreme, we can restart the clock much later, say, at Concestor 1, our split from the chimpanzees, and ask whether the hominids would, in a statistically significant number of reruns given that life had reached Concestor 1, have evolved bipedality, brain enlargement, language, civilisation and baseball. In between, there is a Kauffman question for the origin of the mammals, for the origin of the vertebrates, and any number of other Kauffman questions.

Short of pure speculation, does the history of life, as it actually happened, provide anything approaching a natural Kauffman experiment to guide us? Yes it does. We met several natural experiments throughout our pilgrimage. By happy accidents of prolonged geographical isolation, Australia, New Zealand, Madagascar, South America, even Africa, furnish us with approximate reruns of major episodes of evolution.

These landmasses were isolated from each other, and from the rest of the world, for significant parts of the period after the dinosaurs disappeared, when the mammal group displayed most of its evolutionary creativity. The isolation was not total, but was sufficient to foster the lemurs in Madagascar, and the ancient and diverse radiation of Afrotheria in Africa. In the case of South America, we have distinguished three separate foundations of mammals, with long periods of isolation in between. Australinea provides the most perfect conditions for this kind of natural experiment – its isolation was nearly perfect for much of the period in question, and it began with a very small, possibly single, inoculum of marsupials. New Zealand is an exception, for – alone among these revealing natural experiments – it found itself without mammals during the period in question.

As I look at these natural experiments, mostly I am impressed by how similarly evolution turns out when it is allowed to run twice. We have seen how similar *Thylacinus* is to a dog, *Notoryctes* to a mole, *Petaurus* to flying squirrels, *Thylacosmilus* to the sabretooths (and to various 'false sabretooths' among the placental carnivores). The differences are instructive too. Kangaroos are hopping antelope-substitutes. Bipedal hopping, when perfected at the end of a line of evolutionary progression, may be as impressively fast as quadrupedal galloping. But the two gaits are radically different from each other, in ways that have wrought major changes in the whole anatomy. Presumably, at some ancestral parting of the ways, either of the two 'experimental' lineages could have followed the route of perfecting bipedal hopping, and either could have perfected quadrupedal galloping. As it happens – possibly for almost accidental reasons originally – the kangaroos hopped one way and the antelopes galloped the other. We now marvel at the downstream divergences between the end-products.

The mammals underwent their disparate evolutionary radiations at roughly the same time as each other, on different landmasses. The vacuum left by the dinosaurs freed them to do so. But the dinosaurs in their time had similar evolutionary radiations, although with notable omissions – for example, I can't get an answer to my question of why there seem to have been no dinosaur 'moles'. And before the dinosaurs there were yet other multiple parallels, notably among the mammal-like reptiles, and these too culminated in similar ranges of types.

When I give public lectures I always try to answer questions at the end. The commonest question by far is, 'What might humans evolve into next?' My interlocutor always seems touchingly to imagine it is a fresh and original question, and my heart sinks every time. For it is a question that any prudent evolutionist will evade. You cannot, in detail, forecast the future evolution of any species, except to say that statistically the great majority of species have gone extinct. But although we cannot forecast the future of any species, say, 20 million years hence, we can forecast the general range of ecological types that will be around. There will be herbivores and carnivores, grazers and browsers, meat eaters, fish eaters and insect eaters. These dietary forecasts themselves presuppose that in 20 million years there will still be foods corresponding to the definitions. Browsers presuppose the continued existence of trees. Insectivores presuppose insects, or anyway small, leggy invertebrates – doodoos, to employ that useful technical term from Africa. Within each category, herbivores, carnivores and so on, there will be a range of sizes. There will be runners, fliers, swimmers, climbers and burrowers. The species won't be exactly the same as the ones we see today, or the parallel ones that evolved in Australia or South America, or the dinosaur equivalents, or the mammal-like reptile equivalents. But there will be a similar range of types, making their livings in a similar range of ways.

If, during the next 20 million years, there is a major catastrophe and a mass extinction comparable to the end of the dinosaurs, we can expect the range of ecotypes to be drawn from new ancestral starting points, and – notwithstanding my speculation about rodents at Rendezvous 10 – it might be quite hard to guess which of today's animals will provide those starting points. A Victorian cartoon (see plate 49) shows Professor Ichthyosaurus discoursing upon a human skull from some remote recycling past. If, in the time of the dinosaurs, Professor Ichthyosaurus had mooted their catastrophic end, it would have been quite hard for him to forecast that their place would be taken by the descendants of the mammals, which were then small, insignificant, nocturnal insectivores.

Admittedly, all this concerns quite recent evolution, not so prolonged a rerun as Kauffman imagined. But these recent reruns can surely teach us some lessons about the inherent reproducibility of evolution. If early evolution ran along similar lines to later evolution, those lessons might amount to general principles. My hunch is that the principles we learn from recent evolution since the decease of the dinosaurs probably hold good at least back to the Cambrian, and probably back to the origin of the eukaryotic cell. I have a hunch that the parallelism of mammal radiations in Australia, Madagascar,

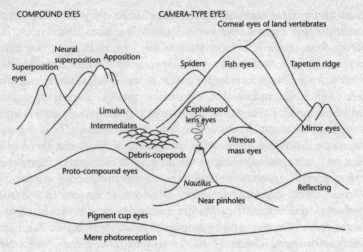

The Fortyfold Path to Enlightenment. Landscape of eye evolution by Michael Land.

South America, Africa and Asia may provide a sort of template for answering Kauffman questions for much older starting points, such as the one he chose, the origin of the eukaryotic cell. Earlier than that landmark event, confidence evaporates. My colleague Mark Ridley, in *Mendel's Demon,* suspects that the origin of eukaryotic complexity was a massively improbable event, perhaps even more improbable than the origin of life itself. Influenced by Ridley, my bet is that most rerun thought experiments that start with the origin of life will not make it into the eukaryocracy.

We don't have to rely on geographical separation as in the Australian natural experiment to study convergence. We can think of the experiment of evolution being rerun, not from the same starting point in different geographical areas, but from different starting points – very possibly in the same geographical area: convergence in animals so unrelated to each other that what they tell us has nothing to do with geographical separation. It has been estimated that 'the eye' has evolved independently between 40 and 60 times around the animal kingdom. This inspired my chapter called 'The Fortyfold Path to Enlightenment' in *Climbing Mount Improbable,* so I won't repeat myself here, except to say that Professor Michael Land of Sussex University, our leading expert on the comparative zoology of eyes, recognises nine independent principles of optical mechanism, each of which has evolved more than once. He was kind enough to prepare for that book the landscape reprinted opposite, in which separate peaks represent independent evolutions of eyes.

It seems that life, at least as we know it on this planet, is almost indecently eager to evolve eyes (see plate 50). We can confidently predict that a statistical sample of Kauffman reruns would culminate in eyes. And not just eyes, but compound eyes like those of an insect, a prawn or a trilobite, and camera eyes like ours or a squid's, with colour vision and with mechanisms for fine-tuning the focus and the aperture. Also very probably parabolic reflector eyes like those of a limpet, and pinhole eyes like those of *Nautilus,* the latter-day ammonite-like mollusc in its floating coiled shell, whom we met at Rendezvous 26. And if there is life on other planets around the universe, it is a good bet that there will also be eyes, based on the same range of optical principles as we know on this planet. There are only so many ways to make an eye, and life as we know it may well have found them all.

We can do the same kind of count for other adaptations. Echolocation – the trick of emitting sound pulses and navigating by accurate timing of the echoes – has evolved at least four times: in bats, toothed whales, oilbirds and cave swiftlets. Not as many times as the eye, but still often enough to make us think it not too unlikely that, if the conditions are right, it will evolve. Very probably, too, reruns of evolution would rediscover the same specific principles: the same tricks for confronting difficulties. Once again, I shall not repeat my exposition from a previous book,* but will simply summarise what we might predict for reruns of evolution. Echolocation should repeatedly evolve using very high-pitched cries (for better resolution of detail than low-pitched). The cries in at least some species are likely to be frequency-modulated, sweeping down or up in pitch during the course of each cry (accuracy is improved because early parts of each echo are distinguishable from late parts by their pitch). The computational apparatus used to analyse the echoes might very well make (subconscious) calculations based on Doppler shifts in frequency of echoes, for the Doppler effect is certainly universally present on any planet where there is sound, and bats make sophisticated use of it.

How do we know that something like the eye or echolocation has evolved independently? By looking at the family tree. Relatives of

---

* *The Blind Watchmaker* in this case.

oilbirds and of cave swiftlets don't do echolocation. Oilbirds and cave swiftlets have separately taken up life in caves. We know they have evolved the technology independently of bats and whales, since nothing else in the surrounding family tree does it. Different groups of bats may have evolved echolocation more than once independently. We don't know how many more times echolocation evolved. Some shrews and seals have a rudimentary form of the skill (and some blind human individuals have learned it). Did pterodactyls do it? Since there is a good living to be made flying at night, and since bats weren't around in those days, it is not unlikely. The same goes for ichthyosaurs. They looked very like dolphins, and presumably made their living in a similar way. Since dolphins make heavy use of echolocation, it is reasonable to wonder whether ichthyosaurs did too, in the days before dolphins. There is no direct evidence, and we must remain open-minded. One point against: ichthyosaurs had extraordinarily big eyes – it is one of their most conspicuous features – which might suggest that they relied upon vision instead of echolocation. Dolphins have relatively small eyes, and one of *their* most conspicuous features, the rounded bump or 'melon' above the beak, acts as an acoustic 'lens', focusing sound into a narrow beam projected in front of the animal like a searchlight.

Like any zoologist, I can search my mental database of the animal kingdom and come up with an estimated answer to questions of the form: 'How many times has X evolved independently?' It would make a good research project, to do the counts more systematically. Presumably some Xs will come up with a 'many times' answer, as with eyes, or 'several times', as with echolocation. Others 'only once' or even 'never', although I have to say it is surprisingly difficult to find examples of these. And the differences could be interesting. I suspect that we'd find certain potential evolutionary pathways which life is 'eager' to go down. Other pathways have more 'resistance'. In *Climbing Mount Improbable,* I developed the analogy of a huge museum of all life, both real and conceivable, with corridors going off in many dimensions and representing evolutionary change, again both real and conceivable. Some of these corridors were wide open, almost beckoning. Others were blocked off by barriers that were hard or even impossible to surmount. Evolution repeatedly races down the easy corridors, and just occasionally, and unexpectedly, leaps one of the hard barriers. I'll return to the idea of 'eagerness' and 'reluctance' to evolve when I discuss the 'evolution of evolvability'.

Let's now go quickly through some more examples where it might be worth making a systematic count of how many times X has evolved. The venomous sting (injecting poison hypodermically through a sharp-pointed tube) has evolved at least ten times independently: in jellyfish and their relatives, spiders, scorpions, centipedes, insects,* molluscs (cone shells) snakes, the shark group (stingrays), bony fish (stonefish), mammals (male platypus) and plants (stinging nettles). It's a good bet that venom, including hypodermic injection, would evolve in reruns.

Sound production for social purposes has evolved independently in birds, mammals, crickets and grasshoppers, cicadas, fish and frogs. Electrolocation, the use of weak electric fields for navigation, has evolved several times, as we saw in the Duckbill's Tale. So has the – probably subsequent – use of electric currents as weapons. The physics of electricity is the same on all worlds, and we could bet with some confidence on repeated evolution of creatures that exploit electricity for both navigational and offensive purposes.

True flapping flight, as opposed to passive gliding, or parachuting, seems to have evolved four times: in insects, pterosaurs, bats and birds. Parachuting and gliding of various kinds have evolved many times, maybe hundreds of times independently, and may well be the evolutionary precursor to true flight. Examples include lizards, frogs, snakes, 'flying' fish, squids, colugos, marsupials and rodents (twice). I'd put a lot of money on gliders turning up in Kauffmanian reruns, and a reasonable sum on true flapping fliers.

Jet propulsion may have evolved twice. Cephalopod molluscs do it, at high speed in the case of squids. The other example I can think of is also a mollusc, but it is not high speed. Scallops mostly live on the sea bottom, but occasionally they swim. They rhythmically open and close their two shells, like a pair of snapping castanets. You'd think (I would) that this would propel them 'backwards' in a direction opposite to the snapping. In fact they move 'forwards', as though biting their way into the water. How can this be? The answer is that the snapping movements pump water through a pair of apertures *behind* the hinge. These two jets propel the animal 'forwards'. The effect is so counter-intuitive it is almost comical.

How about things that have evolved only once, or not at all? As we

---

* In the bees, wasps and ants, the sting is a modified egg-laying tube, and only females sting.

learned from the *Rhizobium*'s Tale, the wheel, with a true, freely rotating bearing, seems to have evolved only once, in bacteria, before being finally invented in human technology. Language, too, has apparently evolved only in us: that is to say at least 40 times less often than eyes. It is surprisingly hard to think of 'good ideas' that have evolved only once.

I put the challenge to my Oxford colleague the entomologist and naturalist George McGavin, and he came up with a nice list, but still a short one compared with the list of things that have evolved many times. Bombardier beetles of the genus *Brachinus* are unique, in Dr McGavin's experience, in mixing chemicals to make an explosion. The ingredients are made and held in separate (obviously!) glands. When danger threatens, they are squirted into a chamber near the rear end of the beetle, where they explode, forcing noxious (caustic and boiling-hot) liquid out through a directed nozzle at the enemy. The case is well known to creationists, who love it. They think it is self-evidently impossible to evolve by gradual degrees because the intermediate stages would all explode. I enjoyed demonstrating the error of this argument during my Royal Institution Christmas Lectures for Children, shown on BBC television in 1991. Donning a Second World War helmet, and inviting nervous members of the audience to leave, I mixed hydroquinone and hydrogen peroxide, the two ingredients of the bombardier explosion. Nothing happened. It didn't even get warm. The explosion requires a catalyst. I raised the concentration of catalyst gradually, which steadily increased the hot whoosh to a satisfactory climax. In nature, the beetle provides the catalyst, and would have had no difficulty in gradually, and safely, increasing the dose over evolutionary time.

Next in the McGavin list is the archer fish, family Toxotidae, which may be unique in shooting a missile to knock prey down from a distance. It comes to the surface of the water and spits a mouthful of water at a perched insect, knocking it down into the water, where it eats it. The other possible candidate for a 'knocking down' predator might be an ant lion. Ant lions are insect larvae of the order Neuroptera. Like many larvae, they look nothing like their adults. With their huge jaws, they would be good casting for a horror film. Each ant lion lurks in sand, just below the surface at the base of a conical pit trap which it digs itself. It digs by flicking sand vigorously outwards from the centre – this causes miniature landslides down the sides of the pit, and the laws of physics do the rest, neatly shaping the cone. Prey, usually ants, fall into the pit and slide down the steep sides into the ant lion's jaws. The possible point of resemblance to the archer fish is that prey don't fall only passively. They are sometimes knocked down into the pit by the particles of sand. These are not, however, aimed with the precision of an archer fish's spit, which is guided, with devastating accuracy, by binocularly focused eyes.

Spitting spiders, family Scytodidae, are a bit different again. Lacking the fleetness of a wolf spider or the net of a web spider, the spitting spider chucks a venomous glue some distance towards its prey, pinning it to the ground until the spider arrives and bites it to death. This is different from the archer fish technique of knocking prey down. Various animals, for example venom-spitting cobras, spit defensively, not to catch prey. The bolas spider, *Mastophora,* is different again, and is probably another unique case. It could be said to throw a missile at prey (moths, attracted by the fake sexual scent of a female moth, which the spider synthesises). But the missile, a blob of silk, is attached to a thread of silk which the spider whirls around like a lasso (or bola) and reels in. Chameleons could be said to spit a missile at prey. The missile is the heavy thickening at the tip of the tongue, and the (much thinner) rest of the tongue is a bit like the rope with which a harpoon is retrieved. The tip of the tongue is technically ballistic, meaning it is hurled free, unlike the tip of your tongue. Chameleons are not unique in this respect, however. Some salamanders also hurl the end of the tongue ballistically at prey, and their missile (but not the chameleon's) contains part of the skeleton. It is fired like a melon pip squeezed between your fingers.

McGavin's next candidate for an evolutionary one-off is a beauty. It is the diving bell spider, *Argyroneta aquatica.* This spider lives and hunts entirely underwater but, like dolphins, dugongs, turtles, freshwater snails and other land animals that have returned to water, it needs to breathe air. Unlike all those other exiles, however, *Argyroneta* constructs its own diving bell. It spins it of silk (silk is the universal solution to any spider problem) attached to an underwater plant. The spider goes to the surface to collect air, which it carries in the same way as some water bugs, in a layer trapped by body hairs. But unlike the bugs, which just carry the air like a scuba cylinder wherever they go, the spider takes it to its diving bell, where it unloads it to replenish the supply. The spider sits in the diving bell watching for prey, and it stores and eats prey there, once caught.

But George McGavin's champion example of a one-off is the larva of an African horsefly called *Tabanus*. Predictably in Africa, the pools of water in which the larvae live and feed eventually dry up. Each larva buries itself in the mud and pupates. The adult fly emerges from the baked mud and flies off to feed on blood, eventually to complete the cycle by laying eggs in pools of water when the rains return. The buried larva is vulnerable to a predictable danger. As mud dries out, it cracks, and there is a risk that a crack will tear right across the grub's refuge. It could theoretically save itself if it could somehow engineer a way for any crack that approaches it to be diverted around it instead. And it does indeed achieve this in a truly wonderful and probably unique manner. Before burying itself in its own pupation chamber, it first corkscrews its way down into the mud in a spiral. It then corkscrews its way back to the surface in an opposite spiral. Finally, it dives into the mud straight down the centre between the two spirals, and that is its resting place through the bad times until water returns. Now, you see what this means? The larva is encased in a cylinder of mud whose circular boundary has been weakened in advance by the preliminary spiral burrowing. This means that when a crack snakes across the drying mud, if it hits the edge of the cylindrical column, instead of cutting straight across the middle it goes instead in a curved bypass around the edge of the cylinder, and the larva is spared. It is just like the perforations around a stamp that stop you tearing the stamp across. Dr McGavin believes this ingenious trick is literally unique to this one genus of horsefly.*

But are there any good ideas that have *never* evolved by natural selection? As far as I know, no animal on this planet has ever evolved an organ for the transmission or reception of radio waves for long-distance communication. The use of fire is another example. Human experience shows how powerfully useful it can be. There are some plants whose seeds need fire to germinate, but I don't think that constitutes using it in the same sense as, for example, electric eels use electricity. The use of metal for skeletal purposes is another example of a good idea that has never evolved except in human artefacts. Presumably it is hard to achieve without fire.

This kind of comparative exercise, counting up which things evolve often, which seldom, when we do it alongside the geographical comparisons discussed earlier, might enable us to predict things about life outside this planet, as well as to guess the likely outcome

* The habit was first described by W. A. Lambourn [165].

of Kauffman-style thought experiments in rerunning evolution. We positively expect eyes, ears, wings and electric organs, but perhaps not bombardier beetle explosions or archer fish water bullets.

Those biologists who could be said to take their lead from the late Stephen Jay Gould regard all of evolution, including post-Cambrian evolution, as massively contingent – lucky, unlikely to be repeated in a Kauffman rerun. Calling it 'rewinding the tape of evolution', Gould independently evolved Kauffman's thought experiment. The chance of anything remotely resembling humans on a second rerun is widely seen as vanishingly small, and Gould voiced it persuasively in *Wonderful Life*. It was this orthodoxy that led me to the cautious self-denying ordinance of my opening chapter; led me, indeed, to undertake my backwards pilgrimage, and now leads me to forsake my pilgrim companions at Canterbury and return alone. And yet . . . I have long wondered whether the hectoring orthodoxy of contingency might have gone too far. My review of Gould's *Full House* (reprinted in *A Devil's Chaplain*) defended the unpopular notion of progress in evolution: not progress towards humanity – Darwin forfend! – but progress in directions that are at least predictable enough to justify the word. As I shall argue in a moment, the cumulative build-up of complex adaptations like eyes strongly suggests a version of progress – especially when coupled in imagination with some of the wonderful products of convergent evolution.

Convergent evolution also inspired the Cambridge geologist Simon Conway Morris, whose provocative book *Life's Solution: Inevitable Humans in a Lonely Universe* presents exactly the opposite case to Gould's 'contingency'. Conway Morris means his subtitle in a sense which is not far from literal. He really thinks that a rerun of evolution would result in a second coming of man: or something extremely close to man. And, for such an unpopular thesis, he mounts a defiantly courageous case. The two witnesses that he repeatedly calls are convergence and constraint.

Convergence we have met again and again through this book, including in this chapter. Similar problems call forth similar solutions, not just twice or three times but, in many cases, dozens of times. I thought I was pretty extreme in my enthusiasm for convergent evolution, but I have met my match in Conway Morris, who presents a stunning array of examples, many of which I had not met before. But whereas I usually explain convergence by invoking similar selection pressures, Conway Morris adds the testimony of his second witness, constraint. The materials of life, and the processes of embryonic development, allow only a limited range of solutions to a particular problem. Given any particular evolutionary starting situation, there is only a limited number of ways out of the box. So if two reruns of a Kauffman experiment encounter anything like similar selection pressures, developmental constraints will enhance the tendency to arrive at the same solution.

You can see how a skilled advocate could deploy these two witnesses in defence of the daring belief that a rerun of evolution would be positively likely to converge on a large-brained biped with two skilled hands, forward-pointing camera eyes and other human features. Unfortunately, it has only happened once on this planet, but I suppose there has to be a first time. I admit that I was impressed by Conway Morris's parallel case for the predictability of the evolution of insects.

Among the defining features of insects are the following: an articulated exoskeleton; compound eyes; a characteristic six-legged gait, whereby three of the six walking legs are always on the ground and thereby define a triangle (two legs on one side, one leg on the other) which keeps the animal stable; respiratory tubes known as tracheae that serve to bring oxygen into the interior of the animal via special openings (spiracles) along the side of the body; and, to complete the list of evolutionary peculiarities, the repeated (11 times independently!) evolution of complex eusocial colonies, as in the honey bees. All pretty strange? All one-offs in the great lottery of life? On the contrary, all are convergent.

Conway Morris goes through his list, showing that each item has evolved more than once in different parts of the animal kingdom, in many cases several times, including several times independently in insects themselves. If nature finds it so easy to evolve the component parts of insecthood separately, it is not all that implausible that the whole collection should evolve twice. I am tempted by Conway Morris's belief that we should stop thinking of convergent evolution as a colourful rarity to be remarked and marvelled at when we find it. Perhaps we should come to see it as a norm, exceptions to which are occasions for surprise. For example, true syntactic language seems to be unique to one species, our own. Perhaps – and I shall return to it – this is one thing that a re-evolved brainy biped would lack?

In my opening chapter, The Conceit of Hindsight, I listened to warnings against seeking patterns, rhymes or reasons in evolution, but said that I would cautiously flirt with them. The Host's Return has provided an opportunity to sweep over the whole course of evolution in the forward direction and see what patterns we can descry. The idea that all evolution was aimed at producing *Homo sapiens* was certainly well rejected, and nothing we have seen on our backwards journey reinstates it. Even Conway Morris claims only that something *approximately* similar to our *kind* of animal is one of several outcomes – others being insects, for example – that we would expect to see recurring if evolution were rerun again and again.

Value-free and Value-laden Progress

What other patterns or rhymes do we discern if we look over our long pilgrimage? Is evolution progressive? There is at least one reasonable definition of progress under which I would defend it. I need to work up to this. To begin with, progress can be defined in a weak, minimalist sense with no value judgement, as the predictable continuation into the future of trends from the past. The growth of a child is progressive in that whatever trends we observe in weight, height and other measurements over one year continue in the next. There is no value judgement in this weak definition of progress. Growth of a cancer is progressive in exactly the same weak sense. So also is shrinkage of a cancer under therapy. What, then, would *not* be progressive in the weak sense? Random, aimless fluctuation: the tumour grows a little, shrinks somewhat, grows a lot, shrinks a bit, grows a little, shrinks a lot, and so on. A progressive trend is one in which there are no reversals; or if there are reversals, they are outnumbered and outweighed by movement in the dominant direction. In a sequence of dated fossils, progress in this value-neutral sense would mean simply that whatever anatomical trend you see as you go from early to intermediate, is continued as a trend from intermediate to late.

I now need to clarify the distinction between value-neutral progress and value-laden progress. Progress in the weak sense just defined is value-neutral. But most people think of progress as value-laden. The doctor, reporting shrinkage of the tumour in response to chemotherapy, announces with satisfaction, 'We are making progress.' Doctors do not, on the whole, look at an X-ray of a swelling tumour with numerous secondaries and announce that the tumour is making progress, though they easily could. That would be value-laden but with negative value. 'Progressive', in human political or social affairs, usually refers to a trend in a direction which the speaker considers desirable. We look over human history and regard the following trends as progressive: abolition of slavery; widening of the franchise; reduction of discrimination by sex or race; reduction of disease and poverty; increase in public hygiene; reduction in atmospheric pollution; increase in education. A person of certain political views might see at least some of these trends as negatively value-laden, and yearn nostalgically for the days before women had the vote or were allowed into the club dining room. But the trends are still progressive in more than just the weak, minimalist, value-neutral sense we first defined. They are progressive according to some specified value system, even if it is not a value-system you or I would share.

Astonishingly, it is only a hundred years since the Wright brothers first achieved powered flight in a heavier-than-air machine. The history of aviation since 1903 has been unmistakeably progressive, and at amazing speed. Only 42 years later, in 1945, Hans Guido Mutke of the Luftwaffe broke the sound barrier in a Messerschmitt jet fighter.* Only 24 years further on, men walked on the moon. The fact that they no longer do so, and the fact that the only supersonic passenger service has just been discontinued, are economically-dictated temporary reversals in an overall trend that is unquestionably progressive. Aircraft are getting faster, and they are progressing in all sorts of other ways at the same time. Much of this progress does not chime with everyone's values – for example those unlucky enough to live under a flightpath. And much of the progress in aviation is driven by military needs. But nobody would deny the existence of a coherently expressible set of values, which at least some sane people might hold, according to which even fighters, bombers and guided missiles have progressively improved over the whole century since the Wrights. The same could be said of all other forms of transport, indeed other forms of technology, including, more than anything else, computers.

I must repeat that in calling this value-laden progress, I am not saying the values necessarily have a positive sign, either for you or for me. As I just remarked, much of the technological progress we are

---

* Mutke's claim is disputed. Either way, the first to do it wasn't USAF Major Chuck Yeager in 1947, as patriotic Americans are taught. An American civilian, George Welch, did it two weeks before him.

talking about is driven by, and contributes towards, military purposes. We could reasonably decide that the world was a better place before such inventions were made. In this sense 'progress' is value-laden with a negative sign. But it is still value-laden in an important sense, over and above my original value-free minimalist definition of progress as any trend into the future continued from the past. The development of weapons, from the stone to the spear through the longbow, the flintlock, musket, rifle, machine gun, shell, atomic bomb, through hydrogen bombs of ever increasing megatonnage, represents progress according to *somebody's* value system, even if not yours or mine – otherwise the research and development to produce them would not have been done.

Evolution exhibits progress not just in the weak, value-free sense. There are episodes of progress that are value-laden, according to at least some entirely plausible value systems. Since we are talking armaments, it is a good moment to note that the most familiar examples come out of arms races between predators and prey.

The first use of the 'arms race' listed in the *Oxford English Dictionary* is from *Hansard* (written proceedings of the House of Commons) for 1936:

> This House cannot agree to a policy which in fact seeks security in national armaments alone and intensifies the ruinous arms race between the nations, inevitably leading to war.

The *Daily Express* in 1937, under the headline 'Arms Race Worry', said, 'All were worried at the armament race.' It was not long before the theme found its way into the literature of evolutionary biology. Hugh Cott, in his classic *Adaptive Coloration in Animals,* published in 1940, deep in the Second World War, wrote:

> Before asserting that the deceptive appearance of a grasshopper or butterfly is unnecessarily detailed, we must first ascertain what are the powers of perception and discrimination of the insects' natural enemies. Not to do so is like asserting that the armour of a battlecruiser is too heavy, or the range of her guns too great,* without inquiring into the nature and effectiveness of the enemy's armament. The fact is that in the primeval struggle of the jungle, as in the refinements of civilized warfare, we see in progress a great evolutionary

---

* Tom Lehrer, probably the all-time wittiest composer of comic songs, included the following musical direction at the head of one of his piano scores: 'A little too fast.'

armament race – whose results, for defence, are manifested in such devices as speed, alertness, armour, spinescence, burrowing habits, nocturnal habits, poisonous secretions, nauseous taste, and procryptic, aposematic, and mimetic coloration; and for offence, in such counter-attributes as speed, surprise, ambush, allurement, visual acuity, claws, teeth, stings, poison fangs, and anticryptic and alluring coloration. Just as greater speed in the pursued has developed in relation to increased speed in the pursuer; or defensive armour in relation to aggresssive weapons; so the perfection of concealing devices has evolved in response to increased powers of perception.

My Oxford colleague John Krebs and I took up the whole matter of evolutionary arms races in a paper given at the Royal Society in 1979. We pointed out that the improvements to be seen in an animal arms race are improvements in equipment to survive, not generally improvements in survival itself – and for an interesting reason. In an arms race between attack and defence, there may be episodes during which one side or the other temporarily pulls ahead. But in general, improvements on one side cancel out improvements on the other. There is even something a bit paradoxical about arms races. They are economically costly for both sides, yet there is no net benefit to either, because potential gains on one side are neutralised by gains on the other. From an economic point of view, both sides would be better off coming to an agreement to call off the arms race. As a ludicrous extreme, prey species might sacrifice a tithe of their number in exchange for secure and untroubled grazing for the rest. Neither predators nor prey need to divert valuable resources into muscles for fast running, sensory systems for detection of enemies, vigilance and prolonged hunts that are time-wasting and stressful for both sides. Both sides would benefit if such a trades union agreement could be reached.

Unfortunately, Darwinian theory knows no route by which this could happen. Instead, both sides pour resources into competing with their own side to outrun the other, and individuals of both are forced into difficult economic trade-offs within their own bodily economies. If there were no predators, rabbits could devote all their economic resources, and all their valuable time, to feeding and reproducing more rabbits. Instead, they are forced to devote substantial time to looking out for predators, and substantial economic resources into building up escape equipment. In turn, this forces predators to shift the balance of their economic investment away from the central business of reproducing, and into improving their weaponry for catching prey. Arms races, in animal evolution and human technology alike, show themselves not in improved performance but in increased shifting of economic investment away from alternative aspects of life and into servicing the arms race itself.

Krebs and I recognised asymmetries in arms races that might result in one side shifting more economic resources into the arms race than the other. One such imbalance we dubbed the 'Life/Dinner Principle'. It takes its name from the Aesop Fable in which the rabbit runs faster than the fox because the rabbit is running for his life, while the fox is only running for his dinner. There is an asymmetry in the cost of failure. In the arms race between cuckoos and hosts, every individual cuckoo can confidently look back on an unbroken line of ancestors who literally never failed to fool a foster parent. An individual of the host species, on the other hand, can look back on ancestors, many of whom never even met a cuckoo, and many of whom met one and were fooled by it. Plenty of genes for failing to detect and kill cuckoos have passed successfully down the generations of the host species. But genes that cause cuckoos to fail in fooling hosts have a much more hazardous ride down the generations. This asymmetry of risk fosters another: an asymmetry in resources devoted to the arms race as opposed to other parts of life's economy. To repeat this important point, the cost of failure is harder on the cuckoos than on the hosts. This leads to asymmetries in how the two sides set their balance between competing calls on their time and other economic resources.

Arms races are deeply and inescapably progressive in a way that, for example, evolutionary accommodation to the weather is not. For an individual of any one generation, predators and parasites just make life harder in pretty much the same way as bad weather does. But over evolutionary time there is a crucial difference. Unlike the weather, which fluctuates aimlessly, predators and parasites (and prey and hosts) are themselves evolving in a systematic direction, getting systematically worse from their victims' point of view. Unlike evolutionary tracking of ice ages and droughts, arms race trends from the past can be extrapolated into the future, and those trends are value-laden in the same kind of way as technological improvements in planes and weapons. Predators' eyes get sharper, though not necessarily more effective, because prey get harder to see. Running speeds increase progressively on both sides, though again the benefits are in general cancelled out by parallel improvements on the other side. Sabre teeth get sharper and longer as hides get tougher. Toxins get nastier as biochemical tricks for neutralising them improve.

With the passing of evolutionary time, the arms race progresses. All the features of life that a human engineer would admire as complex and elegant become more complex, more elegant, and more redolent of the illusion of design.* In *Climbing Mount Improbable* I distinguished designed from 'designoid' (pronounced design-oid, not dezzig-noid). Spectacular feats of designoid engineering, such as the eye of a buzzard, the ear of a bat, the musculoskeletal apparatus of a cheetah or a gazelle, are all climactic products of evolutionary arms races between predators and prey. Parasite/host arms races culminate in even more finely meshed, co-adaptive designoid climaxes.

And now for an important point. The evolution of any complex designoid organ in an arms race must have come about in a large number of steps of progressive evolution. Such evolution qualifies as progressive by our definition because each change tends to continue the direction of its predecessors. How do we know there are many steps rather than just one or two? By elementary probability theory. The parts of a complex machine, such as a bat's ear, could be rearranged at random in a million ways before you hit another arrangement that could hear as well as the real thing. It is statistically improbable, not just in the boring sense that any particular arrangement of parts is as improbable, with hindsight, as any other. Very few permutations of atoms are precision auditory instruments. A real bat's ear is one in a million. It works. Something so statistically improbable cannot sensibly be explained as the result of a single stroke of luck. It has to be constructed by some sort of improbability-generating process, ratcheted up by what the philosopher Daniel Dennett calls a 'crane' (as opposed to a 'skyhook': the analogy is to the man-made lifting machine, not the bird). The only cranes known to science (and I would bet the only cranes there have ever been, or ever will be, in the universe) are design and selection. Design explains the efficient complexity of microphones. Natural selection explains the efficient complexity of bat ears. Ultimately, selection explains

---

* Hume said: 'All these various machines, and even their most minute parts, are adjusted to each other with an accuracy which ravishes into admiration all men who have ever contemplated them.'

microphones and everything designed too because the designers of microphones are themselves evolved engineers generated by natural selection. Ultimately, design cannot explain anything because there is an inevitable regression to the problem of the origin of the designer.

Design and natural selection are both processes of gradual, step-by-step, progressive improvement. Natural selection, at least, could not be anything else. In the case of design it may or may not be a matter of principle, but it is an observed fact. The Wright brothers did not have a blinding flash of inspiration and promptly build a *Concorde* or a *Stealth* bomber. They built a creaking, rickety crate that barely lifted off the ground and lurched into a neighbouring field. From Kitty Hawk to Cape Canaveral, every step of the way was built on its predecessors. Improvement is gradual, step by step in the same continued direction, fulfilling our definition of progressive. We could, with difficulty, conceive of a Victorian genius designing a Sidewinder missile fully formed within his Zeusian, side-whiskered head. The notion defies all common sense and all history, but it does not instantly fall foul of the laws of probability in the way we would have to say of the spontaneous evolution of a flying, echolocating, modern bat.

A single macromutational leap from ground-dwelling ancestral shrew to flying, echolocating bat is ruled out just as safely as we can rule out luck when a conjuror successfully guesses the complete order of a shuffled pack of cards. Luck is not literally impossible in either case. But no good scientist would advance such prodigious luck as an explanation. The card-guessing feat has to be a trick – we've all seen tricks that appear just as baffling to the uninitiated. Nature does not set out to fool us, as a conjuror does. But we can still rule out luck, and it was the genius of Darwin to rumble nature's sleight of hand. The echo-ranging bat is the result of an inching series of minor improvements, each adding cumulatively to its predecessors as it propels the evolutionary trend on in the same direction. That is progress, by definition. The argument applies to all complex biological objects that project the illusion of design and are therefore statistically improbable in a specified direction. All must have evolved progressively.

The returning host, now unabashedly sensitive to major themes in evolution, notes progress as one of them. But progress of this kind is not a uniform, inexorable trend from the start of evolution all the way to the present. Rather, to take up the initial quotation from

Mark Twain on history, it rhymes. We notice an episode of progress during the course of an arms race. But that particular arms race comes to an end. Perhaps one side is driven extinct by the other. Or both sides go extinct, maybe in the course of a mass catastrophe of the kind that did for the dinosaurs. Then the whole process starts again, not from scratch, but from some discernibly earlier part of the arms race. Progress in evolution is not a single upward climb but has a rhyming trajectory more like the teeth of a saw. A sawtooth plunged deeply at the end of the Cretaceous, when the last of the dinosaurs abruptly gave way to the mammals' new and spectacular climb of progressive evolution. But there had been lots of smaller sawteeth during the long reign of the dinosaurs. And since their immediate post-dinosaur rise, the mammals too have had smaller arms races followed by extinctions, followed by renewed arms races. Arms races rhyme with earlier arms races in periodic spurts of many-stepped progressive evolution.

Evolvability

That's all I want to say about arms races as drivers of progress. What other messages from the past does the returning host carry back to the present? Well, I must mention the alleged distinction between macroevolution and microevolution. I say 'alleged' because my own view is that macroevolution (evolution on the grand scale of millions of years) is simply what you get when microevolution (evolution on the scale of individual lifetimes) is allowed to go on for millions of years. The contrary view is that macroevolution is something qualitatively different from microevolution. Neither view is self-evidently silly. Nor are they necessarily contradictory. As so often, it depends on what you mean.

Again we can use the parallel of the growth of a child. Imagine an argument about an alleged distinction between macrogrowth and microgrowth. To study macrogrowth, we weigh the child every few months. Every birthday we stand her up against a white doorpost and draw a pencil line to record her height. More scientifically, we could measure various parts of the body, for example the diameter of the head, the width of the shoulders, the length of the major limb bones, and plot them against each other, perhaps logarithmically transformed for the reasons given in the Handyman's Tale. We also note significant events in development such as the first appearance of pubic hair, or the first sign of breasts and menstruation in girls, and of facial hair in boys. These are the changes that constitute macro-growth, and we measure them on a timescale of years or months. Our instruments are not sensitive enough to pick up the daily and hourly changes in the body – microgrowth – which, when summed over months, constitute macrogrowth. Or, oddly, they may be too sensitive. A very accurate weighing machine could in theory pick up hourly growth, but the delicate signal is swamped by blundering increases in weight with every meal, and decreases with every act of elimination. The acts of microgrowth itself, which all consist of cell divisions, make no immediate impact on weight at all, and an undetectable impact on gross body measurements.

So, is macrogrowth the sum of lots of small episodes of micro-growth? Yes. But it is also true that the different timescales impose completely different methods of study and habits of thought. Micro-scopes looking at cells are not appropriate for the study of child development at the whole-body level. And weighing machines and measuring tapes are not suitable for the study of cell multiplication. The two timescales in practice demand radically different methods of study and habits of thought. The same could be said of macroevolu-tion and microevolution. If the terms are used to signify differences in how best to study them, I have no quarrel with a working distinction between microevolution and macroevolution. I do have a quarrel with those people who elevate this rather mundane practi-cal distinction into one of almost – or more than almost – mystical import. There are those who think Darwin's theory of evolution by natural selection explains microevolution, but is in principle im-potent to explain macroevolution, which consequently needs an extra ingredient – in extreme cases a *divine* extra ingredient!

Unfortunately, this hankering after skyhooks has been given aid and comfort by real scientists whose intentions are innocent of any such thing. I have discussed the theory of 'punctuated equilibrium' before, too often and too thoroughly to repeat myself in this book,* so I shall only add that its advocates usually go on to propose a fundamental 'decoupling' between microevolution and macroevolu-tion. This is an unwarranted inference. No extra ingredient needs to be added at the micro level to explain the macro level. Rather, an extra level of explanation *emerges* at the macro level as a *consequence*

---

* My view is that it is an interesting empirical question, which is likely to have a different answer in different particular cases, and which does not deserve its elevation to the status of major principle.

of events at the micro level, extrapolated over unimaginable timespans.

The working distinction between micro- and macroevolution is similar to one that we meet in many other situations. The changes in the map of the world over geological time are due to the effects, summed over millions of years, of plate tectonic events occurring on a timescale of minutes, days and years. But, as with the growth of a child, there is practically no overlap between the methods of study that serve the two timescales. The language of voltage fluctuations is not useful for discussing how a large computer program, such as Microsoft Excel, works. No sensible person denies that computer programs, however complicated, are entirely executed by temporal and spatial patterns of changes between two voltages. But no sensible person attends to that fact while writing, debugging, or using a large computer program.

I have never seen any good reason to doubt the following proposition: macroevolution is lots of little bits of microevolution joined end to end over geological time, and detected by fossils instead of by genetic sampling. Nevertheless, there could be – and I believe are – major events in evolutionary history after which the very nature of evolution itself changes. Evolution itself might be said to evolve. So far in this chapter, progress has meant individual organisms becoming better over evolutionary time at doing what individuals do, which is survive and reproduce. But we can also countenance changes in the phenomenon of evolution itself. Might evolution itself become better at doing something – what evolution does – as history goes by? Is late evolution some kind of improvement on early evolution? Do creatures evolve to improve not just their capacity to survive and reproduce, but the lineage's capacity to evolve? Is there an evolution of evolvability?

I coined the phrase 'Evolution of Evolvability' in a paper published in the *Proceedings of the 1987 Inaugural Conference on Artificial Life*. Artificial life was a newly invented merger of other disciplines, notably biology, physics and computer science, founded by the visionary physicist Christopher Langton, who edited the *Proceedings*. Since my paper, but probably not because of it, the evolution of evolvability has become a much discussed topic among students of both biology and artificial life. Long before I used the phrase, others had proposed the idea. For example, the American ichthyologist Karel F. Liem in 1973 used the phrase 'prospective adaptation' for the revolutionary jaw apparatus of cichlid fishes which enabled them, as their tale describes, so suddenly and explosively to evolve hundreds of species in all the great African lakes. Liem's suggestion goes beyond the idea of pre-adaptation, I should say. A pre-adaptation is something that originally evolves for one purpose and is co-opted to another. Liem's prospective adaptation and my evolution of evolvability carry the suggestion not just of co-option to a new function but of unshackling a new outburst of divergent evolution. I am suggesting a permanent and even progressive trend towards becoming better at evolving.

In 1987, the idea of the evolution of evolvability was somewhat heretical, especially for me as the alleged 'ultra-Darwinist'. I was placed in the odd situation of advocating an idea at the same time as apologising for it to people who couldn't see why it needed any apology. It is now a much discussed topic, and others have taken it further than I ever envisaged, for example, the cell biologists Marc Kirschner and John Gerhart, and the evolutionary entomologist Mary Jane West-Eberhard in her magisterial book *Developmental Plasticity and Evolution*.

What makes an organism good at evolving, over and above being good at surviving and reproducing? An example, first. We have already met the idea that island archipelagos are workshops of speciation. If the islands are near enough to each other to allow occasional immigrations, but far enough apart to allow time for evolutionary divergence between immigrations, we have a recipe for speciation, which is the first step towards evolutionary radiation. But how near is near enough? How far is far enough? That depends on the locomotive powers of the animals. For woodlice, a separation of a few yards is equivalent to a separation of many miles for a flying bird or bat. The Galapagos Islands are spaced just right for the divergent evolution of small birds such as Darwin's finches, not necessarily for divergent evolution generally. For this purpose, separation of islands should be measured not in absolute units but in units of travelability calibrated to the kind of animal we are talking about – as with the Irish boatman who, when my parents asked him the distance of the Great Blasket Island, replied, 'About three miles in fine weather.'

It follows that a Galapagos finch which either decreased *or increased* its flight range in evolution might thereby *decrease* its evolvability. Shortening the range lowers the chance of initiating a new race of descendants on another island. That way round, it is easily understood. Lengthening the range has a less obvious effect in the same direction. Descendants are seeded onto new islands so frequently that there is no time for separate evolution before the next immigrant arrives. To push to the extreme, birds whose flight range is large enough to render the distance between islands trivial, no longer see the islands as separate at all. As far as gene flow is concerned, the whole archipelago counts as one continent. So once again speciation is not fostered. High evolvability, if we choose to measure evolvability as speciation rate, is an inadvertent consequence of intermediate locomotor range, where what counts as intermediate, as opposed to too short or too long, depends upon the spacing of the islands concerned. Of course 'island' in this sort of argument does not have to mean land surrounded by water. As we saw in the Cichlid's Tale, lakes are islands for aquatic animals, and reefs can be islands within lakes. Mountain tops are islands for land-bound animals that cannot easily tolerate low altitudes. A tree can be an island for an animal with a short range. For the AIDS virus, every man is an island.

If an increase or a decrease in travelling range results in an increase in evolvability, would we want to call this an evolved 'improvement'? My ultra-Darwinist hackles start to quiver at this point. My heresy litmus starts to blush. It sounds uncomfortably like evolutionary foresight. Birds evolve an increase or a decrease to their flying range because of natural selection for individual survival. Future effects on evolution are an irrelevant consequence. Nevertheless, we might find with hindsight that the species that fill the world tend to be descended from ancestral species with a talent for evolution. You could say, therefore, that there is a kind of high-level, between-lineage selection in favour of evolvability – an example of what the great American evolutionist George C. Williams called clade selection. Conventional Darwinian selection leads to individual organisms being finely tuned survival machines. Could it be that, as a consequence of clade selection, life itself has increasingly become a set of finely tuned evolving machines? If this is so, we might expect that in Kauffmanian reruns of evolution, the same progressive improvements in evolvability might be rediscovered.

When I first wrote about the evolution of evolvability, I proposed a number of 'watershed events' in evolution, after which evolvability suddenly improved. The most promising example of a watershed event I could think of was segmentation. Segmentation, you remember, is the train-like modularisation of the body, in which parts and systems are repeated serially down the body. It seems to have been independently invented, in its full form, in arthropods, vertebrates and annelid worms (although the universality of Hox genes argues for some sort of fore and aft serial organisation as a predecessor). The origin of segmentation is one of those evolutionary events that cannot have been gradual. Bony fish typically have about 50 vertebrae, but eels have as many as 200. Caecilians (worm-like amphibians) vary between 95 and 285 vertebrae. Snakes differ hugely in vertebral number: the record known to me is 565 for an extinct snake.

Every vertebra of a snake represents one segment with its own pair of ribs, its own muscle blocks, its own nerves sprouting from the spinal cord. You can't have fractional numbers of segments, and the evolution of variable segment numbers must include numerous instances in which a mutant snake differed from its parents in some whole number of segments: at least one, possibly more, in one fell swoop. Similarly, when segmentation originated, there must have been a mutational transition straight from unsegmented parents to a child with two (at least) segments. It is hard to imagine such a freak surviving, let alone finding a mate and reproducing, but it evidently happened because segmented animals are all around us. Very probably the mutation involved Hox genes, like those of the Fruit Fly's Tale. In my 1987 evolvability paper I guessed that

> . . . the individual success, or otherwise, of the first segmented animal during its own lifetime is relatively unimportant. No doubt many other new mutants have been more successful as individuals. What is important about the first segmented animal is that its descendant lineages were champion *evolvers*. They radiated, speciated, gave rise to whole new phyla. Whether or not segmentation was a beneficial adaptation during the individual lifetime of the first segmented animal, segmentation represented a change in embryology that was pregnant with evolutionary potential.

The ease with which whole segments can be added or subtracted from the body is one thing that contributes to enhanced evolvability. So is differentiation among segments. In animals such as millipedes and earthworms, most of the segments are the same as each other. But there is a recurrent tendency, especially among arthropods and vertebrates, for particular segments to become specialised for particular purposes, and hence different from other segments (compare a lobster with a centipede). A lineage that manages to evolve a segmented body plan is immediately able to evolve a whole range of new animals by altering segmental modules, all along the body.

Segmentation is an example of modularity, and modularity in general is a main ingredient in the thinking of more recent writers on the evolution of evolvability. Of the many meanings of module listed in the *Oxford English Dictionary*, the relevant one is:

> One of a series of production units or component parts that are standardized to facilitate assembly or replacement and are usually prefabricated as selfcontained structures.

Modular is the adjective describing an assemblage of modules, and modularity is the corresponding abstract noun, the property of being modular. Other examples of modular construction include many plants (leaves and flowers are modules). But perhaps the best examples of modularity are to be found at the cellular and biochemical level. Cells themselves are modules *par excellence,* and within cells so are protein molecules and, of course, DNA itself.

So the invention of multicellularity is another important watershed event that almost certainly enhanced evolvability. It preceded segmentation by hundreds of millions of years, and segmentation is itself a kind of large-scale reenactment of it, another leap in modularity. What other watersheds have there been? The dedicatee of this book, John Maynard Smith, collaborated with his Hungarian colleague Eörs Szathmàry on *The Major Transitions in Evolution.* Most of their 'major transitions' would fit under my heading of 'watershed events' – major improvements in evolvability. This obviously includes the origin of replicating molecules, for without them there could be no evolution at all. If, as Cairns-Smith and others have suggested, DNA usurped the key role of replicator from some less proficient predecessor, bridged by intermediate stages, each one of those stages would constitute a leap forward in evolvability.

If we accept the RNA World theory, there would have been a major transition or watershed, when a world of RNA serving as both replicator and enzyme gave over to a separation between DNA in the replicator role and proteins in the enzyme role. Then there was the clubbing together of replicating entities ('genes') in cells with walls, which prevented the gene products leaking away and kept them together with the products of other genes with which they could collaborate in cellular chemistry. A very major transition, and very probably a watershed of evolvability, was the birth of the eukaryotic cell by the commingling of several prokaryotic cells. So was the origin of sexual reproduction, which coincided with the origin of the species itself, with its own gene pool, and all that that implied for future evolution. Maynard Smith and Szathmàry go on to list the origin of multicellularity, the origin of colonies such as ant and termite nests, and the origin of human societies with language. There is a rhyming similarity between at least several of these major transitions: they often involve the coming together of previously independent units in a larger grouping at a higher level, with concomitant loss of independence at the lower level.

To their list I have already added segmentation, and I would stress another, which I call bottlenecking. Once again, to spell it out in full would be to repeat from previous books (especially the final chapter, 'Rediscovering the Organism', of The Extended Phenotype). Bottlenecking refers to a type of life history in multicellular organisms. In bottlenecking, the life cycle regularly returns to a single cell, from which a multicellular body is grown anew. The alternative to a bottlenecked life cycle might be a hypothetical straggling water plant that reproduces by breaking off small, multicellular chunks of itself which drift off, grow and then break off more small chunks. Bottlenecking has three important consequences, all of which are certainly good candidates for improvements in evolvability.

First, evolutionary innovations can be reinvented from the bottom up, rather than as remouldings of existing structures – the equivalent of beating swords into ploughshares. An improvement in, say, a heart, has a better chance of being a clean improvement if genetic changes can alter the whole course of development from a single cell. Imagine the alternative: take the existing heart and modify it by differential tissue growth within its continuously beating fabric. This on-the-trot remodelling would impair the working of the heart and compromise the would-be improvement.

Second, by continually resetting to a consistent starting point in a recurrent life cycle, bottlenecking provides a 'calendar' by which embryological events may be timed. Genes may be turned on or off at key points in the growth cycle. Our hypothetical straggling chunk-extruder lacks a recognisable timetable to regulate such switchings on and off.

Third, without bottlenecking, different mutations would accumulate in different parts of the straggling chunk-extruder. The incentive among cells to co-operate would be reduced. In effect, sub-populations of cells would be tempted to behave as cancers, to increase their chance of contributing genes to the extruded chunks. With bottlenecking, since every generation starts out as a single cell, the whole body has a good chance of being made of a uniform genetic population of co-operating cells, all descended from that single cell. Without bottlenecking, the cells of the body might have, from a genetic point of view, 'divided loyalties'.

Related to bottlenecking is another important landmark event in evolution, and one that may well have contributed to evolvability and might be rediscovered in Kauffman reruns. This is the separation of the germ line from the soma, first clearly understood by the great German biologist August Weismann. As we saw at Rendezvous 31, what happens in the developing embryo is that a portion of the cells are set aside for reproduction (germ-line cells) while the rest are destined to make the body (somatic cells). Germ-line genes are potentially immortal, with a prospect of direct descendants millions of years into the future. Somatic genes are destined for a finite, if not always predictable, number of cell divisions to make the body tissues, after which their line will come to an end and the organism will die. Plants often violate the separation, most obviously when they practise vegetative reproduction. This could constitute an important difference between the ways plants and animals evolve. Before the evolutionary invention of the separate soma, all living cells were potentially the ancestors of an indefinite line of descendants, as sponge cells still are.

The invention of sex is a major watershed, which is superficially confusable both with bottlenecking and with the separation of the germ line, but is logically distinct from both. In its most general form, sex is the partial mixing of genomes. We are familiar with a particular, highly regimented version of it in which every individual gets 50 per cent of its genome from each of two parents. We are used to the idea that there are two kinds of parent, female and male, but this is not a necessary part of sexual reproduction. Isogamy is a system in which two individuals, not distinguished as male and female, combine half their genes to make a new individual. The male/female divide is best seen as a further watershed event, which came after the origin of sex itself. Regimented sex of this kind is accompanied in every generation by a 'reduction division', in which each individual donates 50 per cent of its genome to each offspring.

Without this reduction, genomes would double in size with every generation.

Bacteria practise a haphazard form of sexual donation that is sometimes described as sex but which is really very different, having more in common with the cut-and-paste, or copy-and-paste, functions of a computer program. Fragments of a genome are copied or cut from one bacterium and pasted into another, which does not have to be a member of the same 'species' (though the very meaning of species is in doubt for bacteria). Because genes are software subroutines that perform cellular operations, a 'pasted' gene can immediately go to work in its new environment, doing the same task as it did before.*

What is in it for the donating bacterium? That may be the wrong question. The right question might be, what is in it for the donated gene? And the answer is that genes that successfully get themselves donated, and then successfully help the recipient bacterium to survive and pass them on, thereby increase the number of copies of themselves in the world. It is not clear whether our regimented eukaryotic sex has evolved from bacterial 'cut-and-paste' sex, or whether it was an entirely new watershed event. Both must have had a huge impact on subsequent evolution and are candidates for discussion under the heading of the evolution of evolvability. Regimented sex, as we saw in the Rotifer's Tale, has a dramatic effect upon future evolution because it makes possible the very existence of species with their gene pools.

The positioning of the apostrophe in *The Ancestor's Tale* indicates a singular. I admit that the motive was partly stylistic. Nevertheless, through the millions – probably billions – of individual ancestors whose lives we touched along our Pilgrims' Way, one singular hero has recurred in the minor, like a Wagnerian leitmotiv: DNA. Eve's Tale showed that genes have ancestors, no less than individuals. The Neanderthal's Tale applied the lesson to the question of whether that maligned species perished without any legacy to soften the blow. The

---

* This is why transgenic manipulation in modern agricultural breeding works, for example, the legendary importing of 'antifreeze' genes from Arctic fish into tomatoes. It works for the same reason as a computer subroutine, copied from one program into another, can be relied upon to deliver the same result. The case of GM crops isn't quite so straightforward. But the example serves to allay fears of the 'unnaturalness' of importing, say, fish genes into tomatoes, as though some kind of fishy 'flavour' goes too. A subroutine is a subroutine, and DNA's language of programming is identical in fish and tomatoes.

Gibbon's Tale warmed to the theme of 'majority votes' among genes clamouring to assert their different views of ancestral history. The Lamprey's Tale identified the analogy between gene duplication and speciation, each at its own level – an analogy so close that separate family trees can be drawn up for genes, which parallel, but do not coincide with, the conventional family trees of phylogeny. The leitmotiv in the field of taxonomy echoes, but is distinct from, the major theme of the 'selfish gene' in the understanding of natural selection.

The Host's Farewell

If, as returning host, I reflect on the whole pilgrimage of which I have been a grateful part, my overwhelming reaction is one of amazement. Amazement not only at the extravaganza of details that we have seen; amazement, too, at the very fact that there are any such details to be had at all, on any planet. The universe could so easily have remained lifeless and simple – just physics and chemistry, just the scattered dust of the cosmic explosion that gave birth to time and space. The fact that it did not – the fact that life evolved out of nearly nothing, some 10 billion years after the universe evolved out of literally nothing – is a fact so staggering that I would be mad to attempt words to do it justice. And even that is not the end of the matter. Not only did evolution happen: it eventually led to beings capable of comprehending the process, and even of comprehending the process by which they comprehend it.

This pilgrimage has been a trip, not just in the literal sense but in the countercultural sense I met when a young man in California in the 1960s. The most potent hallucinogen on sale in Haight or Ashbury or Telegraph Avenue would be tame by comparison. If it's amazement you want, the real world has it all. Not to stray outside the covers of this book, think of Venus's girdle, migrating jellyfish and tiny harpoons; think of the platypus's radar and the electric fish; of the horsefly larva with the apparent foresight to pre-empt cracks in the mud; think redwood; think peacock; think starfish with its piped hydraulic power; think cichlids of Lake Victoria, evolving *how* many orders of magnitude faster than *Lingula, Limulus* or *Latimeria*? It is not pride in my book but reverence for life itself that encourages me to say, if you want a justification for the latter, open the former anywhere, at random. And reflect on the fact that although this book has been written from a human point of view, another book could have been written in parallel for any of 10 million starting pilgrims. Not only is life on this planet amazing, and deeply satisfying, to all whose senses have not become dulled by familiarity: the very fact that we have evolved the brain power to understand our evolutionary genesis redoubles the amazement and compounds the satisfaction.

'Pilgrimage' implies piety and reverence. I have not had occasion here to mention my impatience with traditional piety, and my disdain for reverence where the object is anything supernatural. But I make no secret of them. It is not because I wish to limit or circumscribe reverence; not because I want to reduce or downgrade the true reverence with which we are moved to celebrate the universe, once we understand it properly. 'On the contrary' would be an understatement. My objection to supernatural beliefs is precisely that they miserably fail to do justice to the sublime grandeur of the real world. They represent a narrowing-down from reality, an impoverishment of what the real world has to offer.

I suspect that many who call themselves religious would find themselves agreeing with me. To them I would only quote a favourite remark that I overheard at a scientific conference. A distinguished elder statesman of my subject was having a long argument with a colleague. As the altercation came to an end, he twinkled and said, 'You know, we really do agree. It's just that you *say* it wrong!'

I feel I have returned from a true pilgrimage.

FURTHER READING

Numbers in square brackets refer to sources listed in the Bibliography.

Barlow, George (2002) *The Cichlid Fishes: Nature's Grand Experiment in Evolution.* Perseus Publishing, Cambridge, Mass.

Diamond, Jared (1997) *Guns, Germs and Steel: A Short History of Everybody for the Last 13,000 Years.* Chatto & Windus, London.

Fortey, Richard (1997) *Life: An Unauthorised Biography.* HarperCollins, London.

Fortey, Richard (2004) *The Earth: An Intimate History.* HarperCollins, London.

Leakey, Richard (1994) *The Origin of Humankind: Unearthing Our Family Tree.* Science Masters series, Basic Books, New York.

Maynard Smith, John & Szathmàry, Eörs (1999). *The Origins of Life: From the Birth of Life to the Origin of Language.* Oxford University Press, Oxford. (*See also* [189] for a more detailed treatment.)

Quammen, David (1996) *The Song of the Dodo: Island Biogeography in an Age of Extinctions.* Hutchinson, Oxford.

Ridley, Mark (2000) *Mendel's Demon: Gene Justice and the Complexity of Life.* Weidenfeld & Nicolson, London.

Ridley, Matt (1999) *Genome: The Autobiography of a Species in 23 Chapters.* Fourth Estate, London.

Southwood, Richard (2003) *The Story of Life.* Oxford University Press, Oxford.

Tudge, Colin (2000) *The Variety of Life: A Survey and a Celebration of all the Creatures that Have Ever Lived.* Oxford University Press, Oxford.

Weiner, Jonathan (1994) *The Beak of the Finch: A Story of Evolution in Our Time.* Jonathan Cape, London.

Wilson, E. O. (1992) *The Diversity of Life.* Harvard University Press, Cambridge, Mass.

ADVANCED READING

Brusca, Richard C. & Brusca, Gary J. (2002) *Invertebrates.* 2nd ed. Sinauer Associates Inc, Sunderland, Mass.

Carroll, Robert L. (1988) *Vertebrate Paleontology and Evolution.* W. H. Freeman, New York.

Macdonald, David (2001) *The New Encyclopedia of Mammals.* Oxford University Press, Oxford.

Ridley, Mark (2004) *Evolution.* 3rd ed. Blackwell, Oxford.

NOTES TO THE PHYLOGENIES AND RECONSTRUCTIONS

Yan Wong

Numbers in square brackets refer to sources listed in the Bibliography.

PHYLOGENY DIAGRAMS

The following notes outline the scientific basis for the phylogenies in this book, particularly in areas of major recent taxonomic revision and current debate. A good, relatively recent phylogenetic survey is given in Colin Tudge's *The Variety of Life* [289].

RENDEZVOUS 0 The Americas are omitted because evidence points to humans having arrived there recently from Asia. Concestor 0 must logically be at least as recent as any gene MRCA (such as Y-chromosome 'Adam'), and even low levels of interbreeding are enough to result in a very recent MRCA of all humans [45], hence the recent date used here.

RENDEZVOUS 1 & 2 Phylogeny (as for the rest of the trees, the majority 'vote' among genes – *see* the Gibbon's Tale) supported by morphology [102] and molecules [20]. Divergence dates based on the molecular clock [105, 230].

RENDEZVOUS 3 Phylogeny and divergence dates based on morphological, fossil, and molecular data [102, 105, 273].

RENDEZVOUS 4 Gibbon phylogeny is unsure: this tree is based upon mtDNA data [246, fig 2c], supplemented by molecular clock dates for the Concestor and *Symphalangus/Hylobates* nodes [105].

RENDEZVOUS 5 Conventional phylogeny. Divergence dates given by molecular and fossil data [105].

RENDEZVOUS 6 Phylogeny and dates taken directly or inferred from [105]. The position of the Aotinae is not very secure, and may change in the future.

RENDEZVOUS 7 Placement and dating [105] of the tarsier family agrees with molecular [254] and morphological data.

RENDEZVOUS 8 Within strepsirhines, lemur interrelationships are disputed, although the aye-aye is often considered basal. Order and dating of the four other families is from molecules [322], scaled to place basal primate divergence at 63 Mya [105, 207]. However, other calculations place this divergence at 80 Mya [281], moving Rendezvous 9, 10, and 11 backwards by up to 15 million years.

RENDEZVOUS 9 Placement of colugos and tree shrews is highly controversial (see accompanying tale), and is here based on recent molecular data [207]. Basal date then constrained by surrounding nodes to 63–75 Mya.

RENDEZVOUS 10 Placement of Glires from robust molecular evidence [207]. Rendezvous date constrained by molecular clock dating of Rendezvous 11 [207, 137], but may be up to 10 Mya or earlier [271]. Lagomorph placement uncontroversial [137, 207]. Rodent phylogeny debated. Hystricognath rodents (Hystricidae, Phiomorpha, Caviomorpha) generally accepted. Otherwise, 4 groups often found in molecular studies [e.g. 137, 202]: Muridae+Dipodidae, Aplodontidae+Sciuridae+Gliridae, Ctenodactylidae+hystricognaths, Heteromyidae+Geomyidae. Branching order and rough dating of these groups from mtDNA and rDNA [202], but order is not robust [e.g. *see* 137].

RENDEZVOUS 11 & 12 Phylogeny and dating from recent revolutionary molecular studies [207, 271].

RENDEZVOUS 13 Phylogeny and dating from molecular data [207, 271]. Morphology [177] and some molecules [205] agree on elephant/sirenian/hyrax split. However, there is uncertainty in the placement of the aardvark [205, 271], and morphological data may still conflict with the position of the Afrosoricida [177].

RENDEZVOUS 14 Rendezvous supported by old and recent data [208]. Placental-marsupial divergence at 140 Mya consistent with fossils and late molecular dates [7, 144]. Molecular studies find didelphids, then paucituberculates as sister to other marsupials [212, 272], consistent with morphology [251]. Other branches variably supported by molecular data [212, 272]: position of monito del monte particularly uncertain, here interpreted as sister to Diprotodontia [251]. Divergence dates based on molecular clock data, but also constrained by Gondwanan biogeography [212].

RENDEZVOUS 15 Phylogeny and dating from recent molecular, morphological, and fossil data [208].

RENDEZVOUS 16 Date estimates for Rendezvous 16 average around 310 Mya [112], other early branch dates from fossil data [40]. Now-conventional branching within snakes and lizards [228]. Bird branching order from genetic studies [293] with dates from DNA hybridisation [265]: many orders grouped as Neoaves due to uncertain relationships.

RENDEZVOUS 17 Although disputed by some palaeontologists [40], molecular and morphological data strongly support lissamphibian monophyly, and hint at order of branching shown here [325]. Basal date from palaeontological evidence [4], others from maximum likelihood trees of mtDNA [325].

RENDEZVOUS 18 & 19 Phylogeny and dating from molecular [294] and morphological/palaeontological [326] studies.

RENDEZVOUS 20 Rendezvous date generally accepted [209]. Ray-finned fish phylogeny is currently in a state of flux [141, 199], although the traditional view followed here [209] is broadly supported. Divergence dates based on fossil data [40, 209]. Some groups deliberately omitted for simplicity, as phylogeny not robust.

RENDEZVOUS 21 Phylogeny based on morphological data [75, 263] [263]. Divergence dates based on fossil data [209, 252].

RENDEZVOUS 22 Agnathan grouping based on genetic data [97, 279] which contradicts most fossil-based phylogenies (but these specialised groups show secondarily character lost, making morphological data difficult to use). Rendezvous date tightly constrained by fossil data [264]. Lamprey-hagfish divergence time suggested by molecular maximum likelihood trees [279].

RENDEZVOUS 23 Molecular clock data [315] places lancelet split close to basal deuterostome divergences here, estimated at 570 Mya according to medium-fuse dating of Cambrian Explosion (see the Velvet Worm's Tale).

RENDEZVOUS 24 Rendezvous date constrained by surrounding nodes. Possibly closer to ambulacrarians than to lancelets [315].

RENDEZVOUS 25 Ambulacrarian grouping and basal divergences from recent genetic data [32, 97, 315], assuming medium fuse Cambrian explosion. Genetic studies also give deep-branching Xenoturbellida [28], although exact placement not robust. Echinoderm phylogeny and dating from genetic, morphological, and fossil data [176, 297].

RENDEZVOUS 26 Rendezvous date (about 590 Mya) from recent molecular clock studies [8, 10], and broadly consistent with fossil data [291]. Protostome phylogeny recently revised [3]: here a single broad scheme has been followed [103], based on genetics and morphology. Three branches consist of several phyla grouped together. These are: Cephalorhyncha [103], Gnathifera [162] (including Acanthocephala and Myzostomida), and Brachiozoa (phoronids and brachiopods). Edysozoan phylogeny relatively robust [103]: main uncertainties are the onychophore/arthropod grouping and basal inclusion of chaetognaths, here placed according to morphological/genetic data [224]. Many ecdysozoan dates constrained by 'small-shelly' onychophore fossils (see the Velvet Worm's Tale). Lophotrochozoa branching order much more uncertain: annelid/mollusc/sipunculid group robust [224], nemerteans probably sister to this [290], branching order of others unsure.

RENDEZVOUS 27 Phylogeny based on molecular data [247, 283]. These often weakly support a paraphyletic Acoela, but morphological data strongly supports acoelomorph monophyly; divergence date thus arbitrary. Rendezvous date based on genetic distance estimates [247, 283], assuming protostome/deuterostome split of 590 Mya and bilaterian/cnidarian split of 700 Mya.

RENDEZVOUS 28 & 29 Order of branching of cnidarians and ctenophores is still uncertain [35]. Certain molecular data weakly support the order used here [191]. Within cnidarian phylogeny now conventional, dates from genetic studies [50] calibrated to timescale used here.

RENDEZVOUS 30 Trichoplax placement unsure [35], but possibly near the base of the Metazoa [Peter Holland, pers. comm.].

RENDEZVOUS 31 Sponges generally interpreted as basal metazoans, although occasionally molecular data hint that they might be paraphyletic [191]. Rendezvous date of 800 Mya based on molecular clock data [211], recalibrated using protostomedeuterostome divergence of 590 Mya; this conflicts with absence of fossilised sponge spicules before the latest Precambrian, although these may represent a derived character.

RENDEZVOUS 32 & 33 Rendezvous dates roughly estimated from molecular trees [166, 191], assuming Rendezvous 31 at 800 and 34 at 1100 Mya. Position of Mesomycetozoea (Ichthyosporea) [231] based on mtDNA sequences [166], rather than (less extensive) rRNA [191].

RENDEZVOUS 34 Rendezvous date of roughly 1100 Mya commonly argued [91, 244] (but may not be particularly robust). Revised molecular studies now place Microsporidia in Fungi [149], possibly at the base [13]. Morphology and genetics place Ascomycota and Basidiomycota as closest relatives, rDNA additionally identifies Glomeromycota as sister to both [256], with previous 'zygomycetes' two (as shown here), or more paraphyletic branches. Divergence dates from molecular clock [133] rescaled to fit rendezvous date used here.

RENDEZVOUS 35 Grouping of most amoebas and slime moulds as sister group to Metazoa+Fungi has substantial molecular support [13, 43], although unconventional rooting of the eukaryotic tree may collapse Rendezvous 34, 35, 36, and 37 into one [43]. Divergence date arbitrarily placed halfway between two surrounding nodes.

RENDEZVOUS 36 Ribosomal RNA data grouping plants with animals and fungi now recognised as erroneous [13, 113]. As explained in the text of Rendezvous 37, the position of the plants in the eukaryotic phylogeny is uncertain, and the scheme adopted here is somewhat arbitrary. Rendezvous date constrained by 1200 Mya fossils [38, but see 42]: 1300 Mya broadly consistent with molecular clock studies [e.g. 91]. Within plants, phylogeny and relative dates from molecular data [203], although inclusion of red algae sometimes disputed [214].

RENDEZVOUS 37 Branching order and divergence dates of major eukaryote groups uncertain [13] (hence polytomy shown). Ribosomal RNA studies erroneously place different groups as early branching lineages due to long branch attraction; amended trees only able to place eukaryotic branches far from the Archaea [113], implying much later divergence than Rendezvous 38: dates of Rendezvous 37–39 estimated to nearest 500 My.

RENDEZVOUS 38 Rendezvous date uncertain; molecular clock data suggests roughly 2 billion years ago [e.g. 91, but see 42]. Divergence dates and (conventional) phylogeny estimated from rRNA studies [e.g. 16].

RENDEZVOUS 39 Tree inherently difficult to root because there is no outgroup, and changes in mutation rate along different lineages obscure the 'centre' of the tree. It is often rooted between Archaea and Eubacteria (cross A), but other possibilities exist [42] (cross B), [113], and so is presented unrooted. Changes in rooting will affect overall branch lengths, so these cannot truly represent time and are thus somewhat arbitrary. Eubacterial phylogeny based on robust biochemical characteristics (e.g. cell wall glyco-proteins) and rare genomic events (e.g. indels) [42, 117]; rRNA trees can have long branch attraction problems, but indicate that the divergences within the bacteria are deep [113]. Bacterial DNA exchange problematic for building a single tree, unless a core of unswapped genes exists [64].

CONCESTOR RECONSTRUCTIONS
Concestor reconstructions by Malcolm Godwin.

Reconstructions are intended to give a general impression of the probable appearance and habitat of each concestor, based on current scientific know-ledge. Non-skeletal features (e.g. colour of fur or skin) are inevitably a matter of considerable conjecture. Henry Bennett-Clark, Tom Cavalier-Smith, Hugh Dickinson, William Hawthorne, Peter Holland, Tom Kemp, Anna Nekaris, Marcello Ruta, Mark Sutton, and Keith Thomson provided various advice for the reconstructions. However, they bear no responsibility for the final pictures: any errors in interpretation are solely my responsibility.

Concestor 3 Large arboreal quadrupedal ape [20], which probably lived in Asia [273]. The face protrudes less than in orang utans, with rounder, more widely-spaced orbits (inferred from the Miocene ape *Ankarapithe-cus*). Forelimbs are suspensory, although less so than in orang utans; locomotion similar to the proboscis monkey *Nasalis*. Note also the brow ridges, prominent glabella, relatively high degree of encephalisation, pre-dominantly fruit-based diet, and (relative to gibbons and Old World monkeys) the enlarged mammary glands and more bowed radius bone [116].

Concestor 18 Informed by the Lower Devonian rhipidistian *Styloichthys* [326]. Note the fin lobation, the headshield, the lateral line, and the heterocercal tail.

Concestor 23 Similar to lancelets, but notochord does not reach the rostrum, and specialised wheel organ absent. Note the pigment spot eye, gill bars, notochord, myomeres (V-shaped muscle blocks) and atrium (enclosed space below the main body).

Concestor 31 Thought to have been a hollow ball consisting of outward-pointing choanocyte cells [248] (similar to a sponge embryo). Cilia used for locomotion and for wafting food particles into choanocyte 'collar'. Note also cellular specialisation: sexual reproduction is via egg cells and free-swimming sperm. Concestor reconstructed with a pelagic lifestyle, similar to sponge embryos.

Concestor 36 Typical single-celled eukaryote, hence with a pervasive microtubular cytoskeleton, cilium (eukaryotic 'flagella') associated with a centriole (basal body) acting as a microtubule organising centre, a nucleus with pore structure surrounded by perforated sheets of rough ER which graduate into the cytosol, and a grainy appearance caused by tiny ribosomes. Note also the mitochondria with tubular cristae, small numbers of peroxisomes and other cellular vesicles, and movement via a combination of cilium and short pseudopods. Concestor depicted engulfing a food particle (note localised cytoskeleton build-up).

BIBLIOGRAPHY

Numbers in square brackets are used in the captions, notes and index for reference.

[1] Adams, D. (1987) *Dirk Gently's Holistic Detective Agency*. William Heinemann, London.

[2] Adams, D. & Carwardine, M. (1991) *Last Chance to See*. Pan Books, London, 2nd edn.

[3] Aguinaldo, A. M. A., Turbeville, J. M., Linford, L. S., *et al.* (1997) Evidence for a clade of nematodes, arthropods and other moulting animals. *Nature* **387**: 489–493.

[4] Ahlberg, P. E. & Milner, A. R. (2000/1994) The origin and early diversification of tetrapods. In *Shaking the Tree: Readings from Nature in the History of Life* (Gee, H., ed.), University of Chicago Press, Chicago (Originally published *Nature* **368**: 507–514).

[5] Alexander, R. D., Hoogland, J. L., Howard, R. D., *et al.* (1979) Sexual dimorphisms and breeding systems in pinnipeds, ungulates, primates, and humans. In *Evolutionary Biology and Human Social Behavior: An Anthropological Perspective* (Chagnon, N. A. & Irons, W., eds.), pp. 402–435, Duxbury Press, North Scituate, Mass.

[6] *Arabian Nights, The* (1885) (Burton, R. F., trans.). The Kamashastra Society, Benares.

[7] Archibald, J. D. (2003) Timing and biogeography of the eutherian radiation: Fossils and molecules compared. *Molecular Phylogenetics and Evolution* **28**: 350–359.

[8] Aris-Brosou, S. & Yang, Z. (2003) Bayesian models of episodic evolution support a late Precambrian explosive diversification of the Metazoa. *Molecular Biology and Evolution* **20**: 1947–1954.

[9] Arrese, C. A., Hart, N. S., Thomas, N., *et al.* (2002) Trichromacy in Australian marsupials. *Current Biology* **12**: 657–660.

[10] Ayala, F. J., Rzhetsky, A., & Ayala, F. J. (1998) Origin of the metazoan phyla: Molecular clocks confirm paleontological estimates. *Proceedings of the National Academy of Sciences of the USA* **95**: 606–611.

[11] Bada, J. L. & Lazcano, A. (2003) Prebiotic soup – revisiting the Miller experiment. *Science* **300**: 745–746.

[12] Bakker, R. (1986) *The Dinosaur Heresies: A Revolutionary View of Dinosaurs*. Longman Scientific and Technical, Harlow.

[13] Baldauf, S. L. (2003) The deep roots of eukaryotes. *Science* **300**: 1703–1706.

[14] Baldwin, J. M. (1896) A new factor in evolution. *American Naturalist* **30**: 441–451.

[15] Barlow, G. W. (2002) *The Cichlid Fishes: Nature's Grand Experiment in Evolution.* Perseus Publishing, Cambridge, Mass.

[16] Barns, S. M., Delwiche, C. F., Palmer, J. D., & Pace, N. R. (1996) Perspectives on archaeal diversity, thermophily and monophyly from environmental rRNA sequences. *Proceedings of the National Academy of Sciences of the USA* **93**: 9188–9193.

[17] Bateson, P. P. G. (1976) Specificity and the origins of behavior. In *Advances in the Study of Behavior* (Rosenblatt, J., Hinde, R. A., & Beer, C., eds.), vol. 6, pp. 1–20, Academic Press, New York.

[18] Bateson, W. (1894) *Materials for the Study of Variation Treated with Especial Regard to Discontinuity in the Origin of Species.* Macmillan and Co, London.

[19] Bauer, M. & von Halversen, O. (1987) Separate localization of sound recognizing and sound producing neural mechanisms in a grasshopper. *Journal of Comparative Physiology A* **161**: 95–101.

[20] Begun, D. R. (1999) Hominid family values: Morphological and molecular data on relations among the great apes and humans. In *The Mentalities of Gorillas and Orangutans* (Parker, S. T., Mitchell, R. W., & Miles, H. L., eds.), chap. 1, pp. 3–42, Cambridge University Press, Cambridge.

[21] Bell, G. (1982) *The Masterpiece of Nature: The Evolution and Genetics of Sexuality.* Croom Helm, London.

[22] Belloc, H. (1999) *Complete Verse.* Random House Children's Books, London.

[23] Betzig, L. (1995) Medieval monogamy. *Journal of Family History* **20**: 181–216.

[24] Blackmore, S. (1999) *The Meme Machine.* Oxford University Press, Oxford.

[25] Blair, W. F. (1955) Mating call and stage of speciation in the *Microhyla olivacea – M. carolinensis* complex. *Evolution* **9**: 469–480.

[26] Bloch, J. I. & Boyer, D. M. (2002) Grasping primate origins. *Science* **298**: 1606–1610.

[27] Bonner, J. T. (1993) *Life Cycles: Reflections of an Evolutionary Biologist.* Princeton University Press, Princeton, N.J.

[28] Bourlat, S. J., Nielsen, C., Lockyer, A. E., *et al.* (2003) *Xenoturbella* is a deuterostome that eats molluscs. *Nature* **424**: 925–928.

[29] Brasier, M. D., Green, O. R., Jephcoat, A. P., *et al.* (2002) Questioning the evidence for earth's oldest fossils. *Nature* **416**: 76–81.

[30] Briggs, D., Erwin, D., & Collier, F. (1994) *The Fossils of the Burgess Shale.* Smithsonian Institution Press, Washington, D.C.

[31] Briggs, D. E. G. & Fortey, R. A. (in press) Wonderful strife – systematics, stem groups and the phylogenetic signal of the Cambrian radiation. *Paleobiology.*

[32] Bromham, L. & Degnan, B. M. (1999) Hemichordates and deuterostome evolution: Robust molecular phylogenetic support for a hemichordate + echinoderm clade. *Evolution and Development* **1:** 166–171.

[33] Bromham, L. & Penny, D. (2003) The modern molecular clock. *Nature Reviews Genetics* **4:** 216–224.

[34] Bromham, L., Woolfit, M., Lee, M. S. Y., & Rambaut, A. (2002) Testing the relationship between morphological and molecular rates of change along phylogenies. *Evolution* **56:** 1921–1930.

[35] Brooke, N. M. & Holland, P. W. H. (2003) The evolution of multicellularity and early animal genomes. *Current Opinion in Genetics & Development* **13:** 599–603.

[36] Brunet, M., Guy, F., Pilbeam, D., *et al.* (2002) A new hominid from the Upper Miocene of Chad, central Africa. *Nature* **418:** 145–151.

[37] Buchsbaum, R. (1987) *Animals Without Backbones.* University of Chicago Press, Chicago, 3rd edn.

[38] Butterfield, N. J. (2001) Paleobiology of the late Mesoproterozoic (ca. 1200 Ma) Hunting Formation, Somerset Island, arctic Canada. *Precambrian Research* **111:** 235–256.

[39] Cairns-Smith, A. G. (1985) *Seven Clues to the Origin of Life.* Cambridge University Press, Cambridge.

[40] Carroll, R. L. (1988) *Vertebrate Paleontology and Evolution.* W. H. Freeman, New York.

[41] Catania, K. C. & Kaas, J. H. (1997) Somatosensory fovea in the star-nosed mole: Behavioral use of the star in relation to innervation patterns and cortical representation. *Journal of Comparative Neurology* **387:** 215–233.

[42] Cavalier-Smith, T. (2002) The neomuran origin of archaebacteria, the negibacterial root of the universal tree and bacterial megaclassification. *International Journal of Systematic and Evolutionary Microbiology* **52:** 7–76.

[43] Cavalier-Smith, T. & Chao, E. E. Y. (2003) Phylogeny of Choanozoa, Apusozoa, and other Protozoa and early eukaryote megaevolution. *Journal of Molecular Evolution* **56:** 540–563.

[44] Censky, E. J., Hodge, K., & Dudley, J. (1998) Overwater dispersal of lizards due to hurricanes. *Nature* **395:** 556.

[45] Chang, J. T. (1999) Recent common ancestors of all present-day individuals. *Advances in Applied Probability* **31:** 1002–1026.

[46] Chaucer, G. (2000) *Chaucer: The General Prologue on CD-ROM* (Solopova, E., ed.). Cambridge University Press, Cambridge.

[47] Clack, J. (2002) *Gaining Ground: The Origin and Evolution of Tetrapods.* Indiana University Press, Bloomington.

[48] Clarke, R. J. (1998) First ever discovery of a well-preserved skull and associated skeleton of *Australopithecus. South African Journal of Science* **94:** 460–463.

[49] Cleveland, L. R. & Grimstone, A. V. (1964) The fine structure of the flagellate Mixotricha paradoxa and its associated micro-organisms. *Proceedings of the Royal Society of London: Series B* **159:** 668–686.

[50] Collins, A. G. (2002) Phylogeny of medusozoa and the evolution of cnidarian life cycles. *Journal of Evolutionary Biology* **15:** 418–432.

[51] Conway-Morris, S. (1998) *The Crucible of Creation: The Burgess Shale and the Rise of Animals.* Oxford University Press, Oxford.

[52] Conway-Morris, S. (2003) *Life's Solution: Inevitable Humans in a Lonely Universe.* Cambridge University Press, Cambridge.

[53] Cooper, A. & Fortey, R. (1998) Evolutionary explosions and the phylogenetic fuse. *Trends in Ecology and Evolution* **13:** 151–156.

[54] Coppens, Y. (1994) East side story: The origin of humankind. *Scientific American* **271** (May): 88–95.

[55] Cott, H. B. (1940) *Adaptive Coloration in Animals.* Methuen, London.

[56] Crick, F. H. C. (1981) *Life Itself: Its Origin and Nature.* Macdonald, London.

[57] Crockford, S. (2002) *Dog Evolution: A Role for Thyroid Hormone Physiology in Domestication Changes.* Johns Hopkins University Press, Baltimore.

[58] Cronin, H. (1991) *The Ant and the Peacock: Altruism and Sexual Selection from Darwin to Today.* Cambridge University Press, Cambridge.

[59] Darwin, C. (1860/1859) *On The Origin of Species by Means of Natural Selection.* John Murray, London.

[60] Darwin, C. (1987/1842) *The Geology of the Voyage of HMS Beagle: The Structure and Distribution of Coral Reefs.* New York University Press, New York.

[61] Darwin, C. (2002/1839) *The Voyage of the Beagle.* Dover Publications, New York.

[62] Darwin, C. (2003/1871) *The Descent of Man.* Gibson Square Books, London.

[63] Darwin, F. (ed.) (1888) *The Life And Letters of Charles Darwin.* John Murray, London.

[64] Daubin, V., Gouy, M., & Perrière, G. (2002) A phylogenomic approach to bacterial phylogeny: evidence for a core of genes sharing common history. *Genome Research* **12:** 1080–1090.

[65] Davies, P. (1998) *The Fifth Miracle: The Search for the Origin of Life.* Allen Lane, The Penguin Press, London.

[66] Dawkins, R. (1982) *The Extended Phenotype.* W. H. Freeman, Oxford.

[67] Dawkins, R. (1986) *The Blind Watchmaker.* Longman, London.

[68] Dawkins, R. (1989) The evolution of evolvability. In *Artificial Life* (Langton, C., ed.), pp. 201–220, Addison-Wesley, New York.

[69] Dawkins, R. (1989) *The Selfish Gene.* Oxford University Press, Oxford, 2nd edn.

[70] Dawkins, R. (1995) *River Out of Eden.* Weidenfeld & Nicolson, London.

[71] Dawkins, R. (1996) *Climbing Mount Improbable.* Viking, London.

[72] Dawkins, R. (1998) *Unweaving the Rainbow.* Penguin, London.

[73] Dawkins, R. (2003) *A Devil's Chaplain.* Weidenfeld & Nicolson, London.

[74] Dawkins, R. & Krebs, J. R. (1979) Arms races between and within species. *Proceedings of the Royal Society of London: Series B* **205**: 489–511.

[75] de Carvalho, M. R. (1996) Higher-level elasmobranch phylogeny, basal squaleans, and paraphyly. In *Interrelationships of Fishes* (Stiassny, M. L. J., Parenti, L. R., & Johnsson, G. D., eds.), pp. 35–62, Academic Press, San Diego.

[76] de Morgan, A. (2003/1866) *A Budget of Paradoxes.* The Thoemmes Library, Poole, Dorset.

[77] de Waal, F. (1995) Bonobo sex and society. *Scientific American* **272** (March): 82–88.

[78] de Waal, F. (1997) *Bonobo: The Forgotten Ape.* University of California Press, Berkeley.

[79] Dennett, D. (1991) *Consciousness Explained.* Little, Brown, Boston.

[80] Dennett, D. (1995) *Darwin's Dangerous Idea: Evolution and the Meaning of Life.* Simon & Schuster, New York.

[81] Deutsch, D. (1997) *The Fabric of Reality.* Allen Lane, The Penguin Press, London.

[82] Diamond, J. (1991) *The Rise and Fall of the Third Chimpanzee.* Radius, London.

[83] Dixon, D. (1981) *After Man: A Zoology of the Future.* Granada, London.

[84] Drayton, M. (1931–1941) *The Works of Michael Drayton.* Blackwell, Oxford.

[85] Dudley, J. W. & Lambert, R. J. (1992) Ninety generations of selection for oil and protein in maize. *Maydica* **37**: 96–119.

[86] Dulai, K. S., von Dornum, M., Mollon, J. D., & Hunt, D. M. (1999) The evolution of trichromatic color vision by opsin gene duplication in New World and Old World primates. *Genome Research* **9**: 629–638.

[87] Durham, W. H. (1991) *Coevolution: Genes, Culture and Human diversity.* Stanford University Press, Stanford.

[88] Dyson, F. J. (1999) *Origins of Life.* Cambridge University Press, Cambridge, 2nd ed.

[89] Edwards, A. W. F. (2003) Human genetic diversity: Lewontin's fallacy. *BioEssays* **25**: 798–801.

[90] Eigen, M. (1992) *Steps Towards Life: A Perspective on Evolution.* Oxford University Press, Oxford.

[91] Feng, D.-F., Cho, G., & Doolittle, R. F. (1997) Determining divergence times with a protein clock: Update and reevaluation. *Proceedings of the National Academy of Sciences of the USA* **94**: 13028–13033.

[92] Ferrier, D. E. K. & Holland, P. W. H. (2001) Ancient origin of the Hox gene cluster. *Nature Reviews Genetics* **2**: 33–38.

[93] Ferrier, D. E. K., Minguillón, C., Holland, P. W. H., & Garcia-Fernàndez, J. (2000) The amphioxus Hox cluster: Deuterostome posterior flexibility and Hox14. *Evolution and Development* 2: 284–293.

[94] Fisher, R. A. (1999/1930) *The Genetical Theory of Natural Selection: A Complete Variorum Edition.* Oxford University Press, Oxford.

[95] Fogle, B. (1993) *101 Questions Your Dog Would Ask Its Vet.* Michael Joseph, London.

[96] Fortey, R. (1997) *Life: An Unauthorised Biography: A Natural History of the First Four Thousand Million Years of Life on Earth.* HarperCollins, London.

[97] Furlong, R. F. & Holland, P. W. H. (2002) Bayesian phylogenetic analysis supports monophyly of Ambulacraria and of cyclostomes. *Zoological Science* 19: 593–599.

[98] Furnes, H., Banerjee, N. R., Muehlenbachs, K., *et al.* (2004) Early life recorded in Archean pillow larvas. *Science* 304: 578–581.

[99] Garstang, W. (1951) *Larval forms and other zoological verses by the late Walter Garstang* (Hardy, A. C., ed.). Blackwell, Oxford.

[100] Geissmann, T. (2002) Taxonomy and evolution of gibbons. *Evolutionary Anthropology* 11, **Supplement** 1: 28–31.

[101] Georgy, S. T., Widdicombe, J. G., & Young, V. (2002) The pyrophysiology and sexuality of dragons. *Respiratory Physiology & Neurobiology* 133: 3–10.

[102] Gibbs, S., Collard, M., & Wood, B. (2002) Soft-tissue anatomy of the extant hominoids: a review and phylogenetic analysis. *Journal of Anatomy* 200: 3–49.

[103] Giribet, G. (2002) Current advances in the phylogenetic reconstruction of metazoan evolution. A new paradigm for the Cambrian explosion? *Molecular Phylogenetics and Evolution* 24: 345–357.

[104] Gold, T. (1992) The deep, hot biosphere. *Proceedings of the National Academy of Sciences of the USA* 89: 6045–6049.

[105] Goodman, M., Porter, C. A., Czelusniak, J., *et al.* (1998) Toward a phylogenetic classification of primates based on DNA evidence complemented by fossil evidence. *Molecular Phylogenetics and Evolution* 9: 583–598.

[106] Gould, S. J. (1977) *Ontogeny and Phylogeny.* The Belknap Press of Harvard University Press, Cambridge, Mass.

[107] Gould, S. J. (1985) *The Flamingo's Smile: Reflections in Natural History.* Norton, New York.

[108] Gould, S. J. (1989) *Wonderful Life: The Burgess Shale and the Nature of History.* Hutchinson Radius, London.

[109] Gould, S. J. & Calloway, C. B. (1980) Clams and brachiopods: ships that pass in the night. *Paleobiology* 6: 383–396.

[110] Grafen, A. (1990) Sexual selection unhandicapped by the Fisher process. *Journal of Theoretical Biology* 144: 473–516.

[111] Grant, P. R. (1999/1986) *Ecology and Evolution of Darwin's Finches.* Princeton University Press, Princeton, N.J., revised ed.

[112] Graur, D. & Martin, W. (2004) Reading the entrails of chickens: Molecular timescales of evolution and the illusion of precision. *Trends in Genetics* **20:** 80–86.

[113] Gribaldo, S. & Philippe, H. (2002) Ancient phylogenetic relationships. *Theoretical Population Biology* **61:** 391–408.

[114] Gribbin, J. & Cherfas, J. (1982) *The Monkey Puzzle.* The Bodley Head, London.

[115] Gribbin, J. & Cherfas, J. (2001) *The First Chimpanzee: In Search of Human Origins.* Penguin, London.

[116] Groves, C. P. (1986) Systematics of the great apes. In *Systematics, Evolution, and Anatomy* (Swindler, D. R. & Erwin, J., eds.), *Comparative Primate Biology,* vol. 1, pp. 186–217, Alan R. Liss, New York.

[117] Gupta, R. S. & Griffiths, E. (2002) Critical issues in bacterial phylogeny. *Theoretical Population Biology* **61:** 423–434.

[118] Hadzi, J. (1963) *The Evolution of the Metazoa.* Pergamon Press, Oxford.

[119] Haeckel, E. (1866) *Generelle Morphologie der Organismen.* Georg Reimer, Berlin.

[120] Haeckel, E. (1899–1904) *Kunstformen der Natur.*

[121] Haig, D. (1993) Genetic conflicts in human pregnancy. *The Quarterly Review of Biology* **68:** 495–532.

[122] Haldane, J. B. S. (1952) Introducing Douglas Spalding. *British Journal for Animal Behaviour* **2:** 1.

[123] Haldane, J. B. S. (1985) *On Being the Right Size and Other Essays* (Maynard Smith, J., ed.). Oxford University Press, Oxford.

[124] Halder, G., Callaerts, P., & Gehring, W. J. (1995) Induction of ectopic eyes by targeted expression of the eyeless gene in *Drosophila. Science* **267:** 1788–1792.

[125] Hallam, A. & Wignall, P. B. (1997) *Mass Extinctions and their Aftermath.* Oxford University Press, Oxford.

[126] Hamilton, W. D. (2001) *Narrow Roads of Gene Land,* vol. 2. Oxford University Press, Oxford.

[127] Hamilton, W. D. (in prep.) *Narrow Roads of Gene Land,* vol. 3. Oxford University Press, Oxford.

[128] Hamrick, M. W. (2001) Primate origins: Evolutionary change in digital ray patterning and segmentation. *Journal of Human Evolution* **40:** 339–351.

[129] Harcourt, A. H., Harvey, P. H., Larson, S. G., & Short, R. V. (1981) Testis weight, body weight and breeding system in primates. *Nature* **293:** 55–57.

[130] Hardy, A. (1965) *The Living Stream.* Collins, London.

[131] Hardy, A. C. (1954) The escape from specialization. In *Evolution as a Process* (Huxley, J., Hardy, A. C., & Ford, E. B., eds.), Allen and Unwin, London, 1st edn.

[132] Harvey, P. H. & Pagel, M. D. (1991) *The Comparative Method in Evolutionary Biology*. Oxford University Press, Oxford.

[133] Heckman, D. S., Geiser, D. M., Eidell, B. R., *et al.* (2001) Molecular evidence for the early colonization of land by fungi and plants. *Science* **293:** 1129–1133.

[134] Heesy, C. P. & Ross, C. F. (2001) Evolution of activity patterns and chromatic vision in primates: Morphometrics, genetics and cladistics. *Journal of Human Evolution* **40:** 111–149.

[135] Home, E. (1802) A description of the anatomy of the *Ornithorhynchus paradoxus*. *Philosophical Transactions of the Royal Society of London* **92:** 67–84.

[136] Hou, X.-G., Aldridge, R. J., Bergstrom, J., *et al.* (2004) *The Cambrian Fossils of Chengjiang, China: The Flowering of Early Animal Life*. Blackwell Science, Oxford.

[137] Huchon, D., Madsen, O., Sibbald, M. J. J. B., *et al.* (2002) Rodent phylogeny and a timescale for the evolution of Glires: Evidence from an extensive taxon sampling using three nuclear genes. *Molecular Biology and Evolution* **19:** 1053–1065.

[138] Hume, D. (1957/1757) *The Natural History of Religion* (Root, H. E., ed.). Stanford University Press, Stanford.

[139] Huxley, A. (1939) *After Many a Summer*. Chatto and Windus, London.

[140] Huxley, T. H. (2001/1836) *Man's Place in Nature*. Random House USA, New York.

[141] Inoue, J. G., Masaki, M., Tsukamoto, K., & Nishida, M. (2003) Basal actinopterygian relationships: A mitogenomic perspective on the phylogeny of the 'ancient fish'. *Molecular Phylogenetics and Evolution* **26:** 110–120.

[142] Jeffery, W. R. & Martasian, D. P. (1998) Evolution of eye regression in the cavefish *Astyanax*: Apoptosis and the *Pax-6* gene. *American Zoology* **38:** 685–696.

[143] Jerison, H. J. (1973) *Evolution of the Brain and Intelligence*. Academic Press, New York.

[144] Ji, Q., Luo, Z.-X., Yuan, C.-X., *et al.* (2002) The earliest known eutherian mammal. *Nature* **416:** 816–822.

[145] Johanson, D. C. & Edey, M. A. (1981) *Lucy: The Beginnings of Humankind*. Grenada, London.

[146] Jones, S. (1993) *The Language of the Genes: Biology, History, and the Evolutionary Future*. HarperCollins, London.

[147] Judson, O. (2002) *Dr. Tatiana's Sex Advice to all Creation*. Metropolitan Books, New York.

[148] Kauffman, S. A. (1985) Self-organization, selective adaptation, and its limits. *In Evolution at a Crossroads* (Depew, D. J. & Weber, B. H., eds.), pp. 169–207, MIT Press, Cambridge, Mass.

[149] Keeling, P. J. & Fast, N. M. (2002) Microsporidia: Biology and evolution of highly reduced intracellular parasites. *Annual Review of Microbiology* **56:** 93–116.

[150] Kemp, T. S. (1982) *Mammal-like reptiles and the origin of mammals.* Academic Press, London.

[151] Kemp, T. S. (1982) The reptiles that became mammals. *New Scientist* **93:** 581–584.

[152] Kimura, M. (1994) *Population Genetics, Molecular Evolution and the Neutral Theory* (Takahata, N., ed.). University of Chicago Press, Chicago.

[153] Kingdon, J. (1990) *Island Africa.* Collins, London.

[154] Kingdon, J. (2003) *Lowly Origin: Where, When and Why our Ancestors First Stood Up.* Princeton University Press, Princeton/Oxford.

[155] Kingsley, C. (1995/1863) *The Water Babies.* Puffin, London.

[156] Kipling, R. (1995/1906) *Puck of Pook's Hill.* Penguin, London.

[157] Kirschner, M. & Gerhart, J. (1998) Evolvability. *Proceedings of the National Academy of Sciences of the USA* **95:** 8420–8427.

[158] Kittler, R., Kayser, M., & Stoneking, M. (2003) Molecular evolution of *Pediculus humanus* and the origin of clothing. *Current Biology* **13:** 1414–1417.

[159] Klein, R. G. (1999) *The Human Career: Human Biological and Cultural Origins.* Chicago University Press, Chicago/London, 2nd ed.

[160] Kortlandt, A. (1972) *New Perspectives on Ape and Human Evolution.* Stichting voor Psychobiologie, Amsterdam.

[161] Krings, M., Stone, A., Schmitz, R. W., *et al.* (1997) Neanderthal DNA sequences and the origin of modern humans. *Cell* **90:** 19–30.

[162] Kristensen, R. M. (2002) An introduction to Loricifera, Cycliophora, and Micrognathozoa. *Integrative and Comparative Biology* **42:** 641–651.

[163] Kruuk, H. (2003) *Niko's Nature.* Oxford University Press, Oxford.

[164] Lack, D. (1947) *Darwin's Finches.* Cambridge University Press, Cambridge.

[165] Lambourn, W. A. (1930) The remarkable adaptation by which a dipterous pupa (Tabanidae) is preserved from the dangers of fissures in drying mud. *Proceedings of the Royal Society of London: Series B* **106:** 83–87.

[166] Lang, B. F., O'Kelly, C., Nerad, T., *et al.* (2002) The closest unicellular relatives of animals. *Current Biology* **12:** 1773–1778.

[167] Laskey, R. A. & Gurdon, J. B. (1970) Genetic content of adult somatic cells tested by nuclear transplantation from cultured cells. *Nature* **228:** 1332–1334.

[168] Leakey, M. (1987) The hominid footprints: Introduction. In *Laetoli: A Pliocene Site in Northern Tanzania* (Leakey, M. D. & Harris, J. M., eds.), pp. 490–496, Clarendon Press, Oxford.

[169] Leakey, M., Feibel, C., McDougall, I., & Walker, A. (1995) New four-

million-year-old hominid species from Kanapoi and Allia Bay, Kenya. *Nature* **376**: 565–571.

[170] Leakey, R. (1994) *The Origin of Humankind*. Basic Books, New York.

[171] Leakey, R. & Lewin, R. (1992) *Origins Reconsidered: In Search of What Makes us Human*. Little, Brown, London.

[172] Leakey, R. & Lewin, R. (1996) *The Sixth Extinction: Biodiversity and its Survival*. Weidenfeld & Nicolson, London.

[173] Lewis-Williams, D. (2002) *The Mind in the Cave*. Thames and Hudson, London.

[174] Lewontin, R. C. (1972) The apportionment of human diversity. *Evolutionary Biology* **6**: 381–398.

[175] Liem, K. F. (1973) Evolutionary strategies and morphological innovations: cichlid pharyngeal jaws. *Systematic Zoology* **22**: 425–441.

[176] Littlewood, D. T. J., Smith, A. B., Clough, K. A., & Emson, R. H. (1997) The interrelationships of the echinoderm classes: Morphological and molecular evidence. *Biological Journal of the Linnean Society* **61**: 409–438.

[177] Liu, F. G. R., Miyamoto, M. M., Freire, N. P., *et al.* (2001) Molecular and morphological supertrees for eutherian (placental) mammals. *Science* **291**: 1786–1789.

[178] Lorenz, K. (2002) *Man Meets Dog*. Routledge Classics, Routledge, London.

[179] Lovejoy, C. O. (1981) The origin of man. *Science* **211**: 341–350.

[180] Luo, Z.-X., Cifelli, R. L., & Kielan-Jaworowska, Z. (2001) Dual origin of tribosphenic mammals. *Nature* **409**: 53–57.

[181] Manger, P. R. & Pettigrew, J. D. (1995) Electroreception and feeding behaviour of the platypus (*Ornithorhychus anatinus:* Monotrema: Mammalia). *Philosophical Transactions of the Royal Society of London: Biological Sciences* **347**: 359–381.

[182] Marcus, G. F. & Fisher, S. E. (2003) FOXP2 in focus: what can genes tell us about speech and language? *Trends in Cognitive Sciences* **7**: 257–262.

[183] Margulis, L. (1981) *Symbiosis in Cell Evolution*. W. H. Freeman, San Francisco.

[184] Mark Welch, D. & Meselson, M. (2000) Evidence for the evolution of bdelloid rotifers without sexual reproduction or genetic exchange. *Science* **288**: 1211–1219.

[185] Martin, R. D. (1981) Relative brain size and basal metabolic rate in terrestrial vertebrates. *Nature* **293**: 57–60.

[186] Mash, R. (2003/1983) *How to Keep Dinosaurs*. Weidenfeld & Nicholson, London.

[187] Maynard Smith, J. (1978) *The Evolution of Sex*. Cambridge University Press, Cambridge.

[188] Maynard Smith, J. (1986) Evolution – contemplating life without sex. *Nature* **324**: 300–301.

[189]   Maynard Smith, J. & Szathmàry, E. (1995) *The Major Transitions in Evolution.* Oxford University Press, Oxford.

[190]   Mayr, E. (1985/1982) *The Growth of Biological Thought.* Harvard University Press, Cambridge, Mass.

[191]   Medina, M., Collins, A. G., Silberman, J. D., & Sogin, M. L. (2001) Evaluating hypotheses of basal animal phylogeny using complete sequences of large and small subunit rRNA. *Proceedings of the National Academy of Sciences of the USA* **98:** 9707–9712.

[192]   Menotti-Raymond, M. & O'Brien, S. J. (1993) Dating the genetic bottleneck of the African cheetah. *Proceedings of the National Academy of Sciences of the USA* **90:** 3172–3176.

[193]   Milius, S. (2000) Bdelloids: No sex for over 40 million years. *Science News* **157:** 326.

[194]   Miller, G. (2000) *The Mating Mind: How Sexual Choice Shaped the Evolution of Human Nature.* Heinemann, London.

[195]   Miller, K. R. (1999) *Finding Darwin's God: A Scientist's Search for Common Ground Between God and Evolution.* Cliff Street Books (HarperCollins), New York.

[196]   Miller, K. R. (2004) The flagellum unspun: the collapse of 'irreducible complexity'. In *Debating Design: From Darwin to DNA* (Ruse, M. & Dembski, W., eds.), Cambridge University Press, Cambridge.

[197]   Mills, D. R., Peterson, R. L., & Spiegelman, S. (1967) An extracellular Darwinian experiment with a self-duplicating nucleic acid molecule. *Proceedings of the National Academy of Sciences of the USA* **58:** 217–224.

[198]   Milner, A. R. & Sequeira, S. E. K. (1994) The temnospondyl amphibians from the Viséan of East Kirkton. *Transactions of the Royal Society of Edinburgh, Earth Sciences* **84:** 331–361.

[199]   Miya, M., Takeshima, H., Endo, H., *et al.* (2003) Major patterns of higher teleostean phylogenies: A new perspective based on 100 complete mitochondrial DNA sequences. *Molecular Phylogenetics and Evolution* **26:** 121–138.

[200]   Mollon, J. D., Bowmaker, J. K., & Jacobs, G. H. (1984) Variations of colour vision in a New World primate can be explained by polymorphism of retinal photopigments. *Proceedings of the Royal Society of London: Series B* **222:** 373–399.

[201]   Monod, J. (1972) *Chance and Necessity: Essay on the Natural Philosophy of Modern Biology.* Collins, London.

[202]   Montgelard, C., Bentz, S., Tirard, C., *et al.* (2002) Molecular systematics of Sciurognathi (Rodentia): The mitochondrial cytochrome b and 12S rRNA genes support the Anomaluroidea (Pedetidae and Anomaluridae). *Molecular Phylogenetics and Evolution* **22:** 220–233.

[203]   Moreira, D., Le Guyader, H., & Philippe, H. (2000) The origin of red algae and the evolution of chloroplasts. *Nature* **405:** 32–33.

[204]   Morgan, E. (1997) *The Aquatic Ape Hypothesis.* Souvenir Press, London.

[205] Murata, Y., Nikaido, M., Sasaki, T., *et al.* (2003) Afrotherian phylogeny as inferred from complete mitochondrial genomes. *Molecular Phylogenetics and Evolution* **28**: 253–260.

[206] Murdock, G. P. (1967) *Ethnographic Atlas.* University of Pittsburgh Press, Pittsburgh.

[207] Murphy, W. J., Eizirik, E., O'Brien, S. J., *et al.* (2001) Resolution of the early placental mammal radiation using Bayesian phylogenetics. *Science* **294**: 2348–2351.

[208] Musser, A. M. (2003) Review of the monotreme fossil record and comparison of palaeontological and molecular data. *Comparative Biochemistry and Physiology Part A* **136**: 927–942.

[209] Nelson, J. S. (1994) *Fishes of the World.* John Wiley, New York, 3rd edn.

[210] Nesse, R. M. & Williams, G. C. (1994) *The Science of Darwinian Medicine.* Orion, London.

[211] Nikoh, N., Iwabe, N., Kuma, K.-I., *et al.* (1997) An estimate of divergence time of Parazoa and Eumetazoa and that of Cephalochordata and Vertebrata by aldolase and triose phosphate isomerase clocks. *Journal of Molecular Evolution* **45**: 97–106.

[212] Nilsson, M. A., Gullberg, A., Spotorno, A. E., *et al.* (2003) Radiation of extant marsupials after the K/T boundary: Evidence from complete mitochondrial genomes. *Journal of Molecular Evolution* **57**: S3–S12.

[213] Norman, D. (1991) *Dinosaur!* Boxtree, London.

[214] Nozaki, H., Matsuzaki, M., Takahara, M., *et al.* (2003) The phylogenetic position of red algae revealed by multiple nuclear genes from mitochondria-containing eukaryotes and an alternative hypothesis on the origin of plastids. *Journal of Molecular Evolution* **56**: 485–97.

[215] Ohta, T. (1992) The nearly neutral theory of molecular evolution. *Annual Review of Ecology and Systematics* **23**: 263–286.

[216] Oparin, A. I. (1938) *The Origin of Life.* Macmillan, New York.

[217] Orgel, L. E. (1998) The origin of life – a review of facts and speculations. *Trends in Biochemical Sciences* **23**: 491–495.

[218] Pagel, M. & Bodmer, W. (2003) A naked ape would have fewer parasites. *Proceedings of the Royal Society of London: Biological Sciences (Suppl.)* **270**: S117–S119.

[219] Panchen, A. L. (2001) Étienne Geoffroy St.-Hilaire: Father of 'evo-devo'? *Evolution and Development* **3**: 41–46.

[220] Parker, A. (2003) *In the Blink of an Eye: The Cause of the Most Dramatic Event in the History of Life.* Free Press, London.

[221] Partridge, T. C., Granger, D. E., Caffee, M. W., & Clarke, R. J. (2003) Lower Pliocene hominid remains from Sterkfontein. *Science* **300**: 607–612.

[222] Penfield, W. & Rasmussen, T. (1950) *The Cerebral Cortex of Man: A Clinical Study of Localization of Function.*Macmillan, New York.

[223] Perdeck, A. C. (1957) The isolating value of specific song patterns in two sibling species of grasshoppers. *Behaviour* **12**: 1–75.

[224] Peterson, K. J. & Eernisse, D. J. (2001) Animal phylogeny and the ancestry of bilaterians: Inferences from morphology and 18S rDNA gene sequences. *Evolution and Development* 3: 170–205.

[225] Pettigrew, J. D., Manger, P. R., & Fine, S. L. B. (1998) The sensory world of the platypus. *Philosophical Transactions of the Royal Society of London: Biological Sciences* 353: 1199–1210.

[226] Pinker, S. (1994) *The Language Instinct: The New Science of Language and Mind.* Allen Lane, The Penguin Press, London.

[227] Pinker, S. (1997) *How the Mind Works.* Norton, New York.

[228] Pough, F. H., Andrews, R. M., Cadle, J. E., & Crump, M. (2001) *Herpetology.* Prentice Hall, Upper Saddle River, N.J., 2nd ed.

[229] Pullman, P. (2001) *His Dark Materials Trilogy.* Scholastic Press, London.

[230] Purvis, A. (1995) A composite estimate of primate phylogeny. *Philosophical Transactions of the Royal Society: Biological Sciences* 348: 405–421.

[231] Ragan, M. A., Goggin, C. L., Cawthorn, R. J., *et al.* (1996) A novel clade of protistan parasites near the animal-fungal divergence. *Proceedings of the National Academy of Sciences of the USA* 93: 11907–11912.

[232] Reader, J. (1988) *Man on Earth.* Collins, London.

[233] Reader, J. (1998) *Africa: A Biography of the Continent.* Penguin, London.

[234] Rebek, J. (1994) Synthetic self-replicating molecules. *Scientific American* 271 (July): 48–55.

[235] Rees, M. (1999) *Just Six Numbers.* Science Masters, Weidenfeld & Nicolson, London.

[236] Reno, P. L., Meindl, R. S., McCollum, M. A., & Lovejoy, C. O. (2003) Sexual dimorphism in Australopithecus afarensis was similar to that of modern humans. *Proceedings of the National Academy of Sciences of the USA* 100: 9404–9409.

[237] Richardson, M. K. & Keuck, G. (2002) Haeckel's ABC of evolution and development. *Biological Reviews* 77: 495–528.

[238] Richmond, B. G., Begun, D. R., & Strait, D. S. (2001) Origin of human bipedalism: The knuckle-walking hypothesis revisited. *Yearbook of Physical Anthropology* 44: 70–105.

[239] Ridley, Mark (1983) *The Explanation of Organic Diversity – The Comparative Method and Adaptations for Mating.* Clarendon Press/Oxford University Press, Oxford.

[240] Ridley, Mark (1986) Embryology and classical zoology in Great Britain. In *A History of Embryology: The Eighth Symposium of the British Society for Developmental Biology* (Horder, T. J., Witkowski, J., & Wylie, C. C., eds.), pp. 35–67, Cambridge University Press, Cambridge.

[241] Ridley, Mark (2000) *Mendel's Demon: Gene Justice and the Complexity of Life.* Weidenfeld & Nicolson, London.

[242] Ridley, Matt (1993) *The Red Queen: Sex and the Evolution of Human Nature.* Viking, London.

[243] Ridley, Matt (2003) *Nature Via Nurture: Genes, Experience and What Makes Us Human*. Fourth Estate, London.

[244] Rodríguez-Trelles, F., Tarrío, R., & Ayala, F. J. (2002) A methodological bias toward overestimation of molecular evolutionary time scales. *Proceedings of the National Academy of Sciences of the USA* 99: 8112–8115.

[245] Rokas, A. & Holland, P. W. H. (2000) Rare genomic changes as a tool for phylogenetics. *Trends in Ecology and Evolution* 15: 454–459.

[246] Roos, C. & Geissmann, T. (2001) Molecular phylogeny of the major hylobatid divisions. *Molecular Phylogenetics and Evolution* 19: 486–494.

[247] Ruiz-Trillo, I., Paps, J., Loukota, M., *et al.* (2002) A phylogenetic analysis of myosin heavy chain type II sequences corroborates that Acoela and Nemertodermatida are basal bilaterians. *Proceedings of the National Academy of Sciences of the USA* 99: 11246–11251.

[248] Ruppert, E. E. & Barnes, R. D. (1994) *Invertebrate Zoology*. Saunders College Publishing, Fort Worth, 6th ed.

[249] Sacks, O. (1996) *The Island of the Colour-blind and Cycad Island*. Picador, London.

[250] Saffhill, R., Schneider-Bernloer, H., Orgel, L. E., & Spiegelman, S. (1970) *In vitro* selection of bacteriophage Q ribonucleic acid variants resistant to ethidium bromide. *Journal of Molecular Biology* 51: 531–539.

[251] Sànchez-Villagra, M. R. (2001) The phylogenetic relationships of argyrolagid marsupials. *Zoological Journal of the Linnean Society* 131: 481–496.

[252] Sansom, I. J., Smith, M. M., & Smith, M. P. (2001) The Ordovician radiation of vertebrates. In *Major Events in Early Vertebrate Evolution* (Ahlberg, P. E., ed.), chap. 10, Taylor and Francis, London.

[253] Schluter, D. (2000) *The Ecology of Adaptive Radiation*. Oxford University Press, Oxford.

[254] Schmitz, J., Ohme, M., & Zischler, H. (2001) SINE insertions in cladistic analyses and the phylogenetic affiliations of *Tarsius bancanus* to other primates. *Genetics* 157: 777–784.

[255] Schopf, J. W. (1999) *Cradle of Life – The Discovery of Earth's Earliest Fossils*. Princeton University Press, Princeton, N.J.

[256] Schussler, D., Schwarzott, C., & Walker, A. (2001) A new phylum, the Glomeromycota: Phylogeny and evolution. *Mycological Research* 105: 1413–1421.

[257] Scotese, C. R. (2001) *Atlas of Earth History*, vol. 1, Palaeography. PALEOMAP project, Arlington, Texas.

[258] Seehausen, O. & van Alphen, J. J. M. (1998) The effect of male coloration on female mate choice in closely related Lake Victoria cichlids (*Haplochromis nyererei* complex). *Behavioral Ecology and Sociobiology* 42: 1–8.

[259] Senut, B., Pickford, M., Gommery, D., *et al.* (2001) First hominid from

the Miocene (Lukeino Formation, Kenya). *Comptes Rendus de l'Academie des Sciences, Series IIA – Earth and Planetary Science* **332:** 137–144.

[260]   Sepkoski, J. J. (1996) Patterns of Phanerozoic extinction: A perspective from global databases. In *Global Events and Event Stratigraphy in the Phanerozoic* (Walliser, O. H., ed.), pp. 35–51, Springer-Verlag, Berlin.

[261]   Shapiro, B., Sibthorpe, D., Rambaut, A., *et al.* (2002) Flight of the dodo. *Science* **295:** 1683.

[262]   Sheets-Johnstone, M. (1990) *The Roots of Thinking.* Temple University Press, Philadelphia.

[263]   Shirai, S. (1996) Phylogenetic interrelationships of neoselachians (Chondrichthyes: Euselachii). In *Interrelationships of Fishes* (Stiassny, M. L. J., Parenti, L. R., & Johnsson, G. D., eds.), pp. 9–34, Academic Press, San Diego.

[264]   Shu, D.-G., Luo, H.-L., Conway-Morris, S., *et al.* (1999) Lower Cambrian vertebrates from south China. *Nature* **402:** 42–46.

[265]   Sibley, C. G. & Monroe, B. L. (1990) *Distribution and Taxonomy of Birds of the World.* Yale University Press, New Haven.

[266]   Simpson, G. G. (1980) *Splendid Isolation: The Curious History of South American Mammals.* Yale University Press, New Haven.

[267]   Slack, J. M. W., Holland, P. W. H., & Graham, C. F. (1993) The zootype and the phylotypic stage. *Nature* **361:** 490–492.

[268]   Smith, D. C. (1979) From extracellular to intracellular: The establishment of a symbiosis. In *The Cell as a Habitat* (Richmond, M. H. & Smith, D. C., eds.), Royal Society of London, London.

[269]   Smolin, L. (1997) *The Life of the Cosmos.* Weidenfeld & Nicolson, London.

[270]   Southwood, T. R. E. (2003) *The Story of Life.* Oxford University Press, Oxford.

[271]   Springer, M. S., Murphy, W. J., Eizirik, E., & O'Brien, S. J. (2003) Placental mammal diversification and the Cretaceous–Tertiary boundary. *Proceedings of the National Academy of Sciences of the USA* **100:** 1056–1061.

[272]   Springer, M. S., Westerman, M., Kavanagh, J. R., *et al.* (1998) The origin of the Australasian marsupial fauna and the phylogenetic affinities of the enigmatic monito del monte and marsupial mole. *Proceedings of the Royal Society of London, Series B* **265:** 2381–2386.

[273]   Stewart, C. B. & Disotell, T. R. (1998) Primate evolution – in and out of Africa. *Current Biology* **8:** R582–R588.

[274]   Stringer, C. (2003) Human evolution – out of Ethiopia. *Nature* **423:** 692–695.

[275]   Sumper, M. & Luce, R. (1975) Evidence for *de novo* production of self-replicating and environmentally adapted RNA structures by bacteriophage Qβ replicase. *Proceedings of the National Academy of Sciences of the USA* **72:** 162–166.

[276] Sutherland, J. L. (1933) Protozoa from Australian termites. *Quarterly Journal of Microscopic Science* **76**: 145–173.

[277] Swift, J. (1733) *Poetry, A Rhapsody.*

[278] Syed, T. & Schierwater, B. (2002) *Trichoplax adhaerens:* discovered as a missing link, forgotten as a hydrozoan, rediscovered as a key to metazoan evolution. *Vie et Milieu* **52**: 177–187.

[279] Takezaki, N., Figueroa, F., Zaleska-Rutczynska, Z., & Klein, J. (2003) Molecular phylogeny of early vertebrates: Monophyly of the agnathans as revealed by sequences of 35 genes. *Molecular Biology and Evolution* **20**: 287–292.

[280] Tamm, S. L. (1982) Flagellated endosymbiotic bacteria propel a eukaryotic cell. *Journal of Cell Biology* **94**: 697–709.

[281] Tavaré, S., Marshall, C. R., Will, O., *et al.* (2002) Using the fossil record to estimate the age of the last common ancestor of extant primates. *Nature* **416**: 726–729.

[282] Taylor, C. R. & Rowntree, V. J. (1973) Running on two or four legs: which consumes more energy? *Science* **179**: 186–187.

[283] Telford, M. J., Lockyer, A. E., Cartwright-Finch, C., & Littlewood, D. T. J. (2003) Combined large and small subunit ribosomal RNA phylogenies support a basal position of the acoelomorph flatworms. *Proceedings of the Royal Society of London: Biological Sciences* **270**: 1077–1083.

[284] Templeton, A. R. (2002) Out of Africa again and again. *Nature* **416**: 45–51.

[285] Thomson, K. S. (1991) *Living Fossil: The Story of the Coelacanth.* Hutchinson Radius, London.

[286] Trivers, R. L. (1972) Parental investment and sexual selection. In *Sexual Selection and the Descent of Man* (Campbell, B., ed.), pp. 136–179, Aldine, Chicago.

[287] Trut, L. N. (1999) Early canid domestication: The farm-fox experiment. *American Scientist* **87**: 160–169.

[288] Tudge, C. (1998) *Neanderthals, Bandits and Farmers: How Agriculture Really Began.* Weidenfeld & Nicolson, London.

[289] Tudge, C. (2000) *The Variety of Life.* Oxford University Press, Oxford.

[290] Turbeville, J. M. (2002) Progress in nemertean biology: Development and phylogeny. *Integrative and Comparative Biology* **42**: 692–703.

[291] Valentine, J. W. (2002) Prelude to the Cambrian explosion. *Annual Review of Earth and Planetary Sciences* **30**: 285–306.

[292] van Schaik, C. P., Ancrenaz, M., Borgen, G., *et al.* (2003) Orangutan cultures and the evolution of material culture. *Science* **299**: 102–105.

[293] van Tuinen, M., Sibley, C. G., & Hedges, S. B. (2000) The early history of modern birds inferred from DNA sequences of nuclear and mitochondrial genomes. *Molecular Biology and Evolution* **17**: 451–457.

[294] Venkatesh, B., Erdmann, M. V., & Brenner, S. (2001) Molecular synapomorphies resolve evolutionary relationships of extant jawed

vertebrates. *Proceedings of the National Academy of Sciences of the USA* **98:** 11382–11387.

[295] Verheyen, E., Salzburger, W., Snoeks, J., & Meyer, A. (2003) Origin of the superflock of cichlid fishes from Lake Victoria, East Africa. *Science* **300:** 325–329.

[296] Vine, F. J. & Matthews, D. H. (1963) Magnetic anomalies over oceanic ridges. *Nature* **199:** 947–949.

[297] Wada, H. & Satoh, N. (1994) Phylogentic relationships among extant classes of echinoderms, as inferred from sequences of 18S rDNA, coincide with relationships deduced from the fossil record. *Journal of Molecular Evolution* **38:** 41–49.

[298] Wake, D. B. (1997) Incipient species formation in salamanders of the *Ensatina* complex. *Proceedings of the National Academy of Sciences of the USA* **94:** 7761–7767.

[299] Walker, G. (2003) *Snowball Earth: The Story of the Great Global Catastrophe That Spawned Life As We Know It.* Bloomsbury, London.

[300] Ward, C. V., Walker, A., & Teaford, M. F. (1991) *Proconsul* did not have a tail. *Journal of Human Evolution* **21:** 215–220.

[301] Weinberg, S. (1993) *Dreams of a Final Theory.* Hutchinson Radius, London.

[302] Weiner, J. (1994) *The Beak of the Finch.* Jonathan Cape, London.

[303] Wesenberg-Lund, C. (1930) Contributions to the biology of the Rotifera. Part II. The periodicity and sexual periods. *Det Kongelige Danske Videnskabers Selskabs Skrifter* **9, II:** 1–230.

[304] West, G. B., Brown, J. H., & Enquist, B. J. (2000) The origin of universal scaling laws in biology. In *Scaling in Biology* (Brown, J. H. & West, G. B., eds.), Oxford University Press, Oxford.

[305] West-Eberhard, M. J. (2003) *Developmental Plasticity and Evolution.* Oxford University Press, New York.

[306] Westoll, T. S. (1949) On the evolution of the Dipnoi. In *Genetics, Paleontology and Evolution* (Jepsen, G. L., Mayr, E., & Simpson, G. G., eds.), pp. 121–188, Princeton University Press, Princeton, N.J.

[307] Wheeler, J. A. (1990) Information, physics, quantum: The search for links. In *Complexity, Entropy, and the Physics of Information* (Zurek, W. H., ed.), pp. 3–28, Addison-Wesley, New York.

[308] White, T. D., Asfaw, B., DeGusta, D., *et al.* (2003) Pleistocene Homo sapiens from Middle Awash, Ethiopia. *Nature* **423:** 742–747.

[309] White, T. D., Suwa, G., & Asfaw, B. (1994) *Australopithecus ramidus*, a new species of early hominid from Aramis, Ethiopia. *Nature* **371:** 306–312.

[310] Whiten, A., Goodall, J., McGrew, W. C., *et al.* (1999) Cultures in chimpanzees. *Nature* **399:** 682–685.

[311] Williams, G. C. (1975) *Sex and Evolution.* Princeton University Press, Princeton, N.J.

[312] Williams, G. C. (1992) *Natural Selection: Domains, Levels and Challenges.* Oxford University Press, Oxford.

[313] Wilson, E. O. (1992) *The Diversity of Life.* Harvard University Press, Cambridge, Mass.

[314] Wilson, H. V. (1907) On some phenomena of coalescence and regeneration in sponges. *Journal of Experimental Zoology* **5**: 245–258.

[315] Winchell, C. J., Sullivan, J., Cameron, C. B., *et al.* (2002) Evaluating hypotheses of deuterostome phylogeny and chordate evolution with new LSU and SSU ribosomal DNA data. *Molecular Biology and Evolution* **19**: 762–776.

[316] Woese, C. R., Kandler, O., & Wheelis, M. L. (1990) Towards a natural system of organisms: Proposal for the domains Archaea, Bacteria, and Eucarya. *Proceedings of the National Academy of Sciences of the USA* **87**: 4576–4579.

[317] Wolpert, L. (1991) *The Triumph of the Embryo.* Oxford University Press, Oxford.

[318] Wolpert, L., Beddington, R., Brockes, J., *et al.* (1998) *Principles of Development.* Current Biology/Oxford University Press, London/Oxford.

[319] Wray, G. A., Levinton, J. S., & Shapiro, L. H. (1996) Molecular evidence for deep Precambrian divergences among metazoan phyla. *Science* **274**: 568–573.

[320] Xiao, S. H., Yuan, X. L., & Knoll, A. H. (2000) Eumetazoan fossils in terminal Proterozoic phosphorites? *Proceedings of the National Academy of Sciences of the USA* **97**: 13684–13689.

[321] Yeats, W. B. (1984) *The Poems* (Finneran, R. J., ed.). Macmillan, London.

[322] Yoder, A. D. & Yang, Z. (2004) Divergence dates for Malagasy lemurs estimated from multiple gene loci: Geological and evolutionary context. *Molecular Ecology* **13**: 757–773.

[323] Zahavi, A. & Zahavi, A. (1997) *The Handicap Principle.* Oxford University Press, Oxford.

[324] Zardoya, R. & Meyer, A. (2000) Mitochondrial evidence on the phylogenetic position of caecilians (Amphibia: Gymnophiona). *Genetics* **155**: 765–775.

[325] Zardoya, R. & Meyer, A. (2001) On the origin of and phylogenetic relationships among living amphibians. *Proceedings of the National Academy of Sciences of the USA* **98**: 7380–7383.

[326] Zhu, M. & Yu, X. (2002) A primitive fish close to the common ancestor of tetrapods and lungfish. *Nature* **418**: 767–770.

[327] Zuckerkandl, E. & Pauling, L. (1965) Evolutionary divergence and convergence in proteins. In *Evolving genes and proteins* (Bryson, V. & Vogels, H. J., eds.), pp. 97–166, Academic Press, New York.

ILLUSTRATION CREDITS

25 DEEP SEA SNIPE EEL, Peter Parks/imagequestmarine.com; DEEP SEA BLACK SWALLOWER, Peter Herring/imagequestmarine.com; PLAICE CAMOUFLAGED ON SAND, © Linda Pitkin/NHPA; OCEAN SUNFISH, Masa Ushioda/imagequestmarine.com; DEEP-SEA GULPER, © Norbert Wu/Minden Pictures.

26 UNROOTED HAPLOTYPE NETWORK OF THE HAPLOCHROMINE SUPERFLOCK from Verheyen *et al.* Reprinted with permission from [295], fig 3c. Copyright 2003 AAAS.

27 TWO WAYS OF BEING A FLATFISH, by L. Ward, from R. Dawkins [71], fig. 4.7.

28 GREAT HAMMERHEAD SHARK, James, D. Watt/imagequest-marine.com; SMALLTOOTH SAWFISH, © Yves Lanceau/NHPA.

29 ELEPHANT SHARK OR PLOUGH-NOSED CHIMAERA, © ANT Photo Library/NHPA.

30 CONCESTOR 23, © Moonrunner Design.

31 SEA SQUIRTS, © B. Jones & M. Shimlock/NHPA.

32 LEAF CUTTER ANTS, © Michael & Patricia Fogden/Minden Pictures.

33 PHOTOGRAPH OF CONDOLEEZZA RICE, COLIN POWELL, PRESIDENT BUSH AND DONALD H. RUMSFELD, Associated Press.

34 *DROSOPHILA MELANOGASTER* from P. A. Lawrence, *The Making of a Fly*, Blackwell Science (1992), plate 5.1.

35 LIGHT MICROGRAPH OF A ROTIFER, John Walsh/Science Photo Library.

36 VELVET WORM, Dr Morley Read/Science Photo Library.

37 *HALKIERIA* reprinted by kind permission of Simon Conway-Morris, University of Cambridge. From S. Conway-Morris and J. S. Peel, 'Articulated Halkieriids from the Lower Cambrian of North Greenland and Their Role in Early Protostome Evolution', *Phil. Trans. Roy. Soc.* Lond. (1995), B 347: 305–358, figs. 49a, b, c.

38 *DICKINSONIA COSTATA*, all rights reserved by J. G. Gehling, South Australian Museum.

39 *PORPITA PORPITA*, Peter Parks/imagequestmarine.com.

40 MASS OF *MASTIGIAS* JELLYFISH AT SURFACE PALAU, WESTERN PACIFIC ISLANDS, Michael Pitts/naturepl.com.

41 AERIAL VIEW OF HERON ISLAND (CORAL ISLAND), GREAT BARRIER REEF MARINE PARK, QUEENSLAND, AUSTRALIA, © Gerry Ellis/Minden Pictures.

42 CLEANER WRASSE CLEANING ROSY GOATFISH, Georgette Douwma/naturepl.com.

43 VENUS'S GIRDLE, A TYPE OF COMB JELLY, Sinclair Stammers/Science Photo Library.

44 GLASS SPONGE, © The Natural History Museum, London.

45 STINKHORN MUSHROOMS, Vaughan Fleming/Science Photo Library.

46 CONCESTOR 36, © Moonrunner Design.

47 SEQUOIA TREE, SEQUOIA NATIONAL PARK, Tony Craddock/Science Photo Library.

48 TRANSFER RNA MOLECULE, COMPUTER ARTWORK, Alfred Pasieka/Science Photo Library.

49 AWFUL CHANGES © Department of Geology, National Museum of Wales.

50 Clockwise: PHOTOGRAPH OF A SIDE VIEW OF NAUTILUS POMPILIUS showing 'pin-hole' eye, © The Natural History Museum, London; COLOURED SCANNING ELECTRON MICROGRAPH OF TRILOBITE EYE FOSSIL, VVG/Science Photo Library; SCANNING ELECTRON MICROSCOPE IMAGE SHOWING THE COMPOUND EYE OF A SIMULIAN BLACKFLY, © The Natural History Museum, London; STOPLIGHT PARROTFISH EYE, © B. Jones & M. Shimlock/NHPA; EYE OF A GREAT HORNED OWL, Simon Fraser/Science Photo Library.

Page

39 CROWD IMAGE reprinted by kind permission of Professor Robert Winston.

50 VICTORIA FAMILY TREE by Yan Wong.

61 A NEW MODEL OF HUMAN EVOLUTION from A. R. Templeton [284], fig. 1. Reproduced with kind permission of Nature Publishing Group.

81 SCATTER GRAPH OF LOG BRAIN WEIGHT AGAINST LOG BODY WEIGHT, from R. D. Martin *et al.* [185], fig 1. Reproduced with kind permission of Nature Publishing Group.

88 EQ OR 'BRAININESS' INDEX by Yan Wong.

100 TOUMAI CRANIUM, © M.P.F.T. (Mission Paléoanthropologique Franco-Tchadienne).

121 A SYNTHETIC HYPOTHESIS OF CATARRHINE PRIMATE EVOLUTION. Reprinted from C. B. Stewart & T. R. Disotell [273], fig. 2. Copyright 1998, with permission from Elsevier.

129, 130 UNROOTED TREES by Yan Wong; ROOTED CLADOGRAMS by Yan Wong.

135, 136, 138 CHAUCER DIAGRAMS by Yan Wong based on data from [46].

140 'NONCOMMUNICATORY' DATA DIAGRAM from T. Geissmann [100], fig. 1.6. Reprinted by permission of Wiley-Liss, Inc., a subsidiary of John Wiley & Sons, Inc.

141 UNROOTED MAXIMUM LIKELIHOOD TREE. Reprinted from C. Roos & T. Geissmann [246], fig. 2c. Copyright 2001, with permission from Elsevier.

165 TARSIER SKELETON by Stephen D. Nash, from *Primate Adaptation and Evolution* by John G. Fleagle, 2nd ed., Academic Press, San Diego (1999), Fig. 4.25.

213 GRAPH based on data from R. D. Alexander *et al.* [5].

217 TESTIS MASS AGAINST BODY MASS from P. H. Harvey & M. D. Pagel [132], fig. 1.3.

241 HENKELOTHERIUM by Elke Gröning, from B. Krebs, 'Das Skelt von Henkelotherium guimarotae gen. et. sp. nov (*Eupantotheria, Mammalia*) aus dem Oberen Jura von Portugal', *Berliner Geowiss* (1991), Abh. A 133: 1–110.

244 Penfield brain diagram from [222].

246 PLATYPUS ISO-SENSITIVITY LINES from P. R. Manger & J. D. Pettigrew [181], fig. 13e.

250 STAR-NOSED MOLE, Rod Planck/Science Photo Library.

252 MOLUNCULUS from K. C. Catania & J. H. Kaas [41]. Reprinted by permission of Wiley-Liss, Inc., a subsidiary of John Wiley & Sons, Inc.

256 EXTINCTION RATES DIAGRAM from J. J. Sepkoski [260].

288 RICHARD OWEN, BRITISH ZOOLOGIST, WITH A GIANT MOA, George Bernard/Science Photo Library.

300 SEA FLOOR SPREADING DIAGRAM redrawn by Ken Wilson.

316 BELL CURVE DIAGRAM AND DISTRIBUTION DIAGRAM by Richard Dawkins.

333 PHYLOGENETIC POSITION OF THE CAECILIAN from R. Zardoya & A. Meyer [324], fig. 4c.

395 BRACHIOPOD, © Natural History Museum, London.

403 UPSIDE-DOWN CATFISH, © Slip Nicklin/Minden Pictures; APOLLO 11 IMAGE OF MOON CRATERS, NASA/DvR/Science Photo Library.

411 PHOTOGRAPH OF COLIN POWELL AND PRESIDENT DANIEL ARAP MOI, Associated Press.

448 *THAUMATOXENA ANDREINII SILVESTRI*, from R.H.L. Disney and D. H. Kistner, 'Revision of the termitophilous Thaumatoxeninae (Diptera: Phoridae)', *Journal of Natural History* (1992) 26: 953–991.

449 *MEGASELIA SCALARIS* (*Loew*) by Arthur Smith, © Natural History Museum, London.

451 *HALLUCIGENIA* RECONSTRUCTION, by kind permission of Dr Derek J. Siveter, Oxford University Museum of Natural History. From [136].

452 RECONSTRUCTION OF *ANIMALOCARIS*, by kind permission of Dr Derek J. Siveter, Oxford University Museum of Natural History. From [136].

454 *MYLLOKUNMINGIA* SPECIMEN AND DIAGRAM from D-G Shu *et al.* [264], figs. 2a and 3. Reproduced with kind permission of Nature Publishing Group.

478 CNIDARIAN HARPOON, courtesy of BIODIDAC (biodidac.bio.uottawa.ca).

501 DIAGRAM OF PORTION OF SPONGE WALL from R. C. Brusca & G. J. Brusca, *Invertebrates*, 2nd ed., Sinauer Associates, Inc. (2003). Drawn from original by Tim Brown.

502 CHOANOFLAGELLATE STALKED SPECIES from J. N. Farmer, *The Protozoa*, C. V. Mosby Co., St Louis (1980). University of Birmingham.

521 DEEP GREEN DIAGRAM reprinted courtesy of Brent Mishler, Department of Integrative Biology, University of California, Berkeley.

523 KLIEBER'S LAW SCATTER PLOT from West, Brown & Enquist [304], fig. 2. From diagram in Hemmingsen, 'Energy meta as related to body size', *Rep. Steno. Mem. Hosp.*, Copenhagen: 1–110 (1960). Drawn from original by Tim Brown.

524 CROSS-SECTION OF CAULIFLOWER by Yan Wong.

540 CONSENSUS PHYLOGENY OF EUKARYOTES © S. L. Baldauf. Updated from original in [13], fig 1.

547 DIAGRAM OF *MIXOTRICHA* from L. R. Cleveland & A. V. Grimstone [49].

561 BACTERIAL FLAGELLAR MOTOR from D. Voet & J. G. Voet, *Biochemistry*, John Wiley & Sons, Inc. (1994), p. 1259. Drawn from original by Tim Brown.

569 MODERN VIEW OF THE TREE OF LIFE. Modified from S. Gribaldo & H. Philippe [113], fig. 5. Copyright 2002, with permission from Elsevier.

602 EYE LANDSCAPE drawn by L. Ward after original by M. F. Land, from R. Dawkins [71].

INDEX

Key:
> *208* image or diagram
> [**208**] bibliography entry
> (plate 1) coloured plates

Aardvark, 224
Abd-al-Rahman, 287
Abdominal-A, 429, 447
Abdominal-B, 429
Abnormality, genetic, 72, 425–426,
    429–430. *See also* Mutation
Acanthostega, 306 [**47**]
Acoel flatworm (Acoela), 395, 474. *See also*
    Acoelomorph flatworm
Acoelomorph flatworm, 472–476, 505, 634
Acorn worm (Enteropneusta), 382,
    399–400, 432
Actinopterygian. *See* Ray-finned fish
Adam, Y chromosome, 55–56, 62
Adams, Douglas, 171, 286n [**2**]; 284 [**1**]
Adaptation, 64, 97–98, 197, 292n, 358, 379,
    553, 602–610, 620–621
    to life in caves, 354–356
Adaptive radiation, 348 [**253**]
Adzebill (*Aptornis*), 286
Aegyptopithecus, 144
Aepyornis. *See* Elephant bird
Africa, 45–46, 48, 78, 117–123, 143, 147–148,
    171, 219, 221, 223, 224, 290–296,
    298–299, 330, 599, 600, 601
African porcupine (*Hystrix*), 147
Afropithecus, 117
Afrotheres (Afrotheria), 40, 173, 174, 184,
    224–229, 238, 600, 633
Agouti (*Dasyprocta*), 127, 188
Agricultural Revolution, 26–35, 186
Agriculture. *See* Domestication;
    Agricultural Revolution
Alcohol intolerance, 34
Alexander, Richard D., 213 [**5**]
Alexis of Russia, Tsarevitch, 52
Algae, 452, 482, 485, 514, 550–551
    brown algae, 520, 541; *Fucus*, 541; kelp,
    541

dinoflagellates, 475, 541; zooxanthellae,
    485, 486
    green algae, 520; *Tetraselmis convolutae*,
    475; *Volvox*, 505
    red algae (Rhodophyta), 471, 520
Alien abductions, 44
Allele, 34, 49, 51, 52–53, 156, 196, 413, 464
Alligator, 333
Allometry, 426 [**123, 304**]
Allopatric speciation. *See under* Speciation
Alpaca (*Lama pacos*), 222
Altaic, 22
Altruism, 96n, 559
Alu, 132, 159, 162
Alveolate, 541
Amber, preservation in, 12, 20, 437, 442
Ambulacrarian (Ambulacraria), 382–386,
    634
Amino acid, 19, 72, 73, 132, 369, 463, 468n,
    580, 585–586, 589–590
Ammonite, 394
Amniote, 302, 304, 305
Amoeba, 515–517, 541, 568
Amoebozoan, 515, 635
Amphibians, 257, 302–308, 370, 509, 623,
    633
    reproduction, 304–305
    skin not waterproof, 305
Amphioxus. *See* Lancelet
Ampullae of Lorenzini, 247
Andrewsarchus, 199, 206n
Anemone, sea, 460, 475, 477, 480, 483,
    504
Animal, 153, 387, 471, 509, 520
    brain size, 79–87
    communication, 198
    domestic. *See under* Domestication
    new definition of, based on Hox genes,
    436 [**267**]

Naked mole rat (*Heterocephalus glaber*), 188n. *See also* Eusocial animals

Nakedness. *See* Hairlessness, evolution of; Clothes, invention of

Nanotechnology, 563

Natural selection
 carves gene pools, 445–446
 distinguishing products of, *cf.* designed artefacts, 358, 563–564
 only explanation for function and apparent design, 464
 at lower level producing illusion of harmonious whole, 489–490
 power of cumulative, 593

Naturalistic fallacy, 126n

Nautilus, 394, 603

Neanderthal, 20, 66–68
 DNA, 20, 67 [161]

Necrolestes, 234n

Neighbour-joining, 134, 139

Nematocyst. *See* Cnidocyte

Nemertodermatids (Nemertodermatida), 474

Neogene Period, 143, (plate 1)

Neolithic revolution. *See* Agricultural Revolution

Neomura, 553

Neoteny, 102–103, 325, 379

Nerve cord/tube, 368, 372, 377, 399–402, 405, 477

Nervous system, 385, 477, 491, 497. *See also* Nerve cord

Nesomyinae, 172

Nesse, Randolph, 499 [210]

Neutral theory, 465–466, 467 [152]

New Guinea, 27, 171n, 183, 232, 233, 237, 286, 289, 291

New Testament, 18

New Zealand, 232, 286, 288–289, 291, 292, 294, 599, 600

Newt. *See under* Salamander

Nipples, existence of male, 271

Nitrogen fixation, 550n, 566

Nomascus. See under Gibbon

Norman, David, 263n [213]

Nostratic, 22

Notochord, 366, 368, 372, 377

Notoungulate, 202, 223

Nuclear deterrence, 150

Nudibranch. *See under* Snails and slugs

Nyerere, Julius, 349

Obdurodon, 240

Octopus, 393

Ohta, Tomoko, 463, 467 [215]

Oilbird (*Steatornis caripensis*), 603–604

Old Red Sandstone, 14

Old Testament, 18

Oligocene Epoch, 143, (plate 1)

Omomyid, 166

Onychophora (velvet worm), 392, 451, 460
 Peripatopsis moseleyi, (plate 36)
 Peripatus, 392, 450–451

Opabinia, 452

Oparin, A. I., 578, 579, 581 [216]

Opossum (Didelphidae), 151n, 222, 233, 234

Opossum, shrew (Paucituberculata), 222

Opsins, 155–161

Oral/aboral axis, 432. *See also* Mouth

Orang utan (*Pongo*), 94, 108, 111, 114, 115, 116, 120, 121, 185, 218, 276, 632

Ordovician Period, 338, 360, 453, (plate 1)

Oreopithecus, 118, 121

Orgel, Leslie, 563. [*See* 56]; 581 [217]; 593 [250]

Origin of Species, The, 115, 193, 205, 425, 464, 573–574 [59]

Ornithischian. *See under* Dinosaur

Orrorin tugenensis, 99–102, 103–104, 278

Orsten, 452n

Osteolepiforms (Osteolepiformes), 307

Ostracoderm, 364, 366

Ostrich, 94, 287, 289, 292, 294–295, 327

Otter, sea (*Enhydra lutris*), 206, 401

Ouranopithecus, 118, 122–123

Out of Africa theory, 58–62, *61*, 68

Outgroup, 116, 130

Owen, Richard, 113, 205n, *288*

Owl or night monkey (*Aotus*), 151, 156, 163

Owlet nightjars (Aegothelidae), 163n

Oxygen, 95, 313, 330, 344, 369, 484, 524, 544n, 550, 551–552, 578–579, 597, 598, 610

Oyster, 394

Paddlefish (*Polyodon spathula*), 246–247, (plate 15)

Paedomorphosis, 326–328

Pagel, Mark, 273–275 [218]

Pakicetus, 205, 207

Palaeocene Epoch, 163, 204, 206, 290, (plate 1)